P384 ✓
P397-402 ✓ CAYLEY NOS.
P412 ✓

P165 ✓ MOTION
P3-4 ✓ continue Metric

THE GEOMETRY OF GEODESICS

Herbert Busemann

DOVER PUBLICATIONS, INC.
Mineola, New York

Bibliographical Note

This Dover edition, first published in 2005, is an unabridged republication of the work originally published in 1955 by Academic Press, Inc., New York.

Library of Congress Cataloging-in-Publication Data

Busemann, Herbert, 1905–
The geometry of geodesics / Herbert Busemann.
p. cm.
Originally published: New York : Academic Press, 1955, in series: Pure and applied mathematics ; 6.
Includes bibliographical references and index.
ISBN 0-486-44237-3 (pbk.)
1. Geodesics (Mathematics) 2. Geometry, Differential. 3. Curves on surfaces. I. Title.

QA649.B79 2005
516.3'62—dc22

2004065694

Manufactured in the United States of America
Dover Publications, Inc., 31 East 2nd Street, Mineola, N.Y. 11501

PREFACE

Content and Methods

"A geometric approach to qualitative problems in intrinsic differential geometry" would be a brief description of the present book, which extends the methods developed in the author's "Metric Methods in Finsler Spaces and in the Foundations of Geometry," Princeton, 1942.

In the earlier book the emphasis was on spaces in which the geodesic through two points is unique, whereas it is now on the general case in which the geodesics have only local uniqueness properties. Consequently, the relations to the foundations of geometry are no longer very relevant and Finsler spaces are the principal subject.

For two reasons the latter do not appear in the title: Firstly, "Geometry of Geodesics" expresses clearly the method of attack and the intrinsic aspect. Secondly, the term Finsler Space means to many not only a type of space but also a definite approach: The space is considered as a set of line elements to which euclidean metrics are attached. The main problems are connected with parallelism. In spite of the great success of Finsler's thesis, the later development of this aspect lacks simple geometric facts to the extent that their existence in non-Riemannian geometry has been doubted.

Here we take the attitude that Finsler spaces are point spaces of an essentially new character which we must approach directly without being prejudiced by euclidean or Riemannian methods. Of course, it is impossible to live up to this program. The euclidean tradition is much too strong to be shaken off readily, and it may take several generations of mathematicians to gain real freedom.

However, the direct approach has produced many new results and has materially generalized many known phenomena,[1] thereby revealing two startling facts: much of Riemannian geometry is not truly Riemannian and much of differential geometry requires no derivatives.

---

[1] See page 413. All notes to the text appear on pages 413—415.

Actually only the first fact is startling; the second is to a certain extent a consequence of the first. If all theorems were omitted from the present volume which have no bearing on, or become trivial for, Riemann spaces, and the remaining facts were applied only to the latter, there would still remain an interesting book, because it contains — besides many facts which are new also for Riemannian spaces — numerous well-known results of Hadamard, Cartan, Cohn-Vossen, Nielsen and others. Thus the non-existence of concrete geometric facts in Finsler spaces as compared to Riemann spaces proves to be a myth and there emerges the highly important problem of gaining a clear understanding of the true realm of Riemannian geometry, i. e., to recognize the character of the theorems for which it is essential that the local unit spheres be ellipsoids rather than arbitrary convex surfaces with center. At present we have unfortunately not the slightest idea how such a characterization might look.

It is now clear why differentiability hypotheses are often unnecessary: our geometric axioms for the existence of geodesics (see the definition of *G*-space on page **37**) extract from Finsler spaces an essential part of their qualitative geometric properties, but not from Riemann spaces. The difference between the two types of spaces is of an essentially infinitesimal nature, which explains why there are hardly any Riemannian theorems proved without differentiability.[2]

The present methods also offer the esthetical and methodological advantage of a geometric approach to geometric problems and they extend to differential geometry the trend of replacing calculations by syllogisms which has so deeply permeated modern mathematics.

Some differential geometers find this tendency unnatural. We need not take them seriously: the term "differential geometry" carries no obligation to use derivatives, just as names like "calculus of variations" or "algebra" do not confine us to operating within their original meanings. Nor is there any reason why reduction of hypotheses (and their exact formulation!) should be undesirable in differential geometry, but meritorious in all other parts of mathematics.[3]

Moreover, once *G*-spaces have been introduced, our basic assumptions remain the same throughout the book. These form, therefore, an adequate basis for many geometric theories, and it does not seem wholly unreasonable to expect that the concept of *G*-space may eventually be generally accepted in the same way as, for example, the concept of field in algebra.

Prerequisites

Large parts of the book can be understood with a very moderate mathematical background. The first three chapters use only the elements of point set topology, a little dimension theory and a few facts on convex bodies. The remaining part, except for the last four sections, is accessible to anyone who knows the idea of covering space and its relations to the first fundamental group. The last four sections use topological and Lie groups. However, in order to grasp the significance of the problems, some familiarity with non-euclidean geometry and classical differential geometry is necessary. The book "Projective Geometry and Projective Metrics" by the author and P. J. Kelly (1953), which appeared in this series, will prove very useful as preparation for the present work.

Acknowledgments

It had been the author's intention to "sometime" write a comprehensive introduction to his geometric approach to Finsler spaces. Professor H. G. Forder persuaded him to do so at the present time, read those parts which were written in New Zealand, and made many suggestions. Sincere thanks go to him as well as to the Fulbright authorities, who made the association with Professor Forder possible.

For assistance in proofreading the author is greatly indebted to Dr. Flemming P. Pedersen and to Mr. Eugene Zaustinsky.

H. B.

Department of Mathematics
University of Southern California

TABLE OF CONTENTS

CHAPTER I

THE BASIC CONCEPTS

1. Introduction

The spaces occurring in this book are *metric* and share two essential properties with the spaces usually considered in differential geometry: the first guarantees that the space behaves in many respects like a finite-dimensional space which cannot be extended without increasing the dimension; in differential geometry it is frequently called completeness or normality. It is equivalent to the Bolzano-Weierstrass Theorem: a bounded infinite set has an accumulation point. We call a metric space with this property *finitely compact*; the term seems to go back to M. Morse. The second property is the arrangement of the points into *geodesics*, and is obtained here by imposing conditions on the existence and the uniqueness of solutions of the triangle equality.

In Sections 2–5 we study the implications of finite compactness alone. For the benefit of those who approach the subject from a background of classical differential geometry we show first that certain standard theorems in coordinate spaces about uniform continuity, extrema, and selection of subsequences follow from finite compactness. For others it will suffice to simply check on the terminology which varies slightly with different authors.

In Section 3 we define *convergence of sets* once for ever in terms of Hausdorff's closed limit. Otherwise convergence would have to be defined in each case as it arises, for instance, for the hyperplanes in E^n. Nowadays we are not satisfied with the mere definition of a limit, we wish to ascertain the nature of the topology induced by this limit. We show that the closed subsets of a finitely compact space may be metrized such that they become again a finitely compact space and the resulting limit is equivalent to the closed limit. Hausdorff had proved the corresponding result for compact spaces.

Again, in order to obviate special agreements in individual cases the next Section 4 derives a similar result for *motions*, i. e., distance preserving mappings of the space on itself. It is shown that the motions of a finitely compact space form with a suitable right-invariant metric themselves a finitely compact space, in which convergence is equivalent to pointwise convergence.

We then discuss *length of curves.* The definition and properties of length are very much the same as in elementary spaces.

The length of any curve is at least the distance of its endpoints. If the two numbers are equal we call the curve a *segment,* because it is then isometric to a segment in ordinary space. In the intrinsic differential geometry of surfaces a segment is a geodesic arc which is a shortest connection of its endpoints. A consistent use of segments would shorten many arguments in differential geometry, but as the literature shows, the concept is not indispensable. In the present setup it is basic. It is the subject of Section 6, where *Menger's convexity* is introduced as a sufficient condition for the existence of segments.

A *geodesic* (Section 7) is a curve which extends indefinitely in both directions and behaves locally like a segment. This definition merely abstracts the process of continuing indefinitely a solution of the differential equations for a geodesic. To guarantee the existence of geodesics it must be postulated that *prolongation is locally possible.*

So far questions of uniqueness have been entirely left aside. In differential geometry the assumptions are always such that a line element determines a geodesic, hence continuation of a solution can be effected in one way only. Our final postulate is therefore *uniqueness of prolongation.* The spaces which satisfy all five postulates, viz., that they are *metric, finitely compact, and convex in Menger's sense and that prolongation is locally possible and unique, will be called G-spaces.* They are the primary subject of this book.

Section 8 discusses the most elementary properties of these spaces. We then (Section 9) define the *multiplicity* for a point of a geodesic and treat the simplest types of geodesics. In Section 10 we show that a *two-dimensional G-space is always a manifold.* Whether this is true for higher dimensions is an interesting, but at present quite inaccessible, problem.

Particularly simple and important are those *G*-spaces where prolongation is possible in the large. We call such spaces *straight,* because each geodesic is isometric to an ordinary straight line and is uniquely determined by any two of its points. In the language of the calculus of variations they are the *simply connected spaces without conjugate points.* A two-dimensional straight space is homeomorphic to the plane. Each geodesic is an open Jordan curve which tends to infinity if traversed in either direction, and there is exactly one curve through two given points. In Section 11 we prove: if a system of Jordan curves in the plane with these two properties is given, then the plane can be metrized as a straight space with the given curves as geodesics. *For plane metrics without conjugate points we thus solve the inverse problem of the calculus of variations in the large.* This is a first example where the present methods not only reestablish a classical result without differentiability hypotheses, but go far beyond it by providing a solution in the large.

The result is used for the construction of non-Desarguesian spaces which illustrate Chapter II.

2. Compact and finitely compact metric spaces

The basic concept underlying all the investigations of this book is that of the *distance* between two points of a space. The properties of distance will depend to some extent on the space considered, but we shall always assume that distance satisfies the following conditions:

(1) the distance of a point from itself is zero,

(2) the distance of two distinct points is positive, and the distance of the first point from the second is the same as that of the second from the first,

(3) distance is shortest distance; if we interpolate a point between two given points, the distance between the given points is less than the sum of their distances to the interpolated point.

These assumptions can be stated a little more exactly by formulae. We consider a set R of elements $a, b, \ldots, y, z$; we call the elements, "*points*", and the set R, a "*space*".

Our assumptions are:

A real valued function xy is defined on all pairs x, y of points in R which has the properties:

(1) $$xx = 0,$$

(2) $$xy = yx > 0, \quad \text{if} \quad x \neq y,$$

(3) $$xy + yz \geqslant xz \qquad (\textit{the triangle inequality}).$$

A space with these properties is called "*metric*".

As a minor point, we may mention that these conditions can be replaced by

(1′) $$xy = 0 \text{ if and only if } x = y,$$

(2′) $$xy + zy \geqslant xy.$$

? Last Term xz

Comments on the axioms

These properties are too general to serve as a basis for an interesting geometric theory; for they are satisfied by the following trivial space R_1: Take any set of elements and define $xy = 0$ when $x = y$, $xy = 1$, when $x \neq y$.

Nevertheless the conditions exclude many interesting spaces, for example, spaces which occur in the calculus of variations frequently do not satisfy $xy = yx$; for a practical illustration, if the distance between two points x, y on the surface of the earth be measured by the time it takes to walk from one to the other, then xy and yx will differ if one direction is uphill, and the other downhill.

Much of the early work in this book has been extended to non-symmetric distance[1], but the most interesting later results are often meaningless or have not been carried over. Thus the extension is, at least at present, not justified by the results.

The condition $xy > 0$ for $x \neq y$ fails in the spaces considered in the theory of relativity.

Finally, if we omit the triangle inequality, distance loses its meaning of shortest distance, and so much of its geometric interest. Investigations based on (1), (2) only have therefore a different aim and trend. Partly they concern spaces in which (3) holds, but is not assumed[2], because it implies the following very strong *continuity property* of distance:

$$|pq - p'q'| \leqslant pp' + qq' \tag{2.1}$$

For $\quad pq \leqslant pp' + p'q' + q'q \quad$ and $\quad p'q' \leqslant p'p + pq + qq'$

The basic definitions

If A is a non-empty set, the "*distance from p to A*" is denoted by pA and defined as the greatest lower bound of px as x traverses A:

$$pA = \inf_{x \in A} px, \tag{2.2}$$

$$|pA - qA| \leqslant pq. \tag{2.3}$$

For, if $x \in A$ then $pA \leqslant px \leqslant pq + qx$.

Hence $pA \leqslant pq + \inf qx = pq + qA$. Similarly $qA \leqslant pq + pA$. Hence (2.3).

If $\rho > 0$ the "*ρ-neighborhood* of the set A", denoted by $S(A, \rho)$, consists of all points x for which $xA < \rho$.

If A consists of a single point c, we call $S(c, \rho)$ the "(open) *sphere with center c and radius* ρ".

Thus $S(c, \rho)$ is the set of all points x such that $cx < \rho$. If $\cup$ denotes union of sets, then

$$(2.4) \qquad S(A, \rho) = \bigcup_{x \in A} S(x, \rho)$$

(2.5) *If* $q \in S(A, \rho)$, *then* $S(q, \rho - qA) \subset S(A, \rho)$.

For, if $\quad x \in S(q, \rho - qA)$, then $xq < \rho - qA$;

hence by (2.3), $xA \leqslant xq + qA < \rho$.

The space R_1 shows us that metric space is far too general to yield an interesting theory. We consider additional assumptions of two different types: the first is essentially topological and centers round the notion of compactness; the second is more geometrical, and concerns the detailed ordering of points on "lines". The first type is discussed in sections 2 to 5. Those of the second type are treated in sections 6 to 9.

Although we assume that the elementary topological properties of metric spaces are known, we discuss those briefly which we need in order to insure agreement on the terminology.

A subset M of the metric space R is "*open*" if each point p of M is the center of a sphere contained entirely in M. In symbols: given $p \in M$, there is a positive number $\delta(p)$ such that

$$S(p, \delta(p)) \subset M.$$

The point p is an "*accumulation point*" of M if, for each positive number ρ, the common part, or intersection $M \cap S(p, \rho)$ of M and $S(p, \rho)$ contains infinitely many points of M.

The "*closure of M*" denoted by $\overline{M}$, consists of M and its accumulation points. M is "*closed*", if it coincides with its closure.

If $\{p_\nu\}$ denotes a sequence of points $p_1, p_2, \ldots$, not necessarily distinct, then p is an "*accumulation point*" of the sequence, if $S(p, \varepsilon)$, for each $\varepsilon > 0$, contains infinitely many elements of the sequence. It must be noticed that, with these definitions, if M is the set traversed by p_ν, an accumulation point of $\{p_\nu\}$ need not be such as point of M; for, if M is a finite set, it has no accumulation point, but $\{p_\nu\}$ has at least one.

When there is a point p such that $S(p, \varepsilon)$, for each $\varepsilon > 0$, contains all but a finite number of the points of $\{p_\nu\}$, we say that $\{p_\nu\}$ "*converges to p*", and that p is its "*limit point*", we write, $p = \lim p_\nu$, or $p_\nu \to p$. This is equivalent to

$p_\nu p \to 0$. A set M is "*bounded*" if there is a number α such that we have $xy < \alpha$ for any pair of points x, y of M. Then, if p is any point of R, there is a number β such that $M \subset S(p, \beta)$. Conversely, if there is a point p, and a number β such that $M \subset S(p, \beta)$, then M is bounded.

A sequence $\{p_\nu\}$ of points is "bounded", if the point set which it traverses is bounded; or if $p\, p_\nu < \beta$ for some p, β and all ν.

(2.6) *A converging sequence is bounded and has just one accumulation point. In a finitely compact space (see below) the converse holds.*

Compact and finitely compact sets

A set C is "*compact*" if every infinite subset of C has an accumulation point, and this lies in C.

Thus in our terminology, a compact set is always closed.

A compact set C satisfies the *Heine-Borel Theorem*:

(2.7) If C is contained in the union of a (not necessarily denumerable[2]) collection of open sets, then it is contained in the union of a finite subcollection.

Apply this to the set of all spheres $S(p, \eta)$ with η fixed, centers p in C and we have

(2.8) A compact set C contains, for each $\eta > 0$, a finite set of points C_η such that $C \subset S(C_\eta, \eta)$.

We make an important deduction: Since each $S(p, \eta)$ is bounded, so is $S(C_\eta, \eta)$, hence C is bounded. Take $\eta = 1, 1/2, 1/3, \ldots$; altogether we get a countable set of points which is dense in C. In the usual terminology C is "*separable*".

(2.9) *A compact set (in a metric space) is bounded, closed, and separable.*

For our geometric investigations it would not suffice to consider only compact spaces, for this would exclude even Euclidean space. But the Bolzano-Weierstrass Theorem holds for that space and for all those which we will consider, namely:

(2.10) *A bounded infinite set has at least one accumulation point.*

If (2.10) holds in a set, with the accumulation point in the set, we call the set "*finitely compact*" [finitely, because (2.10) makes a statement only for bounded sets].

(2.11) A closed subset of a finitely compact set is finitely compact. A bounded closed subset of a finitely compact set is compact.

(2.12) *A finitely compact set F is closed and separable.*

To show separability, let $S_\nu = F \cap S(p, \nu)$ then by (2.11), $\overline{S}_\nu$ is compact, and hence by (2.8), it contains a finite subset T_ν with $S(T_\nu, \nu^{-1}) \supset \overline{S}_\nu \supset F \cap S(p, \nu)$. Now the set $\overset{\infty}{\underset{\nu=1}{\cup}} T_\nu$ is countable and it is dense in F, since $\cup\, S(p, \nu) \supset F$.

If R, R^1 be two metric spaces, and $M \subset R$, we say M is "mapped *in* R^1" if there is a correspondence, or function $f(p)$ such that if $p \in M$, then $f(p) \in R^1$.

If $N \subset M$ then $f(N)$ denotes the set of points $f(p)$ in R^1 obtained when p traverses N, and we say f "*maps N on f(N)*".

If $M' \subset R'$ then $f^{-1}(M')$ is the "*original*" of M', consisting of the points p in M for which $f(p) \in M'$. This set may be empty.

If the mapping $p' = f(p)$ is one-to-one, then $f^{-1}(p')$ is a single point for each $p' \in f(M)$, and $p = f^{-1}(p')$ is a mapping *of* M' on M.

The mapping $p' = f(p)$ of M in R' is "*continuous*", if $f(p_\nu) \to f(p)$ whenever $p_\nu, p \in M$ and $p_\nu \to p$. Thus if $p \in M$ and $\varepsilon > 0$, then a number δ, depending on ε, p exists such that

$$pq < \delta(\varepsilon, p) \qquad \text{implies} \qquad f(p)\, f(q) < \varepsilon.$$

(2.13) *A continuous mapping of a compact set C is uniformly continuous.*

This means that, given $\varepsilon > 0$, a positive number δ exists such that $f(x)\, f(y) < \varepsilon$ whenever $x, y \in C$ and $xy < \delta$; this follows from the Heine-Borel Theorem. For with the above meaning of $\delta(\varepsilon, p)$, there is a finite number of spheres $S\left(p_\nu, \frac{1}{2}\delta(\varepsilon/2, p_\nu)\right)$ which cover C. Let δ be the least of $\delta(\varepsilon/2, p_\nu)/2$ and let $xy < \delta$. Then if $x \in S\left(p_\mu, \frac{1}{2}\delta(\varepsilon/2, p_\mu)\right)$, we have

$$y\, p_\mu \leqslant yx + x\, p_\mu < \delta + \tfrac{1}{2}\delta(\varepsilon/2, p_\mu) \leqslant \delta(\varepsilon/2, p_\mu);$$

and by the definition of $\delta(\varepsilon/2, p_\mu)$,

$$f(x)\, f(y) \leqslant f(x)\, f(p_\mu) + f(p_\mu)\, f(y) < \varepsilon/2 + \varepsilon/2 = \varepsilon.$$

(2.14) *If the mapping $f(p) = p'$ of a compact set C is continuous, then $f(C)$ is compact. If, further, $f(p)$ is one-to-one, then $f^{-1}(p')$ is continuous.*

For, if $\{q_\nu'\} = \{f(q_\nu)\}$ is any sequence of points in $f(C)$, it has an accumulation point q in C; hence $\{q_\nu\}$ has a subsequence $\{q_\mu\}$ such that $q_\mu \to q$. As the mapping is continuous, $f(q_\mu) \to f(q) \in f(C)$.

If $f(p)$ is one-to-one and $q_\nu' = f(q_\nu) \to q' = f(q)$ then no subsequence $\{q_\mu\}$ of $\{q_\nu\}$ can converge to a point $p \neq q$, because then would $f(q_\mu) \to f(p) \neq f(q) = q'$. But as $\{q_\nu\}$ is bounded, it has an accumulation point; hence by (2.6) $q_\nu \to q$.

A one-to-one mapping $p' = f(p)$ of a subset M of R on a subset M' of R' for which both $f(p)$ and $f^{-1}(p)$ are continuous is called "*topological*", and two sets M, M' for which such a mapping exists are "*homeomorphic*".

Thus a one-to-one continuous mapping of a compact set is topological.

Selection Theorems

The reader will have noticed the analogy between (2.13) and an elementary theorem in analysis. Certain standard selection theorems in analysis also hold for finitely compact spaces, and this is one reason for their importance. We first apply Cantor's diagonal process.

(2.15) Let the mappings $f_\nu(p)$ $(\nu = 1, 2, \ldots)$ of the denumerable subset N of R in the finitely compact metric space R' have the property that $\{f_\nu(p)\}$ $(\nu = 1, 2, \ldots)$ is bounded for each $p \in N$. Then a suitable subsequence of $\{f_\nu(p)\}$ converges for all p of N.

For let $p_1, p_2, \ldots$ be the points of N. Since $\{f_\nu(p_1)\}$ is bounded and R' is finitely compact, some subsequence $\{f_{\nu,1}(p_1)\}$ converges. Similarly the sequence $\{f_{\nu,1}(p_2)\}$ has a convergent subsequence $\{f_{\nu,2}(p_2)\}$, and of course $\{f_{\nu,2}(p_1)\}$ converges. So continuing, we find a sequence $\{f_{\nu,\varkappa}(p)\}$ converging for $p = p_1, p_2, \ldots, p_\varkappa$, and we can take $\varkappa = 1, 2, \ldots$ indefinitely. The diagonal sequence $\{f_{\nu,\nu}(p)\}$ converges for all $p_\varkappa \in N$, since for $\nu \geqslant \varkappa$ it is a subsequence of $\{f_{\nu,\varkappa}(p)\}$.

Various forms of Ascoli's Theorem are based on this process:

A sequence of mappings $\{f_\nu(p)\}$ of M in R' is "*equicontinuous*" if, when $\varepsilon > 0$ is given, there is some $\delta > 0$ such that

(2.16) $\quad f_\nu(p)\, f_\nu(q) < \varepsilon$ for all ν, and for any p, q in M with $pq < \delta$.

This is sometimes called "uniform equicontinuity".

(2.17) *If R' is finitely compact and $\{f_\nu(p)\}$ is bounded for each $p \in M$, and M contains a countable dense set N, and if the $f_\nu(p)$ are equicontinuous in M, then some subsequence $\{f_\mu(p)\}$ of $\{f_\nu(p)\}$ tends to a uniformly continuous mapping of M in R'.*

Proof: By (2.15) there is a subsequence $\{\mu\}$ of $\{\nu\}$ such that $\{f_\mu(p)\}$ converges for all p in N. Let ε be any given positive number and choose δ so that, if $p, q \in M$ and $pq < \delta$, then $f_\mu(p)\, f_\mu(q) < \varepsilon/3$. Now if p is any point of M choose q in N with $pq < \delta$, then

(2.18) $$f_{\mu_1}(p)\, f_{\mu_2}(p) \leqslant f_{\mu_1}(p)\, f_{\mu_1}(q) + f_{\mu_1}(q)\, f_{\mu_2}(q) + f_{\mu_2}(q)\, f_{\mu_2}(p).$$

Since $\{f_\mu(q)\}$ converges there is a positive number β such that $f_{\mu_1}(q)\, f_{\mu_2}(q) \leqslant \varepsilon/3$ if $\mu_1, \mu_2 > \beta$. Hence $f_{\mu_1}(p)\, f_{\mu_2}(p) < \varepsilon$. As R' is finitely compact $\{f_\mu(p)\}$ converges. But as β depends on q and thus on p, the convergence need not be uniform in general. Finally, $f(p) = \lim f_\mu(p)$ is uniformly continuous, since, if $pq < \delta$, then $f_\nu(p)\, f_\nu(q) < \varepsilon$, and $f(p)\, f(q) \leqslant \varepsilon$.

(2.19) *The uniform convergence of a suitable subsequence* $\{f_\mu(p)\}$ *can be proved when* M *is compact.*

Proof: Let $f_\nu(p)$ be equicontinuous mappings of the compact set C in the finitely compact space R'. As N take the set $\cup C_{1/\nu}$ of (2.8), then by (2.17) there is a subsequence $\{f_\mu(p)\}$ which tends to a continuous mapping $f(p)$.

To show that the convergence is uniform, choose δ again so that $f_\mu(p)\,f_\mu(q) < \varepsilon/3$ when $pq < \delta$. Consider a fixed $\nu_0 > 1/\delta$; the set C_{1/ν_0} contains a point q with $pq < \delta$. Now β exists such that $f_{\mu_1}(q)\,f_{\mu_2}(q) < \varepsilon/3$ for all q in C_{1/ν_0}, when $\mu_1, \mu_2 > \beta$. Thus β is independent of p, and (2.18) gives $f_{\mu_1}(p)\,f_{\mu_2}(p) < \varepsilon$ when $\mu_1, \mu_2 > \beta$. For $\mu_2 \to \infty$, we find $f_{\mu_1}(p)\,f(p) \leqslant \varepsilon$, when $\mu_1 > \beta$. This shows uniformity of the convergence.

Real valued functions

Two special cases of mappings of R in R' which are of great importance occur when either R or R' is the real axis T. We now take the second case. We shall always define "*distance*" on T in the natural way: the distance of the numbers τ_1, τ_2 shall be $|\tau_1 - \tau_2|$. If $R' = T$, then $f(p)$ is a real valued function of p. From (2.14):

(2.20) *A real continuous function* $f(p)$ *defined on a compact set* C *attains its maximum and minimum.*

For the image $f(C)$ is a compact, hence closed bounded set on T and includes a greatest and a smallest number.

An important example of a continuous function in R is the distance pA of a variable point p from a fixed set A (See 2.3). If there is a point f of A such that $pf = pA$ then f is a "*foot*" of p on A. If A is compact then by (2.20) a foot always exists, but it need not be unique; e. g. in ordinary space, each point of a spherical surface is a foot for its center.

(2.21) *If* H *is finitely compact (and not empty), then each point* p *has a foot on* H.

For if $\alpha = pH = \inf_{x \in H} px$, let $px_\nu \to \alpha$, where $\{x_\nu\}$ is a sequence of points on H. Then $\{x_\nu\}$ is bounded and, since H is finitely compact, contains a subsequence $\{x_\mu\}$ which converges to a point f of H. By (2.1), $px_\mu \to pf = \alpha$.

(2.22) If f_ν is a foot of x_ν on A and $x_\nu \to x, f_\nu \to f \in A$, then f is a foot of x on A.

For if g is any point on A, then

$$xg - xf \geqslant x_\nu g - xx_\nu - (xx_\nu + x_\nu f_\nu + f_\nu f) \geqslant -2\,xx_\nu - f_\nu f \to 0$$

hence $xg \geqslant xf$.

(2.23) *If H is finitely compact and each point x of M has a unique foot $f(x)$ on H then $f(x)$ depends continuously on x;* thus $f(x)$ is a continuous map of M in H.

For let $x_\nu \to x$, where $x_\nu, x \in M$. By (2.3) $\{f(x_\nu)\}$ is a bounded sequence; $xf(x_\nu) \leqslant xx_\nu + x_\nu f(x_\nu) = xx_\nu + x_\nu H \leqslant 2\ xx_\nu + xH$. By (2.22) each accumulation point of $\{f(x_\nu)\}$ is a foot of x, and hence coincides with $f(x)$; an accumulation point exists on H, since H is finitely compact (compare (2.6)).

3. Convergence of point sets

We shall introduce that notion of convergence which Hausdorff called the "*closed limit*", and as we use no other, we omit the word "closed". Since the notion will be applied to unbounded sets, it will be necessary to understand clearly the meaning of statements of which the following is a simple example: in the Cartesian (ξ, η)-plane, the circle

$$(\xi - \nu)^2 + \eta^2 = \nu^2 \tag{3.1}$$

tends as $\nu \to \infty$ to the line $\xi = 0$.

Now let $\{M_\nu\}$ be any sequence of non-empty subsets of a metric space R. We define the upper and lower limits of the sequence as follows:

The "*upper limit*", $\limsup M_\nu$, consists of all accumulation points of all sequences $\{p_\nu\}$ when $p_\nu \in M_\nu$:

Thus $p \in \limsup M_\nu$ if, and only if, each $S(p, \varepsilon)$, $\varepsilon > 0$, contains points of infinitely many M_ν.

The "*lower limit*", $\liminf M_\nu$, consists of all points p such that each $S(p, \varepsilon)$ for any $\varepsilon > 0$ contains points of all but a finite number of the M_ν. Obviously

$$\liminf M_\nu \subset \limsup M_\nu.$$

If these limits coincide we say $\{M_\nu\}$ converges:

$$\lim M_\nu = M \quad \text{means} \quad M = \liminf M_\nu = \limsup M_\nu.$$

Consequences of the definitions

The upper and the lower limits of $\{\overline{M}_\nu\}$ are those of $\{M_\nu\}$. For $M_\nu \subset \overline{M}_\nu$, and a sphere which contains a point of $\overline{M}_\nu$ contains points of M_ν.

Further, *the upper and lower limits of $\{M_\nu\}$ are closed sets.* For if p is the limit of points p^ν in $\limsup M_\nu$, there is a subscript $\mu_\nu > \nu$, and a point

$p^{\nu}_{\mu_\nu} \in M_{\mu_\nu}$ such that $p^{\nu} p^{\nu}_{\mu_\nu} < 1/\nu$; hence $p \in \limsup M_\nu$. Similarly for the lower limit.

Example: Consider the sequence of straight lines $\xi = \nu$, $\nu = 0, \pm 1, \pm 2, \ldots$ in the (ξ, η) plane. According to the definition given, this converges to the empty set; but if to each line we adjoin a fixed point p to form a new set, these sets have p as limit. The lines, as it were, "disappear to infinity". The same may be said of the semicircles, $\xi \geqslant \nu$ in (3.1); the semicircles $\xi < \nu$ "tend" in an intuitive fashion to $\xi = 0$. Those for $\xi \geqslant \nu$ count less and less, so to speak, as we increase ν.

These remarks on convergence of sets suffice for reading this book. The remainder of this section serves to elucidate the notion of convergence, and to make precise the intuitive statements above.

Our aim is to define a distance $\delta(M, N)$ between sets M, N in a metric space R such that when R is finitely compact $\delta(M_\nu, M) \to 0$, if and only if $M_\nu \to \overline{M}$. Hausdorff considered this problem for the bounded case only.

DISTANCE OF SETS

Guided by our example, we select a point p and define the *"distance"* of the (non-empty) sets M, N as follows:

$$\delta_p(M, N) = \sup_{x \in R} |xM - xN| \, e^{-px} \tag{3.2}$$

(px is, of course, the distance between p and x). Thus

$$\delta_p(M, N) \geqslant |pM - pN|.$$

This definition diminishes the importance of points x at a large distance from p. The weighting factor e^{-px} is chosen for convenience; many other functions which decrease fast enough would do.

We note for reference

$$\tau e^{-\tau} < 1 \quad \text{for real } \tau, \tag{3.3}$$

$$e^{-px} \leqslant e^{-px} \cdot e^{qp + px - qx} = e^{-qx} e^{pq}. \tag{3.4}$$

The number (3.2) is always finite, since, by (2.3)

$$|xM - xN| \leqslant xM + xN \leqslant pM + pN + 2px.$$

Hence by (3.3):

$$\delta_p(M, N) \leqslant pM + pN + 2. \tag{3.5}$$

We show that with this distance, the closed sets M in R constitute a metric space. Clearly $\delta_p(M, N) = \delta_p(N, M) \geqslant 0$.

We next prove the triangle inequality: $\delta_p(L, M) + \delta_p(M, N) \geqslant \delta_p(L, N)$. If $\varepsilon > 0$ is given, there is a point z such that

$$\delta_p(L, N) - \varepsilon < |zL - zN|\, e^{-pz}$$
$$\leqslant (|zL - zM| + |zM - zN|)\, e^{-pz} \leqslant \delta_p(L, M) + \delta_p(M, N).$$

(3.6) $\delta_p(M, N) = 0$ if, and only if, $\overline{M} = \overline{N}$.

(We cannot expect that $\delta_p(M, N) = 0$ if, and only if, $M = N$).

Proof: $\delta_p(M, N) = 0$ is equivalent to $xM = xN$ for all x in R. Now for any set L and any point q

$$qL = q\overline{L}. \tag{3.7}$$

Hence $\overline{M} = \overline{N}$ implies $xN = xM$ for all x in R, and $\delta_p(M, N) = 0$. But if a point y lies in $\overline{M}$ and not in $\overline{N}$, then $y\overline{N} = yN > 0$, $y\overline{M} = yM = 0$; and hence

$$0 < |yM - yN|\, e^{-py} \leqslant \delta_p(N, M).$$

We have shown:

(3.8) *With our distance* (3.2) *the non-empty closed sets of a metric space themselves form a metric space.*

We next show that the exceptional role of the point p is really immaterial. We introduce a notion of general importance. If $\delta_1(p, q)$ and $\delta_2(p, q)$ be two metrics defined for the same set (of points or of any elements), they are "*topologically equivalent*" if each of the relations $\delta_1(p_\nu, q) \to 0$, $\delta_2(p_\nu, q) \to 0$ implies the other. Thus any notions which involve convergence only coincide for the two metrics.

We easily show that the metrics δ_p, δ_q are topologically equivalent. Using (3.4),

$$\delta_q(M, N) = \sup_{x \in R} |xM - xN|\, e^{-qx} \leqslant \sup_{x \in R} |xM - xN|\, e^{-px} \cdot e^{pq} = \delta_p(M, N)\, e^{pq}.$$

Similarly $\delta_p(M, N) \leqslant \delta_q(M, N)\, e^{pq}$.

LIMIT AND DISTANCE, FOR SETS

(3.9) *If $\delta_p(M_\nu, M) \to 0$, as $\nu \to \infty$, then lim $M_\nu = \overline{M}$.*

Proof: Let $x \in \overline{M}$, then, noting (3.7), $xM = x\overline{M} = 0$,

$$\delta_p(M_\nu, M) \geqslant |xM_\nu - xM|\, e^{-px} = xM_\nu \cdot e^{-px}.$$

Hence $xM_\nu \to 0$; in M_ν there is a point y_ν such that $xy_\nu \to 0$. Thus $\overline{M} \subset \liminf M_\nu$.

Next let $y \in \limsup M_\nu$. Then there is a subsequence $\{\mu\}$ of $\{\nu\}$ such that there are points $y_\mu \in M_\mu$ with $y_\mu \to y$ and

$$e^{py}\, \delta_p(M_\mu, M) \geqslant |y\overline{M} - yM_\mu| \geqslant y\overline{M} - y_\mu M_\mu - yy_\mu = y\overline{M} - yy_\mu.$$

Hence $y\overline{M} = 0$, $\limsup M_\nu \subset \overline{M}$.

These two conclusions give the result.

Note: The converse of (3.9) is not true for general metric spaces. Consider as an example a space consisting of a denumerable number of distinct points $p_1, p_2, \ldots$ with the trivial distance $p_\nu p_\mu = 1$ for $\nu \neq \mu$. If $M_\nu = R - p_\nu$ then lim $M_\nu = R$, because any $S(p_{\nu_0}, \varepsilon)$ contains for arbitrary positive ε the point p_{ν_0} of M_ν, if $\nu \neq \nu_0$. But $\delta_{p_1}(M_\nu, R) = p_\nu M_\nu \cdot e^{-p_1 p_\nu} = e^{-1}$ for $\nu > 1$. However for finitely compact spaces, this converse is true.

(3.10) *If the space R is finitely compact, and* lim $M_\nu = M = \overline{M} \neq 0$ *then*

$$\delta_p(M_\nu, M) \to 0 \qquad as \qquad \nu \to \infty.$$

Proof: If not, there is a subsequence $\{\mu\}$ of $\{\nu\}$ for which $\delta_p(M_\mu, M) \geqslant 4\alpha > 0$. Thus there is a sequence $\{x_\mu\}$ of points such that, for large μ,

$$|x_\mu M_\mu - x_\mu M|\, e^{-px_\mu} > 3\alpha. \tag{3.11}$$

If $z \in M$, then M_μ contains a point z_μ with $z_\mu \to z$, and

$$|x_\mu M_\mu - x_\mu M| \leqslant x_\mu z_\mu + x_\mu z \leqslant 2\, px_\mu + pz_\mu + pz.$$

By (3.11),

$$(2\, px_\mu + pz_\mu + pz)\, e^{-px_\mu} > 3\alpha.$$

Hence $\{px_\mu\}$ is bounded. By (3.11) there is either a subsequence $\{\lambda\}$ of $\{\mu\}$ with

$$x_\lambda M_\lambda - x_\lambda M > 3\alpha \tag{3.12}$$

or a subsequence $\{\theta\}$ of $\{\mu\}$ with

$$x_\theta M - x_\theta M_\theta > 3\alpha. \tag{3.13}$$

These cases we treat separately. In the first, choose y_λ in M with $x_\lambda y_\lambda < x_\lambda M + \alpha$. Thus $\{y_\lambda\}$ is bounded and has an accumulation point y. Since M is closed we have $y \in M$. But

$$y_\lambda M_\lambda + \alpha \geqslant x_\lambda M_\lambda - x_\lambda y_\lambda + \alpha > x_\lambda M_\lambda - x_\lambda M > 3\alpha.$$

Hence for large λ, $yM_\lambda > y_\lambda M_\lambda - \alpha > \alpha$ which contradicts $y \in \liminf M_\lambda$. In the second case (3.13), choose $u_\theta \in M_\theta$ with $u_\theta x_\theta < x_\theta M_\theta + \alpha$. Then since $\{x_\theta\}$, $\{z_\theta\}$ are bounded so is $\{u_\theta\}$. By (3.13),

$$u_\theta M + \alpha \geqslant x_\theta M - u_\theta x_\theta + \alpha > x_\theta M - x_\theta M_\theta > 2\alpha.$$

Thus there would be an accumulation-point u of $\{u_\theta\}$ which was not in M, contrary to $u \in \limsup M_\nu = M$.

This theorem, together with a result of Hausdorff, leads to the following statement which, although not explicitly, underlies many of our arguments.

(3.14) *If R is finitely compact and $\{M_\nu\}$ a sequence of non-empty sets for which $\delta_p(M_\nu, M_\mu)$ is bounded then $\{M_\nu\}$ contains a subsequence $\{M_\lambda\}$ which converges to a non-empty set M.*

For as R is separable, Hausdorff's result [2, p. 147] assures us that $\{M_\nu\}$ contains a subsequence $\{M_\lambda\}$ whose limit is a set M in R. By (3. 2)

$$\delta_p(M_\lambda, M_1) \geqslant |pM_\lambda - pM_1|.$$

Hence $\{pM_\lambda\}$ is bounded, and M_λ contains a point z_λ such that $\{pz_\lambda\}$ is bounded. $\{z_\lambda\}$ has an accumulation point $z \in \limsup M_\lambda = M$. Thus M is not empty and (3.14) follows from (3.10).

We now can state more than (3.8). By (3. 9, 10, 14):

(3.15) ***Theorem:*** *The non-empty closed sets of a finitely compact metric space R form with the metric* (3.2) *again a finitely compact metric space.*

If the space R is bounded the factor e^{-px} may be omitted in (3.2) and distance then coincides with Hausdorff's distance.[1]

4. Motion and Isometry

The concept of motion as a transformation of points which preserves distances is of paramount importance in geometry and particularly applicable to metric spaces.

A mapping $p' = f(p)$ of the set M in the space R on a set M' of R' is "*isometric*", or an "*isometry*", if it preserves distances:

(4.1) *$f(p)\,f(q) = pq$ for any p, q in M.*

The mapping is topological, and it has an inverse which is an isometry, since, if $pq > 0$ then $f(p)\,f(q) > 0$, and so $f(p) \neq f(q)$.

Two sets M, M' of points are "*congruent*" if they can be related by an isometry.

A "*motion*" of a metric space R may be defined as an isometry of R on itself.

Notice that $f(p)$ must traverse *all* points of R when p does so. Thus the mapping $\tau' = f(\tau) = \tau + 1$ of the non negative real axis $\tau \geqslant 0$ is not a motion of $\tau \geqslant 0$, since the point $f(\tau)$ does not traverse the part $0 \leqslant \tau < 1$.

In this section we treat properties of motions which hold for general, or at any rate for finitely compact spaces.

Motions will be denoted by capital Greek letters, usually Φ, Ψ, with or without subscripts. When Φ is the motion $p' = f(p)$ we write $p' = p\,\Phi$, and $f(M) = M\,\Phi$ for sets. If Ψ is the motion $p' = g(p)$, then the product $\Phi\Psi$ is defined by $p' = p\,\Phi\Psi = g(f(p))$. Thus products of motions are read from left to right.

The point p is a "*fixed point*" of Φ if $p = p\,\Phi$.

(4.2) *The fixed points of a motion form a closed set.*

For if $p_1, p_2, \ldots$ are fixed points, and $p_\nu \to p$, then $pf(p) = \lim p_\nu f(p) = \lim f(p_\nu)\,f(p) = \lim p_\nu p = 0$.

The motion Φ "*leaves the set M invariant*", if $M = M\,\Phi$, i. e. if Φ maps M on itself; it "*leaves M point-wise invariant*", or "*fixed*", if $p\,\Phi = p$ for each point p of M. We have obviously:

(4.3) The totality of motions which a metric space can carry form a group.

(4.4) Of this group, the totality of motions which leaves a set M invariant form a subgroup. So do those motions which leave M point-wise invariant.

(4.5) *If a motion Ψ exists which maps M on M', then the group of motions which leaves M (pointwise) invariant is isomorphic to the group of motions leaving M' (pointwise) invariant.*

For if $M\Phi = M$ then since $M\Psi = M'$, we have $M'\Psi^{-1}\Phi\Psi = M'$. Thus $\Psi' = \Psi^{-1}\Phi\Psi$ leaves M' invariant. Conversely if Φ' is any motion leaving M' invariant then $\Psi\Phi'\Psi^{-1}$ leaves M invariant. Thus the groups correspond by an isomorphism. Similarly if the sets are pointwise fixed.

(4.6) *Isometric spaces have isomorphic groups of motions.*

For if Ω is an isometry of space R on R' and Φ is a motion of R then $\Phi\,\Omega$ is a motion of R', and every motion of R' can be written in this form. The isomorphism desired is obtained by making $\Phi, \Phi\,\Omega$ correspond.

Isometric spaces have the same geometric properties. For example, if f is a foot of x on a set A in R, then $f\Omega$ is a foot of $x\Omega$ in the set $A\Omega$ in R'.

Distance between motions

In geometric discussions, we often use terms which presuppose that notions of continuity or of limit have been defined for motions: e. g. if we speak of a continuous rotation about a point, or say that the group of covering motions of a space is discrete. In these cases the group is explicitly given and it would be easy to define limits within the group.

It is not at all obvious, and probably not true, that a general definition of limit could be given which would always coincide with that used in the special cases; but it is feasible for the finitely compact spaces which we treat.

We have a problem and a difficulty like that encountered when we considered the convergence of sets. If the space R is bounded, then a satisfactory definition for the distance between the motions Φ and Ψ would be $\sup_{x \in R} x\Phi\, x\Psi$. But this would be infinite even for such simple motions as rotations in a plane. Guided by (3.2) we define for any metric space R, a distance $\delta_p(\Phi, \Psi)$ between motions Φ, Ψ, by taking an arbitrary fixed point p and putting

$$\delta_p(\Phi, \Psi) = \sup_{x \in R} x\Phi x\Psi\, e^{-px}. \tag{4.7}$$

Now $x\Phi\, x\Psi \leqslant x\Phi\, p + p\, x\Psi = xp\Phi^{-1} + p\Psi^{-1}x \leqslant 2\, xp + pp\Phi^{-1} + pp\Psi^{-1}$. Hence by (3.3)

$$\delta_p(\Phi, \Psi) \leqslant 2 + pp\Phi^{-1} + pp\Psi^{-1} \tag{4.8}$$

and thus is *finite*. Also, $\delta_p(\Phi, \Psi)$ is bounded if R is bounded.

Of the metric axioms, two are obvious: $\delta_p(\Phi, \Psi) = \delta_p(\Psi, \Phi) \geqslant 0$ and $\delta_p(\Phi, \Psi) = 0$ if, and only if, $x\Phi\, x\Psi = 0$ for all x, i. e. if $\Phi = \Psi$.

For the triangle inequality,

$$\delta_p(\Phi, \Psi) + \delta_p(\Psi, \Omega) \geqslant \delta_p(\Phi, \Omega), \tag{4.9}$$

we need only observe that for a suitable point z

$$\delta_p(\Phi,\Omega)-\varepsilon\leqslant z\Phi z\Omega\, e^{-pz}\leqslant(z\Phi z\Psi+z\Phi z\Omega)\, e^{-pz}\leqslant\delta_p(\Phi,\Psi)+\delta_p(\Psi,\Omega).$$

If we replace the point p by another q we obtain an equivalent distance, since by (3.4)

$$\delta_q(\Phi,\Psi)=\sup_{x\in R} x\Phi x\Psi\, e^{-qx}\leqslant\sup_{x\in R} x\Phi x\Psi\, e^{-px}\, e^{pq}=\delta_p(\Phi,\Psi)\, e^{pq}.$$

Hence and similarly

(4.10) $$\delta_p(\Phi,\Psi)\, e^{-pq}\leqslant\delta_q(\Phi,\Psi)\leqslant\delta_p(\Phi,\Psi)\, e^{+pq}.$$

The distance $\delta_p(\Phi,\Psi)$ *is right-invariant,* that is, for any three motions Φ, Ψ, Ω.

(4.11) $$\delta_p(\Phi,\Psi)=\delta_p(\Phi\Omega,\Psi\Omega).$$

This is so, since $x\Phi\Omega\; x\Psi\Omega = x\Phi\; x\Psi$ for all x. But distance is, in general, not left-invariant. Nevertheless we have

(4.12) $$\delta_p(\Phi,\Psi)=\delta_{p\Omega^{-1}}(\Omega\Phi,\Omega\Psi).$$

For $y=x\Omega^{-1}$ traverses all of R when x does. Hence

$$\sup_x x\Phi x\Psi\, e^{-px}=\sup_x x\Omega^{-1}\Omega\Phi\; x\Omega^{-1}\Omega\Psi\, e^{-p\Omega^{-1}\, x\Omega^{-1}}=\sup_y y\Omega\Phi\; y\Omega\Psi\, e^{-p\Omega^{-1}y}$$

Convergence of motions

(4.13) *If* $\delta_p(\Phi_\nu,\Phi)\to 0$, *then* $x\Phi_\nu\to x\Phi$ *for each point* x of R.

This follows at once from $e^{px}\,\delta_p(\Phi_\nu,\Phi)\geqslant x\Phi_\nu\, x\Phi$. The converse in general, fails. For consider again the space consisting of a denumerable set of distinct points $p_1, p_2, \ldots$ with distance $p_\nu p_\mu = 1$ for $\nu\neq\mu$. Let Φ_ν be the motion which interchanges p_ν, $p_{\nu+1}$, and leaves the other points fixed.

Then $\lim_\nu p_\mu\Phi_\nu = p_\mu$ for every μ, but if E is the identity,

$$\delta_p(\Phi,E)=p_\nu\Phi_\nu\, p_\nu\, e^{-1}=e^{-1}\qquad\text{for}\qquad \nu>1.$$

However:

(4.14) *If* R *is finitely compact, and if* $x\Phi_\nu\to x\Phi$ *for each* x *in* R, *then* $\delta_p(\Phi_\nu,\Phi)\to 0$.

For, otherwise there would be a sequence $\{y_\mu\}$ of points, and a subsequence $\{\Phi_\mu\}$ of $\{\Phi_\nu\}$, with

$$y_\mu\Phi_\mu\, y_\mu\Phi\, e^{-py_\mu}>\alpha>0.$$

Because R is finitely compact, there is either a subsequence $\{\lambda\}$ of $\{\mu\}$ with $py_\lambda\to\infty$ or a subsequence $\{\theta\}$ of $\{\mu\}$ such the y_θ converges to a point y.

The first case is impossible, since

$$y_\lambda\Phi_\lambda\, y_\lambda\Phi\, e^{-py_\lambda} \leqslant (y_\lambda\Phi_\lambda\, p\Phi_\lambda + p\Phi_\lambda\, p\Phi + p\Phi\, y_\lambda\Phi)\, e^{-py_\lambda}$$
$$\leqslant 2\, py_\lambda\, e^{-py_\lambda} + p\Phi_\lambda\, p\Phi \to 0.$$

The second case is also impossible, since

$$y_\theta\Phi_\theta\, y_\theta\Phi\, e^{-py_\theta} \leqslant y_\theta\Phi_\theta\, y\Phi_\theta + y\Phi_\theta\, y\Phi + y\Phi\, y_\theta\Phi = 2\, yy_\theta + y\Phi_\theta\, y\Phi \to 0.$$

We now prove the important fact:

(4.15) **Theorem:** *The group of all motions which a finitely compact space possesses is a finitely compact space when the metric is defined by* (4.7). *The group is compact when the space is compact.*

Proof. Let $\{\Phi_\nu\}$ be a bounded sequence of motions, that is, we suppose that, for some α, $x\Phi_\nu\, x\Phi_\mu < \alpha\, e^{px}$.

Hence $\{x\Phi_\nu\}$ is bounded for every x, moreover the mappings Φ_ν are equicontinuous, since $xy < \varepsilon$ implies

$$x\Phi_\nu\, y\Phi_\nu = xy < \varepsilon.$$

By (2.17) there is a subsequence $\{\lambda\}$ of $\{\mu\}$ such that $x\Phi_\lambda$ converges to a point $x\Phi$ for every x, and

$$x\Phi\, y\Phi = \lim x\Phi_\lambda\, y\Phi_\lambda = xy.$$

Hence the correspondence of x to $x\Phi$ is an isometric mapping of R in itself, but to show that it is a motion we must also prove that for each point y of R there is a point y' with $y'\Phi = y$. Now the sequence $\{y\Phi_\lambda^{-1}\}$ is bounded, since $y\Phi_\lambda^{-1}\, p = yp\Phi_\lambda \to yp\Phi$. Hence there is a subsequence $\{\theta\}$ of $\{\lambda\}$ such that the points $y\Phi_\theta^{-1}$ converge to a point y'. Then $y = y'\Phi$ since

$$yy'\Phi = \lim yy'\Phi_\theta = \lim y\Phi_\theta^{-1}\, y' = 0.$$

We have shown that $x\Phi_\lambda \to x\Phi$ for each x. Hence by (4.14) $\delta_p(\Phi_\lambda, \Phi) \to 0$, as desired. If R is compact then any $\{\Phi_\nu\}$ is bounded.

(4.16) *If* $\mathfrak{G}$ *is a group of motions of a metric space* R *then its closure* $\overline{\mathfrak{G}}$ (under the metric (4.7)) *is a group.*

Proof. We must show that if $\Phi, \Psi \in \overline{\mathfrak{G}}$, then $\Phi\Psi^{-1} \in \overline{\mathfrak{G}}$. Now $\Phi \in \overline{\mathfrak{G}}$ means there are motions $\Phi_\nu \in \overline{\mathfrak{G}}$ with $\delta_p(\Phi_\nu, \Phi) \to 0$. Similarly for Ψ. Then, using (4.10, 11, 12), we have

$$\delta_p(\Phi_\nu\Psi_\nu^{-1}, \Phi\Psi^{-1}) = \delta_p(\Phi_\nu, \Phi\Psi^{-1}\Psi_\nu) \leqslant \delta_p(\Phi_\nu, \Phi) + \delta_p(\Phi, \Phi\Psi^{-1}\Psi_\nu)$$
$$= \delta_p(\Phi_\nu, \Phi) + \delta_q(\Psi, \Psi_\nu)$$

where $q = p\,\Phi\Psi^{-1}$. Hence $\delta_p(\Phi_\nu\Psi_\nu^{-1}, \Phi\Psi^{-1}) \to 0$ and $\Phi\Psi^{-1} \in \overline{\mathfrak{G}}$.

5. Curves and their lengths

A continuous mapping $x(\tau)$ of a closed interval $[\alpha, \beta]$, of the real τ-axis, in the metric space R is called a "*curve*" in R ($[\alpha, \beta]$ stands for the set of those τ which satisfy $\alpha \leqslant \tau \leqslant \beta$). The curve may cross itself, or repeat itself. Thus distinct τ may correspond to points that coincide geometrically.

The theory of the length of a curve resembles that in ordinary space.

Let Δ stand for a partition $\alpha = \tau_0 \leqslant \tau_1 \leqslant \ldots \leqslant \tau_\varkappa = \beta$ of $[\alpha, \beta]$ and put

$$\lambda(x, \Delta) = \sum_{\nu=1}^{\varkappa} x(\tau_{\nu-1})\, x(\tau_\nu).$$

Define "*the length*" $\lambda(x)$ of $x(\tau)$, $\alpha \leqslant \tau \leqslant \beta$, as the least upper bound of $\lambda(x, \Delta)$ for all possible partitions Δ of $[\alpha, \beta]$:

(5.1) $$\lambda(x) = \sup_{\Delta} \lambda(x, \Delta).$$

The curve is "*rectifiable*" when $\lambda(x)$ is finite. As a method of evaluating this length, we have the following analogue to the usual procedure. For brevity define for each partition Δ:

$$||\Delta|| = \max_{\nu} |\tau_{\nu+1} - \tau_\nu|.$$

Then

(5.2) *If the curve $x(\tau)$ is rectifiable and ε is any positive number, then a number δ exists such that $||\Delta|| < \delta$ implies*

$$\lambda(x) - \lambda(x, \Delta) < \varepsilon.$$

But if $\lambda(x) = \infty$, then for any $\gamma > 0$, a number δ exists such that $||\Delta|| \leqslant \delta$ implies $\lambda(x, \Delta) > \gamma$.

Proof. We note that in our definition of partition, some partition points may coincide; it will clearly be sufficient to prove (5.2) when they are all distinct. Then by the definition of $\lambda(x)$, there is a partition

$$\Delta' : \alpha = \sigma_0 < \sigma_1 < \ldots < \sigma_\eta = \beta$$

with

(5.3) $$\lambda(x, \Delta') > \lambda(x) - \varepsilon/2 \quad \text{or} \quad \lambda(x, \Delta') > \gamma + \varepsilon/2.$$

By (2.13) $x(\tau)$ is uniformly continuous. Hence δ' exists such that

(5.4) $$x(\tau')\, x(\tau'') < \varepsilon/4\eta \quad \text{if} \quad |\tau' - \tau''| < \delta'.$$

Define $\delta(\varepsilon)$ or $\delta(\gamma)$ in the two cases, by

$$\delta = \min\left\{\delta', \frac{1}{3}(\sigma_1-\sigma_0), \frac{1}{3}(\sigma_2-\sigma_1), \ldots, \frac{1}{3}(\sigma_\eta-\sigma_{\eta-1})\right\} \tag{5.5}$$

and let Δ be any partition $\alpha = \tau_0 < \tau_1 < \ldots < \tau_\varkappa = \beta$ for which $\|\Delta\| < \delta$.

Denote by τ_μ' one of the two possible τ_ν with least distance from $\sigma_\mu, \mu = 1, \ldots, \eta$.

Then $|\tau_\mu' - \sigma_\mu| < \delta$, and by (5.5), $\tau_\mu' < \tau_{\mu+1}'$.

By the triangle inequality,

$$x(\tau_\mu')\, x(\tau_{\mu+1}') \geqslant x(\sigma_\mu)\, x(\sigma_{\mu+1}) - x(\sigma_\mu)\, x(\tau_\mu') - x(\sigma_{\mu+1})\, x(\tau_{\mu+1}').$$

(This shows, by the way, that omitting points in a partition does not increase $\lambda(x, \Delta)$).

Hence by (5.4, 3)

$$\lambda(x,\Delta) \geqslant \lambda(x,\Delta') - 2\sum_{\mu=1}^{\eta-1} x(\sigma_\mu)\, x(\tau_\mu') > \lambda(x,\Delta') - 2\eta\varepsilon/4\eta > \lambda(x) - \varepsilon \quad \text{or} \quad \gamma.$$

The following two statements are corollaries of (5.2):

(5.6) *If $\{\Delta_\varkappa\}$ is a sequence of partitions with $\|\Delta_\varkappa\| \to 0$ then $\lambda(x, \Delta_\varkappa) \to \lambda(x)$.*

(5.7) *Additivity of length. If $\lambda_\gamma^\delta(x)$ denotes the length of the curve $x(\tau), \gamma \leqslant \tau \leqslant \delta$ and if $\alpha < \gamma_1 < \gamma_2 < \ldots < \gamma_\varkappa < \beta$ then*

$$\lambda(x) = \lambda_\alpha^\beta(x) = \lambda_\alpha^{\gamma_1}(x) + \lambda_{\gamma_1}^{\gamma_2}(x) + \ldots + \lambda_{\gamma_\varkappa}^\beta(x).$$

The following is simple but very important:

(5.8) *Semicontinuity of length. If $x_\nu(\tau)$ and $x(\tau)$, $\alpha \leqslant \tau \leqslant \beta$, $\nu = 1, 2, \ldots$ are curves in R and $x_\nu(\tau) \to x(\tau)$ as $\nu \to \infty$ for each τ in $[\alpha, \beta]$, then*

$$\lambda(x) \leqslant \liminf \lambda(x_\nu).$$

Proof. Given $\varepsilon > 0$ or $\gamma > 0$, choose a partition $\Delta: \alpha = \tau_0 \leqslant \tau_1 \leqslant \ldots \leqslant \tau_\eta = \beta$ such that $\lambda(x, \Delta) > \lambda(x) - \varepsilon$ or $\lambda(x, \Delta) > \gamma$.

Then by (5.1)

$$\lambda(x_\nu) \geqslant \lambda(x_\nu, \Delta) = \sum_{\mu=0}^{\eta-1} x_\nu(\tau_\mu)\, x_\nu(\tau_{\mu+1}) \to \sum_{\mu=0}^{\eta-1} x(\tau_\mu)\, x(\tau_{\mu+1}) = \lambda(x, \Delta).$$

As ε, γ were arbitrary, (5.8) follows.

Note. It is well known that $\lambda(x) = \lim \lambda(x_\nu)$ need not hold in the Cartesian (x, y) plane even for uniformly converging analytic curves. For example, let

$$y = \frac{1}{\nu} \cos \nu^2 x, \qquad \nu = 1, 2, \ldots, \qquad 0 \leqslant x \leqslant \pi.$$

These curves tend uniformly to the x-axis between 0 and π. But

$$\int_0^\pi [1 + (\nu \sin \nu^2 x)^2]^{1/2} dx \geqslant \nu \int_0^\pi |\sin \nu^2 x| \, dx = \nu^3 \int_0^{\pi/\nu^2} \sin \nu^2 x \, dx = 2\nu \to \infty.$$

Arc length as parameter

For a rectifiable curve it is natural to introduce as a parameter, the arc-length from a fixed point of the curve. We need the lemma:

(5.9) *For a rectifiable curve* $x(\tau)$, $\alpha \leqslant \tau \leqslant \beta$, *the arc length* $\lambda_\alpha^\gamma(x)$ $(\alpha \leqslant \gamma \leqslant \beta)$ *is a continous function of* γ.

By the additivity property (5.6) we need only prove:

(5.10) *Given* $\varepsilon > 0$, *then* $\delta > 0$ *exists such that* $\lambda_{\gamma_1}^{\gamma_2}(x) < \varepsilon$ *whenever* $0 < \gamma_2 - \gamma_1 < \delta$.

For choose $\delta > 0$ so that $||\Delta|| < \delta$ implies $\lambda(x) < \lambda(x, \Delta) + \varepsilon/2$ and so that $x(\gamma_1)\, x(\gamma_2) < \varepsilon/2$ for $0 < \gamma_2 - \gamma_1 < \delta$.

Then choose Δ so that γ_1, γ_2 occur as successive points of the partition, say $\gamma_1 = \tau_\mu$, $\gamma_2 = \tau_{\mu+1}$. Then

$$\sum_{\nu=1}^{\varkappa} x(\tau_\nu)\, x(\tau_{\nu+1}) = \lambda(x, \Delta) \geqslant \lambda_\alpha^{\gamma_1} + \lambda_{\gamma_1}^{\gamma_2} + \lambda_{\gamma_2}^{\beta} - \varepsilon/2$$

$$\geqslant \sum_{\nu=1}^{\mu} x(\tau_{\nu-1})\, x(\tau_\nu) + \lambda_{\gamma_1}^{\gamma_2} + \sum_{\nu=\mu+2}^{\varkappa} x(\tau_{\nu-1})\, x(\tau_\nu) - \varepsilon/2.$$

Hence $\varepsilon/2 > x(\gamma_1)\, x(\gamma_2) = x(\tau_\mu)\, x(\tau_{\mu+1}) \geqslant \lambda_{\gamma_1}^{\gamma_2} - \varepsilon/2$ which proves (5.10).

The standard representation of a rectifiable curve is now defined as follows: $y(\sigma)$ represents that point $x(\tau)$ for which $\lambda_\alpha^\tau(x) = \sigma$. The point $x(\tau)$ is uniquely determined by the arc-length σ, but in general τ is not. Indeed, all the points of an interval of the τ-axis may correspond to the same point on the curve, and then $\lambda_\alpha^\tau(x)$ is constant as τ describes the interval. But in all cases σ is a non-decreasing function of τ, continuous by (5.10).

The triangle inequality and the definition of arc-length show

(5.11) *For any curve* $x(\tau)$, $\alpha \leqslant \tau \leqslant \beta$ *we have* $x(\alpha)\, x(\beta) \leqslant \lambda(x)$.

Hence if $\sigma_1 < \sigma_2$ and $y(\sigma_1) = x(\tau_1)$, $y(\sigma_2) = x(\tau_2)$, then

(5.12) $$y(\sigma_1)\, y(\sigma_2) = x(\tau_1)\, x(\tau_2) \leqslant \lambda_{\tau_1}^{\tau_2}(x) = \sigma_2 - \sigma_1.$$

Thus $y(\sigma)$, $0 \leqslant \sigma \leqslant \lambda(x)$, is a curve in accord with our definition.

We shall show that

(5.13) *On* $y(\sigma)$, *the parameter* σ *is the arc-length.*

If $\sigma(\tau_1) = \sigma_1 < \sigma_2 = \sigma(\tau_2)$ then $\lambda_{\sigma_1}^{\sigma_2}(y) = \sigma_2 - \sigma_1 = \lambda_{\tau_1}^{\tau_2}(x)$. For, if Δ_ν: $\tau_1 = \tau_{\nu,0} < \tau_{\nu,1} < \ldots < \tau_{\nu,\varkappa_\nu} = \tau_2$, $\nu = 1, 2, \ldots$ be any sequence of partitions of $[\tau_1, \tau_2]$ with $||\Delta_\nu|| \to 0$ as $\nu \to \infty$ and we put $\sigma_{\nu,\mu} = \sigma(\tau_{\nu,\mu})$ then, by uniform continuity, the partition Δ_ν': $\sigma_1 = \sigma_{\nu,0} \leqslant \sigma_{\nu,1} \leqslant \cdots \leqslant \sigma_{\nu,\varkappa_\nu} = \sigma_2$ satisfies $||\Delta_\nu'|| \to 0$ as $\nu \to \infty$. Then by (5.6)

$$\lambda_{\sigma_1}^{\sigma_2}(y) = \lim \sum_\mu y(\sigma_{\nu,\mu-1})\, y(\sigma_{\nu,\mu}) = \lim \sum_\mu x(\tau_{\nu,\mu-1})\, x(\tau_{\nu,\mu}) = \lambda_{\tau_1}^{\tau_2}(x) = \sigma_2 - \sigma_1,$$

as desired.

We shall say quite generally that σ is the "*arc length*" on the rectifiable curve $x(\sigma)$, $\alpha \leqslant \sigma \leqslant \beta$, or that $x(\tau)$ is "*represented by* $x(\sigma)$ *in terms of arc length*", if $\lambda_{\sigma_1}^{\sigma_2}(x) = \sigma_2 - \sigma_1$ holds for all $\sigma_1 \leqslant \sigma_2$ in $[\alpha, \beta]$.

If a rectifiable curve in ordinary space has no special differentiability properties, then the theorem, "arc divided by chord tends to one" is only true almost everywhere. Although differentiability is here not defined, the statement is still true for our general rectifiable curves.[1]

(5.14) *If* σ *is the arc-length on* $y(\sigma)$, $\alpha \leqslant \sigma \leqslant \beta$ *then for almost all* σ *in* $[\alpha, \beta]$ *we have*

$$\lim \frac{y(\sigma_1)\, y(\sigma_2)}{\sigma_2 - \sigma_1} = 1 \quad \text{if} \quad \sigma_1 < \sigma < \sigma_2 \quad \text{and} \quad \sigma_1, \sigma_2 \to \sigma,$$

$$\lim \frac{y(\sigma)\, y(\sigma + h)}{|h|} \to 1 \quad \text{when} \quad h \to 0.$$

Proof. Put

$$w(\sigma) = \liminf_{\sigma_1 < \sigma < \sigma_2,\ \sigma_1, \sigma_2 \to \sigma} \frac{y(\sigma_1)\, y(\sigma_2)}{\sigma_2 - \sigma_1}.$$

Since, by (5.12) $y(\sigma_1)\, y(\sigma_2) \leqslant \sigma_2 - \sigma_1$ the first part will follow if $w(\sigma) = 1$ almost everywhere.

Now $w(\sigma)$ is measurable, since $y(\sigma_1)\, y(\sigma_2)$ is a continuous function of σ_1, σ_2. If $w(\sigma) < 1$ on a set of positive measure, there is a set M, of measure $2\,\theta > 0$ and a number $\eta > 0$ such that $w(\sigma) \leqslant 1 - 2\,\eta$ on M.

Choose δ such that for a partition Δ of $[\alpha, \beta]$ with $\|\Delta\| \leqslant \delta$ we have $\lambda(y, \Delta) > \lambda(y) - \eta\,\theta/2$.

For each σ in M, there is a pair σ_1, σ_2 such that $\sigma_1 < \sigma < \sigma_2$ and

$$y(\sigma_1)\, y(\sigma_2) < (1 - \eta)\, (\sigma_2 - \sigma_1).$$

Hence a finite set of non-overlapping intervals $[\sigma_1^\nu, \sigma_2^\nu]$ exists such that $y(\sigma_1^\nu)\, y(\sigma_2^\nu) < (1 - \eta)\, (\sigma_2^\nu - \sigma_1^\nu)$, $\Sigma(\sigma_2^\nu - \sigma_1^\nu) > \theta$, $\sigma_2^\nu - \sigma_1^\nu < \delta$. Let Δ be a partition of $[0, \lambda(y)]$ in which $\sigma_1^\nu, \sigma_2^\nu$ occur as consecutive elements and $\|\Delta\| < \delta$ apply (5.12) to the pairs of consecutive elements not of form $\sigma_1^\nu, \sigma_2^\nu$ then

$$\lambda(y) - \eta\,\theta/2 < \lambda(y, \Delta) \leqslant \lambda(y) - \sum_\nu{}' (\sigma_2^\nu - \sigma_1^\nu) + \sum_\nu{}' y(\sigma_1^\nu)\, y(\sigma_2^\nu)$$

$$< \lambda(y) - \eta \sum_\nu{}' (\sigma_2^\nu - \sigma_1^\nu) \leqslant \lambda(y) - \eta\,\theta$$

a contradiction.

The second part is a consequence. Let $0 < \sigma < \lambda(y)$. We have

$$\frac{y(\sigma - h)\, y(\sigma + h)}{|h|} \leqslant \frac{y(\sigma - h)\, y(\sigma)}{|h|} + \frac{y(\sigma)\, y(\sigma + h)}{|h|} \leqslant 2.$$

Since $w(\sigma) = 1$ the left-hand side tends to 2 as $h \to 0$. But as both terms on the right hand have values at most unity each tends to one.

Geometric Curves

Our definition of a curve involved a definite parametrization. This parametrization introduces properties which are not all geometrically significant. The point set carrying the curve is one of the important geometric features of the curve, though it does not in general carry all the important properties, for instance:

In the Cartesian (x, y)-plane the curves $(x, 0)$, $0 \leqslant x \leqslant 1$ and $(\sin \pi\, x, 0)$, $0 \leqslant x \leqslant 1$ are carried by the same set of points, namely the segment $[0, 1]$ of the x axis, but the curves differ in the important feature of length; the second arc covers the segment twice and has length 2.

Which properties of a parametrization are considered essential depends on the investigation. For our purposes the standard representation expresses all the essential properties, and two rectifiable curves with the same standard representations are called "*equivalent*". Because of (5.13) any curve $x(\tau)$ is equivalent to the curve given by its standard representation $y(\sigma)$.

Accordingly, we may define a "*rectifiable curve in the geometric sense*" as a class $\mathfrak{c}$ of equivalent curves. Each element of the class is a parametrization of $\mathfrak{c}$; the properties common to all parametrizations are the properties of $\mathfrak{c}$. Thus the common length of all parametrizations is the length $\lambda(\mathfrak{c})$ of $\mathfrak{c}$, their common initial and final points are the initial and final point of $\mathfrak{c}$, the common standard representation is the standard representation of $\mathfrak{c}$.

(5.15) *If $x(\tau)$, $\alpha \leqslant \tau \leqslant \beta$, and $z(\tau')$, $\alpha' \leqslant \tau' \leqslant \beta'$, are rectifiable curves and if there is a topological mapping of $[\alpha, \beta]$ on $[\alpha', \beta']$ such that α corresponds to α' and hence β to β' and $x(\tau) = z(\tau')$ then the curves $x(\tau)$, $z(\tau')$ are equivalent.*

For let $\alpha < \gamma < \beta$ and consider any sequence $\{\Delta_\varkappa\}$ of partitions of $[\alpha, \gamma]$ with $||\Delta_\varkappa|| \to 0$. Then the given mapping defines a sequence of partitions $\{\Delta_\varkappa'\}$ of $[\alpha', \beta']$ with $||\Delta_\varkappa'|| \to 0$, by uniform continuity. Since $x(\tau) = z(\tau')$ corresponding pairs in the two partitions have the same distance apart. Thus $\lambda_\alpha^\gamma(x) = \lambda_{\alpha'}^{\gamma'}(z)$ and $x(\gamma) = z(\gamma')$; the standard representations of the two curves are identical.

We can now establish some important geometric properties of our geometric curves.

(5.16) **Theorem:** *If the lengths $\lambda(\mathfrak{c}_\nu)$ of a sequence of rectifiable curves $\mathfrak{c}_\nu$ in a finitely compact set H, are bounded and the initial points a_ν of $\mathfrak{c}_\nu$ form a bounded set then $\{\mathfrak{c}_\nu\}$ contains a subsequence $\{\mathfrak{c}_\mu\}$ which converges uniformly to a curve $\mathfrak{c}$ in H and*

(5.17) $$\lambda(\mathfrak{c}) \leqslant \liminf \lambda(\mathfrak{c}_\mu).$$

More precisely: there is a subsequence $\{\mathfrak{c}_\mu\}$ of $\{\mathfrak{c}_\nu\}$ and a rectifiable curve $\mathfrak{c}$ and parametrizations $x_\mu(\tau)$ of $\mathfrak{c}_\mu$ and $x(\tau)$ of $\mathfrak{c}$, $0 \leqslant \tau \leqslant 1$, such that $x_\mu(\tau)$ tends uniformly to $x(\tau)$.

Proof. This depends on (5.8) and (2.17). Select a subsequence $\{a_\theta\}$ of $\{a_\nu\}$ which converges to a point a. If $\lambda(\mathfrak{c}_\nu) = 0$, then $\mathfrak{c}_\nu$ is a single point. Hence if $\lambda(\mathfrak{c}_\mu) = 0$ for a subsequence $\{\mathfrak{c}_\mu\}$ of $\{\mathfrak{c}_\theta\}$, the theorem is trivial.

Assume then $\lambda(\mathfrak{c}_\theta) > 0$ and let $y_\theta(\sigma)$ be the standard representation of $\mathfrak{c}_\theta$, then we define a new parametrization of $\mathfrak{c}_\theta$ by $x_\theta(\tau) = y_\theta(\sigma)$ with $\tau = \sigma/\lambda(\mathfrak{c}_\theta)$, $0 \leqslant \tau \leqslant 1$.

By hypothesis, $\lambda(\mathfrak{c}_\theta) < \beta$ for a suitable β, hence by (5.11)

$$a\, x_\theta(\tau) = a\, y_\theta(\sigma) \leqslant a\, a_\theta + a_\theta\, y_\theta(\sigma) \leqslant a\, a_\theta + \beta$$

whence $\{x_\theta(\tau)\}$ is bounded for all θ, τ.

Using (5.11) again

$$x_\theta(\tau_1)\, x_\theta(\tau_2) = y_\theta(\sigma_1)\, y_\theta(\sigma_2) \leqslant |\sigma_1 - \sigma_2| = \lambda(\mathfrak{c}_\theta)\, |\tau_1 - \tau_2| < \beta\, |\tau_1 - \tau_2|$$

so that the sequence $\{x_\theta(\tau)\}$ is equicontinuous. The assertion now follows from (2.17, 19).

We use this to show an existence theorem of great importance.

If a, b are two points of a set H, a "*shortest join*" of a and b in H is one whose length is not greater than that of any other curve from a to b in H.

(5.18) *If in a finitely compact set H the points a, b can be connected by a rectifiable curve, then a shortest join of a and b exists in H.*

Proof: If α is the infimum (finite) of the lengths of curves joining a to b, there is a sequence $\{\mathfrak{c}_\nu\}$ of such curves whose lengths tend to α. This sequence satisfies the hypothesis of the last theorem and hence contains a subsequence $\{\mathfrak{c}_\mu\}$ which converges to a curve $\mathfrak{c}_0$ from a to b, and

$$\lambda(\mathfrak{c}_0) \leqslant \liminf \lambda(\mathfrak{c}_\mu) = \alpha.$$

But $\lambda(\mathfrak{c}_0) \geqslant \alpha$ by the definition of α; hence $\lambda(\mathfrak{c}_0) = \alpha$.

Note: The reader might easily overlook the width and strength of this result. We give an illustration in Cartesian (ξ, η, ζ)-space. Suppose $\varphi(\xi, \eta)$ is continuous in all the (ξ, η)-plane or in a finitely compact domain thereon. Then the points of the surface $\zeta = \varphi(\xi, \eta)$ form a finitely compact subset of the space. If the points a, b can be connected on the surface by a curve of finite length in the usual sense, then there is a shortest join on the surface from a to b.

In more usual terms we have established the existence of *geodesic joins*.

ARCS. A curve is called an "*arc*" if it has a parametrization such that $x(\tau_1) \neq x(\tau_2)$ when $\tau_1 \neq \tau_2$.

The standard representation of a shortest join has this property.

(5.19) *A (rectifiable) shortest join $\mathfrak{c}$ of two points a, b is an arc.*

Proof: If it were not, then it would contain a loop which could be removed. We put this in precise form:

Let $y(\sigma)$ be the standard representation of a curve from a to b which is not an arc, and thus $y(\sigma_1) = y(\sigma_2)$ for some σ_1, σ_2 with $\sigma_1 < \sigma_2$. Then

$$z(\tau) = \begin{cases} y(\tau) & \text{for} \quad 0 \leqslant \tau \leqslant \sigma_1 \\ y(\tau + \sigma_2 - \sigma_1) & \text{for} \quad \sigma_1 \leqslant \tau \leqslant \lambda(\mathfrak{c}) - \sigma_2 + \sigma_1 \end{cases}$$

defines a curve from a to b with

$$\lambda(z) = \lambda_0^{\sigma_1}(y) + \lambda_{\sigma_2}^{\lambda(\mathfrak{c})} < \lambda(y).$$

Orientation of curves

If $x(\tau), \alpha \leqslant \tau \leqslant \beta$, is a curve, then the curve $x^{-1}(\tau) = x(-\tau), -\beta \leqslant \tau \leqslant -\alpha$ is a curve from $x(\beta)$ to $x(\alpha)$, *"the curve $x(\tau)$ traversed in the opposite sense"*.

We may call this curve the *"inverse of $x(\tau)$"*. Clearly $\lambda(x) = \lambda(x^{-1})$ and if $y(\sigma)$ is the standard representation of $x(\tau)$ then

$$y'(\sigma) = y(\lambda(x) - \sigma), \qquad 0 \leqslant \sigma \leqslant \lambda(x) = \lambda(x^{-1})$$

is the standard representation of $x^{-1}(\tau)$.

Thus equivalent curves have equivalent inverses, and we may speak of the inverse $\mathfrak{c}^{-1}$ of $\mathfrak{c}$.

(5.20) If $\mathfrak{c}$ is a shortest join of a, b, then $\mathfrak{c}^{-1}$ is a shortest join of b, a.

(5.21) *If $\mathfrak{c}_1, \mathfrak{c}_2$ are rectifiable arcs carried by the same point set, then either $\mathfrak{c}_1 = \mathfrak{c}_2$, or $\mathfrak{c}_1 = \mathfrak{c}_2^{-1}$.*

Thus the point set determines the arc apart from its orientation.

Proof: Let $x_i(\tau), \alpha_i \leqslant \tau \leqslant \beta_i, i = 1, 2$ be a parametrization of $\mathfrak{c}_i$ with

$$x_i(\tau_1) \neq x_i(\tau_2) \text{ for } \tau_1 \neq \tau_2.$$

For each τ in $[\alpha_1, \beta_1]$ there is a unique τ' in $[\alpha_2, \beta_2]$ such that $x_1(\tau) = x_2(\tau')$. Since $[\alpha_i, \beta_i]$ is compact, it follows by (2.14) that the correspondence of τ to τ' is a topological mapping of $[\alpha_1, \beta_1]$ on $[\alpha_2, \beta_2]$.

If α_1 is mapped on α_2, then $x_1(\tau)$ and $x_2(\tau)$ are equivalent by (5.15) and $\mathfrak{c}_1 = \mathfrak{c}_2$. If α_1 is mapped on β_2 then $x_1(\tau)$ and $x_2^{-1}(\tau)$ are equivalent, and $\mathfrak{c}_1 = \mathfrak{c}_2^{-1}$.

6. Segments

A curve $\mathfrak{c}$ from a to b for which

$$\lambda(\mathfrak{c}) = a\,b$$

is called an "*oriented segment*" from a to b, and is denoted by $\mathfrak{s}(a, b)$, or by $T^+(a, b)$. By (5.11) we have for any rectifiable curve from a to b

$$\lambda(\mathfrak{c}) \geqslant a\,b; \tag{6.1}$$

hence a segment is a shortest join and by (5.19) an arc. Also, $\mathfrak{s}^{-1}(a, b)$ is a segment $\mathfrak{s}(b, a)$ from b to a.

The point set which carries the segment $\mathfrak{s}(a, b)$ will be denoted by $T(a, b)$. By (5.21) it determines the segment up to orientation, and it is called the "*non-oriented*" segment when the distinction is emphasized.

(6.2) *Any subarc of a segment is a segment.*

For let $y(\sigma)$ be the standard representation of $\mathfrak{s}(a, b)$, and γ, δ be any numbers such that $0 \leqslant \gamma < \delta \leqslant a\,b$, then

$$a\,b \leqslant a\,y(\gamma) + y(\gamma)\,y(\delta) + y(\delta)\,y(a\,b) \leqslant \lambda_\alpha^\gamma(y) + \lambda_\gamma^\delta(y) + \lambda_\delta^{ab}(y) = \lambda(y) = a\,b.$$

Hence

$$\gamma - \delta = \lambda_\gamma^\delta(y) = y(\gamma)\,y(\delta) \tag{6.3}$$

and the assertion is proved.

The last equation gives the additional information that the correspondence of $y(\sigma)$ to σ is an isometry of $T(a, b)$ on the segment $[0, ab]$ of the real axis.

Any representation of $\mathfrak{s}(a, b)$ or of $\mathfrak{s}^{-1}(a, b)$ in terms of arc length as defined after (5.13) will be called a *representation* of $T(a, b)$. It satisfies

$$z(\sigma_1)\,z(\sigma_2) = |\sigma_1 - \sigma_2|, \qquad \alpha \leqslant \sigma_i \leqslant \beta = \alpha + a\,b \tag{6.4}$$

and it gives an isometric mapping of $T(a, b)$ on a segment of the real axis.

Conversely any set T in a metric space which is congruent to a segment of the real axis, is a segment. For we have a mapping of the interval $[\alpha, \beta]$ of the real axis in which σ in the interval corresponds to $z(\sigma)$ in T, and for which (6.4) holds. Then if $\alpha = \sigma_0 \leqslant \sigma_1 \leqslant \ldots \leqslant \sigma_\varkappa = \beta$ be any partition Δ,

$$\lambda(z, \Delta) = \sum z(\sigma_\nu)\,z(\sigma_{\nu+1}) = \sum (\sigma_{\nu+1} - \sigma_\nu) = \beta - \alpha = z(\alpha)\,z(\beta).$$

Hence $\lambda(z) = z(\alpha)\,z(\beta)$.

Examples. Segments, of course, need not exist, except for the trivial case when the segment consists of one point only. The space R_1 at the beginning of Section 2 gives an example.

A more interesting one is furnished by the points of a sphere $\xi^2+\eta^2+\zeta^2=\rho^2$ in ordinary space with the Euclidean distance. However, if distance is measured along great circles, we have a space in which any two points can be connected by a segment, with our definition. But a segment joining two points need not be unique; two antipodal points can be joined by an infinite number of segments.

A more drastic breakdown of uniqueness occurs in the space R_2 consisting of all ν-tuples $\xi = (\xi_1, \ldots, \xi_\nu)$ of real numbers with distance $d(\xi, \eta)$ defined by $\sum_{\mu=1}^{\nu} |\xi_\mu - \eta_\mu|$. Then any curve, $\xi(\tau)$ $0 \leqslant \tau \leqslant 1$, for which $\xi(0) = \xi$, $\xi(1) = \eta$ and for which each component $\xi_\varkappa(\tau)$ is monotone, is a segment from ξ to η. This follows since $|\alpha - \beta| + |\beta - \gamma| = |\alpha - \gamma|$ if β lies between α and γ.

M-CONVEX SETS AND BETWEENNESS

A set M in ordinary space is called "convex", if any two of its points can be joined by a segment lying entirely in M.

For general metric spaces Menger used the word "convex" to denote a somewhat weaker property. To formulate it we introduce a new notation:

If x, y, z be any points, $(x\,y\,z)$ shall mean that x, y, z are distinct and that

$$xy + yz = xz. \tag{6.5}$$

Clearly $(x\,y\,z)$ implies $(z\,y\,x)$. Moreover,

(6.6) *If* $(w\,x\,y)$ *and* $(w\,y\,z)$, *then* $(x\,y\,z)$ *and* $(w\,x\,z)$; *and conversely.*

This follows from

$$wz = wy + yz = wx + xy + yz \geqslant wx + xz \geqslant wz.$$

But it must be carefully noted that:

$$(wxy) \quad \text{and} \quad (xyz) \quad \text{do } \textit{not} \text{ imply} \quad (wxz) \quad \text{or} \quad (wyz)$$

as they would, for example, on a euclidean straight line. For consider points on a circle, with distance measured along the circle, and let w, y be diametrically opposite. Then for any other point x on the circle (wxy) is true and there are points z with (xyz), but neither (wyz) nor (wxz) holds.

From (6.6) we deduce:

(6.7) *If* (xyz) *and* T' *is a segment from* x *to* y *and* T'' *a segment from* y *to* z *then* $T' \cup T''$ *is a segment from* x *to* z.

We define:

A set H in a metric space is convex in Menger's sense or briefly *M-convex*, if it contains with any two distinct points x, z a point y with (xyz).

We do not follow Menger [1] in simply calling H convex, because it will prove useful to reserve the term convex for sets which are more nearly related to convex sets in the usual sense. The rational points on the τ-axis with the distance $|\tau_1 - \tau_2|$ form an M-convex set. However, Menger proved that any complete M-convex set contains with any two points x, y a segment $T(x, y)$. The following weaker statement suffices for our purposes:

(6.8) *Any two points of a finitely compact M-convex set* H *can be joined by a segment in* H.

Proof. Let x, y be two distinct points of H, then the set B of points z which satisfy $xz + zy = xy$ is bounded and closed, and hence by (2.11) compact. The function min (zx, zy) is continuous on B, hence attains on B its maximum β at some point m, and

$$\min (mx, my) = \beta \leqslant (1/2)\, xy.$$

If there is a segment joining x, y, then $\beta = (1/2)\, xy$. Suppose $\beta = mx < my$, hence $\beta < (1/2)\, xy$. Let z satisfy (mzy) then, since (xmy), lemma (6.6) implies (xzy) and (xmz), and so $zy \leqslant \beta$, otherwise would $xz > \beta$ and $yz > \beta$. Thence $mz \geqslant xy - 2\,\beta > 0$, so that the points z with (mzy) together with y form a compact set on which mz reaches a positive minimum at some point z_0. By the convexity assumption, there is a point z' in H with $(mz'\,z_0)$; then $mz' < mz_0$, and by (6.6) $(mz'y)$, and this contradicts the minimum property of z_0. Rename x, y as x_0, x_1.

Thus if points x_0, x_1 are given in H, we have shown that in H they have a mid-point $x_{1/2}$ Now if $\alpha = x_0\, x_1$ we map x_0, x_1 on the points $0, \alpha$ of the real axis, and $x_{1/2}$ on the point $\alpha/2$. Similarly, midpoints $x_{1/4}, x_{3/4}$ of the respective pairs $x_0, x_{1/2}$ and $x_{1/2}, x_1$, are to be mapped on the points $\alpha/4$, $3\,\alpha/4$. By (6.6) the mapping is isometric, and we may continue the bisection process, getting a denumerable set of points x_θ, where θ has the form $\varkappa\, 2^{-n}$, $0 \leqslant \varkappa \leqslant 2^n$ and the set is isometric to the points $\varkappa\, 2^{-n}\, \alpha$ of the interval $[0, \alpha]$.

If $\lim \theta_\nu = \theta$ where θ_ν are numbers of form $\varkappa\, 2^{-n}$, then since $x_{\theta_\nu} x_{\theta_\mu} = |\theta_\nu - \theta_\mu|$ and H is finitely compact, the points x_{θ_ν} converge to a point x_θ of H, and as, clearly, $x_{\theta'} x_{\theta''} = \alpha \, |\theta' - \theta''|$ for any θ', θ'' in $[0, 1]$, the mapping of x_θ on $[0, \alpha]$ is isometric.

Hence the set x_θ forms a segment from x_0 to x_1.

A LEMMA ON SPHERES

(6.9) *If* $x, y \in S(p, \rho)$, *then* $T(x, y) \subset S(p, 2\rho)$.

For if (xzy) then $\mu = \min(xz, yz) \leqslant \frac{1}{2} xy \leqslant \frac{1}{2}(xp + py) < \rho$ and if $\mu = xz$, then $zp \leqslant zx + xp < 2\rho$.

The reader would expect the conclusion $T \subset S(p, \rho)$, but *spheres need not be convex*. For example on a spherical surface of radius ρ in ordinary space, no circle with radius $\pi\rho/2 + \varepsilon$, $0 < \varepsilon < \pi\rho/2$ is convex. Let p be the center of the circle. Then $px = \frac{1}{2}\pi\rho$ represents the equator with p as one pole. A great circle through p cuts the given circle in two points which can be joined by a segment on the sphere of length less than $\pi\rho$ and passing through the antipodal point of p. Thus this segment does not fall in the circle.

7. Geodesics

The subject of geodesics is the central theme of this book. A geodesic may be regarded as arising from a segment by indefinite extension in both directions. Thus we must introduce some postulate of prolongation, and, to obtain a non-trivial theory, the prolongation must be unique.

In this section we first discuss an example which exhibits the meaning of geodesic and the kind of prolongation needed. We shall then define geodesics and prove that they exist. Uniqueness of prolongation will be discussed in the following section.

EXAMPLE

In Cartesian (ξ, η, ζ)-space, consider the cylinder $\xi^2 + \eta^2 = 1$; its geodesics are the helices.

(7.1) $$\xi = \cos\tau, \qquad \eta = \sin\tau, \qquad \lambda\zeta + \mu\tau + \nu = 0, \qquad \lambda^2 + \mu^2 > 0.$$

We write the equations in this rather unusual form so that they include the degenerate cases, namely

the circles $\zeta = \text{const.}$ when $\mu = 0$, and

the generators $\xi = \text{const.}$, $\eta = \text{const.}$ when $\lambda = 0$.

The parameter τ in (7.1) now runs from $-\infty$ to ∞, so that this geodesic is not a curve as defined in Section 5, but the part traversed for $\alpha \leqslant \tau \leqslant \beta$ is a rectifiable curve, although τ is its arc length, only in the case of the circles $\mu = 0$. If $\lambda \neq 0$, put $\mu_1 = \mu/\lambda$ and we have for the element of arc

$$d\sigma^2 = (1 + \mu_1^2)\, d\tau^2.$$

Hence if $x(\sigma)$ stands for the point $(\xi(\sigma), \eta(\sigma), \zeta(\sigma))$ then $x(\sigma) = (\cos \varkappa\, \sigma, \sin \varkappa\, \sigma, \mu_1 \varkappa\, \sigma)$ where $\varkappa = (1 + \mu_1^2)^{-1/2}$, is a representation of the helix in terms of arc length; when $\lambda = 0$, then $x(\sigma) = (\cos \nu/\mu, -\sin \nu/\mu, \sigma)$ is a representation of a generator in terms of arc length.

If points p, q do not lie on the same circle $\zeta =$ const., there are infinitely many geodesics through them, and, unless p, q lie on diametrically opposite generators, there is just one "steepest" helix through them, that is, one for which $|\mu/\lambda|$ is maximal (we include the value ∞, which gives a generator). The arc pq of this helix is the shortest join of p, q on the cylinder. If p, q be on diametrically opposite generators, then there are two shortest joins.

When points p, q lie on the same circle $\zeta =$ const., the infinitely many helical arcs which join two points in general position, degenerate into the arcs joining p, q which are obtained by traversing the circle an arbitrary number of times.

If we define the distance between two points p, q on the cylinder as the length of their shortest join (on the cylinder), then the cylinder becomes a finitely compact convex metric space. To construct the geodesics by the means we have developed we may proceed as follows:

Let $T(p_0, q)$, $p_0 \neq q$, be any segment. If points x with $(p_0\, q x)$ exist then $T(p_0, q)$ can be prolonged beyond q to yield a segment $T(p\ x)$ (see 6.7), but unless $T(p_0, q)$ lies on a generator, this will end when we reach the point p_2 on the generator diametrically opposite to p_0. Let p_1 be the mid-point of the segment $T(p_0, p_2)$ thus obtained. The subsegment $T(p_1, p_2)$ of $T(p_0, p_2)$ may be prolonged until we reach a point p_3 on the generator opposite to p_1. The segment $T(p_1, p_3)$ thus obtained overlaps $T(p_0, p_2)$ along $T(p_1, p_2)$. So proceeding, we obtain a segment $T(p_2, p_4)$, with p_4 on the same generator as p_0, which overlaps $T(p_1, p_3)$ along $T(p_2, p_3)$. This process may be repeated indefinitely; similarly we may prolong the original segment beyond p_0 instead of beyond q. In this way we construct the whole geodesic which contains $T(p_0, q)$.

We note that each point of a geodesic is the mid-point of a segment which lies on it. We utilize this in the definition of a geodesic in general.

Definition of a geodesic in a metric space R

A *"geodesic"* is a locally isometric map $\tau \to x(\tau)$ of the *whole* real τ-axis in R. This means:

For any real number τ_0, there is a positive number $\varepsilon(\tau_0)$ such that

$$(7.2) \qquad |\tau_0 - \tau_i| \leqslant \varepsilon(\tau_0), \quad i = 1, 2, \quad \text{implies} \quad x(\tau_1)\, x(\tau_2) = |\tau_1 - \tau_2|.$$

By this condition, $x(\tau)$ is a continuous map of the real τ-axis. The parameter τ is the arc-length in the following sense:

(7.3) *If $\lambda_{\tau_1}^{\tau_2}(x)$ denotes the length of the arc $x(\tau)$, $\tau_1 \leqslant \tau \leqslant \tau_2$ on the geodesic $x(\tau)$ then $\lambda_{\tau_1}^{\tau_2}(x) = \tau_2 - \tau_1$.*

Proof. By the Heine-Borel-Theorem, there is a finite set of values τ_i' with $\tau_1 = \tau_1' < \tau_2' < \ldots < \tau_\varkappa' = \tau_2$ such that $\tau_{\nu+1}' < \tau_\nu' + \varepsilon(\tau_\nu')$. Then $x(\tau)$ with $\tau_\nu' \leqslant \tau \leqslant \tau_{\nu+1}'$ represents a segment in terms of length by (7.2).
Hence

$$\lambda_{\tau_1}^{\tau_2}(x) = \sum_{\nu=1}^{\varkappa-1} \lambda_{\tau_\nu'}^{\tau_{\nu+1}'} = \sum_{\nu=1}^{\varkappa-1} (\tau_{\nu+1}' - \tau_\nu') = \tau_2 - \tau_1.$$

To reach a truly geometric concept, we must eliminate the special parameter τ; but as this has already the geometric meaning of length the only possible changes are a shift of the point on the geodesic from which the length is measured, and a reversal of the direction in which we traverse the geodesic. This suggests that we consider that two locally isometric maps $x(\tau)$ and $y(\tau)$ of the real axis represent the same geodesic $\mathfrak{g}$ when for suitable numbers $\eta = \pm 1$ and β, we have $y(\tau) = x(\eta\tau + \beta)$ for all τ. To make this precise we notice:

If $x(\tau)$ is a locally isometric map of the τ-axis, then so is

$$x(\tau') \qquad \textit{when} \qquad \tau' = \eta\tau + \beta, \ (\eta = \pm 1).$$

We now define two locally isometric maps $x(\tau)$, $y(\tau)$ of the real axis to be "equivalent", if η, β exist with $\eta = \pm 1$ such that $y(\tau) = x(\eta\tau + \beta)$.

Since the transformations $\tau' = \eta\tau + \beta$ of the real axis on itself form a group, this equivalence is reflexive, symmetric, transitive. A (geometric) geodesic $\mathfrak{g}$ is a class of equivalent $x(\tau)$, $-\infty < \tau < \infty$, with property (7.2). Each element of $\mathfrak{g}$ is called a *"representation"* of the geodesic.

Geodesics need not exist in general metric spaces, nor even in finitely compact convex metric spaces as e. g., in a closed convex domain in an ordinary plane. What is needed is the following axiom:

AXIOM OF LOCAL PROLONGABILITY

(7.4) *To each point p corresponds a positive number ρ_p such that if x, y be any two distinct points inside the sphere $S(p, \rho_p)$ then there is a point z with (xyz).*

Though we require nothing regarding the dependence of ρ_p on p, this generality is apparent only, since we can show from the axiom that a continuous function $\rho(p)$ exists satisfying (7.4).

As this kind of conclusion will recur, we formulate the underlying principle generally for any property $P(x, y)$ of an ordered pair of distinct points (in the present case, $P(x, y)$ means that z exists with (xyz)).

(7.5) *For each point p of a metric space R let a positive number ω_p exist such that $P(x, y)$ holds whenever $x, y \in S(p, \omega_p)$.*

Let $\omega(p)$ be the least upper bound of all ω such that $P(x, y)$ holds in $S(p, \omega)$. Then $P(x, y)$ holds in all $S(p, \omega(p))$. (If $\omega(p) = \infty$ this is the whole space.)

Further, either $\omega(p) = \infty$ for all p, or $\omega(p)$ is finite and positive for all p and

$$(7.6) \qquad |\omega(p) - \omega(q)| \leqslant pq.$$

Proof. Let $x, y \in S(p, \omega(p))$ and $x \neq y$, then for some $\omega < \omega(p)$ we have $x, y \in S(p, \omega)$, and hence $P(x, y)$ holds.

If $\omega(p) = \infty$ for one p, then $P(x, y)$ holds for any distinct x, y. Hence $\omega(q) = \infty$ for all q.

$\omega_p > 0$, implies $0 < \omega(p) < \infty$ unless $\omega(p) = \infty$. Suppose $\omega(p) > \omega(q)$, then (7.6) is obvious if $p\,q \geqslant \omega(p)$.

Assume then $p\,q < \omega(p)$. Then (compare 2.5) $S(p, \omega(p)) \supset S(q, \omega(p) - p\,q)$; hence $\omega(p) - p\,q \leqslant \omega(q)$.

(7.7) *If moreover R is finitely compact, then $\inf_{p \in M} \omega(p) > 0$ for points p in any bounded set M.* (In particular in R, if R is compact.)

For $\overline{M}$ is bounded and closed, hence compact, and $\omega(p)$ attains a minimum in $\overline{M}$, and this is positive since $\omega(p) > 0$ everywhere.

Apply this result to the property $P(x, y)$ in (7.4). Now let $\rho(p)$ be the least upper bound of those ρ for which $P(x, y)$ holds in $S(p, \rho)$. Then by (7.5):

Either $\rho(p) = \infty$ for all p or $0 < \rho(p) < \infty$ and $|\rho(p) - \rho(q)| \leqslant p\,q$.

This meaning of $\rho(p)$ is used throughout the book. We need a further lemma.

(7.8). *Let the space be finitely compact and M-convex. If $0 < q\,p < \rho(p)$ and α is any positive number less than $\rho(p)$ then a point q' exists with $(q\,p\,q')$ and $p\,q' = \alpha$.*

Proof. The set of points x with $q\,p + p\,x = q\,x$ and $p\,x \leqslant \alpha$ is compact. Hence px attains its maximum at some point z_0 in that set. If $pz_0 = \alpha$ take $z_0 = q'$. Suppose $pz_0 < \alpha$. Since q, z_0 lie in $S(p, \rho(p))$, there is a point z_1, with $(q\,z_0\,z_1)$. Thence by (6.6) $(q\,p\,z_1)$ and $(p\,z_0\,z_1)$, unless $p = z_0$. Since $p\,z_1 > p\,z_0$, we have $p\,z_1 > \alpha$. By (6.8) there is a segment $T(p, z_1)$; this contains a point q' with $p\,q' = \alpha$, and since $(q\,p\,z_1)$, $(p\,q'\,z_1)$ we have $(q\,p\,q')$.

It should be noted that we have not said, neither can we prove, that q' is unique, compare R_2 in the last section.

Existence of geodesics

(7.9) **Theorem:** *Let R be a finitely compact, M-convex metric space which satisfies the axiom of local prolongability, and let $x(\tau)$, $\alpha \leqslant \tau \leqslant \beta$, $\alpha < \beta$, represent a segment. Then a geodesic $\mathfrak{g}$ exists which has a representation $y(\tau)$ such that $y(\tau) = x(\tau)$ when $\alpha \leqslant \tau \leqslant \beta$.*

Thus the geodesic contains the given segment.

Proof. Put $\sigma(p) = \min(\rho(p)/2, 1)$. Then $\sigma(p)$ is still continuous and positive. Write α_0, β_0 for α, β, and $y(\tau)$ for $x(\tau)$ when $\alpha_0 \leqslant \tau \leqslant \beta_0$.

Choose τ_0 with $\alpha_0 < \tau_0 < \beta_0$, $\beta_0 - \tau_0 < \sigma(y(\beta_0))$. Then (7.8) guarantees the existence of a point q_1, with $(y(\tau_0)\,y(\beta_0)\,q_1)$ and $y(\beta_0)\,q_1 = \sigma(y(\beta_0))$. Let $y(\tau)$, $\beta_0 \leqslant \tau \leqslant \beta_1$, where $\beta_1 = \beta + \sigma(y(\beta_1))$, represent a segment $T(y(\beta), q_1)$. Then, by (6.7), the curve $y(\tau)$, $\alpha_0 \leqslant \tau \leqslant \beta_1$ represents a segment for $\tau_0 \leqslant \tau \leqslant \beta_1$.

Proceed similarly with $y(\beta_1)$. Choose τ_1 with $\beta_0 < \tau_1 < \beta_1$, $\beta_1 - \tau_1 < \sigma(y(\beta_1))$ and q_2 with $(y(\tau_1)\,y(\beta_1)\,q_2)$ and $y(\beta_1)\,q_2 = \sigma(y(\beta_1))$. Using (7.8) construct a segment $T(y(\beta_1), q_2)$ of the form $y(\tau)$, $\beta_1 \leqslant \tau \leqslant \beta_1 + \sigma(y(\beta_1)) = \beta_2$. By (6.7) the curve $y(\tau)$, $\alpha_0 \leqslant \tau \leqslant \beta_2$, represents a segment for $\tau_1 \leqslant \tau \leqslant \beta_2$.

Continuing the process, we find a sequence

$$\alpha_0 < \tau_0 < \beta_0 < \tau_1 < \beta_1 < \tau_2 < \beta_2 < \dots$$

with $\beta_{\nu+1} = \beta_0 + \sum_{\varkappa=1}^{\nu} \sigma(y(\beta_\varkappa))$ and $y(\beta_\nu)\,y(\beta_{\nu+1}) = \beta_{\nu+1} - \beta_\nu = \sigma(y(\beta_\nu))$. Now $\beta_\nu \to \infty$. For otherwise $\sigma(y(\beta_\nu)) \to 0$; but if $\mu > \nu$ then $y(\beta_\nu)\,y(\beta_\mu) \leqslant \beta_\mu - \beta_\nu$, and so $y(\beta_\nu)$ would converge to a point p; but then $\sigma(y(\beta_\nu)) \to \sigma(p)$ and $\sigma(p) > 0$.

Hence $y(\tau)$ is defined for all $\tau \geqslant \alpha$, and it represents a segment when $\tau_\nu \leqslant \tau \leqslant \beta_{\nu+1}$.

Similarly we can construct a sequence $\beta_0 > \tau_0' > \alpha_0 > \tau_1' > \alpha_1 > \tau_2' > \alpha_2'$ with $\alpha_{\nu+1} = \alpha_0 - \sum_{\varkappa=1}^{\nu} \sigma(y(\alpha_\varkappa)) \to -\infty$ and a function $y(\tau)$ for $-\infty < \tau \leqslant \beta$

which coincides with $y(\tau)$ for $\alpha_0 \leqslant \tau \leqslant \beta_0$, and which represents a segment when $\tau_\nu' \leqslant \tau \leqslant \alpha_{\nu-1}$.

It is clear that $y(\tau)$ satisfies (7.2) and hence it represents a geodesic.

Our definition of $\sigma(p)$ enabled us to include the case when $\rho(p) = \infty$. This case is very important, and when it obtains, the theorem can be strengthened.

STRAIGHT LINES: If the equation $x(\tau_1)\, x(\tau_2) = |\tau_1 - \tau_2|$ holds for all τ_1, τ_2, where $x(\tau)$ is a representation of a geodesic, then $x(\tau)$ is an *isometric* map of the *whole* real axis.

The equation then holds for all representations of the geodesic, and the geodesic is called a „*straight line*".

(7.10) *If the space is finitely compact and M-convex, and if for any distinct points x, y a point z exists with (xyz) (which means $\rho(p) = \infty$), then if $x(\tau)$, $\alpha \leqslant \tau \leqslant \beta$, $\alpha < \beta$, represents a segment, there is a straight line with a representation $y(\tau)$ such that $y(\tau) = x(\tau)$ for $\alpha \leqslant \tau \leqslant \beta$.*

The segment can be prolonged to a straight line.

For (7.8) asserts the existence of points q_ν ($\nu = 0, \pm 1, \pm 2, \ldots$) such that $q_0 = y(\alpha)$, $q_1 = y(\beta)$; $(q_{-1} q_0 q_1)$, $q_{-1} q_0 = 1$; $(q_{-1} q_1 q_2)$, $q_1 q_2 = 1$, $(q_{-2} q_{-1} q_2)$, $q_{-2} q_{-1} = 1$ and so on. Define $y(\tau)$, $-\infty < \tau < \infty$, such that

$y(\tau)$, $\beta + \nu - 1 \leqslant \tau \leqslant \beta + \nu$, represents a segment $T(q_\nu, q_{\nu+1})$, $\nu = 1, 2, \ldots$,

$y(\tau)$, $-\nu + \alpha \leqslant \tau \leqslant -\nu + 1 + \alpha$, represents $T(q_{-\nu}, q_{-\nu+1})$, $\nu = 1, 2, \ldots$.

Then, by (6.7), these $y(\tau)$ together with $y(\tau) = x(\tau)$ for $\alpha \leqslant \tau \leqslant \beta$ represent a straight line.

The example R_2 in the last section shows that the straight line $y(\tau)$ which coincides with $x(\tau)$ for $\alpha \leqslant \tau \leqslant \beta$ need not be unique.

GREAT CIRCLES: In analogy to geodesics on a sphere, we shall call any geodesic a "*great circle*" of radius β when it is an isometric map of a circle of radius β in the (ξ, η)-plane, with distances measured along the circle. More explicitly, the equation $x(\tau) = (\beta \cos \beta^{-1} \tau, \beta \sin \beta^{-1} \tau), -\infty < \tau < \infty$ represents the circle $\xi^2 + \eta^2 = \beta^2$ in terms of arc length (the circle being traversed an infinity of times), and the distance $x(\tau_1)\, x(\tau_2)$ of $x(\tau_1)$, $x(\tau_2)$ measured along the shorter arc of the circle is $\min |\tau_1 - \tau^\nu|$ where τ^ν traverses all parameter values for which the point $x(\tau^\nu)$ coincides with $x(\tau_2)$. Hence

$$x(\tau_1)\, x(\tau_2) = \min_{\nu = 0, \pm 1, \pm 2, \ldots} |\tau_1 - \tau_2 + 2\nu\pi\beta|. \tag{7.11}$$

Accordingly: a geodesic is a *"great circle"* of radius $\beta > 0$ if one representation $x(\tau)$, (and hence each representation), of the geodesic satisfies (7.11).

If the distance of two points on the great circle is maximal, namely $\pi\beta$, we call the points *"conjugate"*; the great circle carries two segments joining such a pair; for other pairs, only one.

Note. The example of the cylinder shows that straight lines, great circles, and geodesics which are neither, may occur in the same space. The latter may be more complicated than suggested by those on the cylinder, such a geodesic may intersect itself infinitely often, and it may be everywhere dense in the space. Examples will be given later.

8. G-spaces

In our considerations we have not yet taken up the question of the uniqueness either of the join of two points, or of the prolongation of a segment. The example we gave of the metric $d(\xi, \eta) = \sum_{\mu=1}^{\nu} |\xi_\mu - \eta_\mu|$ shows that our requirements of local compactness, convexity and local prolongability do not imply any uniqueness theorem.

If our geometry is to resemble differential geometry we must adjoin some uniqueness properties. Now in those geometries the geodesics, and more generally the extremals in the calculus of variations, are given by differential equations of the second order, and under the hypotheses usually made in those fields, there is just one solution through a given line element. Thus a geodesic has a unique prolongation, though the shortest geodesic arc joining two points even on simple surfaces such as the sphere, need not be unique. This suggests the assumption:

Uniqueness of prolongation

(8.1) *If* (xyz_1), (xyz_2) *and* $yz_1 = yz_2$ *then* $z_1 = z_2$.

This however is a condition in the large and so may not fit into the framework of differential geometry. But (8.1) is equivalent to a local condition:

(8.2) *If any two points of the metric space R can be joined by a segment, then* (8.1) *follows from the following local requirement*:

(8.3) *Every point* p *has a neighborhood* $S(p, \delta(p))$, $\delta(p) > 0$ *such that, if* (xpy_1), (xpy_2) *and* $py_1 = py_2 < \delta(p)$, *then* $y_1 = y_2$.

For, if (8.1) were false, there would exist points x', p', y_1', y_2' with $(x'p'y_1')$, $(x'p'y_2')$, $p'y_1' = p'y_2'$ but $y_1' \neq y_2'$. Now there are segments $T(x', p')$, $T(p', y_1')$, $T(p', y_2')$ and $T(x', p') \cup T(p', y_i')$ is a segment T_i from x' to y_i'. Let these segments T_i be represented by $y_i(\tau)$ with $y_i(0) = x'$. There will be a last value τ_0 of τ such that $y_1(\tau) = y_2(\tau)$ for $0 \leqslant \tau \leqslant \tau_0$ and since $y_1' \neq y_2'$, we have $x'p' \leqslant \tau_0 < x'y_i'$. Put $p = y_1(\tau_0) = y_2(\tau_0)$. Let $\tau_0' = \min(\tau_0, \delta(p)/2)$ and $x = y_i(\tau_0 - \tau_0')$. Thus $xp < \delta(p)$. By the definition of τ_0, there is a τ_1 with $\tau_1 > \tau_0$, $\tau_1 < \tau_0 + \min(\delta(p)/2, py_i')$ such that $y_1(\tau_1) \neq y_2(\tau_1)$. But this is impossible by (8.3), since $xp < \delta(p)$, $(xpy_1(\tau_1))$, $(xpy_2(\tau_1))$ and $py_1(\tau_1) = = py_2(\tau_1) < \delta(p)$.

Definition of G-space

A finitely compact M-convex metric space which satisfies the axiom of local prolongability and for which prolongation is unique is called a "*G-space*". The symbol G is chosen to suggest that geodesics have all the usual properties apart from differentiability.

G-spaces are the primary subject of this book, and nearly all spaces considered from now on will be G-spaces. It will therefore be convenient to list the complete list of axioms explicitly for reference. The axioms are:

I *The space is metric* with distance xy.

II *The space is finitely compact*, i. e., a bounded infinite set has at least one accumulation point.

III *Given two distinct points x, z then a point y with* (xyz), i. e., different from x and z and with $xy + yz = xz$, *exists.*

IV *To every point p of the space there corresponds a positive ρ_p, such that for any two distinct points x, y in* $S(p, \rho_p)$ (i. e. $xp < \rho_p$, $yp < \rho_p$) *a point z with (xyz) exists.*

V *If (xyz_1), (xyz_2), and $yz_1 = yz_2$ then $z_1 = z_2$.*

The simplest general properties of G-spaces are treated in this section and the next. Then we turn to two-dimensional G-spaces. Chapter II deals with special G-spaces, the general theory is developed further in Chapters III and IV.

First consequences

We first note that the proof of (8.3) yields also the following improvement of (7.9).

(8.4) ***Theorem:*** *If $x(\tau)$, $\alpha \leqslant \tau \leqslant \beta$, $\alpha < \beta$, represents a segment in a G-space, then there is exactly one representation $y(\tau)$ of a geodesic such that $y(\tau) = x(\tau)$ for $\alpha \leqslant \tau \leqslant \beta$.*

For (7.9) guarantees the existence of at least one such $y(\tau)$. If there were two distinct representations of this kind, $y_1(\tau)$ and $y_2(\tau)$, then we should have either a last $\tau_0 \geqslant \beta$ such that $y_1(\tau) = y_2(\tau)$ for $\alpha \leqslant \tau \leqslant \tau_0$, or a first $\tau_0 \leqslant \alpha$ such that $y_1(\tau) = y_2(\tau)$ for $\tau_0 \leqslant \tau \leqslant \beta$. Choose a positive ε such that $y_1(\tau)$ represents a segment for $|\tau - \tau_0| \leqslant \varepsilon$. Taking the first case, for example, there would be a τ_1 with $\tau_0 < \tau_1 < \tau_0 + \varepsilon$, $y_1(\tau_1) \neq y_2(\tau_1)$ whereas $(y_i(\tau_0 - \varepsilon)\ y_i(\tau_0)\ y_i(\tau_1))$. We next have the extremely important fact:

(8.5) *If $\rho(p) = \infty$ for all p, then all geodesics are straight lines. If $\rho(p) < \infty$ and $y(\tau)$ represents a geodesic, then $y(\tau)$ represents a segment for $|\tau - \tau_0| \leqslant \rho(y(\tau_0))$.*

Proof. The first part is a corollary of (8.4) and (7.10). In the second case, there is a positive number ε such that $y(\tau)$ represents a segment for $|\tau - \tau_0| \leqslant \varepsilon$. Put $p = y(\tau_0)$. If $\varepsilon < \rho(p)$ then, by (7.8), for $0 < \alpha < \rho(p)$ there is a point q with $qp = \alpha$ and $(qpy(\tau_0 + \varepsilon))$. If $x(\tau)$ represents a segment from q to $y(\tau_0 + \varepsilon)$ which coincides with $y(\tau)$ for $\tau_0 \leqslant \tau \leqslant \tau_0 + \varepsilon$, then $x(\tau)$ is part of a representation of a geodesic, hence (8.4) implies $x(\tau) = y(\tau)$ for $\tau_0 - \alpha \leqslant \tau \leqslant \tau_0 + \varepsilon$. Using (7.8) again, we deduce that a point q' exists with (qpq') and $pq' = \alpha$. The same argument shows that $y(\tau)$ represents a segment for $|\tau - \tau_0| \leqslant \alpha$. As this holds for any $\alpha < \rho(p)$ it also holds for $\alpha = \rho(p)$.

The spaces in which $\rho(p) = \infty$ play a great role; the euclidean, hyperbolic and more generally all simply connected spaces with non-positive curvature are examples for $\rho(p) = \infty$. In the terminology of the calculus of variations, they are the simply connected spaces without conjugate points (see Section 25). We introduce the brief term "*straight space*" for a G-space with $\rho(p) = \infty$, which is suggested by the fact that all geodesics in such a space are straight lines.

Without mentioning it every time explicitly, we assume that the underlying space is a G-space and show next:

(8.6) *If (xyz), then the segments $T(x, y)$ and $T(y, z)$ are unique, and hence $T(x, y) \cup T(y, z)$ is the only segment from x to z that contains y.* (A good illustration of this theorem is given by the geodesics on a sphere.)

For, if there were two distinct segments from y to z, they would contain, respectively, distinct points z_1, z_2 with $yz_1 = yz_2$ and (6.6) gives (xyz_i) contrary to (8.1). A first application of this yields:

(8.7) *If two segments T_1, T_2 have more than two common points then $T_1 \cap T_2$ is a segment; but if they have just two common points these are the end points of both segments.*

Proof. Suppose a, b, c are common to T_1 and T_2 and choose the names so that (abc). Then the segment $T(b, c)$ is unique, and hence lies on both T_1 and T_2. Choose representations $y_i(\tau)$, $\alpha_i \leqslant \tau \leqslant \beta_i$ of T_i such that $y_i(\tau)$, $0 \leqslant \tau \leqslant bc$, represents $T(b, c)$. Since the $y_i(\tau)$ can be extended to representations of geodesics, we have $y_1(\tau) = y_2(\tau)$ in $[\alpha_1, \beta_1] \cap [\alpha_2, \beta_2]$ by (8.4). This shows the first part.

If, however, T_1 and T_2 have just two common points a, b there can be no point c with (abc), since $T(a, b)$ would then be unique and part of both T_1 and T_2. Thus b must be an end point of both segments.

Another consequence of (8.6) is that $T(x, y)$ is unique when $x, y \subset S(p, \rho(p))$. Hence *joins are unique in the small.* Then from (7.5) we have:

(8.8) **Theorem:** *Let $\rho_1(p)$ be the least upper bound of those ρ for which $T(x, y)$ is unique when $x, y \in S(p, \rho)$. Then the segment $T(x, y)$ is also unique when $x, y \in S(p, \rho_1(p))$ and either $\rho_1(p) = \infty$ for all p, or $\rho(p) \leqslant \rho_1(p) < \infty$ and $|\rho_1(p) - \rho_1(q)| \leqslant pq$.*

We shall say that the segment T "*lies on the geodesic* $\mathfrak{g}$", if for one (and hence for every), representation $x(\tau)$ of $\mathfrak{g}$ an interval $[\alpha, \beta]$ exists in which $x(\tau)$ represents T.

This would appear to leave open the possibility that each point of T was a point of $\mathfrak{g}$ and yet T did not "lie on" $\mathfrak{g}$. This cannot happen, for T lies on some geodesic anyway and we shall show:

(8.9) *Two distinct geodesics have either no common point or only a countable number of common points.*

Proof. For suppose the geodesics $\mathfrak{g}$, $\mathfrak{g}'$ have a non-empty, uncountable set N of common points. We shall prove that the geodesics coincide.

For let $x(\tau)$, $x'(\tau)$ be representations of $\mathfrak{g}$, $\mathfrak{g}'$ respectively. Choose numbers $\tau_1, \tau_2, \ldots$ such that the intervals $|\tau - \tau_\nu| \leqslant \rho(x(\tau_\nu))$ cover the τ-axis. Then by (8.5) $x(\tau)$ represents a segment T_ν for $|\tau - \tau_\nu| \leqslant \rho(x(\tau_\nu))$.

At least one T_ν, say T_λ, must contain a (non-empty and) non-countable subset of points in N. Now choose $\tau_1', \tau_2', \ldots$ so that the intervals $|\tau - \tau_\nu'| \leqslant \rho(x'(\tau_\nu'))$ cover the τ-axis. Then one of the segments T_ν' thus given, say T_μ', contains a non-countable subset of $T_\lambda \cap N$. By (8.7) $T_\lambda \cap T_\mu'$ is a segment T. Let it be represented by $y(\tau)$, $0 \leqslant \tau \leqslant \gamma$, $\gamma > 0$. Then since $x(\tau)$ and $x'(\tau)$ represent T_λ and T_μ' in $|\tau - \tau_\lambda| \leqslant \rho(x(\tau_\lambda))$, $|\tau - \tau_\mu'| \leqslant \rho(x'(\tau_\mu'))$ respectively, we can find numbers $\eta, \eta' = \pm 1$ and numbers β, β' such that $x(\eta\tau + \beta) = y(\tau)$, $x'(\eta'\tau + \beta') = y(\tau)$ for $0 \leqslant \tau \leqslant \gamma$. But then $y(\tau) = x(\eta\tau + \beta)$ and $y'(\tau) = x'(\eta'\tau + \beta')$ are representations of $\mathfrak{g}$ and $\mathfrak{g}'$ which coincide in $[0, \gamma]$. Hence $\mathfrak{g}, \mathfrak{g}'$ coincide by (8.4).

Special neighborhoods

For many applications, it is important to have neighborhoods $S(p, \rho)$ of a given point p such that, for some given integer $\nu \geqslant 2$, every segment $T(x, y)$ with endpoints in $S(p, \rho)$ can be imbedded in a segment $T(x_1, y_1)$ with the same mid-point as $T(x, y)$, but whose length is at least $2\nu\rho$, and hence at least ν times the length of $T(x, y)$.

To construct such spheres take $\eta < 1/3$; if $x, y \in S(p, \eta\,\rho(p))$ with $\rho(p)$ as in (7.4), then $\rho(x) \geqslant \rho(p) - \eta\,\rho(p)$, $xy \leqslant 2\,\eta\,\rho(p)$. Represent the (unique) segment $T(x, y)$ by $z(\tau)$, such that $z(0) = x$ and $z(xy) = y$. Then $z(\tau)$ is part of a representation $z(\tau)$ of a geodesic, which by (8.5) represents a segment for $|\tau| \leqslant \rho(x)$ and hence also for $|\tau - xy/2| \leqslant \rho(x) - xy/2$. The latter segment, say $T(x_1, y_1)$ has the same mid-point as $T(x, y)$, and length $2\,\rho(x) - - xy \geqslant \rho(p)\,(2 - 4\,\eta) \geqslant 2\nu\eta\rho(p)$, if we take η so that $\eta \leqslant (\nu + 2)^{-1}$.

Hence we have

(8.10) *If x, y are any two distinct points in $S(p, \rho(p)/(\nu + 2))$, $\nu \geqslant 2$, then there is a segment which contains the segment $T(x, y)$, has the same mid-point and has length $2\nu\rho(p)/(\nu + 2) \geqslant \nu \cdot xy$.*

Convergence of geodesics

The discussion is based on the lemma:

(8.11) *Let $x(\tau)$ and $x_\nu(\tau)$ represent geodesics, $\nu = 1, 2, \ldots$. If $\tau_\varkappa \to \tau_0$ and $\lim_{\nu \to \infty} x_\nu(\tau_\varkappa) = x(\tau_\varkappa)$ for $\varkappa = 1, 2, \ldots$ then $x_\nu(\tau_0) \to x(\tau_0)$ and $x_\nu(\tau_\nu) \to x(\tau_0)$.*

Proof.

$$x(\tau_0)\,x_\nu(\tau_0) \leqslant x(\tau_0)\,x(\tau_\varkappa) + x(\tau_\varkappa)\,x_\nu(\tau_\varkappa) + x_\nu(\tau_\varkappa)\,x_\nu(\tau_0) \leqslant 2\,|\tau_0 - \tau_\varkappa| + x(\tau_\varkappa)\,x_\nu(\tau_\varkappa),$$

using (7.3). Hence, given $\varepsilon > 0$, we may first choose $\varkappa$ so that $|\tau_0 - \tau_\varkappa| < \varepsilon/3$ and then an integer m such that $x(\tau_\varkappa)\,x_\nu(\tau_\varkappa) < \varepsilon/3$ for $\nu > m$. This gives the first statement. The second follows since

$$x_\nu(\tau_\nu)\,x(\tau_0) \leqslant x_\nu(\tau_\nu)\,x_\nu(\tau_0) + x_\nu(\tau_0)\,x(\tau_0) \leqslant |\tau_\nu - \tau_0| + x_\nu(\tau_0)\,x(\tau_0).$$

(8.12) ***Theorem:*** *Let $x_\nu(\tau)$ represent a geodesic, $\nu = 1, 2, \ldots$. If $\{x_\nu(\tau)\}$ converges for each τ in $[\alpha, \beta]$, $\alpha < \beta$, then the sequence $\{x_\nu(\tau)\}$ converges, uniformly on any bounded set W of the τ-axis, to a limit $x(\tau)$ which represents a geodesic.*

Proof. Put $\gamma = (\alpha + \beta)/2$; then, by hypothesis, $q_\nu = x_\nu(\gamma)$ tends to a point q, and $x_\nu(\tau)$ represents a segment for

$$|\tau - \gamma| \leqslant \rho(q_\nu). \quad \text{(Replace } \rho(q_\nu) \text{ by 1 if } \rho(p) \equiv \infty.)$$

Now $x_\nu(\tau_1)\, x_\nu(\tau_2) = |\tau_1 - \tau_2|$ when $|\tau_i - \gamma| \leqslant \frac{1}{2}\rho(q) < \rho(q_\nu)$. Hence, if $x(\tau) = \lim x_\nu(\tau)$ when $|\tau - \gamma| \leqslant \frac{1}{2} \min (\rho(q), \beta - \alpha)$, then $x(\tau)$ represents a segment in that interval of τ, which we call $[\alpha', \beta']$. There is a representation of a geodesic which coincides with $x(\tau)$ in this interval. Call this representation $x(\tau)$.

Consider the set V of all values $\tau' \geqslant \gamma$ for which $x_\nu(\tau) \to x(\tau)$ when $\gamma \leqslant \tau \leqslant \tau'$. Then V contains β', and if it contains τ_1', then it contains all τ' in $[\gamma, \tau_1']$. By (8.11) V is a closed set. But it is also open in $\tau \geqslant \gamma$. For, if $\tau' \geqslant \beta'$, then $x_\nu(\tau)$ represents a segment for $|\tau - \tau'| \leqslant \rho(x(\tau'))$.

If $\tau_1 = \min (\alpha', \tau' - \rho(x(\tau'))/2)$ then, since $\rho(x_\nu(\tau')) \to \rho(x(\tau'))$, we have, when ν is large,

$$(x_\nu(\tau_1)\, x_\nu(\tau')\, x_\nu(\tau)) \quad \text{for} \quad 0 < \tau - \tau' \leqslant \rho(x(\tau'))/2;$$

thence each accumulation point y of $x_\nu(\tau)$ satisfies the relations $(x(\tau_1)\, x(\tau')\, y)$ and $x(\tau')\, y = \tau - \tau'$. Since also $(x(\tau_1)\, x(\tau')\, x(\tau))$ and $x(\tau')\, x(\tau) = \tau - \tau'$ we have by (8.1), that $y = x(\tau)$, and hence $x_\nu(\tau) \to x(\tau)$.

Thus V coincides with $\tau \geqslant \gamma$. Similarly $x_\nu(\tau) \to x(\tau)$ when $\tau \leqslant \gamma$. And so $x_\nu(\tau)$ converges to $x(\tau)$, which represents a geodesic.

If the convergence were not uniform on a bounded set W of the τ-axis, it would not be so on its closure, and we could find points τ_ν in the closure with $x_\nu(\tau_\nu)\, x(\tau_\nu) \geqslant \eta > 0$. If $\tau_\nu \to \tau$ we get a contradiction to (8.11) since $x(\tau)\, x(\tau_\nu) \leqslant |\tau - \tau_\nu|$. (8.12) implies:

(8.13) *Under the hypotheses of* (8.12), *if G_ν and G are the point sets traversed by $x_\nu(\tau)$ and $x(\tau)$ respectively, then* $\bar{G} \subset \liminf G_\nu$. In spite of the strength of (8.12) this is the most that can be said in general.

For example, the sequence of helices on the cylinder in (7.1) given by $\nu = 0$, $\lambda = 1$, $\mu = 1/\varkappa$ has, for $\varkappa \to \infty$, as lower limit the whole cylinder and the case is worse in spaces which contain an everywhere dense geodesic.

Another consequence of (8.12) is:

(8.14) *If $x_\nu(\tau)$ represents a geodesic, $\nu = 1, 2, \ldots$ and the sequence $\{x_\nu(\tau_0)\}$ is bounded, then $\{x_\nu(\tau)\}$ contains a subsequence $\{x_\lambda(\tau)\}$ which converges (uniformly in every bounded set of the τ-axis) to a representation $x(\tau)$ of a geodesic.*

Proof. Choose a subsequence $\{\mu\}$ of $\{\nu\}$ such that $x_\mu(\tau_0)$ converges to a point q, say. Then $x_\mu(\tau)$ represents a segment for $|\tau - \tau_0| \leqslant (\frac{1}{2})\ \rho(q)$, when μ is large enough. Now choose a subsequence $\{\lambda\}$ of $\{\mu\}$ such that $x_\lambda(\tau_0 + (\frac{1}{2})\rho(q))$ converges to a point r, say. The uniqueness of $T(q, r)$ implies as in the proof of (8.12) that $x_\nu(\tau)$ tends to a representation $x(\tau)$ of $T(q, r)$ for $0 \leqslant \tau \leqslant \rho(q)/2$. The rest follows by (8.12).

PRODUCTS OF G-SPACES

From two G-spaces R, R' a new G-space may be obtained by forming their topological product and metrizing it properly. A one-parameter family of such metrizations which will prove useful later on is given in the following theorem:

(8.15) **Theorem:** *If R is a G-space with points $x, y, z, \ldots$ and R' a G-space with points $x', y', z', \ldots$, then the pairs (x, x') in the product $R \times R'$ form with the metric*

$$(x, x')\,(y, y') = [\,xy^\alpha + x'y'^\alpha\,]^{1/\alpha}, \qquad \alpha > 1,$$

a G-space $[R \times R']_\alpha$.

If R and R' are straight, then $[R \times R']_\alpha$ is straight.

Proof. $(x, x')\,(y, y') = 0$ only for $x = y$ and $x' = y'$ is obvious, and so is the symmetry of distance. The triangle inequality is a consequence of Minkowski's Inequality

$$\text{(8.16)} \qquad \left[\sum_{i=1}^{n} (a_i + b_i)^\alpha\right]^{1/\alpha} \leqslant \left[\sum_{i=1}^{n} a_i^\alpha\right]^{1/\alpha} + \left[\sum_{i=1}^{n} b_i^\alpha\right]^{1/\alpha}$$

$$a_i \geqslant 0, \qquad b_i \geqslant 0, \qquad \alpha > 1,$$

with equality only when all a_i vanish, or all b_i vanish or the sets $a_1, \ldots, a_n$ and $b_1, \ldots, b_n$ are proportional[1].

(8.16) yields for any three distinct pairs (x, x'), (y, y'), (z, z') that

$$\text{(8.17)} \qquad (x, x')\,(y, y') + (y, y')\,(z, z') \geqslant [(xy + yz)^\alpha + (x'y' + y'z')^\alpha\,]^{1/\alpha}$$
$$\geqslant (xz^\alpha + x'z'^\alpha)^{1/\alpha}$$

with the equality signs if and only if $xy + yz = xz$, $x'y' + y'z' = x'z'$ and either $xy = yz = 0$ or $x'y' = y'z' = 0$ or $xy : yz = x'y' : y'z'$.

It follows that the space is convex. For if (x, x') and (z, z') are given and y, y' are chosen such that $xy = yz = xz/2$, $x'y' = y'z' = x'z'/2$, then $(x, x')(y, y') = (y, y')(z, z') = (x, x')(z, z')/2$.

Local prolangability is seen in the same way: we show that $\rho(p, p') = \min(\rho(p), \rho(p'))$ satisfies Axiom IV. If $(x, x') \neq (y, y')$ and

$$(p, p')(x, x') < \rho(p, p'), \qquad (p, p')(y, y') < \rho(p, p')$$

then

$$px < \rho(p), \qquad py < \rho(p), \qquad p'x' < \rho(p'), \qquad p'y' < \rho(p').$$

If $x = y$ then $x' \neq y'$ and z' with $(x'y'z')$ exists, and the relation

$$[(x, x')(y, y')(z, z')] \tag{8.18}$$

clearly holds for $z = y$.

If $x \neq y$ and $x' \neq y'$, then points z and z_0' with (xyz) and $(x'y'z_0')$ exist. Assume $xy : yz \geqslant x'y' : y'z_0'$. Then there is a z' on $T(y', z_0')$ such that $xy : yz = x'y' : y'z'$. It follows from (6.6) that $(x'y'z')$ and from the conditions for the equality sign in (8.17) that (8.18) holds.

The finite compactness of $[R \times R']_\alpha$ is clear, so that it remains only to show that prolongation is unique. If

$$[(x, x')(y, y')(z_i, z_i')], \quad i = 1, 2, \quad \text{and} \quad (y, y')(z_1, z_1') = (y, y')(z_2, z_2'),$$

then we conclude first that $xy + yz_i = xz_i$, $i = 1, 2$, and distinguish several cases

1) $xy = yz_1 = 0$. Then $x'y' > 0$ and $y'z_1' > 0$ otherwise $(x, x') = (y, y')$ or $(y, y') = (z_1, z_1')$. Now $y'z_1' = [yz_2^\alpha + y'z_2'^\alpha]^{1/\alpha}$ and $yz_2 = 0$, because $yz_2 > 0$ and $0 = xy : yz_2 = x'y' : y'z_2'$ would imply $x'y' = 0$. Thus $y = z_2$ and uniqueness of prolongation in R' guarantees $z_1' = z_2'$ and $(z_1, z_1') = (z_2, z_2')$.

The cases $xy = yz_2 = 0$, $x'y' = y'z_1' = 0$, and $x'y' = y'z_2' = 0$ are treated in the same way.

2) If none of these four cases enter, then

$$xy : yz_1 = x'y' : y'z_1' \qquad \text{and} \qquad xy : yz_2 = x'y' : y'z_2'$$

show together that none of the eight numbers vanish. Therefore $yz_1 > yz_2$ would imply $y'z_1' > y'z_2'$ which contradicts $(y, y')(z_1, z_1') = (y, y')(z_2, z_2')$.

Since $\rho(p, p') = \infty$, if $\rho(p) = \infty$ and $\rho(p') = \infty$, $[R \times R]_\alpha$ is straight with R and R'.

9. Multiplicity, Geodesics without multiple points

As differentials are not defined in general G-spaces, we cannot speak of lineal elements in the usual sense. However, the following definition provides a good substitute.

A "*lineal element*" with mid-point p is a segment with mid-point p and length $\sigma(p) = \min(\rho(p)/2, 1)$.

The 1 introduced enables us to include the case when $\rho(p) = \infty$; the $\frac{1}{2}$ is inserted for convenience.

Two lineal elements either have at most one point in common, or they have a common subsegment, by (8.7).

Hence by (8.7) and $|\rho(x) - \rho(y)| \leqslant xy$, we have.

(9.1) *Two lineal elements which have p as a common point, and other common points as well, lie on a segment with mid-point p and length* $2\,\rho(p)$ (or length 2, if $\rho(p) = \infty$).

MULTIPLICITY OF $\mathfrak{g}$ AT A POINT: If p is a point on the geodesic $\mathfrak{g}$ the cardinal number of the set of lineal elements, with mid-point p, lying on $\mathfrak{g}$ is the multiplicity of p as a point of $\mathfrak{g}$. If the multiplicity of p is *one* then p is a "*simple*" point of $\mathfrak{g}$; in all other cases it is a "*multiple*" point of $\mathfrak{g}$.

(9.2) *The multiplicity of a geodesic at any of its points is countable.*

This follows from the trivial fact:

(9.3) *Let the segments T_1, T_2 lie on the geodesic* $\mathfrak{g}$ *and have at most one common point, then if $x(\tau)$ is a representation of* $\mathfrak{g}$, *and T_i is given when τ is in $[\alpha_i, \beta_i]$, then $[\alpha_1, \beta_1]$, $[\alpha_2, \beta_2]$ have no common interior point.*

Then (9.2) follows since (*i*) two lineal elements on $\mathfrak{g}$ with the same mid-point have no other common point and (*ii*) there can be only a countable number of non-overlapping intervals of length $\rho(p)$ on the τ-axis.

(9.4) *A geodesic has at most a denumerable number of multiple points.*

We may assume $\rho(p) < \infty$. It suffices to prove that a given lineal element L on $\mathfrak{g}$ contains a countable number of multiple points of $\mathfrak{g}$ if any.

Now if q is the mid-point of L, there is a segment T with the same mid-point, which contains L, and has length $2\,\rho(q)$.

By the definition of $\sigma(p)$, we have for any point p of L,

$$\sigma(p) - \tfrac{1}{2}\,\rho\,(q) \leqslant \tfrac{1}{2}\,\rho(p) - \tfrac{1}{2}\,\rho(q) \leqslant \tfrac{1}{2}\,\sigma(q) = \tfrac{1}{4}\,\rho(q).$$

Hence $\sigma(p) \leqslant 3\,\rho(q)/4$, and if x is any point of a lineal element with mid-point p, then

$$xq \leqslant xp + pq \leqslant \sigma(p)/2 + p\,q \leqslant 3\,\rho(q)/8 + \rho(q)/4 < 3\,\rho(q)/4.$$

If now p_1, p_2 are multiple points of $\mathfrak{g}$ on L, then by the last inequality, one lineal element with mid-point p_i lies on T and there is at least one other, L_i, which does not lie on T.

The elements L_1 and L_2 cannot have more than one common point, for if they had, then they would be subsegments of a segment T', see (9.1), which would meet L in p_1, p_2 only, whereas $T(p_1, p_2)$ is unique.

Thence the theorem follows like (9.2).

GEODESICS WITHOUT MULTIPLE POINTS

(9.5) *A geodesic* $\mathfrak{g}$ *represented by* $x(\tau)$ *is without multiple points if, and only if,* $x(\tau_1) = x(\tau_2)$ *implies* $x(\tau_1 + \tau) = x(\tau_2 + \tau)$ *for all* τ.

Proof: (1) *Sufficiency.* Let L_1 and L_2 be lineal elements of $\mathfrak{g}$ with the same mid-point p, and let $x(\tau)$ represent L_i for $|\tau - \tau_i| \leqslant \frac{1}{2}\,\sigma(p)$. Then $x(\tau_1) = x(\tau_2)$; hence, by hypothesis $x(\tau_1 + \tau) = x(\tau_2 + \tau)$ for all τ, and in particular $L_1 = L_2$.

(2) *Necessity.* Let $\mathfrak{g}$ have no multiple points. If $x(\tau_1) = x(\tau_2) = p$ and $\tau_1 < \tau_2$, then $x(\tau)$ for $|\tau - \tau_i| \leqslant \frac{1}{2}\,\sigma(p)$ represents a lineal element L_i. Since p is a simple point, we have $L_1 = L_2$, and hence either

$$x(\tau_1 + \tau) = x(\tau_2 + \tau) \text{ or } x(\tau_1 + \tau) = x(\tau_2 - \tau) \text{ for } |\tau - \tau_i| \leqslant \sigma(p)/2.$$

Now, $x(\tau_2 + \tau)$ and $x(\tau_2 - \tau)$ represent geodesics and hence by (8.4) both the last relations hold for all τ. But the second is impossible, for, if we put $\tau_0 = \frac{1}{2}\,(\tau_1 + \tau_2)$, it gives.

$$x(\tau_0 + \tau) = x(\tau_1 + \tfrac{1}{2}\,(\tau_2 - \tau_1) + \tau) = x(\tau_2 - \tfrac{1}{2}\,(\tau_2 - \tau_1) - \tau) = x(\tau_0 - \tau)$$

which contradicts the property (7.2) of the representation of geodesics.

Note: If the representation $x(\tau)$ of any geodesic $\mathfrak{g}$ is such that numbers τ_1, τ_2 exist with $\tau_1 < \tau_2$, $x(\tau_1 + \tau) = x(\tau_2 + \tau)$ for all τ, then a standard argument shows that there is a smallest positive number η such that $x(\tau + \eta) = x(\tau)$ for all τ. This η is the same for all representations of $\mathfrak{g}$, and we call $\mathfrak{g}$ a "*closed geodesic*" or a "*periodic geodesic*" with "*period*" η. A great circle of length η on a sphere is an example; on the torus there are in addition to great circles, other closed geodesics, without multiple points. Among periodic geodesics the great circle is characterised by the property that two points on it can be joined by a segment contained in it.

(9.6) **Theorem:** *If a geodesic is such that any two of its points can be joined by a segment lying on it, it is a straight line or a great circle.*

Proof. First, a geodesic $\mathfrak{g}$ with this property cannot have a multiple point p. For suppose that $L_1 = T(q_1, q_1')$ and $L_2 = T(q_2, q_2')$ were two distinct lineal elements with mid-point p, then the segment T joining q_1 to q_2 is unique and does not contain p; moreover, if x is any point on T, then by (6.9) $xp \leqslant \sigma(p) < \rho(p)$, and hence the segments $T(p, x)$, with x on T, are unique and no two have any common point save p. But, on the other hand, T and hence all the segments $T(p, x)$ would have to lie on $\mathfrak{g}$, so that p would be a point of $\mathfrak{g}$ with non-countable multiplicity.

Let $x(\tau)$ represent $\mathfrak{g}$. We consider two cases:

(1) $x(\tau_1) \neq x(\tau_2)$ for $\tau_1 < \tau_2$. Then $x(\tau)$, $\tau_1 \leqslant \tau \leqslant \tau_2$, is the only arc from $x(\tau_1)$ to $x(\tau_2)$ on the geodesic hence, by hypothesis, is a segment. By (6.3), $x(\tau_1)\, x(\tau_2) = |\tau_1 - \tau_2|$, which is the characteristic property of a straight line.

(2) Suppose $x(\tau_1) = x(\tau_2)$ for some $\tau_1 < \tau_2$; then by (9.5) and the ensuing discussion, there is a smallest positive number η such that $x(\tau + \eta) = x(\tau)$ for all τ. When τ_1, τ_2 are given, we determine the integer ν such that $|\tau_1 - \tau_2 + \nu\,\eta|$ is minimal; then $|\tau_1 - \tau_2 + \nu\,\eta| \leqslant \eta/2$. Take the case when $\tau_1 - \tau_2 + \nu\,\eta \leqslant 0$. Since $x(\tau_2) = x(\tau_2 - \nu\,\eta)$, the arcs $\tau_1 \leqslant \tau \leqslant \tau_2 - \nu\,\eta$ and $\tau_2 - \nu\,\eta \leqslant \tau \leqslant \tau_1 + \eta$ are the only arcs on $\mathfrak{g}$ from $x(\tau_1)$ to $x(\tau_2)$. By hypothesis, at least one is a segment. Since

$$\tau_2 - \nu\,\eta - \tau_1 \leqslant \eta/2 \qquad \text{and} \qquad \tau_1 + \eta - \tau_2 + \nu\,\eta \geqslant \eta - \eta/2 = \eta/2$$

the first arc is by (6.3) a segment, and $x(\tau_1)\, x(\tau_2) = |\tau_1 - \tau_2 + \nu\,\eta|$, which is the characteristic property of a great circle. Hence the theorem.

Applications

This theorem has several important consequences, two of which we now give.

(9.7) *A G-space of one dimension is a straight line or a great circle.*

We shall use the notion of dimension in the sense defined by Menger and Urysohn, but only a few of the simplest theorems will be used here, and they are obvious for the intuitive notion of dimension. For the present theorem, then, we assume as known that a one-dimensional space contains more than one point, but not a subset homeomorphic to the surface of a triangle or to a two-simplex.

The first of these facts implies that there is at least one geodesic in the space; and the second that no point can be the mid-point of two distinct lineal elements L_1, L_2 (not assumed to be necessarily on the same geodesic). For if this was the case, then the segments $T(p, x)$ of the last proof would form a set homeomorphic to a two-simplex, as we show in the Lemma which succeeds this proof.

Hence, if there are two geodesics $\mathfrak{g}$, $\mathfrak{g}'$, they cannot meet. Now if p, p' were points of $\mathfrak{g}$, $\mathfrak{g}'$ respectively, then $T(p, p')$ would lie on a geodesic meeting both $\mathfrak{g}$ and $\mathfrak{g}'$. Hence there can be only one geodesic $\mathfrak{g}$ in the space, and as this is a G-space, if two points be given on $\mathfrak{g}$, then $\mathfrak{g}$ contains a segment joining them. Hence the G-space is a straight line or great circle.

This theorem shows that G-spaces are interesting only if their dimension exceeds one. We shall, when convenient, assume tacitly that this is the case.

In the proof we used:

(9.8) *Let $x(\tau)$, $\alpha \leqslant \tau \leqslant \beta$, $\alpha < \beta$, be a curve with $x(\tau_1) \neq x(\tau_2)$ whenever $\tau_1 \neq \tau_2$. Suppose the curve does not contain the point p, that $T_\tau = T(p, x(\tau))$ is unique for each τ in $[\alpha, \beta]$, and that p is the only point common to T_{τ_1} and T_{τ_2} when $\tau_1 \neq \tau_2$. Then $\bigcup_{\alpha \leqslant \tau \leqslant \beta} T_\tau$ is homeomorphic to a two-simplex.*

Proof. Take the point $q = q(\tau, \sigma)$ on T_τ where $pq/px(\tau) = \sigma$, and map it in the (ξ, η)-plane on the point

$$\xi = \sigma\tau, \qquad \eta = 1 - \sigma.$$

The image points of these points q form a simplex with vertices $(\alpha, 0)$, $(\beta, 0)$, $(0, 1)$; the mapping is one-to-one, since two distinct T_τ meet in p only.

By (2.14) we need only show that the correspondence between (ξ, η) and (τ, σ) is continuous, that is, that the correspondence between (τ, σ) and $q(\tau, \sigma)$ is continuous when $\sigma \neq 0$; the continuity is trivial when $\sigma = 0$ (at p).

Let $(\tau_\nu, \sigma_\nu) \to (\tau_0, \sigma_0)$, $\sigma_0 \neq 0$. The points $q_\nu = q(\tau_\nu, \sigma_\nu)$ form a bounded set, let q_0 be an accumulation point, and $q_\lambda = q(\tau_\lambda, \sigma_\lambda) \to q_0$. Then $(pq_\lambda x(\tau_\lambda))$ and $pq/px(\tau_\lambda) = \sigma_\lambda$ imply $(pq_0\, x(\tau_0))$ and $pq_0/px(\tau_0) = \sigma_0$. As the segment T_{τ_0} is unique there can only be one point q_0 with these properties, and hence $q_0 = q(\tau_0, \sigma_0)$.

(9.9) *If a G-space is such that there is only one geodesic through each given pair of points, then each geodesic is a straight line or great circle.*

For as any segment lies on some geodesic, and the geodesic through two points is unique, any geodesic must contain the segments which join two of its points.

Later in this book (Section 31) we shall prove by deeper methods, based on covering spaces, that the hypothesis of (9.9) implies that either *all* geodesics are straight lines or *all* are great circles of the same length. We had called the former type of G-space straight and will use the term "*space of the elliptic type*" for the latter, i. e., for a *G-space* in which the geodesic through two points is unique and all geodesics are great circles.

In these spaces convergent sequences of geodesics converge also as point-sets.

(9.10) *Let the geodesic* $\mathfrak{g}(a, b)$ *through any two distinct points* a, b *of a G-space be unique. Then if* $a_\nu \to a_0$, $b_\nu \to b_0 \neq a_0$ *and* G_ν *denotes the point set carrying* $\mathfrak{g}(a_\nu, b_\nu)$, *then* $\lim G_\nu = G_0$, *and we have* $x_\nu(\tau) \to x_0(\tau)$ *for a suitable representation* $x_\nu(\tau)$ *of* $\mathfrak{g}(a_\nu, b_\nu)$.

Proof. If $x_\nu'(\tau)$ is one representation of $\mathfrak{g}(a_\nu, b_\nu)$ with $x_\nu'(0) = a_\nu$, then the only other representation is $x_\nu''(\tau) = x_\nu'(-\tau)$. By (8.14) any subsequence of $\{x_\nu'(\tau), x_\nu''(\tau)\}$ itself contains a subsequence which converges to a representation $x_0(\tau)$ of $\mathfrak{g}(a_0, b_0)$ with $x_0(0) = a_0$.

Hence if $\eta_\nu = \pm 1$ is suitably chosen, $x_\nu(\tau) = x_\nu'(\eta_\nu \tau) \to x_0(\tau)$. If $\{\lambda\}$ is a subsequence of $\{\nu\}$ for which $p_\lambda \in G_\lambda$ and $p_\lambda \to p$ choose $\tau_\lambda = \pm a_\lambda p_\lambda$ such that $x_\lambda(\tau_\lambda) = p_\lambda$. Since $a_\lambda p_\lambda \to a_0 p$ there is a subsequence $\{\tau_\Theta\}$ of $\{\tau_\lambda\}$ which converges to $\eta\, a_0 p$, $\eta = \pm 1$, and (9.1) gives:

$$p = \lim p_\theta = \lim x_\theta(\tau_\theta) = x(\eta a_0 p) \in G_0.$$

Hence $\limsup G_\nu \subset G_0$; and by (8.13), $\lim G_\nu = G_0$.

In view of the deeper theorem already mentioned, the following is trivially true, but we need it in the proof.

(9.11) Under the hypotheses of (9.10) let λ_ν be the length of $\mathfrak{g}(a_\nu, b_\nu)$ if that geodesic is a great circle, otherwise put $\lambda_\nu = \infty$; then $\lambda_\nu \to \lambda_0$.

Proof. For any $\beta > 0$, let $\tau_\nu = \min(\beta, \lambda_\nu/2)$. Then $a_\nu x(\tau_\nu) = \tau_\nu$, hence $\lambda_0/2 \geqslant a_0\, x(\limsup \tau_\nu) = \limsup \tau_\nu = \min(\beta, \tfrac{1}{2} \limsup \lambda_\nu)$ which shows $\lambda_0 \geqslant \limsup \lambda_\nu$.

The relation $\lim \lambda_\theta < \lambda_0$ for a subsequence, would lead to the contradiction.

$$\lim x_\theta(0)\, x_\theta(\lambda_0/2) < \lambda_0/2 = x_0(0)\, x_0(\lambda_0/2).$$

On several occasions we will need the following lemma which is another consequence of (9.6).

(9.12) *If* $\mathfrak{g}_1$ *and* $\mathfrak{g}_2$ *are distinct geodesics (in a G-space), each of which contains with any two points a segment connecting them, then* $\mathfrak{g}_1$ *and* $\mathfrak{g}_2$ *have not more than two common points. If they have two common points* p, q *then* $\mathfrak{g}_1$ *and* $\mathfrak{g}_2$ *are great circles of the same length and* p, q *are conjugate on both* $\mathfrak{g}_1$ *and* $\mathfrak{g}_2$.

For if $\mathfrak{g}_1$ and $\mathfrak{g}_2$ have p and q in common, then $\mathfrak{g}_i$ contains a segment T_i from p to q by hypothesis, and $T_1 \neq T_2$ because $\mathfrak{g}_1 \neq \mathfrak{g}_2$. Hence no point r with (pqr) can exist. Since we know from (9.6) that $\mathfrak{g}_i$ is a straight line or great circle, it follows that $\mathfrak{g}_i$ is a great circle and that p and q are conjugate on $\mathfrak{g}_i$. Hence $\mathfrak{g}_1$ and $\mathfrak{g}_2$ both have length $2\,pq$. Finally, $\mathfrak{g}_1$ and $\mathfrak{g}_2$ cannot have a third common point r, because $T(p, r)$ is unique and would have to lie on both $\mathfrak{g}_1$ and $\mathfrak{g}_2$, which is possible only for $\mathfrak{g}_1 = \mathfrak{g}_2$.

10. Two-dimensional *G*-spaces

We saw that a *G*-space of one dimension is a topological manifold, and we shall prove the same for those of two-dimensions. Although this is probably true for any *G*-space, the proof (if the conjecture is correct) seems quite inaccessible in the present state of topology.

A HOMOGENEITY PROPERTY OF GENERAL *G*-SPACES

The only fact known in this direction is the following:

(10.1) *If two points* p, q *of a G-space and a positive number* ε *be given, then a topological mapping of* $S(p, \rho(p))$ *on an open subset* Q *of* $S(q, \varepsilon)$ *exists which sends* p *into* q.

Proof. If $q = p$, the theorem is easy. Let $\varepsilon' = \min\big(\varepsilon, \rho(p)\big)$; then we may choose Q as $S(q, \varepsilon')$; map p on itself and any point x of $S(p, \rho(p))$ on the point x' of the (unique) segment $T(p, x)$ such that $px' = px \cdot \varepsilon'/\rho(p)$, then $S(p, \rho(p))$ is mapped topologically on Q.

Next let p, q be distinct with $pq < \frac{1}{2}\rho(q)$. Put $\delta = \min(\varepsilon/3, pq/4)$ and determine r by (pqr) and $qr = \delta$. Since $pq + 2\,\delta < \rho(q)/2 + 3\,\rho(q)/8 = 7\,\rho(q)/8$ and $\rho(r) \geqslant \rho(q) - qr = \rho(q) - \delta \geqslant 7\,\rho(q)/8$ it follows that each point y of $S(r, pq + 2\,\delta)$, distinct from r, lies on a unique segment with origin r, of length $pq + 2\,\delta$. Let V_1, V_2 be the open sets defined by the inequalities

$$0 < r\,x < 2\,\delta \qquad \text{and} \qquad pq < rx < pq + 2\,\delta.$$

Then we can define a mapping Ω of V_2 on V_1 which is easily seen to be topological, by associating x_1 of V_1 and x_2 of V_2 by (rx_1x_2), $rx_1 + pq = rx_2$. Now $S(p, \delta) \subset V_2$, because $px < \delta$ implies

$$pq = rp - \delta < rp - px \leqslant rx \leqslant rp + px < rp + \delta = pq + 2\,\delta.$$

Hence $Q = S(p, \delta)\,\Omega$ is defined and is an open set. Clearly $p\Omega = q$ and $Q \subset V_1 \subset S(r, 2\,\delta) \subset S(q, \varepsilon)$ since, if $\delta \leqslant \varepsilon/3$ then $2\,\delta \leqslant \varepsilon - \delta = \varepsilon - rq$. Now by the first part of the proof, there is a topological mapping Ω_1 of $S(p, \rho(p))$ on $S(p, \delta)$ which leaves p fixed. Then $\Omega_1\Omega$ maps $S(p, \rho(p))$ topologically on Q and sends p into q.

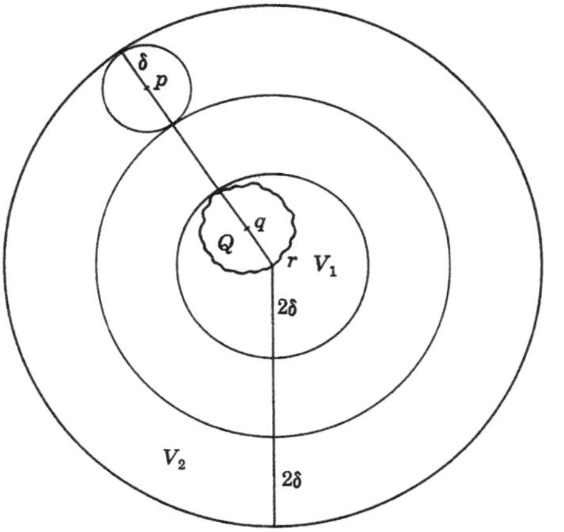

Fig. 1.

We now remove the restriction $pq < \frac{1}{2}\,\rho(q)$: if the theorem is true for p and q, and also for q and s then it holds for p and s. For then there are topological mappings Ω', Ω'' such that Ω' maps $S(p, \rho(q))$ on an open subset U' in $S(q, \rho(q))$ with $p\Omega' = q$; and Ω'' maps $S(q, \rho(q))$ on an open subset U'' of $S(s, \varepsilon)$ with $q\Omega'' = s$. Then $\Omega'\Omega''$ maps p on s and $S(p, \rho(p))$ on an open set U in U'', and $p\Omega'\Omega'' = s$.

Now $\inf \rho(x) > 0$ when $x \in S(p, 2\,pq)$, therefore we can reach q along a segment $T(p, q)$ by means of intermediate points q_i such that $q_iq_{i+1} < \frac{1}{2}\,\rho(q_{i+1})$.

The theorem now follows.

(10.2) *A G-space has the same dimension at each of its points.*

Pasch's axiom.

When in the foundations of geometry we wish to obtain a plane, some axiom must be introduced analogous to Pasch's axiom; the latter may be stated thus:

If a, b, c are not collinear, and (adc), then any line through d meets $T(ab) \cup T(bc)$.

In general G-spaces it would not be reasonable to adjoin an axiom of this type, for the line through d may be everywhere dense on a surface, it may have infinitely many multiple points, and both these properties may hold.

Even when the segments $T(a, b)$ and $T(b, c)$ are unique, and $\mathfrak{g}$ is a straight line through d, the axiom need not hold. For example, on the cylinder (7.1) let $a = (\cos\alpha, \sin\alpha, 0)$, $c = (\cos\alpha, -\sin\alpha, 0)$, $0 < \alpha < \pi/2$, $b = (-1, 0, 1)$, $d = (1, 0, 0)$; then the line $\xi = 1$, $\eta = 0$ through d does not meet $T(a, b) \cup T(b, c)$, although d is on $T(a, c)$.

To be significant for general G-spaces it is thus clear that Pasch's Axiom must be formulated, if at all, only locally.

(10.3) *Local Axiom of Pasch: If a, b, c lie in $S(p, \rho(p)/8)$ and (apc), then any segment of length $\rho(p)$ with mid-point p, contains a point of $T(a, b) \cup T(b, c)$.*

Although we could replace $\rho(p)$, $\rho(p)/8$ by other numbers, we have to make certain that the segment through p is long enough to reach $T(a, b)$ and $T(b, c)$, and we can in general only say that these segments lie in $S(p, \rho(p)/4)$, see (6.9). To see the implications of (10.3), consider any G-space R of dimension greater than one, and let a, b, c be three points in $S(p, \rho(p)/8)$ and not on a segment, and let (apc). Then there is a quadruple a', b', c', p' in the Euclidean plane R' which is *congruent* to a, b, c, p (distance of corresponding pairs of points are the same).

Map $T(a, b)$ isometrically on $T(a', b')$, and $T(b, c)$ on $T(b', c')$ so that in both mappings b goes into b'. This defines a topological mapping of $T(a, b) \cup T(b, c)$ on $T(a', b') \cup T(b', c')$, in which, say, x corresponds to x'. Finally if T_x is the segment with mid-point p, length $\rho(p)$, which contains x, map it isometrically on the segment $T_{x'}$ with midpoint p', length $\rho(p)$ which contains x', so that x' and the image of x lie on the same side of p' on $T_{x'}$. This is consistent with the previous agreement for $x = a, c$ and defines a topological mapping of $\cup T_x$ on the closed circular disc in R' with center p', radius $\frac{1}{2}\rho(p)$. Hence:

If in a G-space R there exists a quadruple a, b, c, p where a, b, c do not lie on one segment, $a, b, c \in S(p, \rho(p)/8)$, and (apc), and if every segment of length $\rho(p)$ with mid-point p meets $T(a, b) \cup T(b, c)$ then R is a two-dimensional manifold.

For then $\cup T_x$ exhausts $S(p, \rho(p)/2)$ and hence there is a circle about p, namely $S(p, \rho(p)/2)$, which is homeomorphic to the interior of a circle of the Euclidean plane.

Now (10.1) shows that R is a two-dimensional manifold. Thus the Local Pasch Axiom (10.3) not only implies that R is two-dimensional, but also that R is a manifold. It contains an intersection property which is not possessed by all two-parameter systems of curves in a plane. But actually (10.3) follows from the fact that the dimension of R is two.

(10.4) ***Theorem:*** *A two-dimensional G-space R satisfies the Local Axiom of Pasch and is therefore a manifold.*

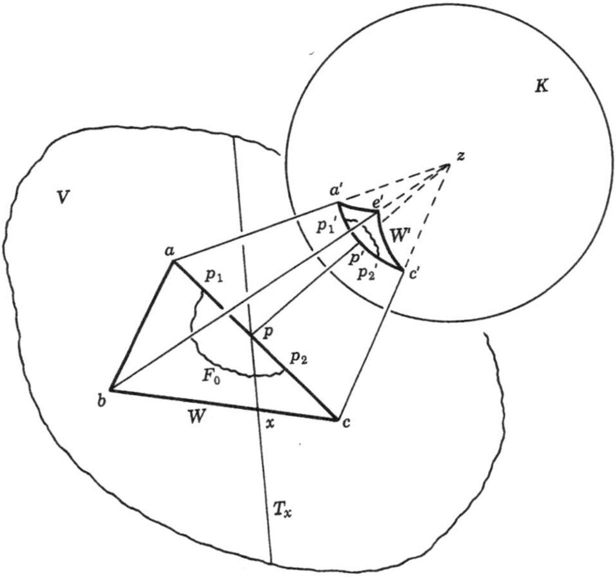

Fig. 2.

Proof. Let a, b, c, p in R satisfy the assumptions of (10.3) and to avoid a trivial case, let a, b, c not lie on a segment. Let $W = T(a, b) \cup T(b, c)$. If $x \in W$ let again T_x be the segment through x with midpoint p and length $\rho(p)$. We have shown that $V = \cup T_x$ is homeomorphic to a closed circular disk. (10.4) will be true if we prove:

(10.5) *If $S(p, \rho(p)/2)$ contains a point z, not in V, then the space R has at least dimension three.*

Proof. Let $\alpha = zV$, then $\alpha > 0$. Let K be the locus of points r with $zr = \alpha$. If $y \in V$, then $T(z, y)$ contains just one point z' in K. Define the operation Ω by $y' = y\Omega$, and call y' the "projection" of y on K. Then Ω is clearly a continuous mapping of V in K; (10.5) will be trivial if Ω is one-to-one. But this is not the case for general G-spaces, since planes need not exist. z does not lie on V, therefore $p' = p\Omega$ will not lie in $W' = W\Omega$. Since W' is closed, we have

$$p'W' = 2\delta > 0. \tag{10.6}$$

We first show that $V' = V\Omega$ has dimension at least two at p'. For this it is enough to prove that if F' is any closed set in $V' \cap S(p', \delta)$ which separates $b' = b\Omega$ from p' then $\dim F' \geqslant 1$. (Cf. Hurewicz and Wallman. [1, p 34]).

Now the set $F = F'\Omega^{-1}$ is closed and $F \cap W = 0$, since $F' \cap W' = 0$ by (10.6). Moreover, F separates p from b in V. For, if not, then there would be a continuum C in V connecting p and b but not meeting F. But this would entail that $C' = C\Omega$ were a continuum connecting p' and b' not meeting F', and then F' would not separate p' from b'.

Thus F separates p from b in V and, as V is homeomorphic to a circular disk, F contains a continuum F_0 which already separates p from b, see Hausdorff [1, p. 343, VI]. Then F_0 must contain points p_1, p_2 with (ap_1p), (pp_2c), for otherwise $T(p, a) \cup T(a, c)$ or $T(p, c) \cup T(c, b)$ would connect p to b, but not meet F_0 since $W \cap F = 0$.

The projection $F_0' = F_0\Omega$ of F_0 is a continuum, since it is a continuous map of F_0, and contains the two distinct points $p_i' = p_i\Omega$. But $F_0' \subset F'$, $\dim F_0' \geqslant 1$. Hence $\dim F' \geqslant 1$.

Thus $\dim V' \geqslant 2$. If u is the mid-point of $T(x', z)$, where $x' \in V'$ then $\bigcup_{x' \in V'} T(x', u)$ is the topological product of V' by a segment of length $\alpha/2$. Hence by a theorem of Hurewiez [1]

$$\dim R \geqslant \dim \bigcup_{x' \in V'} T(x', u) = \dim V' + 1 \geqslant 3.$$

Consequences

(10.7) *If R is a two-dimensional G-space and the geodesic through any pair of points is unique, then R is homeomorphic to the projective plane or to a proper subset of it.*

Proof. Choose a, b, c, p as before. An elliptic plane E with distance $\varepsilon(x', y')$ for which the common length λ' of its lines is at least $2\rho(p)$ contains a quadruple a', b', c', p' congruent to a, b, c, p. Map again $T(a, b)$, $T(b, c)$ isometrically on the (unique) segments $T(a', b')$, $T(b', c')$ so that b corresponds to b' in both maps. This defines a topological mapping of $W = T(a, b) \cup T(b, c)$ on $W' = T(a', b') \cup T(b', c')$. Since each geodesic through p contains a segment of length $\rho(p)$ with mid-point p and each point can be joined to p we have

$$R = \bigcup_{y \in W} \mathfrak{g}(p, y), \qquad E = \bigcup_{y' \in W'} \mathfrak{g}(p', y').$$

We now map, for fixed y, the geodesic $\mathfrak{g}(p, y)$ on $\mathfrak{g}(p', y')$ as follows:

(a) If $\mathfrak{g}(p, y)$ is a great circle of length λ_y, let p_y, p_y' be respectively conjugate to p on $\mathfrak{g}(p, y)$ and to p' on $\mathfrak{g}(p', y')$. Let T_y^1 be the segment from p to p_y which contains $y \in W$ where $y \neq c$, and T_y^2 be the other segment. Define $T_{y'}^1$, $T_{y'}^2$ similarly for $y' \in W'$, $y' \neq c'$. Then map the point x of T_y^i on the point x' of $T_{y'}^i$ for which

$$\varepsilon(p', x') = \frac{\lambda'}{2} \frac{px}{1 + px} \frac{2 + \lambda_y}{\lambda_y}.$$

Observe that the mapping is consistent for $x = p_y$ since $pp_y = \lambda_y/2$, and the right-hand side equals $\frac{1}{2}\lambda'$ so that $x' = p_{y'}'$.

(b) If $\mathfrak{g}(p, y)$ is a straight line let T_y^1 be the ray with origin p which contains y and T_y^2 the opposite ray. Then map the point x of T_y^i on x' of $T_{y'}^i$ for which

$$\varepsilon(p', x') = \frac{\lambda'}{2} \frac{px}{1 + px}$$

That this mapping is topological follows since λ_y is a continuous function of y and $(2 + \lambda_y)/\lambda_y \to 1$ as $\lambda_y \to \infty$, compare (9, 11).

Further applications of (10.3) which will be used frequently are the following: call a set M in a G-space *convex*, if the segment connecting any two points of the closure $\overline{M}$ of M is unique and $x, y \in M$ implies $T(x, y) \subset M$. Then:

(10.8) *Every point of a two-dimensional G-space is interior point of a closed and an open convex set.*

The set $W = \cup\, T(p, x)$, $x \in T(a, b) \cup T(b, c)$ used above is convex. For a given point y a set W can be constructed that contains y as interior point. If the three segments bounding W are omitted we obtain an open set containing y.

(10.9) *Two distinct segments of length $\rho(p)$ with p as common midpoint cross each other at p.*

More precisely: either segment decomposes $S(p, \rho(p)/2)$ into two sets either of which contains one of the half open segments into which p decomposes the other segment. A corollary of (10.9) and (9.10) is:

(10.10) If the geodesic through two points of a two-dimensional G-space is unique, and the geodesics $\mathfrak{g}_1$, $\mathfrak{g}_2$ intersect at p, then two geodesics $\mathfrak{g}_1'$ close to $\mathfrak{g}_1$ and $\mathfrak{g}_2'$ close to $\mathfrak{g}_2$ will also intersect.

(10.11) *In a two-dimensional G-space let p be the midpoint of a and b where $0 < \gamma = ab < \rho(p)/4$. Given two positive numbers α, β with $|\alpha - \beta| < \gamma$, $\alpha + \beta > \gamma$ and $\alpha < \rho(p)/8$, $\beta < \rho(p)/8$, there exists in $S(p, \rho(p)/2)$, on either side of the diameter D carrying $T(a, b)$, exactly one point c with $ac = \beta$ and $bc = \alpha$.*

Proof. The circle K_a: $ax = \beta$ lies in $S(p, \rho(p)/2)$ and is therefore homeomorphic to an ordinary circle, similarly for the circle K_b: $bx = \alpha$. The diameter D divides the circles K_a, K_b into semicircles K_a', K_a'' and K_b', K_b'', where we assume that K_a' and K_b' lie on the same side D' of D.

The semicircle K_a' connects the points q and q_0 on D with (qaq_0) and $qa = aq_0 = \beta$, and K_b' connects the points r, r_0 with (rbr_0) and $rb = br_0 = \alpha$. Because $|\alpha - \beta| < \gamma$ and $\alpha + \beta > \gamma$ the pair q, q_0 separates the pair r, r_0 on D. Hence K_a' and K_b' intersect at a point c.

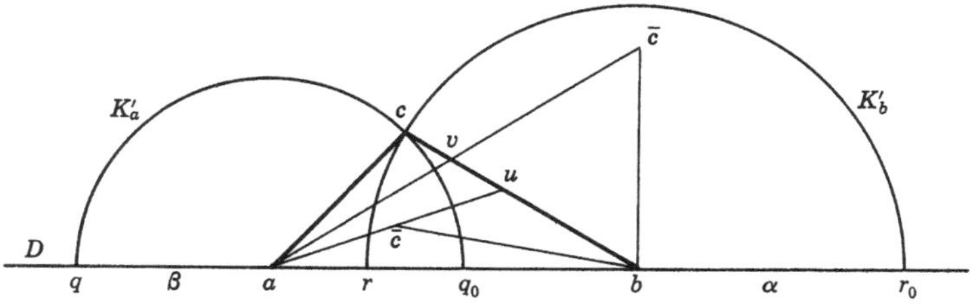

Fig. 3.

To establish the uniqueness of c in D' assume the existence of another point $\bar{c}$ in D' with $a\bar{c} = \beta$, $b\bar{c} = \alpha$. It is clear that $\bar{c}$ cannot lie on $T(a, c) \cup T(c, b)$, nor c on $T(a, \bar{c}) \cup T(\bar{c}, b)$. Therefore either $\bar{c}$ lies in the interior of the triangle abc, or c in the interior of $ab\bar{c}$, or neither. In the first case (10.3) implies the existence of a point u with $(a\bar{c}u)$ and (buc). Then

$$\alpha + \beta = ac + cb = ac + cu + ub > au + ub = a\bar{c} + \bar{c}u + ub >$$
$$> a\bar{c} + \bar{c}b = \alpha + \beta.$$

Interchanging c and $\bar{c}$ in this argument settles the second case.

In the third case either $T(a, \bar{c})$ and $T(b, c)$ or $T(a, c)$ and $T(b, \bar{c})$ intersect. Assume $v = T(a, \bar{c}) \cap T(b, c)$. Then

$$\alpha - cv + v\bar{c} = bv + v\bar{c} > b\bar{c} = \alpha \quad \text{or} \quad v\bar{c} > cv$$

and

$$\beta - v\bar{c} + cv = av + vc > vc = \beta \quad \text{or} \quad cv > v\bar{c}.$$

This proves (10.11).

11. Plane metrics without conjugate points

The proof of (10.7) shows:

(11.1) *A two-dimensional straight space is homeomorphic to the plane.*

Such a space may therefore be regarded as a remetrization of the euclidean plane P. One might then pose the problem of determining all the metrizations of P for which P becomes a straight space. The following considerations will show that the problem is, in this form, too general to be interesting. But asking for all curve systems which can serve as geodesics for such metrizations of P turns out to be fruitful, because two simple, obviously necessary conditions prove to be sufficient:

(11.2) **Theorem:** *In the euclidean plane P, with a euclidean distance $e(x, y)$, let a system Σ of curves be given with the following two properties:*

I. *Each curve in Σ is representable in the form $p(t)$, $-\infty < t < \infty$, such that $p(t_1) \neq p(t_2)$ for $t_1 \neq t_2$ and $e(p(0), p(t)) \to \infty$ for $|t| \to \infty$.*

II. *There is exactly one curve of Σ through two given distinct points of P.*

Then P may be metrized as a straight space such that the curves in Σ are the geodesics.

The proof of the theorem is quite long, but this is justified by the importance of the result. Besides leading to many important examples, the theorem constitutes a solution in the large, of the inverse problem of the calculation of variations[1] for plane metrics without conjugate points, a problem which is quite inaccessible to the standard methods. In addition, the length of the proof is largely due to the weakness of our hypotheses. For instance, it must be established that a curve in Σ varies continuously with two of its points, a fact which is obvious under the usual differentiability hypotheses.

We denote the unique curve of Σ (called *Σ-curve or line*) through the two distinct points x and y by $\mathfrak{g}(x, y)$, and the arc of $\mathfrak{g}(x, y)$ with endpoints x and y by $T(x, y)$, and put $T(x, x) = x$. Moreover if $x \neq y$ then $[xzy]$ denotes that z is an interior point of $T(x, y)$. The points on $T(x, y)$, $x \neq y$, form together with the points z for which $[xyz]$ the ray $R(x, y)$. The proof will be decomposed into several steps.

Topological properties of Σ

(a) If $[apb]$ and p lies on a Σ-curve $\mathfrak{h} \neq \mathfrak{g}(a, b)$, then $\mathfrak{h}$ separates a from b.

For if $[rps]$ and $r, s \in \mathfrak{h}$, then $D = T(a, r) \cup T(r, s) \cup T(s, a)$ is a closed Jordan curve whose interior lies on the same side of $\mathfrak{h}$ as a. If b were also on this side, then $R(p, b) - p$ would lie on this side and intersect D because of the Jordan curve theorem and I. But $R(p, a)$ intersects each of the lines $\mathfrak{g}(r, s)$, $\mathfrak{g}(r, a)$, $\mathfrak{g}(s, a)$, hence $\mathfrak{g}(a, b) = R(p, a) \cup (R(p, b) - p)$ would intersect at least one of these lines twice.

This proof implies:

(b) *The system Σ satisfies the Axiom of Pasch*: if b is not on $\mathfrak{g}(a, c)$ and $[adc]$, then each Σ-curve through d intersects $T(a, b) \cup T(b, c)$.

We call a set *Σ-convex* if it contains with x and y the whole arc $T(x, y)$. If a, b, c do not lie on one Σ-curve then $T(a, b) \cup T(b, c) \cup T(c, a)$ is the boundary of a closed and an open Σ-convex domain. We denote the former by $F(abc)$. If a, b, c lie on one Σ-curve, $F(abc)$ denotes the smallest Σ-segment containing a, b, c. If $x_1, \ldots, x_\sigma$ are any points then

$$C(x) = C(x_1, \ldots, x_\sigma) = \bigcup_{\lambda,\mu,\nu} F(x_\lambda x_\mu x_\nu)$$

is readily seen to be convex. It is obviously the smallest Σ-convex set containing the points $x_1, \ldots, x_\sigma$. Unless all these points lie on one Σ-curve, $C(x)$ is bounded by a Σ-polygon Q of the form $Q = \bigcup_{\nu=1}^{\tau} T(x_{\lambda_\nu}, x_{\lambda_{\nu+1}})$, $x_{\lambda_{\tau+1}} = x_{\lambda_1}$ where no three x_{λ_ν} lie on the same Σ-curve and $C(x)$ is the intersection of the closed half planes bounded by the lines $\mathfrak{g}(x_{\lambda_\nu}, x_{\lambda_{\nu+1}})$ and containing $C(x)$. Thus

(c) *Any finite set of points $x_1, \ldots, x_\sigma$ which is not contained in a Σ-curve has a bounded Σ-convex closure $C(x)$, whose boundary has the form $\bigcup_{\nu=1}^{\tau} T(x_{\lambda_\nu}, x_{\lambda_{\nu+1}})$,*

$x_{\lambda_\tau+1} = x_{\lambda_1}$. *Any point outside of* $C(x)$ *can be separated from the interior of* $C(x)$ *by a suitable* Σ*-curve* $\mathfrak{g}(x_{\lambda_\nu}, x_{\lambda_{\nu+1}})$.

Next, convergence of Σ-segments and Σ-curves will be discussed. The underlying limit concept is here, as always, that of Hausdorff's closed limit defined in Section 3.

(d) If $e(a_\nu, a) \to 0$ and $e(b_\nu, b) \to 0$, then $\lim T(a_\nu, b_\nu) = T(a, b)$.

In the proof we assume $a \neq b$, the modifications necessary for $a = b$ are obvious. If $\limsup T(a_\nu, b_\nu) \subset T(a, b)$ were not correct then a subsequence $\{\mu\}$ of $\{\nu\}$ and points $c_\mu \in T(a_\mu, b_\mu)$ would exist, such that $c = \lim c_\mu \notin T(a, b)$. There is a Σ-curve $\mathfrak{h}$ separating c from $T(a, b)$. This is obvious for $c \in \mathfrak{g}(a, b)$. If $c \notin \mathfrak{g}(a, b)$ then a $\mathfrak{g}(a_1, b_1)$ with $[aa_1c]$, $[bb_1c]$ separates c from $T(a, b)$ because of (a).

For large μ the curve $\mathfrak{h}$ also separates a_μ and b_μ from c_μ. But then $T(a_\mu, c_\mu)$ and $T(c_\mu, b_\mu)$ would both intersect $\mathfrak{h}$, and $\mathfrak{g}(a_\mu, b_\mu)$ would intersect $\mathfrak{h}$ twice.

Now we show that every point c with $[acb]$ is limit of a sequence $c_\nu \in T(a_\nu, b_\nu)$. Let $\mathfrak{h}$ be any Σ-curve through c distinct from $\mathfrak{g}(a, b)$. Then $\mathfrak{h}$ separates a_ν from b_ν for large ν, hence intersects $T(a_\nu, b_\nu)$ in a point c_ν. The first part of this proof shows that every accumulation point of $\{c_\nu\}$ lies on $T(a, b)$. On the other hand, c is the only common point of $\mathfrak{h}$ and $T(a, b)$, hence $\lim c_\nu = c$.

(e) *If* $e(a_\nu, a) \to 0$, $e(b_\nu, b) \to 0$ *and* $a \neq b$ *then* $\lim \mathfrak{g}(a_\nu, b_\nu) = \mathfrak{g}(a, b)$.

For the proof of $\limsup \mathfrak{g}(a_\nu, b_\nu) \subset \mathfrak{g}(a, b)$ it suffices, because of (d), to consider a subsequence $\{\mu\}$ of $\{\nu\}$ for which a point c_μ with $[a_\mu b_\mu c_\mu]$ exists that converges to a point c. If c did not lie on $\mathfrak{g}(a, b)$, a suitable Σ-curve $\mathfrak{h}$ would separate b from a and c, therefore also b_μ from a_μ and c_μ for large μ. Again $\mathfrak{g}(a_\mu, b_\mu)$ would intersect $\mathfrak{h}$ at least twice.

Let an arbitrary point c with $[abc]$ be given. Put $\gamma = \max_{x \in T(a,c)} e(x, a)$. By I there is a point p_ν on $R(a_\nu, b_\nu)$ with $e(a_\nu, p_\nu) = 2\gamma$. If a neighborhood $N : e(c, x) < \varepsilon$ of c and a subsequence $\{\mu\}$ of $\{\nu\}$ existed such that N contains no point of $\mathfrak{g}(a_\mu, b_\mu)$, take a subsequence $\{p_\lambda\}$ of $\{p_\mu\}$ which converges to a point p. The first part of this proof implies that $p \in \mathfrak{g}(a, b)$, and (d) shows $\lim T(a_\lambda, p_\lambda) = T(a, p)$. Because of $[a_\lambda b_\lambda p_\lambda]$ we have $[abp]$ and, since $ap = 2\gamma$, also $[acp]$. But then (d) would show that N contains points of $T(a, c)$ for all large λ.

(f) *For a given point* z *and a given* $\rho > 0$ *a* Σ*-convex polygon exists that contains the disk* $N_\rho : e(z, x) < 2\rho$ *in its interior.*

For a proof consider the circle $K : e(z, x) = 2\rho$. Because of (d) there is a $\delta > 0$ such that for $x, y \in K$ and $e(x, y) < \delta$ the diameter of $T(x, y)$ is less than $\rho/2$. Select points $x_1, \ldots, x_\sigma, x_{\sigma+1} = x_1$ in this order on K such that $e(x_\nu, x_{\nu+1}) < \delta$. Then $\bigcup_{\nu=1}^{\sigma} T(x_\nu, x_{\nu+1})$ is a (not necessarily simple) closed polygon that separates N_ρ from all points y with $e(z, y) > 3\rho$. For these points $x_1, \ldots, x_\sigma$ construct $C(x)$ as under (d). The boundary $\bigcup_{\nu=1}^{\tau} T(x_{\lambda_\nu}, x_{\lambda_{\nu+1}})$ of $C(x)$ satisfies the assertion.

We denote as *simple family* Φ of Σ-curves a set of curves in Σ that covers the plane simply, i. e., every point of P lies on exactly one curve in Φ. If the Σ-curves are the euclidean straight lines, then the families of parallel lines are the only simple families. This fact indicates how to construct simple families in the general case.

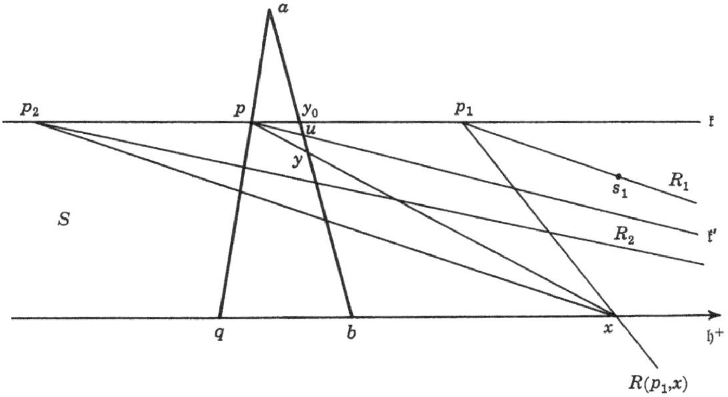

Fig. 4.

(g) *Every curve* $\mathfrak{h}$ *in* Σ *is contained in a simple family of* Σ*-curves* $\Phi(\mathfrak{h})$.

Proof. Denote by $\mathfrak{h}^+$ an orientation of $\mathfrak{h}$. Let $p \notin \mathfrak{h}$ and choose q and b on $\mathfrak{h}^+$ such that b follows q, and any point a such that $[qpa]$. As x traverses $\mathfrak{h}^+$ from b on, in the positive direction, the intersection y of $T(p, x)$ with $T(a, b)$ moves monotonically from b toward a, but does not reach a (because of (a) and (e)). Therefore as x tends on $\mathfrak{h}^+$ to infinity, the point y tends to a limit position y_0 with $[by_0a]$, so that the line $\mathfrak{g}(p, x)$ tends by (e) to the line $\mathfrak{g}(p, y_0)$.

We call $\mathfrak{k} = \mathfrak{g}(p, y_0)$ the *asymptote* to $\mathfrak{h}^+$ through p. The line $\mathfrak{k}$ cannot intersect $\mathfrak{h}$, for then $\mathfrak{g}(p, y)$ would by (a) intersect $\mathfrak{h}$ twice for y on $T(b, y_0)$ close to y_0.

To prove (g) it suffices to see that $\mathfrak{k}$ does not depend on the choice of p on $\mathfrak{k}$. First let p_1 be on $\mathfrak{k}$ on the same side as y_0. Since $T(p_1, x)$ lies for every x in the closed strip S bounded by $\mathfrak{h}$ and $\mathfrak{k}$, the ray $R(p_1, x)$ tends to a ray R_1 in S. If R_1 did not lie on $\mathfrak{k}$, it would contain an interior point s_1 of S. Then the line $\mathfrak{g}(p, s_1)$ would not intersect $\mathfrak{h}$, which contradicts the definition of $\mathfrak{k}$.

Now let p_2 be a point of $\mathfrak{k}$ on the other side of p from y_0. Again $T(p_2, x)$ lies in S, hence $R_2 = \lim T(p_2, x)$ lies in S. If R_2 did not lie on $\mathfrak{k}$, then the asymptote $\mathfrak{k}'$ through p to the oriented line which contains R_2 as positive subray cannot be $\mathfrak{k}$ because $\mathfrak{k}$ intersects R_2 at p_2, hence would have to intersect $T(y_0, y)$ in an interior point u without intersecting $\mathfrak{h}$ which again contradicts the definition of $\mathfrak{k}$.

Construction of the metric

Because of (e) there is a countable set of curves in Σ (for instance, all curves containing two distinct rational points) which is dense in Σ in the sense that every Σ-curve is limit of a sequence of curves in the set. Therefore there is a countable set of simple families $\Phi_1, \Phi_2, \ldots$ such that the lines in $\cup \Phi_i$ are dense in Σ.

With each Φ_i we associate an auxiliary function $\delta_i(x, y)$ as follows. We denote the curve in Φ_i through the point a by L_a^i and select a point z once for ever. Denote the two sides of L_z^i by σ_+^i and σ_-^i and put

$$t_i(L) = \begin{cases} e(z, L) & \text{if } L \subset \sigma_+^i \\ -e(z, L) & \text{if } L \subset \sigma_-^i \\ 0 & \text{if } L = L_z^i \end{cases}$$

If the line L in Φ_i separates the lines L' and L'' in Φ_i then

$$t_i(L') < t_i(L) < t_i(L'') \qquad \text{or} \qquad t_i(L') > t_i(L) > t_i(L'').$$

The function $\delta_i(x, y)$ is defined by

$$\delta_i(x, y) = |t_i(L_x^i) - t_i(L_y^i)|$$

and has the following properties:

(h) $$\delta_i(x, y) = \delta_i(y, x) \geqslant 0,$$

(i) $$\delta_i(x, y) = 0 \quad \text{if and only if} \quad L_x^i = L_y^i,$$

(k) $$\delta_i(x, y) + \delta_i(y, z) \geqslant \delta_i(x, z)$$

and the equality sign holds if and only if L_y^i coincides with L_x^i or L_z^i, or separates L_x^i from L_z^i.

(l) $$\delta_i(a_\nu, a) \to 0 \quad \text{if} \quad e(a_\nu, a) \to 0.$$

For if $e(a_\nu, a) \to 0$, then $L_{a_\nu}^i \to L_a^i$ since otherwise $L_{a_\nu}^i$ would intersect L_a^i for infinitely many ν, and $e(L_{a_\nu}^i, z) \to e(L_a^i, z)$ follows immediately from the definition of $L_{a_\nu}^i \to L_a^i$.

(m) $$\delta_i(x, z) \leqslant e(x, z)$$

because $\delta_i(x, z) = |t_i(L_x^i) - t_i(L_z^i)| = |t_i(L_x^i)| = e(L_x^i, z) \leqslant e(x, z)$.

Put now

$$\delta(x, y) = \sum_{i=1}^{\infty} \delta_i(x, y)\, 2^{-i}.$$

This series converges for every pair x, y because by (k) and (m)

$$\delta_i(x, y) \leqslant \delta_i(x, z) + \delta_i(z, y) \leqslant e(x, z) + e(z, y).$$

Clearly $\delta(x, y) \geqslant 0$. Since for $x \neq y$ a Σ-curve $\mathfrak{k}$ different from $\mathfrak{g}(x, y)$ through a point z with $[xzy]$ separates by (a) x from y and the curves in $\cup \Phi_i$ are dense in Σ, there is a curve L^{i_0} in a suitable Φ_{i_0} which separates x from y. It follows from (i) that $\delta_{i_0}(x, y) > 0$, hence $\delta(x, y) = \delta(y, x) > 0$. Also $\delta(x, y)$ satisfies the triangle inequality by (k). Thus $\delta(x, y)$ satisfies the axioms for a metric space. If the symbol (xyz) has the usual meaning with respect to $\delta(x, y)$ then

(n) $$(xyz) \text{ if and only if } [xyz].$$

For let (xyz). If $[xyz]$ were not true we could separate y from x and z by a Σ-curve (compare the proof (d)), hence also by a curve L^{i_0} in some Φ_{i_0}. Then the remark on the equality sign in (k) shows $\delta_{i_0}(x, y) + \delta_{i_0}(y, z) > \delta_{i_0}(x, z)$, hence $\delta(x, y) + \delta(y, z) > \delta(x, z)$ contrary to the hypothesis.

If $[xyz]$, then the curves L_x^i, L_y^i, L_z^i either coincide or L_y^i separates L_x^i from L_z^i, hence (k) yields $\delta_i(x, y) + \delta_i(y, z) = \delta_i(x, z)$ for every i, and $\delta(x, y) + \delta(y, z) = \delta(x, z)$ follows.

(o) $$\delta(a, a_\nu) \to 0 \text{ if and only if } e(a, a_\nu) \to 0.$$

Let $e(a, a_\nu) \to 0$. Then $\delta_i(a, a_\nu)$ is uniformly bounded because of (k) and (m). Therefore, given $\varepsilon > 0$, μ may be chosen such that

$$\sum_{i=\mu+1}^{\infty} \delta_i(a, a_\nu)/2^i < \varepsilon/2 \quad \text{for all } \nu.$$

Because of (l) we can find λ such that

$$\delta_i(a, a_\nu)/2^\nu < \varepsilon/2^{\mu+1} \quad \text{for} \quad \nu > \lambda \quad \text{and} \quad i = 1, \ldots, \mu.$$

Then $\delta(a, a_\nu) < \varepsilon$ for $\nu > \lambda$.

Let $\delta(a, a_\nu) \to 0$ and assume for an indirect proof that $e(a, a_\mu) > \varepsilon > 0$ for a subsequence $\{\mu\}$ of $\{\nu\}$. We may also assume that the rays $R(a, a_\mu)$ converge. There is a curve L^{i_0} in some Φ_{i_0} separating a_μ from a for large μ. If q is any point on L^{i_0} we have

$$\delta_{i_0}(a, a_\mu) > \delta_{i_0}(a, q) > 0 \quad \text{for large } \mu,$$

hence $\delta(a, a_\mu) > \delta_{i_0}(a, q)\, 2^{-i_0} > 0$.

If the metric $\delta(x, y)$ is finitely compact, then it solves our problem. However, this may in general not be the case. We therefore make the following additional construction. Let N_ν denote the disk $e(z, x) < \nu$. For every positive integer ν construct by (f) a Σ-convex polygon $Q_\nu\colon \bigcup_{\lambda=1}^{\gamma} T(x_\lambda, x_{\lambda+1}),\ x_{\gamma+1} = x_1$ containing N_ν in its interior and put

$$\varphi(\nu) = \max_{x \in Q_\nu} (1 + e(z, x)).$$

For every point p with $e(z, p) > \varphi(\nu)$ there is a line $\mathfrak{g}(x_\lambda, x_{\lambda+1})$ which separates p from N_ν. Define $\psi_\nu(\tau)$ by

$$\psi_\nu(\tau) = \begin{cases} \dfrac{\tau}{\gamma\, 2^\nu} & \text{for} \quad 0 \leqslant \tau \leqslant \nu/2 \\[2ex] \varphi(\nu) + (\tau - \nu)\left(\dfrac{2\varphi(\nu)}{\nu} - \dfrac{1}{\gamma 2^\nu}\right) & \text{for} \quad \tau \geqslant \nu/2 \\[2ex] -\psi_\nu(-\tau) & \text{if} \quad \tau \leqslant 0. \end{cases}$$

Let Ψ_λ be a simple family containing $\mathfrak{g}(x_\lambda, x_{\lambda+1})$ and define $\tau_\lambda(L)$ with respect to this family exactly as $\tau_i(L)$ with respect to Φ_i.

If L_x^λ denotes the curve in Ψ_λ through x we put

$$\varepsilon_\nu(x, y) = \sum_{\lambda=1}^{\gamma} \left|\psi_\nu[\tau_\lambda(L_x^\lambda)] - \psi_\nu[\tau_\lambda(L_y^\lambda)]\right|.$$

Since $\psi_\nu(\tau)$ is a monotone function of τ the above arguments show that $\varepsilon_\nu(x, y) = \varepsilon_\nu(y, x) \geqslant 0$, $\varepsilon_\nu(x, y) + \varepsilon_\nu(y, z) \geqslant \varepsilon_\nu(x, z)$ and that $[xyz]$ implies

$\varepsilon_\nu(x, y) + \varepsilon_\nu(y, z) = \varepsilon_\nu(x, z)$. It follows from the definition of $\psi_\nu(\tau)$ and $\tau_\nu(L_x) \leqslant e(z, x)$ that

$$\varepsilon_\nu(z, x) \leqslant 2^{-\nu} \quad \text{for} \quad 0 \leqslant e(z, x) \leqslant \nu/2.$$

If $e(z, x) > \varphi(\nu) > \nu$ then at least one line $M = \mathfrak{g}(x_\lambda, x_{\lambda+1})$ separates x from N_ν, hence

$$\varepsilon_\nu(z, x) \geqslant |\psi_\nu(\tau_\lambda(M))| > \psi_\nu(\nu) = \varphi(\nu) > \nu.$$

Therefore

$$xy = \delta(x, y) + \sum_{\nu=1}^{\infty} \varepsilon_\nu(x, y)$$

is finite for all pairs x, y and satisfies the condition

$$zx_\nu \to \infty \quad \text{for} \quad e(z, x_\nu) \to \infty.$$

This implies that the plane with the metric xy is finitely compact, so that xy solves our problem.

Remarks

If the system Σ has differentiability properties (in the inverse problem as originally conceived it consists of the solutions of a second order differential equation) then our summations can be replaced by integrations and a distance function xy with adequate differentiability properties is obtained.

The same problem for higher dimensions is quite difficult. It is known that conditions I and II no longer suffice, i. e., geodesics have additional properties of a topological nature. For the classical three-dimensional problem in the small, necessary and sufficient conditions have been found by J. Douglas [1]. The problem is entirely unsolved for higher dimensions than three.

Non-desarguesian systems

It is of importance for the next chapter to remember from the foundations of geometry, that the curves in a system with properties I and II need not satisfy Desargues' Theorem or its converse (compare the Desargues Property in Section 13) even if Σ satisfies strong additional conditions.

A reader not familiar with the well-known example of Hilbert [1, pp. 66–71] need not look it up; the present book provides several other examples of such systems: the two systems constructed to verify (23.5 a, b) furnish examples of non-desarguesian Σ which go into themselves under all translations of the euclidean plane. However, there is no metrization of the plane with either of these systems as geodesics, which is invariant under all translations (this follows from Theorem (50.1)).

A non-desarguesian system Σ which (in cartesian coordinates ξ, η) goes into itself under the translations $\xi' = \xi + \nu$, ν an integer, $\eta' = \eta + \rho$, ρ arbitrary real, and a metrization of the plane as a straight space with the curves of Σ as geodesics which admits these translations as motions are found in Section **33**. Like Hilbert's example this system satisfies the parallel axiom.

The importance of these examples for the next chapter is this: we call a subset L of a G-space R "*flat*", if it is with the metric of R itself a G-space. If L has dimension ρ we call it briefly a "ρ-*flat*". Thus the straight lines and great circles of a G-space are its one-flats. The euclidean, hyperbolic, and elliptic spaces have the property that any $\rho + 1$ points which do not lie in a σ-flat with $\sigma < \rho$ lie in exactly one ρ-flat, in particular, three points, which do not lie on one geodesic, determine a two-flat.

Every two-dimensional straight space has trivially this property for $\rho = 1, 2$. But higher dimensional straight spaces need not possess it. For it is well known, and will be proved again in (14.4), that every 2-flat in the space must satisfy Desargues' Theorem, if any three points lie in a 2-flat.

Therefore, to obtain examples of straight spaces of dimension greater than two, in which 3 points do in general not lie in a 2-flat, it suffices to take a non-desarguesian two-dimensional straight space R and form the product $[R \times E^\nu]_\alpha$, defined in (8.15), of R with the ν-dimensional euclidean space E^ν. Even if R satisfies Desargues' Theorem, $[R \times E^\nu]_\alpha$ need not have the property that any three points lie in a two-flat. For if R is the hyperbolic plane, then $[R \times E^\nu]_2$ is a Riemann space in which the curvature is not constant, hence three given points will, by Beltrami's Theorem, (see Section 15) in general not lie in a 2-flat.

CHAPTER II

DESARGUESIAN SPACES

12. Introduction

G-spaces in which the ordinary lines are the geodesics are the subject of this chapter. More precisely, a G-space R falls in this category if it can be mapped topologically, or imbedded, in a projective space P^n in such a way that each geodesic is mapped in, or lies on, a line of the P^n. Such G-spaces will be called Desarguesian because of the fundamental role played by Desargues' Theorem.

Necessary conditions for R to be Desarguesian are obviously

(1) *That the geodesic through two distinct points is unique.*

(2) *If R is two dimensional that Desargues' Theorem and its converse hold,* whenever the intersections, with which these theorems deal, exist.

(3) *If R is higher dimensional, that any three points lie in a plane,* that is in a two dimensional subset of R which, with the metric of R, is a G-space.

It was pointed out at the end of Chapter I that (2) does not hold in every two-dimensional G-space that satisfies (1). A simple Riemannian example is furnished by the paraboloid $z = xy$ in ordinary space, with the length of the shortest join on the surface as distance [1].

Spaces of dimensions greater than two which show that (1) does not imply (3) in higher dimensions, are found on the preceding page.

In Section 13 it will be proved that (1) and (2) are sufficient for a two-dimensional G-space to be Desarguesian, and in Section 14 it will be seen that (1) and (3) characterize higher dimensional Desarguesian G-spaces. In addition we find in both cases that R, considered as imbedded in P^n (where n is the dimension of R), *either covers all of P^n*, in which case all geodesics are great circles of the same length, *or* leaves out an entire hyperplane of P^n and may therefore be regarded as an *open convex set in the affine space A^n*. That the proofs of these theorems are quite long, will not surprise readers familiar with the foundations of geometry.

In the Riemannian case the only Desarguesian G-spaces are the euclidean, hyperbolic, and elliptic spaces, so that the Desarguesian character implies strong mobility properties. Although the methods are quite foreign to the rest of the book, we prove this fundamental *Theorem of Beltrami* in Section 15, because it is by far the most striking example of a Riemannian theorem without a simple analogue in more general spaces. In fact, it is one of Hilbert's famous problems [2] to characterize the Desarguesian spaces among all G-spaces with certain differentiability properties. Whereas Hamel [1] has given a method for constructing all Desarguesian, sufficiently differentiable G-spaces, no entirely satisfactory infinitesimal characterization of these spaces, in terms of analogues to curvature tensors say, has ever been given. The freedom in the choice of a metric with given geodesics is for non-Riemannian metrics so great, that it may be doubted, whether there really exists a convincing characterization of all Desarguesian spaces. At any rate, without differentiability properties (1) and (2) or (1) and (3) are most likely the simplest conditions.

However, there are important special Desarguesian non-Riemannian spaces, in particular the *Minkowskian spaces, which may be considered as the prototype for all non-Riemannian spaces*: every non-Riemannian space behaves locally like a Minkowski space, in the same way as the local behaviour of a Riemannian case is euclidean (Section 15). It is therefore not surprising that we will meet in the course of this book many different characterizations of Minkowski spaces.

Convex surfaces play an important part in the theory of Minkowski spaces and in other parts of the book. In Section 16 we compile all material on convex surfaces or bodies which is needed later. This will partly be applied in the study of Minkowski spaces which is taken up in Section 17. Only those properties will be discussed which will prove important later. The theory of Minkowski spaces has lately been carried rather far, but in a different direction.

Finally we discuss another Desarguesian space which was discovered by Hilbert. It will furnish valuable examples for the properties of parallels and of the spaces with non-positive curvature. This metric is also used to construct G-spaces in a given, open convex subset of A^n with the ordinary lines as geodesics.

13. Planes with the Desargues Property

This section is closely related to the classical results of the *foundations of geometry*, and the methods of this field are partly used here. We outline briefly the analogies as well as the differences of that work with the present.

In building up plane projective or affine geometry the following points are essential:

I) Incidence conditions, *O) Order relations,* *C) Continuity,*

H) The uniqueness of the fourth harmonic point,

D) Desargues' Theorem, *P) Pappus' Theorem.*

The interdependence between these is as follows:

$$(I, D) \to H, \qquad (I, P) \to D, \qquad (I, H, O) \to D, \qquad (I, H, O, C) \to P.$$

If the lines are the geodesics of a two-dimensional G-space in which the geodesic through two points is unique, then the order and continuity relations on a given line are either those of affine or those of projective geometry. Hamel proved, and therefore *we do not want to assume,* that they are either on all lines of the affine type, or on all lines of the projective type. The order relations for a plane are guaranteed, at least locally, by Pasch's Axiom. Owing to the continuity properties of a G-space P will follow from D or H.

Of the incidence conditions only the existence of a unique line through two points is always true. A priori, there may be lines which intersect every other line and also pairs of lines which do not intersect.

We assume Desargues' Theorem in a weak form in which it follows from the existence of planes in a higher dimensional space. It is tempting to try to introduce ideal elements following the classical work by Hjelmslev and others. We cannot simply refer to this work, because the order relations need not be on every line of the affine type, nor do we know from the outset that Pappus' Theorem holds.

We therefore proceed differently: We establish the existence and uniqueness of the fourth harmonic point in special situations, construct on the basis of this a Moebius net in a small domain Q, use continuity to pass from the net to an imbedding of Q in P^2, and extend the imbedding of Q to cover the whole given G-space.

The following is the form of Desargues' Theorem assumed here:

The Desargues Property

(1) *If the geodesics* $\mathfrak{g}(a_1, a_2)$, $\mathfrak{g}(b_1, b_2)$, $\mathfrak{g}(c_1, c_2)$ *have a common point, and the intersections* $p = \mathfrak{g}(a_1, b_1) \cap \mathfrak{g}(a_2, b_2)$, $q = \mathfrak{g}(b_1, c_1) \cap \mathfrak{g}(b_2, c_2)$ *exist, then, if two of the three intersections* $\mathfrak{g}(p, q) \cap \mathfrak{g}(c_1, a_1)$, $\mathfrak{g}(p, q) \cap \mathfrak{g}(c_2, a_2)$, $\mathfrak{g}(c_1, a_1) \cap \mathfrak{g}(c_2, a_2)$ *exist, they coincide.*

(2) *If the intersections* $\mathfrak{g}(a_1, b_1) \cap \mathfrak{g}(a_2, b_2)$, $\mathfrak{g}(b_1, c_1) \cap \mathfrak{g}(b_2, c_2)$, $\mathfrak{g}(c_1, a_1) \cap \mathfrak{g}(c_2, a_2)$ *exist and are collinear* (i. e., lie on a geodesic) *and if two of the three intersections* $\mathfrak{g}(a_1, a_2) \cap \mathfrak{g}(b_1, b_2)$, $\mathfrak{g}(b_1, b_2) \cap \mathfrak{g}(c_1, c_2)$, $\mathfrak{g}(c_1, c_2) \cap \mathfrak{g}(a_1, a_2)$ *exist then they coincide.*

We assume both parts because two geodesics do not necessarily meet, and the duality principle which reduces (2) to (1) does not hold. [1]

The object of the present section is the proof of the following important theorem:

(13.1) **Theorem:** *Let R be a two dimensional G-space in which the geodesic through two points is unique and in which the Desargues Property holds. Then*

Either all geodesics are great circles of the same length and R can be mapped topologically on the projective plane P^2 *in such a way that each great circle in R goes into a line in* P^2

Or all geodesics are straight lines and R can be mapped topologically on an open convex subset C of the affine plane A^2 *in such a way that each straight line in R goes into the intersection of C with a line in* A^2. [2]

Instead of mapping R on P or C we may then consider R as coinciding with (or imbedded in) P^2, or coinciding with C. We express this by saying that we identify R with P^2 or C.

We decompose the long proof of (13.1) into several steps

(*a*) *Harmonic points*

Let D be any open non-empty convex set, see (10.8). If $a, b, d \in D$ and $(a\, b\, d)$, choose any point $u \in D$ not on $\mathfrak{g}(a, d)$, then v with $(a\, v\, u)$. Because of Pasch's Axiom $T(v, d)$ and $T(u, b)$ intersect at a point w, put $t = T(a, w) \cap T(b, v)$. Then $\mathfrak{g}(u, t)$ intersects $T(a, b)$ in a point c with $(u\, t\, c)$. We call c the *fourth harmonic point to a, b, d in this order.*

That c does not depend on the choice of u and v may be seen by the following modification of the standard proof:

Let $u' \in D$ be not on $\mathfrak{g}(a, d)$ or $\mathfrak{g}(a, u)$. If $(a\, v'\, u')$ and v' is sufficiently close to u', and w', t', c' are constructed as before, then the intersections (see Figure 5) $p = \mathfrak{g}(v, v') \cap \mathfrak{g}(w, w')$, $\mathfrak{g}(t, t') \cap \mathfrak{g}(v, v')$ and $\mathfrak{g}(u, u') \cap \mathfrak{g}(v, v')$ exist (since they fall in u' if $v' = u'$).

The triangles u, v, w and u', v', w' satisfy the hypothesis of the second part of the Desargues Property, hence $\mathfrak{g}(v, v')$, $\mathfrak{g}(u, u')$, $\mathfrak{g}(w, w')$ concur at p. By the same argument $\mathfrak{g}(v, v')$, $\mathfrak{g}(w, w')$, $\mathfrak{g}(t, t')$ concur at p. Now the first part of the Desargues Property applied to the triangles u, v, t and u', v', t'

yields that the intersections c of $\mathfrak{g}(a, b)$ with $\mathfrak{g}(u, t)$ and c' of $\mathfrak{g}(a, b)$ with $\mathfrak{g}(u', t')$ coincide.

If now u_0 is any point in D not on $\mathfrak{g}(a, d)$ and $(a\, v_0\, u_0)$ then v' may be chosen so close to u', that also the intersections $p_0 = \mathfrak{g}(v_0, v') \cap \mathfrak{g}(w_0, w')$, $\mathfrak{g}(t_0, t') \cap \mathfrak{g}(v_0, v')$ and $\mathfrak{g}(u_0, u') \cap \mathfrak{g}(v_0, v')$ exist. Then, if c_0 is the point obtained, $c_0 = c'$ hence $c_0 = c$. We add a simple consequence:

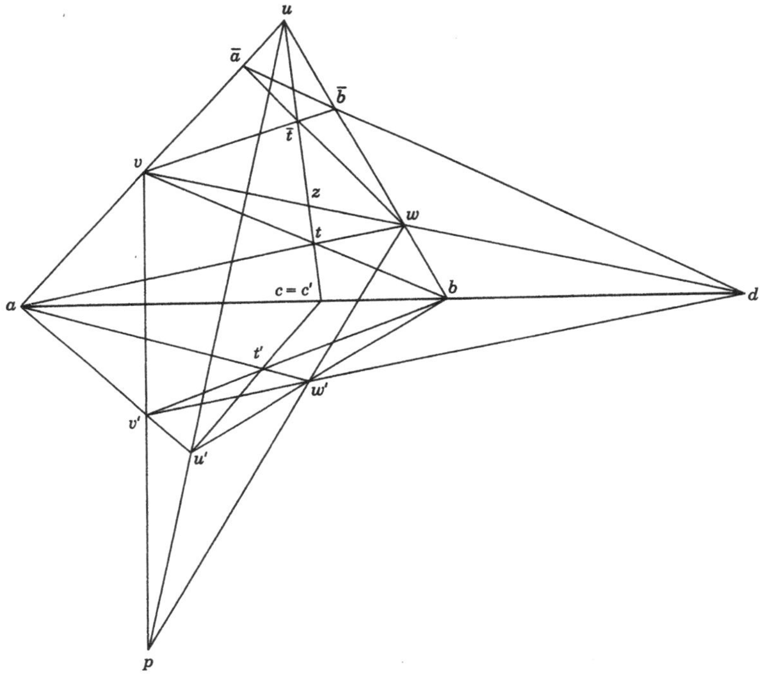

Fig. 5.

If $(v\, \bar{a}\, u)$ and $\bar{b} = T(\bar{a}, d) \cap T(u, b)$, $\bar{t} = T(v, \bar{b}) \cap T(w, \bar{a})$ then the intersection z of $\mathfrak{g}\,(u, \bar{t})$ with $T\,(v, w)$ is, by definition the fourth harmonic point to v, w, d. On the other hand, the Desargues Property applied to a, b, t and $\bar{a}, \bar{b}, \bar{t}$ shows that $u, t, \bar{t}$ are collinear.

Thus the fourth harmonic point z to v, w, d, where $(a\, v\, u)$ and $w = T(b, u) \cap T(v, d)$ lies, for variable v, on a line. Since t lies on this line, it is the fourth harmonic point to the intersections of $T(a, u)$ and $T(b, u)$ with $\mathfrak{g}(d, t)$ and d itself.

(*b*) *The Moebius net*

If a, b, v, w, t, c, z are as before, we put $a = (0,0)$, $b = (1,0)$, $v = (0,1)$, $w = (1,1)$, $t = (^1/_2, {}^1/_2)$, $c = (^1/_2, 0)$, $z = (^1/_2, 1)$.

Then t is harmonic to c, z, u, also, as just observed, t is harmonic to $(0, {}^1/_2) = T(a, u) \cap \mathfrak{g}(t, d)$, $(1, {}^1/_2) = T(b, u) \cap \mathfrak{g}(t, d)$, and d. Similarly put

$$(^1/_4, {}^3/_4) = T[(0, {}^1/_2), z] \cap T(b, v), \qquad (^3/_4, {}^1/_4) = T[c, (1, {}^1/_2)] \cap T(b, v)$$
$$(^1/_4, {}^1/_4) = T[(0, {}^1/_2), c] \cap T(a, w), \qquad (^3/_4, {}^3/_4) = T[z, (1, {}^1/_2)] \cap T(a, w)$$

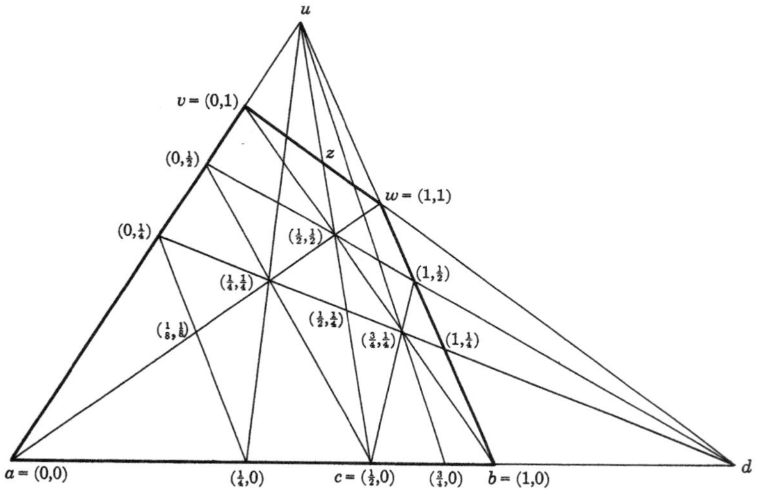

Fig. 6.

The points $(^1/_4, {}^1/_4)$, $(^3/_4, {}^1/_4)$, d are collinear and the line connecting them intersects $T(a, u)$, $T(c, u)$, $T(b, u)$ in points $(0, {}^1/_4)$, $(^1/_2, {}^1/_4)$, $(1, {}^1/_4)$ which are the fourth harmonic points to a, $(0, {}^1/_4)$, u and c, t, u and b, $(1, {}^1/_2)$, u respectively. A similar statement holds for the line connecting $(^1/_4, {}^3/_4)$ and $(^3/_4, {}^3/_4)$: it passes through d, intersects $T(a, u)$ in the point $(0, {}^3/_4)$ harmonic to $(0, {}^1/_2)$, $(0, 1)$ and u, etc. Also $(^1/_4, {}^1/_4)$, $(^1/_4, {}^3/_4)$, u are collinear and their line intersects $T(a, b)$ in the point $(^1/_4, 0)$ harmonic to a, c, d and so forth.

Continuing this process yields points (ρ, σ) for any numbers ρ, σ of the form $\varkappa/2^n$, $0 \leqslant \varkappa \leqslant 2^n$.

We map (ρ, σ) on the point $\xi = \rho$, $\eta = \sigma$ of the unit square Q': $0 \leqslant \xi \leqslant 1$, $0 \leqslant \eta \leqslant 1$, of a cartesian (ξ, η)-plane and denote the image of (ρ, σ) in Q' generally by $(\rho, \sigma)'$. The construction of the (so-called Moebius) net (ρ, σ)

implies that if three points in Q' lie on a line whose slope is 0, ± 1, or ∞ the corresponding points in the net are collinear. Actually any three points of the net are collinear, if and only if the corresponding points in Q' are collinear.

As an example take $(0, {}^1/_2)'$, $({}^1/_2, {}^1/_4)'$, $(1, 0)'$ which lie on $\xi + 2\eta = 1$. Since $(0, {}^1/_4)$ is the fourth harmonic point to a, $(0, {}^1/_2)$, u, the intersection p of the segments $T\,[(0, {}^1/_2), b]$ and $T\,[a, (1, {}^1/_2)]$ must lie on $T\,[(0, {}^1/_4), d]$, and because c is harmonic to a, b, d the point p must also lie on $T(u, c)$. Hence $p = T(u, c) \cap T\,[(0, {}^1/_4), d] = ({}^1/_2, {}^1/_4)$.

A consequence of this construction is: If $\sigma = (\rho_1 + \rho_2)/2$ then $(\sigma, 0)$ is the harmonic conjugate of d with respect to $(\rho_1, 0)$, $(\rho_2, 0)$; for in the net, $T\,[(\rho_1, 0), (\rho_2, 1)]$ and $T\,[(\rho_2, 0), (\rho_1, 1)]$ meet at $(\sigma, {}^1/_2)$, since this is true for the corresponding figure on the square. The line $\mathfrak{g}[(\sigma, 0), (\sigma, {}^1/_2)]$ goes through u and $\mathfrak{g}[(\rho_1, 1), (\rho_2, 1)] = \mathfrak{g}\,(v, w)$ goes through d.

(c) *The mapping*

We want to show that the closed convex set Q in D bounded by the four segments $T(a, b)$, $T(b, w)$, $T(w, v)$, $T(v, a)$ can be mapped topologically on Q' such that collinear points in Q go into collinear points in Q'.

First consider the mapping of the points $(\rho, 0)$ of $T\,(a, b)$ on the points $(\rho, 0)'$ of the side $\eta = 0$ of Q. The construction yields that if $\rho_1 < \rho_2$ then $(\rho_2, 0)$ is between $(\rho_1, 0)$ and d. The mapping of $(\rho, 0)$ on $(\rho, 0)'$ thus preserves order. For any real r, $0 < r < 1$, define the points x and y on $T\,(a, b)$ by

$$a\,x = \limsup_{\rho \leqslant r} a\,(\rho, 0) \qquad a\,y = \liminf_{\rho > r} a\,(\rho, 0)$$

where, as before, ρ is of the form $\varkappa/2^n$, $0 \leqslant \varkappa \leqslant 2^n$. We want to show that $x = y$. Suppose this is not the case, then no point $(\rho_1, 0)$ can lie between x and y since then $a\,(\rho_1, 0) < a\,y$ which contradicts the definition of y. On the other hand let $\rho' \leqslant r$ and $\rho'' \geqslant r$, $\sigma = (\rho' + \rho'')/2$. Then $(\sigma, 0)$, by the preceding remark, is harmonic to $(\rho', 0)$, $(\rho'', 0)$ and d. If z is the fourth harmonic point to x, y and d then $(x\,z\,y)$, and since $(\sigma, 0)$ is close to z, if ρ', ρ'' are close to r, the point $(\sigma, 0)$ is between x and y which we saw was impossible.

This implies: the mapping of the points $p = (\rho, 0)$ of the net on the points $p' = (\rho, 0)'$ of Q' can be extended to a topological mapping $p \to \xi_p$ of $T(a, b)$ on the interval $0 \leqslant \xi \leqslant 1$, $\eta = 0$. Similarly the mapping of the points $q = (0, \sigma)$ on the points $q' = (0, \sigma)'$ on Q', can be extended to a topological mapping $q \to \eta_q$ of $T(a, v)$ on the interval $0 \leqslant \eta \leqslant 1$, $\xi = 0$.

We now map $z = T(p, u) \cap T(q, d)$ of Q on the point (ξ_p, η_q) in Q'. This defines a topological mapping of Q on Q', since $T(p, u)$, $T(q, d)$ depend continuously on p and q.

The points (ρ, σ) are dense in Q and $(\rho, \sigma)'$ in Q', moreover collinearity is preserved in the mapping for these dyadic points. As any line can be approximated by lines through such points [compare (9.10)], it follows that collinearity is preserved generally.

(d) Extension of the mapping

Interpreting ξ and η as non-homogeneous coordinates in P^2, we consider Q' as imbedded in P^2.

Let p be any point in the given space R. Draw two geodesics L_1 and L_2 through p that contain interior points of Q. They intersect Q in proper segments, T_1 and T_2. The latter are mapped on segments T_1', T_2' in Q' which lie on lines L_1' and L_2' in P^2. We map p on $p' = L_1' \cap L_2'$, and are going to show that this defines a topological mapping of R in P^2 which preserves collinearity.

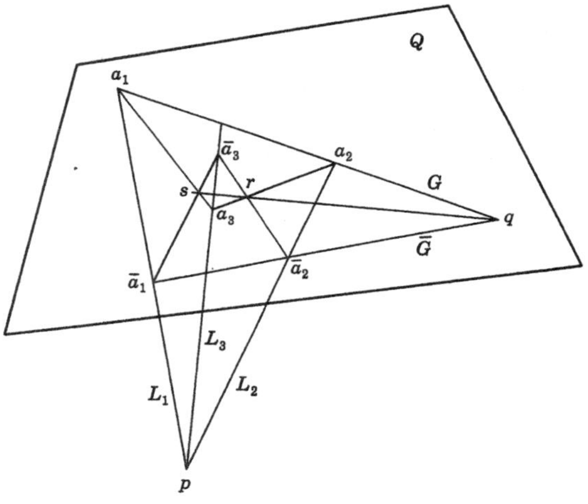

Fig. 7.

First of all it must be shown that p' does not depend on the choice of the lines L_1 and L_2. Let L_3 be any third line through p. Since p may be assumed outside Q, one of the segments in which the L_i intersect Q separates the other two in Q. Connecting two points of these last segments yields a line intersecting all three segments in Q. Therefore two lines G, $\bar{G}$ can be found as in the figure, which intersect in a point q of Q. Then two triangles with vertices a_1, a_2, a_3 and $\bar{a}_1, \bar{a}_2, \bar{a}_3$ exist such that $a_i = L_i \cap G$, $\bar{a}_i = L_i \cap \bar{G}$ for $i = 1, 2$, and corresponding sides intersect at q and two further points r, s of Q. By the Desargues Property the points q, r, s are collinear. The points a_i, $\bar{a}_i$, q, r, s go into points of Q' under preservation of collinearity. It follows from the converse of Desargues' Theorem in P^2 that the lines L_i' corresponding to L_i must concur, hence the point p' defined by L_1 and L_2 is the same as that defined by L_1 and L_3 (or L_2 and L_3). This shows that p' does not depend on the choice of L_1 and L_2.

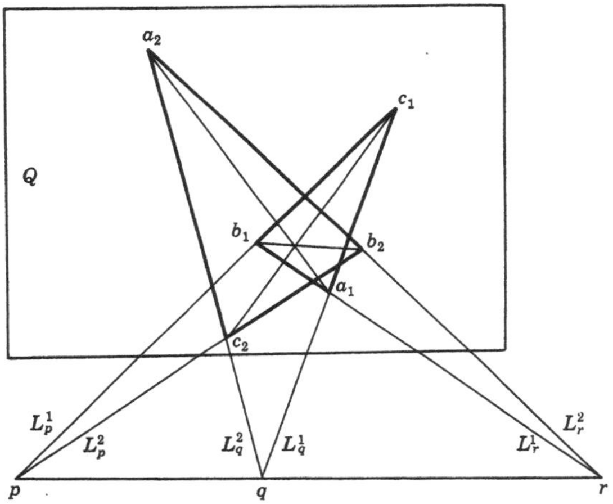

Fig. 8.

Next we show that collinear points p, q, r in R go into collinear points in P^2. Take any interior point m of Q. Since the lines $\mathfrak{g}(p, m)$, $\mathfrak{g}(q, m)$, $\mathfrak{g}(r, m)$ intersect each other, pairs of lines L_p^1, L_p^2 close to $\mathfrak{g}(p, m)$, L_q^1, L_q^2 close to $\mathfrak{g}(q, m)$, L_r^1, L_r^2 close to $\mathfrak{g}(r, m)$ will have the property that either line in any pair intersects in Q either line in any other pair. By interchanging, if necessary, the notation within a pair we obtain triangles

$$c_i = L^i_p \cap L^i_q, \qquad a_i = L^i_q \cap L^i_r, \qquad b_i = L^i_r \cap L^i_p, \qquad i = 1, 2,$$

such that the intersections $\mathfrak{g}(a_1, a_2) \cap \mathfrak{g}(b_1, b_2)$, and $\mathfrak{g}(b_1, b_2) \cap \mathfrak{g}(c_1, c_2)$ exist in Q, and hence coincide by the Desargues Property. The two triangles $a_i b_i c_i$ correspond to triangles in Q' for which the lines connecting corresponding vertices intersect in Q', hence the images p', q', r' must be collinear by Desargues' Theorem in P^2.

The mapping of R in P^2 is obviously topological because the intersection of two distinct lines both in R and P^2 varies continuously with the lines.

(*e*) *Hamel's Theorem*:

We now introduce the standard (2:1) mapping Ω: $\pm x \to [x]$ of the sphere S: $\sum_{i=0}^{2} x_i^2 = 1$ on the projective plane P^2, whose points are the classes $[x]$ of proportional non-zero triples (x_1, x_2, x_3). Denote by R' the set $R\,\Omega^{-1}$, that is the set of all points of S whose image lies in R. By considering the geodesics through a definite point z in R and their images through the corresponding antipodal points z_1', z_2' we see that R' is connected if and only if R contains at least one great circle.

We consider the case where R' is connected first. The theorem of Hamel, already mentioned, states that then R' *must coincide with* S, *or* R *with* P^2 *and that all great circles in* R *have the same length.* We prove this fact now.

About every point p of R we can find a convex neighborhood $U(p)$. To this neighborhood there correspond on R' two disjoint neighborhoods $U'(p_1')$ and $U'(p_2')$ of the antipodal points $p_i' = p\,\Omega^{-1}$. We ascribe to two points in either of these neighborhoods the same distance as their images in $U(p)$ have. This definition of local distance on R' is evidently consistent for pairs in different neighborhoods. We define as length of a curve $x'(\tau)$, $\alpha \leqslant \tau \leqslant \beta$, in R' the greatest lower bound of the numbers $\Sigma\, x'(\tau_i)\, x'(\tau_{i+1})$ for all partitions $\tau_0 = \alpha < \tau_1 < \ldots < \tau_n = \beta$ for which $x'(\tau_i)$ and $x'(\tau_{i+1})$ lie in one neighborhood $U'(p')$. As distance of any two points in R' we define the greatest lower bound of the lengths of all curves connecting them. This is again consistent for pairs of points for which distance has already been defined[3]. The mapping Ω of R' on R is locally isometric because $U'(p')$ is isometric to $U(p'\,\Omega)$.

Since R is finitely compact, so is R'. There are, trivially, curves of finite length connecting two given distinct points p', q' of R', hence they have by (5.18) a shortest join. Any point r' on this join distinct from p' and q' satisfies $(p'\,r'\,q')$ hence R' is convex. Prolongation is locally possible and locally unique because this is true in $U(p)$, hence, compare (8.3), R' is a G-space.

A geodesic $x'(\tau)$ in R' is locally a segment. Because Ω sends a segment in $U'(p')$ into a segment in $U(p)$, the curve $x'(\tau)\Omega$ in R is locally a segment and hence a geodesic. Therefore $x'(\tau)$ must lie entirely on one great circle C' of S. If $x'(\tau)$ left out one point q_i', of C', it would also leave out the antipodal point q_2'. If now r' and q' are two points of $R' \cap C'$ which are separated on C' by q_1' and q_2' and are not antipodal (such points exist because R' is open and contains with any point the antipodal point), then a geodesic $\mathfrak{g}'$ containing

r' and s' cannot lie on C' and would be mapped by Ω on a geodesic $\mathfrak{g}$ through $r = r'\Omega$ and $s = s'\Omega \neq r$. But $\mathfrak{g}$ lies on a projective line, and this line would intersect $C = C'\Omega$ twice.

Thus we have proved that if one geodesic in R is a great circle, all are; therefore R' is the complete sphere S. We have still to show that all great circles have the same length. If p_1', p_2' are any two points on R', which are not antipodal, then the geodesic $\mathfrak{g}'$ through them is unique and hence contains at least one segment. For reasons of continuity $\mathfrak{g}'$ also contains a segment connecting antipodal points. Hence $\mathfrak{g}'$ is by (9.9) a great circle in the metric of R' as well as in that of S. Any two great circles $\mathfrak{g}_1'$, $\mathfrak{g}_2'$ in R' intersect at a pair of antipodal points a_1' and a_2'. By (9.12) the lengths of $\mathfrak{g}_1'$ and $\mathfrak{g}_2'$ are equal.

(f) *The Affine Plane*

To establish Theorem (13.1) completely, it must be shown that when all geodesics in R are straight lines, then R, considered as a subset of P^2, leaves out at least one line in P^2. The set R' in S, considered in (e), is then disconnected. It consists of two disjoint sets R_1' and R_2', the points of one being antipodal in S to the points of the other, and neither containing a pair of antipodal points. Since R_i' is derived from R, if R_i' contains points p', q', it contains the (unique) shorter arc of the great circle on S from p' to q'. Denote this arc by $v(p', q')$. Thus R_1' is what is usually called a "*convex set*" on S and it is known that it therefore has at each boundary point a supporting great circle. An image of this great circle is a line in P^2 containing no points of the open set R. (See Figure 9 on page 77.)

Since we do not presuppose the theory of convex sets on a sphere, we give a proof. Let p' be a boundary point of R_1'. Denote by H' the open hemisphere of S with center p'. Then H' may be considered as an affine plane with the open great semicircles in H' as lines ($H'\Omega$ is the projective plane with one line omitted). The set $R_1' \cap H'$ is a convex subset of H' (as an affine plane) and therefore has a supporting line L_0' at p' [4]. Denote by L' the great circle containing L_0'. If L' contained a point of R_1' this must be an interior point, and then L' contains a point q' of R_1' not antipodal to p'. By construction the arc $v(p', q')$ crosses the boundary of H' at some point s' and the sub-arc $v(p', s')$ of $v(p', q')$ does not lie in R_1'. Take now a point t' close to s' on $v(p', s')$; the point p' will be in the open hemisphere H'' with center s'. But the only possible supporting line L_0'' of $H'' \cap R_1'$ at p' (on H'' regarded as an affine plane) is the arc carried by L' since $v(p', s')$ lies on the boundary of R_1', but L_0'' contains the interior point q' of $H'' \cap R_1'$.

14. Spaces which contain planes

We now discuss the higher-dimensional analogue of (13.1):

(14.1) ***Theorem:*** *Let the geodesic through any two points of an n-dimensional G-space R, $(n \geqslant 3)$, be unique and let any three non-collinear points of R lie in a subset of R which is, with the metric of R, a two-dimensional G-space.*

Then either all geodesics of R are great circles of the same length, and R can be mapped topologically on the n-dimensional projective space P^n such that each geodesic in R goes into a line in P^n;

Or all geodesics of R are straight lines and R can be mapped topologically on an open subset C of the n-dimensional affine space A^n such that each geodesic in R goes into the intersection of C with a line in A^n.

Denoting generally an r-flat (compare p. 64) by L_r we begin with the following frequently useful fact:

(14.2) *If R is a G-space with a unique geodesic through any two points, then $r+1$ points of R which do not lie on any $L_{r'}$ with $r' < r$ lie on at most one L_r.*

Proof: Suppose L' and L'' are two L_r which contain the given $r+1$ points, then $L' \cap L'' = B$ also contains these points. Further it is a flat of dimension r. For dim $B \leqslant$ dim L', and by hypothesis, the $r+1$ points lie in no flat of dimension less than r.

Unless $L' = L''$ one of these flats, say L'', contains a point p not in the other, L'. No line $\mathfrak{g}(p, x)$, where $x \in B$, can intersect B twice, because then $\mathfrak{g}(p, x) \subset B \subset L'$. If p' is any point of B and $(p\,q\,p')$ with $qp' < \rho(p')/2$ it follows from Hurewicz's Theorem in [1] that $V = \cup\, T(q, x)$ with $x \in \overline{S}(p', \rho(p')/2) \cap B$ is $(r+1)$-dimensional. On the other hand $V \subset L''$, a contradiction.

Using the more suggestive terms „plane" for two-flat and „line" for geodesic we conclude from (14.2):

(14.3) Two distinct planes have in common either no points, or one point, or a line.

Desargues' Theorem

(14.4) *If R satisfies the hypothesis of* (14.1) *then the Desargues Property holds in every plane.*

We establish, for example, the first part of Desargues' Theorem. Let a_i, b_i, c_i $(i = 1, 2)$ be given in the plane E of R such that $\mathfrak{g}(a_1, a_2)$, $\mathfrak{g}(b_1, b_2)$, $\mathfrak{g}(c_1, c_2)$ concur at u. Let $\mathfrak{g}(a_1, b_1) \cap \mathfrak{g}(a_2, b_2) = s$, $\mathfrak{g}(b_1, c_1) \cap \mathfrak{g}(b_2, c_2) = q$ and let two of the three intersections $\mathfrak{g}(q, s) \cap \mathfrak{g}(c_1, a_1)$, $\mathfrak{g}(q, s) \cap \mathfrak{g}(c_2, a_2)$,

$\mathfrak{g}(c_1, a_1) \cap \mathfrak{g}(c_2, a_2)$ exist. Choose c_1' close to c_1, and not in E. This is possible since $\dim R > 2$. On the line $\mathfrak{g}(c_1', u)$ we can choose c_2' close to c_2 since $\mathfrak{g}(c_1', u) \to \mathfrak{g}(c_1, u)$ when $c_1' \to c$, [compare (9.10)]. The lines $\mathfrak{g}(b_1, c_1')$ and $\mathfrak{g}(b_2, c_2')$ lie in a plane E' because $\mathfrak{g}(b_1, b_2)$ and $\mathfrak{g}(c_1', c_2')$ intersect at u. Since, by hypothesis, $\mathfrak{g}(b_1, c_1)$ and $\mathfrak{g}(b_2, c_2)$ intersect, it follows from continuity considerations that $\mathfrak{g}(b_1, c_1')$ and $\mathfrak{g}(b_2, c_2')$ intersect at a point q', say, when c_1' is sufficiently near to c_1. For similar reasons, two of the three intersections $\mathfrak{g}(q', s) \cap \mathfrak{g}(c_1', a_1)$, $\mathfrak{g}(q', s) \cap \mathfrak{g}(c_2', a_2)$, $\mathfrak{g}(c_1', a_1) \cap \mathfrak{g}(c_2', a_2)$ exist. The two lines of each pair lie in a plane — those in the first pair because $\mathfrak{g}(q', c_1') \cap \mathfrak{g}(s, a_1) = b_1$, and similarly for the second pair, those in the third pair because $\mathfrak{g}(c_1', c_2') \cap \mathfrak{g}(a_1, a_2) = u$. Because of (14.3) the three planes have not more than one common point; hence the two intersections, which are common to the three planes, must coincide.

This result implies one of the statements of the theorem, namely that all geodesics of R are great circles of the same length or else all are straight lines. For this holds for any two intersecting geodesics because they lie in a plane in which Desargues' Theorem holds; and for any two non-intersecting geodesics, because we can draw a third geodesic which meets both.

If the space is of the elliptic type, standard methods of projective geometry lead to the assertion of (14.1). In the case of a straight space, the methods of the foundations of geometry become very involved owing to the difficulty of defining r-flats. The following method settles both cases at once in a comparatively simple way.

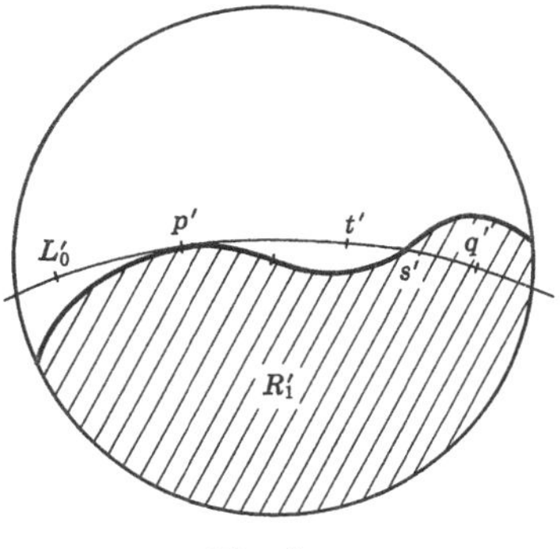

Fig. 9.

THE MAPPING

Let a, b, c, d be non co-planar and contained in $S(a, \rho(a)/8)$. Then the segments connecting any two of the four points are unique and lie in $S(a, \rho(a)/4)$. The four triangles abc, abd, adc, bdc lie then in $S(a, \rho(a)/2)$, where for instance, because planes exist,

$$abc = \bigcup_{x \in T(b,c)} T(a, x) = \bigcup_{x \in T(c,a)} T(b, x) = \bigcup_{x \in T(a,b)} T(c, x)$$

Finally $V = \cup T(d, x)$, $x \in abc$ lies in $S(a, \rho(a))$. The segment connecting two points of V is therefore unique. Moreover V is convex, that is, contains $T(x, z)$ if it contains x and z. For by definition there are points x_d, z_d in abc such that $x \in T(d, x_d)$, $z \in T(d, z_d)$. If $x_d = z_d$ the assertion is trivial. If $x_d \neq z_d$ the plane through d, x_d, z_d intersects the plane through a, b, c in the two points x_d, z_d, hence in the line $\mathfrak{g}(x_d, z_d)$. If now $(x\, y\, z)$, then $\mathfrak{g}(d, y)$ intersects $\mathfrak{g}(x_d, z_d)$ in a point y_d with $(x_d\, y_d\, z_d)$; hence, $y_d \in abc$ and $y \in V$.

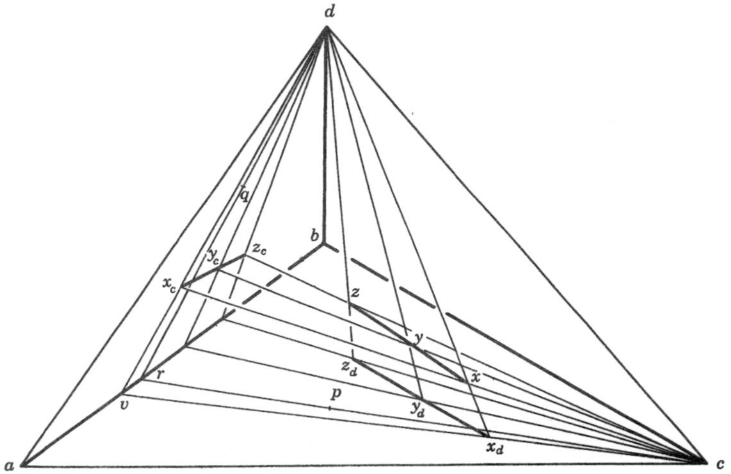

Fig. 10.

We now map a, b, c, d on the points $a' = (0, 0, 0)$, $b' = (1, 0, 0)$, $c' = (0, 1, 0)$, $d' = (0, 0, 1)$ of a cartesian (ξ, η, ζ)-space. Then we map abc topologically under preservation of collinearity on $a'b'c'$. Let p be any arbitrary interior point of abc and — generally — p' the image of p. Let $\mathfrak{g}(c, p)$ intersect $T(a, b)$ in r. Choose in abd an interior point q on $T(d, r)$, and in $a'b'd'$ an interior point q' on $T(d', r')$. Then, by a well known result on collineations, abd can be mapped topologically under preservation of collinearity on $a'b'd'$ such that q goes into q'. Then r is mapped on r', hence — by the fundamental theorem of projective geometry — the two induced mappings of $T(a, b)$ on $T(a', b')$ are identical.

For a given point $x \neq d, c$ in V let $\mathfrak{g}(d, x)$ and $\mathfrak{g}(c, x)$ intersect abc and abd in x_d and x_c respectively. Since $\mathfrak{g}(c, x)$ and $\mathfrak{g}(d, x)$ intersect they lie in a plane which cannot intersect $T(a, b)$ in more than one point. Hence

$\mathfrak{g}(d, x_c)$ and $\mathfrak{g}(c, x_d)$ intersect $T(a, b)$ in the same point v. Then $x_c' \in T(d', v')$, $x_d' \in T(c', v')$ hence c', d', x_c', x_d' are coplanar, so that $T(d', x_d')$ and $T(c', x_c')$ intersect in a point x' which is defined as image of x.

If completed by $d \to d'$, and $c \to c'$ this defines a one-to-one mapping of V on the tetrahedron V' with vertices a', b', c', d'.

This mapping coincides with the previously defined mapping on abc and abd. Since in any G-space with a unique geodesic through two points $a_\nu \to a$ and $b_\nu \to b$ implies $\mathfrak{g}(a_\nu, b_\nu) \to \mathfrak{g}(a, b)$, the mapping is topological.

It also preserves collinearity: if x, y, z lie in V and $(x\,y\,z)$, then x_d, y_d, z_d lie in abc and are collinear, similarly x_c, y_c, z_c lie in abd and are collinear. Hence x_d', y_d', z_d' and x_c', y_c', z_c' are collinear triples in $a'b'c'$ and $a'b'd'$ respectively. The points x', y', z' lie therefore in a plane through d' and x_d', y_d', z_d' and also in a plane through c' and x_c', y_c', z_c' and are collinear.

If any line through an interior point p of V exists which intersects V only in p, then the space contains clearly a product of V and a segment and is therefore at least four-dimensional. The same method produces ultimately a set W which can be mapped topologically and under preservation of collinearity on a non-degenerate n-simplex W' in E^n.

Now every line that contains an interior point of W must intersect W in a proper segment, otherwise the space would be at least $(n+1)$-dimensional. This mapping of W can be extended as in the plane case:

We consider W' as imbedded in P^n. If p is any point of R not in W, there are lines L_1, L_2 through p and interior points of W. They intersect W in proper segments T_1, T_2. The segments T_1' and T_2' in W' corresponding to T_1 and T_2 are coplanar, because T_1 and T_2 are, hence the lines L_i' in P^n containing T_i' intersect at a point p', which is defined as image of p. Then p' does not depend on the choice of L_1 and L_2. For if L_3 is a third line through p and an interior point of W, then L_1, L_2, L_3 are pairwise coplanar and lie therefore in a subset of W with the properties of a 3-flat. In particular there are triangles a_1, a_2, a_3 and $\bar{a}_1, \bar{a}_2, \bar{a}_3$ with $a_i, \bar{a}_i \in L_i$ such that the lines connecting corresponding vertices intersect at a point w of W. The converse of the spatial form of the Theorem of Desargues applied to the points $a_i', \bar{a}_i'$ and w' shows that the line L_3' determined by L_3 also passes through p'.

That the mapping preserves collinearity is shown by the argument of the plane case: if q, r, s are collinear take a plane through these points and an interior point of W, and operate in this plane.

Finally, the topological character of the mapping follows again from (9.10).

It remains to show that R, if straight and considered imbedded in P^n, leaves out an entire hyperplane of P^n. This can be reduced to the two-dimensional case: We consider again the $(2:1)$-mapping Ω of the n-sphere S^n on P^n, and call R' the set corresponding to R. This set consists in the straight case of two disjoint connected open sets R_1' and R_2' with the property that neither contains a pair of antipodal points and either contains the shorter arc of the great circle through two of its points.

If p' is a boundary point of R_1' and H' the open hemisphere of S^n with center p', then $H' \cap R_1'$ is a convex set of H' considered as an A^n. It therefore possesses at p' a supporting hyperplane[1] K_0, which lies in a great S^{n-1} on S^n (a great S^r on S^n is an r-dimensional sphere on S^n with maximal radius), which we denote by K'. If K' contained a point q' of R_1', we would obtain a contradiction to the two-dimensional result if applied to the intersection of S^n with a great circle S^2 through p' and q'.

This establishes Theorem (14.1). It is sometimes convenient to use

(14.5) *The conclusion of* (14.1) *remains valid if the hypothesis that any three non-collinear points lie in a plane, is replaced by the assumption that for a fixed* r, $2 \leqslant r \leqslant n-1$, *any* $r+1$ *points which do not lie in a flat of dimension less than* r, *lie in an* r-*flat.*

A proof may be obtained by induction. Assume (14.5) to be true for $r-1$, where $2 \leqslant r-1 < n-1$ and assume that $r+1$ points, which lie in no flat of dimension less than r, lie in an r-flat. Let $a_1, \ldots, a_r$ be given such that they lie in no flat with dimension less than $r-1$. If we can show that they lie in an $(r-1)$-flat then the inductive assumption proves the theorem.

The intersection of all flats that contain $a_1, \ldots, a_r$ is a flat and hence has dimension $\rho \geqslant r-1$. We show that $\rho = r-1$. Assume $\rho > r-1$ and take any point a_{r+1}. Then $a_1, \ldots, a_{r+1}$ lie in no flat of dimension less than r, hence lie in an r-flat L_r'. Since $n > r$ there is a point b not on L_r'. Then $a_1, \ldots, a_r, b$ again lie in no flat of dimension less than r, hence they lie in an r-flat $L_r'' \neq L_r'$. This contradicts (14.2).

SIGNIFICANCE OF DESARGUES' THEOREM

In the foundations of geometry the validity of Desargues' Theorem for a system of plane curves is shown to be (necessary and) sufficient for the imbedding of the plane in a space with planes.

A two-dimensional Desarguesian G-space R carries a definite metric, and so the question with us is whether it can be imbedded in a higher dimensional Desarguesian space with preservation of the metric in R. A method due to

Hessenberg[2] enables us to answer the question for straight spaces; the problem is open for spaces of the elliptic type.

(14.6) *A two-dimensional straight Desarguesian space R can be imbedded in a three-dimensional Desarguesian space R′* (preserving the metric in R).

Consider R as part of an affine plane A^2 such that each straight line in R lies on a line in A^2. Choose 3 lines L_a, L_b, L_c in A^2 no two of which are parallel. The triangles $p' = \{a, b, c\}$ with vertices in R for which the sides $\mathfrak{g}(b, c)$, $\mathfrak{g}(c, a)$, $\mathfrak{g}(a, b)$ are parallel to L_a, L_b, L_c respectively form the "*points*" of R'. The case when two of the points a, b, c and therefore all three coincide is allowed.

The distance between p_1' and p_2' where $p_i' = \{a_i, b_i, c_i\}$ we define by

$$p_1' p_2' = (a_1 a_2 + b_1 b_2 + c_1 c_2)/3. \tag{14.7}$$

The triangles with coincident vertices form a space isometric to R. It is obvious that the distance (14.7) satisfies the axioms for a metric space and that this space is finitely compact.

The lines $\mathfrak{g}(a_1, a_2)$, $\mathfrak{g}(b_1, b_2)$, $\mathfrak{g}(c_1, c_2)$ which connect corresponding elements of two points p_1', p_2' are either identical or by Desargues' Theorem are concurrent at the same point of A^2 or are parallel. Again by Desargues' Theorem, the totality of points $p' = \{a, b, c\}$ in R' for which $a \in \mathfrak{g}(a_1, a_2)$, $b \in \mathfrak{g}(b_1, b_2)$, $c \in \mathfrak{g}(c_1, c_2)$ form a straight line in R' considered as a metric space, and we can have $(p_1' p' p_2')$, only if $(a_1 a a_2)$, $(b_1 b b_2)$, $(c_1 c c_2)$ and hence if p' lies on the straight line through p_1' and p_2'. Thus R' is a straight space.

The details of the proof that any three points of R' lie in a plane may be found in Hessenberg's book[2].

The argument obviously generalizes. If R is a three dimensional straight space, we may consider it as imbedded in an affine space A^3, such that each straight line in R lies on a line in A^3. The higher dimensional analogues of Desargues Theorem[3] are true, and we may consider a tetrahedron T_0 in A^3 with vertices a_0, b_0, c_0, d_0. The quadruples $p' = \{a, b, c, d\}$ in R for which the sides of the tetrahedron $a\,b\,c\,d$ are parallel to the corresponding sides of T_0 (with vertices corresponding in the order given) may be taken as points in a four dimensional space R' in which R is imbedded. Hence

(14.8) *An m-dimensional straight Desarguesian space R can be imbedded in an n-dimensional straight Desarguesian space R′, for any $n > m$ in such a way that the metric in R is preserved.*

Note. The theorem is also true when $m = 1$, because a straight line is congruent to a euclidean straight line. For $m = 1$ it even holds when the space is a great circle, since that is congruent to a line in a suitable elliptic plane.

15. Riemann and Finsler spaces. Beltramis Theorem

It was mentioned in the introduction to this chapter that requiring the space to be Desarguesian is very much more stringent for Riemann spaces, than for more general spaces, usually called Finsler spaces. To fully comprehend Beltrami's Theorem it is therefore necessary to realize exactly what distinguishes a Riemann space from a Finsler space.

Finsler Spaces

The assumptions in differential geometry are usually of the following type:

(a) *The space R is an n-dimensional manifold of a certain class C^r, $r \geqslant 1$.*

An exact definition of this term is found in many books, for instance in Lefschetz' Algebraic Topology. The gist of the definition is: there is a base of distinguished neighborhoods $U(p)$ with these properties: Each $U(p)$ is a definite topological map $(x_1, \ldots, x_n) = x \to q$ of a connected open set G in cartesian n-space. The numbers $x_1, \ldots, x_n$ are called the coordinates of q in $U(p)$. A point q in the overlap of two distinguished neighborhoods $U(p_1)$, $U(p_2)$ has then two sets of coordinates, say $x_1, \ldots, x_n$ and $y_1, \ldots, y_n$, hence relations of the form

$$x_i = g_i(y), \qquad y_i = f_i(x) \qquad i = 1, \ldots, n$$

hold in $U(p_1) \cap U(p_2)$ where the $f_i(x)$ and $g_i(y)$ are continuous. That R is of class C^r means that $f_i(x)$ and $g_i(y)$ have continuous partial derivatives of orders $1, 2, \ldots, r$. The term manifold implies that the space is connected.

The actual value of r varies from theorem to theorem, depending on the analytical tools used in the proofs. At first we only need that $r = 1$.

(b) *On the contravariant vectors $(x, \xi) = (x_1, \ldots, x_n; \xi_1, \ldots, \xi_n)$ in R, a function $F(x, \xi)$ continuous in the $2n$ variables is defined, which determines the length $\Lambda(x)$ of a given curve $x(\tau) = (x_1(\tau), \ldots, x_n(\tau))$, $\alpha \leqslant \tau \leqslant \beta$, of class C^1 (or D^1) by*

$$\Lambda(x) = \int_\alpha^\beta F(x(\tau), \dot{x}(\tau))\, d\tau,$$

where $\dot{x}(\tau) = \left(\frac{dx_1}{d\tau}, \ldots \frac{dx_n}{d\tau}\right)$.

This number will be an acceptable length only if $F(x, \xi)$ satisfies certain further conditions: length must be positive, hence

(c) $$F(x, \xi) > 0 \quad \textit{if} \quad \xi \neq 0.$$

Length must be a geometric concept, i. e., independent of the parametrization: if $\tau = \varphi(\sigma)$, where $\varphi(\sigma)$ is of class C^1 and $\varphi'(\sigma) > 0$, $\gamma = \varphi(\alpha)$, $\delta = \varphi(\beta)$, then with $y(\sigma) = x(\varphi(\sigma))$

$$\int_{\alpha}^{\beta} F(x(\tau),\, x(\tau))\, d\tau = \int_{\gamma}^{\delta} F(y(\sigma), \dot{y}(\sigma))\, d\sigma.$$

Requiring this relation for any curve and any $\varphi(\sigma)$ proves equivalent to

(d) $$F(x, k\,\xi) = k\,F(x, \xi) \qquad \textit{if} \quad k > 0.$$

This is easy to verify and is found in any book on the calculus of variations.

In differential geometry it is frequently not assumed that the length of a curve is independent of the sense in which a curve is traversed. This must be required if we want to arrive at G-spaces and is obviously equivalent to

(e) $$F(x, -\xi) = F(x, \xi).$$

Notice that (d) and (e) may be written as one condition:

(f) $$F(x, k\,\xi) = |k|\,F(x, \xi) \ \textit{for any real } k.$$

It is now tempting to define as distance in R the number

(g) $$xy = \inf_{x(\tau)} \int_{\alpha}^{\beta} F(x(\tau), \dot{x}(\tau))\, d\tau,$$

where $x(\tau)$ traverses all curves of class C^1 (or D^1) from x to y (i. e., $x(\alpha) = x$, $x(\beta) = y$). It is trivial that xy satisfies the axioms for a metric space, and — owing to the continuity of $F(x, \xi)$ — that the topology defined by xy is equivalent to the original topology in R.

However, if nothing else is assumed for $F(x, \xi)$, then the length $\Lambda(x)$ defined above will *not coincide* with the length $\lambda(x)$ of $x(\tau)$ defined in Section 5 in terms of the distance xy. *A necessary and sufficient condition that* $\Lambda(x) = \lambda(x)$ holds for any curve $x(\tau)$ of class C^1 is:

(h) *The surface* $F(x, \xi) = 1$ *in* ξ*-space is convex for every* x.

A proof of this fact may be found in H. Busemann and W. Mayer [1]. This paper will be referred to as B. M.. The mentioned result is contained in Theorem 2, p. 186.

There is still a long way from a space R of class C^1 with properties (b) to (h) to a G-space. An obviously necessary condition is:

(i) *The space R is with the distance* (g) *finitely compact.*

It then follows readily that the space is convex, so that segments exist, but no general statement regarding existence or uniqueness of prolongation can as yet be made, even when $F(x, \xi) = 1$ is strictly convex (for this terminology compare the next section), see Section 7 in B. M. The relations of the differentiability properties of R and $F(x, \xi)$ to the geometric properties of the space R metrized by (g) have never been seriously studied, but the following conditions are known to imply that the geodesics or extremals are given as solutions of the second order differential equations with the usual properties[1], and hence that the space is a G-space.

(j) *R is of class C^4 and $F(x, \xi)$ is of class C^3.*

(k) *The surface $F(x, \xi) = 1$ in ξ-space for fixed x has everywhere positive Gauss curvature.*

(k) implies, of course, (h) which may therefore be omitted.

The conditions (a), (b), (c), (d), (j), (k) *define, what is usually called a Finsler space.* For theorems in the large the finite compactness (i) — frequently called normality — is added, and the class requirements in (j) are increased, when the proofs make it necessary.

Under the assumptions (a), (b), (c), (f), (h) it can be shown (see B. M. Section 4) that

$$\lim_{x \to p, y \to p} \frac{xy}{F(p, x-y)} = 1 \qquad (x \neq y).$$

This can be improved when (j) and (k) hold. Thus $F(p, x-y)$ is a good approximation of xy in the neighborhood of p. Now $F(p, x-y)$ defines for fixed p a Minkowski metric in $U(p)$. Section 17 discusses these metrics in detail. The metric $F(p, x-y)$ is euclidean, if and only if, $F(p, \xi)$ has the form

$$F^2(p, \xi) = \sum g_{ik}(p)\, \xi_i \xi_k$$

(see Section 17), that is, when the space is Riemannian. In that case $F(p, \xi) = 1$ is an ellipsoid.

Thus the difference between a Riemann space and a Finsler space is that the former behaves locally like a euclidean, the latter locally like a Minkowski space, or analytically, that to an ellipsoid in the Riemannian case, there corresponds an — except for differentiability properties — arbitrary convex surface, which, if (e) holds, has the origin of the ξ-space as centre. The increase in generality by passing from Riemann to Finsler spaces is therefore considerable.

BELTRAMI'S THEOREM

We now turn to Beltrami's Theorem[2]. Like all theorems of classical differential geometry it is a local statement which, however, implies the theorem in the large in which we are at present interested.

(15.1) *Let a connected open set C of the projective plane P^2 be metrized so that the metric is Riemannian and the geodesics lie on projective lines. Then the Gauss curvature is constant.*

It suffices to see that the curvature $K(x)$ is constant for all points x in a neighborhood U of a given point p, or that $K(q) = K(p)$ for all q in U. There are pencils of lines such that p and q lie on the same orthogonal trajectory of the pencil. For instance, a circle $cx = \rho$ with radius $\rho > pq$ whose center c is the midpoint of $T(p, q)$ contains (at least two) points z such that $pz = qz$, (all concepts are used in the sense of the Riemann metric). The circles $zx =$ const. are the orthogonal trajectories of the pencil of lines through z, and p, q lie on the same trajectory.

For a pencil of this type introduce non-homogeneous projective coordinates ξ, v such that $v =$ const. are the lines of the pencil and $p = (0, 0)$. Let the orthogonal trajectory of the pencil through (ξ, v) intersect $v = 0$ at the point $(\xi', 0)$ whose distance from p equals $\pm u$ where sign $u =$ sign ξ'. This defines ξ as a function (u, v) of u and v, and the u, v are geodesic parallel coordinates in which the line element takes the form

$$ds^2 = du^2 + B^2(u, v)\, dv^2.$$

Denoting derivatives by subscripts, the Gauss curvature[3] is

$$K(u, v) = -\frac{B_{uu}}{B}.$$

It suffices to show that B is the product of a function of u alone and a function of v alone. For then $K(u, v)$ will be independent of v and hence constant on $u = 0$ so that $K(p) = K(q)$.

Now if $u = u(v)$ represents a geodesic then for a general line element[4] $ds^2 = E\, du^2 + 2F\, du\, dv + G\, dv^2$

$$-\frac{d^2u}{dv^2}(EG - F^2) + \left(E\frac{du}{dv} + F\right)\left[\left(F_u - \frac{1}{2}E_v\right)\left(\frac{du}{dv}\right)^2 + G_u\frac{du}{dv} + \frac{1}{2}G_v\right]$$
$$-\left(F\frac{du}{dv} + G\right)\left[\frac{1}{2}E_u\left(\frac{du}{dv}\right)^2 + E_v\frac{du}{dv} + \left(F_v - \frac{1}{2}G_u\right)\right] = 0.$$

Hence in our case

$$\frac{d^2u}{dv^2} = \frac{2\,B_u}{B}\left(\frac{du}{dv}\right)^2 + \frac{B_v}{B}\frac{du}{dv} + B\,B_u. \tag{15.2}$$

By hypothesis

$\xi\,(u, v) = D\,v + C$ represents a geodesic for any D, C.

Differentiating this twice yields

$$\xi_u \frac{d^2u}{dv^2} + \xi_{uu}\left(\frac{du}{dv}\right)^2 + 2\,\xi_{uv}\frac{du}{dv} + \xi_{vv} = 0$$

Compare with (15.2), then

$$\frac{\xi_{uu}}{\xi_u} = -\frac{2\,B_u}{B}, \qquad \frac{2\,\xi_{uv}}{\xi_u} = -\frac{B_v}{B},$$

so that $(B^2\,\xi_u)_u = 0$, $(B\,\xi_u{}^2)_v = 0$ and with suitable functions $f(v)$, $q(u)$,

$$B^2\,\xi_u = f^3\,(v), \qquad B\,\xi_u{}^2 = 1/q^3\,(u), \qquad \text{hence} \qquad B^2 = f^4\,(v)\cdot q^2\,(u),$$

and this proves this theorem.

We may express this result by saying that the metric in C is euclidean, hyperbolic, or elliptic[5]. The corresponding result in higher dimensions we deduce from the two dimensional case.

(15.3) ***Beltrami's Theorem:*** *Let a connected open set C of the projective space P^n be metrized so that the metric is Riemannian and the geodesics lie on projective lines. Then the metric in C is euclidean, hyperbolic or elliptic.*

It suffices to consider a convex neighborhood U of a given point in C. We know that the metric is euclidean, hyperbolic or elliptic in the intersection of any plane with U. It remains to show that the metrics in the intersections with different planes P_1, P_2 are both euclidean or both hyperbolic with the same value of the constant k, or both elliptic with the same k.

We can reduce the general case to that in which P_1, P_2 cut in a line which meets U. For if $p_i \in P_i$ $(i = 1, 2)$ are connected by a line L, and L_i is a line in P_i through p_i, then the plane P_i^* determined by L_i and L meets P_i in L_i, and P_1^*, P_2^* meet in L.

Assume therefore that P_1 and P_2 intersect in a line L that contains a point p of U and take on a line L_i normal at p to L and in P_i, a point q_i such that $pq_1 = pq_2 = \alpha > 0$. Also take on $L \cap U$ a point r with $rp = \beta > 0$. Let m be the midpoint of $T(q_1, q_2)$. Then $T(p, m)$ and $T(q_1, q_2)$ are perpendicular, since $p\,q_1\,q_2$ is an isosceles triangle in a euclidean, hyperbolic, or elliptic space.

Also $T(r, m)$ is perpendicular to $T(q_1, q_2)$. For under parallel displacement along $T(p, m)$ the unit vector with origin p and tangent to $T(p, r)$ remains normal to the plane pq_1q_2 and tangent to the plane prm [6].

Therefore the line normal to pq_1q_2 at m lies in the plane prm and thus, as $T(q_1, q_2)$ is perpendicular to two lines in prm through m, it is perpendicular to all lines in prm through m, in particular to $T(r, m)$.

Operating in the plane prm we conclude that $rq_1 = rq_2$ and that rq_i equals one of the following expressions

$$k_i \operatorname{arc\,cos}\left(\cos\frac{\alpha}{k_i}\cos\frac{\beta}{k_i}\right),$$

$$k_i \operatorname{area\,cosh}\left(\cosh\frac{\alpha}{k_i}\cosh\frac{\beta}{k_i}\right),$$

$$(\alpha^2 + \beta^2)^{1/2}.$$

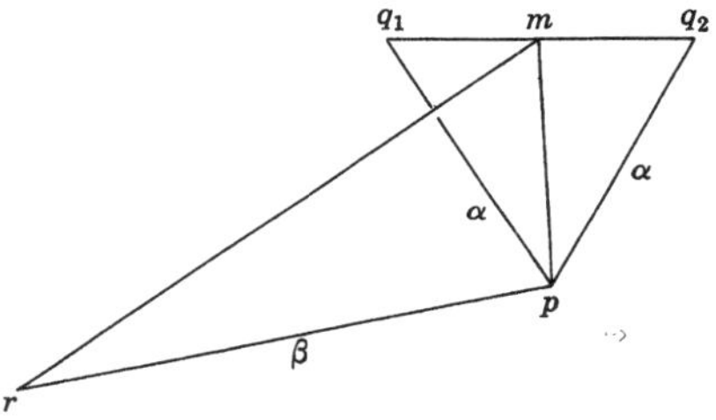

Fig. 11.

But any two such expressions can be equal only if they are of the same type and $k_1 = k_2$; for k arc cos $(\cos \alpha/k \cos \beta/k)$ and k area cosh $(\cosh \alpha/k \cosh \beta/k)$ as $k \to \infty$ tend to $(\alpha^2 + \beta^2)^{1/2}$ by increasing and decreasing values respectively. (We may choose $\alpha = \beta$, then the calculation becomes trivial).

Beltrami's theorem implies

(15.4) *A Desarguesian and Riemannian G-space is a euclidean, hyperbolic or elliptic space.*

That the whole space is covered follows in each case from the finite compactness.

16. Convex sets in affine space

A set in the affine space A^n is called "*convex*" if it contains with any two points x, y the segment $(1 - \theta)\, x + \theta\, y, 0 \leqslant \theta \leqslant 1$. In any euclidean metrization of A^n in which the affine lines are euclidean straight lines, the segment will be the segment $T(x, y)$ in the metric sense. We may therefore use this symbol and any euclidean concept which is the same for these metrizations. In particular $x^\nu \to x$ is always equivalent to $x_i^\nu \to x_i$ for $i = 1, \ldots, n$.

A "*supporting*" (hyper)plane H of a non empty set M is a plane which contains points of $\overline{M}$ but does not separate any two points of M. It then does not separate any two points of $\overline{M}$ so that M and $\overline{M}$ have the same supporting planes. A supporting plane of M contains at least one boundary point of M, but no interior point.

(16.1) *If M is bounded, then a converging sequence of supporting planes of M tends to a supporting plane of M.*

The convergence of hyperplanes may be interpreted in the sense of the closed limit of Hausdorff discussed in Section 2. With the definition of distance of sets given there, we may reformulate (16.1) as follows: The supporting planes of a bounded set form a compact subset of the set of all planes. Convergence of planes is, of course, equivalent to convergence in terms of suitable parameters, for instance the coefficients in the normal form of the equation of a plane.

(16.2) *A convex set possesses a supporting plane through each of its boundary points.*[1]

The converse is not true without restrictions but the following holds:

(16.3) *If every boundary point of a closed set, with interior points, lies in a supporting plane of the set, then the set is convex.*

The set M is „*strictly convex*", if any supporting plane of M contains exactly one point of $\overline{M}$. This is equivalent to requiring that the boundary of M contains no proper segment.

The boundary C of a convex set M with interior points is called a "*convex surface*" provided it is connected. (The only set M excluded is the set bounded by two parallel hyperplanes.) The terminology for convex sets is extended to convex surfaces: a "*supporting plane*" of C is a supporting plane of M, the surface C is "*strictly convex*" if M is, and so forth.

If the convex surface C has at the point p just one supporting plane H, we call C "*differentiable at p*" and H the tangent plane of C at p. As usual C is called "*differentiable*" (simply) if it is differentiable at all its points.

(16.4) *A convex curve (= convex surface in A^2) is differentiable except perhaps at a denumerable set of points*, see (25.1).

(16.5) A convex surface in A^n is differentiable except perhaps at a set of points which, in every euclidean metrization of A^n, has $(n-1)$-dimensional surface measure zero. Therefore the points at which a surface is differentiable are everywhere dense on this surface[2].

Let p be a point of the convex surface C. The intersection of all the closed half spaces bounded by planes through p and containing C (that is, supporting planes of C at p) is a convex set of points. The boundary of this set is a convex cone and is called the "*tangent cone*" of C at p. We may say:

(16.6) *A convex surface C is differentiable at p if, and only if, the tangent cone at p is a plane.*

Convex curves with center

A convex surface is (by definition) closed if it is bounded. A point z inside C is called a "*center*" of C if every chord of C through z has z as affine center. Clearly C has at most one center.

If z is the center of the closed surface C then C possesses parallel supporting planes at the endpoints of any chord through z. The converse is also true, and the higher dimensional case is an obvious consequence of the two-dimensional case. Here we need only the latter:

(16.7) *If a closed convex curve C in A^2 and a point z inside C have the property that C possesses parallel supporting lines at the endpoints of every chord through z, then z is the center of C.*

The author is indebted to Dr. Petty for the following elementary proof: Choose z as origin of a system of polar coordinates r, α in an associated euclidean metric $e(p, q)$. Then C has an equation of the form

$$r = f(\alpha), \text{ where } f(\alpha) \text{ has period } 2\pi.$$

We must show that $f(\alpha)$ actually has period π [3]. Consider the curves D and E defined by

$$D\colon\ r = g(\alpha) = (f(\alpha) + f(\alpha + \pi))/2$$

$$E\colon\ r = f(\alpha + \pi).$$

If C', D', E' denote the compact domains bounded by C, D, E respectively and $|A|$ is the area of A, then

$$(16.8) \qquad 2\,|D'| = \int_0^{2\pi} [f(\alpha) + f(\alpha + \pi)]^2\, 4^{-1}\, d\alpha \leqslant$$

$$\leqslant \int_0^{2\pi} [f^2(\alpha) + f^2(\alpha + \pi)]\, 2^{-1}\, d\alpha = |C'| + |E'|$$

and the equality sign holds only for $f(\alpha) \equiv f(\alpha + \pi)$.

The set D' consists of the points $(c + e)/2$ for $c \in C'$ and $e \in E'$. For a point a of D' has the form (s, α) with $0 \leqslant s \leqslant g(\alpha)$ and is the center of $c = (f(\alpha)\, s/g(\alpha), \alpha)$ and $e = (f(\alpha + \pi)\, s/g(\alpha), \alpha)$. Conversely, let $c \in C'$ and $e \in E'$. The line $L(-e, c)$ through $-e$ and c is parallel to the line $L(z, (c + e)/2)$. The point $-e$ lies in C', hence the chord of maximal length of C or E parallel to $L(-e, c)$ has at least length $e(-e, c) = 2\, e(z, (c + e)/2)$. The hypothesis that C, and hence E, have parallel supporting lines at the endpoints of a given chord through z implies that this chord has maximal length among all chords parallel to it. Thus the chords of both C and E through z have at least length $2\, e(z, (c + e)/2)$. Since D has z as center it follows that $(c + e)/2$ lies in D'.

There is at least one value α_0 such that $f(\alpha_0) = f(\alpha_0 + \pi)$, so that z bisects the chords with direction α_0 of C and E. Therefore C and E have a pair of common parallel supporting lines. Assume that the x-axis is perpendicular to these supporting lines. If H' denotes the set of points of the form $(x, (y_C + y_E)/2)$ where $(x, y_C) \in C'$ and $(x, y_E) \in E'$ then $H' \subset D'$ and

$$|D'| \geqslant |H'| \geqslant \tfrac{1}{2}|C'| + \tfrac{1}{2}|E'|. \tag{16.9}$$

The inequalities (16.8) and (16.9) prove $f(\alpha) \equiv f(\alpha + \pi)$.

Characterization of ellipsoids among convex bodies

We base the discussion on the following lemma due to Loewner and Behrend[4].

(16.10) *Given a closed bounded set F with positive measure and a point z in E^n there is just one (solid) ellipsoid E with center z, and of minimal volume, which contains F.*

That at least one such ellipsoid exists is plain since ellipsoids depend on a finite number of parameters and $|F| > 0$.

If E_1', E_2' be two such ellipsoids we must show they coincide. By an affinity which keeps volumes invariant, we can transform E_1' into a sphere E_1. This affinity maps E_2' on an ellipsoid E_2 and F on a set F'. In proper coordinates E_1 and E_2 have equations of the form $\Sigma x_i^2 \leqslant a^2$ and $\Sigma x_i^2/b_i^2 \leqslant 1$ respectively. Now E_1', E_2' are both minimal, hence $|E_1'| = |E_2'|$. But $|E_1| = \varkappa_n a^n$, $|E_2| = \varkappa_n b_1 b_2 \dots b_n$, $\varkappa_n = \pi^{n/2}\Gamma^{-1}(n/2 + 1)$. Since the transformed sphere and ellipsoid contain F'

$$\sum x_i^2/a^2 \leqslant 1, \quad \sum x_i^2/b_i^2 \leqslant 1 \quad \text{for all } x \in F'.$$

Hence

$$\sum x_i^2(a^{-2}+b_i^{-2})/2 \leqslant 1 \qquad \text{for} \qquad x \in F'$$

and so the ellipsoid $E': \Sigma\, x_i^2(a^{-2}+b_i^{-2})/2 \leqslant 1$ also contains F' and its volume is not less than $|E_1| = |E_2|$. But

$$|E'| = \varkappa_n \prod_{i=1}^{n} \sqrt{2}\, ab_i(a^2+b_i^2)^{-\frac{1}{2}} \leqslant \varkappa_n \Pi(a\, b_i)^{\frac{1}{2}} = |E_1|^{\frac{1}{2}}\, |E_2|^{\frac{1}{2}} = |E_1|.$$

Hence we have equality, and $a^2+b_i^2 = 2\, ab_i$, that is $a = b_i$ for all i; whence $E_1 = E_2$, $E_1' = E_2'$. As an application we have:

(16.11) *Let C be a closed convex surface in A^n and z an interior point of C. If for any two points p and q of C an affinity exists that leaves z fixed, maps C on itself and p on q, then C is an ellipsoid with center z.*

For let E' be the solid ellipsoid with center z of minimal volume which contains C. Because $|E'|$ is minimal, the ellipsoid E which bounds E' contains at least one point p of C. Let q be a given point of C. By hypothesis there is an affinity Φ which maps C on itself; it leaves volumes unchanged, hence it maps E' on an ellipsoid of the same volume; which contains C and has center z. By (16.10) it must coincide with E'. Since Φ maps E on itself, and p on q, the point q lies on E; hence $C = E$.

Note: Our proof holds also for the case when C is star shaped with respect to z and not necessarily convex.

Before applying this theorem, we establish a simple but frequently useful lemma.

(16.12) *If a closed convex surface C in A^n has the property that for a fixed r, $2 \leqslant r \leqslant n-1$, every r-flat through a fixed point p inside C intersects C in an ellipsoid* (ellipse when $r = 2$) *then C is an ellipsoid.*

The case $r = 2$ implies the case $r > 2$. For any 2-flat L_2 through p lies in an r-flat L_r and since $L_r \cap C$ is an ellipsoid the intersection $L_2 \cap (L_r \cap C) = L_2 \cap C$ is an ellipse.

Now let the line $\mathfrak{g}$ through p meet C at q_1 and q_2, and let H_i denote a supporting plane of C at q_i. The $(n-2)$-flat L in which H_1, H_2 intersect (unless they are parallel) cannot have a point in common with C, since the cut of C and a 2-flat through $\mathfrak{g}$ and a point of $L \cap C$ would not then be an ellipse. Consider A^n as imbedded in P^n and let p' be the harmonic conjugate of p for q_1, q_2; choose the hyperplane through $H_1 \cap H_2$ and p' as the plane at infinity. In the new affine space, p is the midpoint of q_1, q_2 and H_1, H_2 are parallel; let H be the plane through p parallel to these.

First let $n = 3$. Then $H \cap C$ is an ellipse E. Each L_2 through $\mathfrak{g}$ meets C in an ellipse E' with tangents T_i in H_i. Since T_1 and T_2 are parallel, p is the center of E'. Moreover, the plane of E' meets H in a line T parallel to T_i, that is, in the diameter conjugate to $\mathfrak{g}$. The tangents to E' at the points $E' \cap E$ are therefore parallel to $\mathfrak{g}$. Moreover p is the mid-point of the chord carried by T_2 hence the center of E. With a suitable euclidean metric, E is a circle, $\mathfrak{g}$ is normal to H, and $pq_1 = pq_2$ equals the radius of E. Then E' appears as circle with diameter q_1q_2, hence C is a sphere.

Use induction: if $n = 4$, then $H \cap C$ is an ellipsoid, from the case $n = 3$. With a suitable euclidean metric, E is a sphere with center p and now $\mathfrak{g}$ will be normal to H and the distance q_1p will equal the radius of E. Hence C is a sphere.

(16.13) *If for every family of parallel chords of a closed convex surface C, the mid-points of the chords lie on a hyperplane, then C is an ellipsoid.*

Let L_2 be any 2-flat which contains points inside C. Any family F of parallel chords of $E = C \cap L_2$ belongs to a family of parallel chords of C. By hypothesis, their mid-points lie on a hyperplane H, hence the mid-points of the chords of F lie on the line $H \cap L_2$. If (16.13) has been shown for curves, then E is an ellipse, and hence by (16.12) C is an ellipsoid.

Assume therefore that the mid-points of a family of parallel chords of the closed plane convex curve C are collinear. Let F_1 be any family of parallel chords of C and $\mathfrak{g}_1$ the line carrying their mid-points, moreover $\mathfrak{g}_2$ the line through the mid-points of the family F_2 of chords parallel to $\mathfrak{g}_1$. Let Φ_i be the affinity which maps each point of $\mathfrak{g}_i$ on itself and interchanges the ends of one non-degenerate (and so of each) chord in F_i.

Φ_1 maps C on itself, and sends every chord of F_2 into another chord in F_2, moreover the mid-point of the first into the mid-point of the second. Hence Φ_1 maps the line $\mathfrak{g}_2$ containing the mid-points of chords of F_2 on itself. A similar statement holds for Φ_2. Take affine coordinates in which $\mathfrak{g}_1$ and $\mathfrak{g}_2$ have equations $x_1 = 0$ and $x_2 = 0$; the affinity Φ_i is then the transformation $x_i' = -x_i$, $x_j' = x_j$ $(j \neq i)$. Since $\Phi_1\Phi_2$ maps C on itself and is the affinity $x_1' = -x_1$, $x_2' = -x_2$ it follows that $z = \mathfrak{g}_1 \cap \mathfrak{g}_2$ is the center of C.

If now p and q are any two distinct points of C and F_1 is the family of chords parallel to $L(p, q)$, where $L(p, q)$ denotes generally the line in A^2 through p and q, then the affinity Φ_1, defined above, maps C on itself, p on q and leaves z fixed. Hence by (16.11) C is an ellipse.

We call a line L a *supporting line* of the closed convex surface C, if it contains a point on C but no point inside C, and prove the following important theorem due to Blaschke [1].

(16.14) ***Theorem:*** *If the closed convex surface C in A^n $(n \geqslant 3)$ has the property that the points common to C and a given family of parallel supporting lines of C lie in a hyperplane, then C is an ellipsoid.*

For, because of (16.12) it suffices to show that every 3-flat L_3 containing interior points of C intersects C in an ellipsoid. Let $\mathfrak{g}$ be any line in L_3 and $Z'(\mathfrak{g})$ the set of all points of C which lie on the supporting lines of C parallel to $\mathfrak{g}$. A supporting line of C in L_3 is also a supporting line of $K = C \cap L_3$ and conversely. Hence $Z(\mathfrak{g}) = Z'(\mathfrak{g}) \cap L_3$ is the set of points on the supporting lines of K parallel to $\mathfrak{g}$. Since $Z'(\mathfrak{g})$ lies in a hyperplane H, the set $Z(\mathfrak{g})$ lies in $H \cap L_3$ that is, in a 2-flat. Therefore K satisfies the hypothesis of the theorem and it suffices to establish the theorem for a closed convex surface K in A^3.

We observe that K is strictly convex. For if the supporting line $\mathfrak{g}$ of K intersected K in a proper segment, then the set $Z(\mathfrak{g})$ could not lie in a plane.

Consider any two parallel supporting planes H_1 and H_2 of K touching K at p_1 and p_2, and let $\mathfrak{h} = L(p_1, p_2)$. For any line $\mathfrak{g}$ in H_1 through p_1 the curve $Z(\mathfrak{g})$ goes through p_1 and p_2 and lies, by hypothesis, in a plane $H(\mathfrak{g})$. Any plane P parallel to H_1, H_2 and lying between them intersects K in a convex curve A, and cuts $Z(\mathfrak{g})$ in two points q_1, q_2. Their join $T(q_1, q_2)$ lies in the plane $H(\mathfrak{g})$ containing $\mathfrak{h}$, and hence meets $\mathfrak{h}$ in the point $c = P \cap \mathfrak{h}$. At the points q_i the curve A has supporting lines parallel to $\mathfrak{g}$ and hence to each other.

Varying $\mathfrak{g}$ it follows that A has parallel supporting lines at the ends of any chord through c; hence c is the center of A, see (16.7).

Changing the position of P between H_1 and H_2 yields that $T(p_1, p_2)$ is the locus of the mid-points of a family of parallel chords of $Z(\mathfrak{g})$. If now $Z(\mathfrak{g})$ is kept fixed and p_1 is allowed to vary on $Z(\mathfrak{g})$ it is seen that every family of parallel chords of $Z(\mathfrak{g})$ has collinear mid-points. Hence by (16.13), $Z(\mathfrak{g})$ is an ellipse.

We cannot immediately apply lemma (16.12), since the planes $H(\mathfrak{g})$ are not yet known to be concurrent. With the same notation as before consider the ellipse $Z(\mathfrak{h})$. Each ellipse $Z(\mathfrak{g})$ (with $\mathfrak{g}$ in H_1 through p_1) intersects $Z(\mathfrak{h})$ in two points r_1, r_2 at which the tangents to $Z(\mathfrak{g})$ are parallel to $\mathfrak{h}$; hence $T(r_1, r_2)$ is the diameter of $Z(\mathfrak{g})$ conjugate to $\mathfrak{h}$. Consequently the tangents to $Z(\mathfrak{g})$ at p_1 and p_2 are parallel to $T(r_1, r_2)$. Thus H_1 and H_2 are parallel to the plane of $Z(\mathfrak{h})$ and the situation is the same as in the proof (16.12).

17. Minkowskian Geometry

A metric xy defined in the whole space A^n is called "*Minkowskian*" if it is equivalent, on each affine line, to the (natural) topology of A^n and if the affine mid-point z of two points x, y is also the mid-point in the sense of the metric xy, that is, if $xz = zy = xy/2$.

We have assumed as little as possible in this definition; in particular, we have not assumed that xy is topologically equivalent to A^n as a whole, because this will follow. The equivalence relation cannot be omitted altogether for, as there are discontinuous solutions of the functional equation

$$f\left(\frac{x+y}{2}\right) = \tfrac{1}{2}\,(f(x) + f(y)),$$

the mid-point condition is not enough.

In the present discussion it will prove useful to call a euclidean metric $e(x, y)$ defined in A^n, "*associated*" with A^n if it is equivalent to A^n and the straight lines defined by $e(x, y)$ coincide with the affine lines. Such a euclidean metric is a particular Minkowski metric.

(17.1) *The metric xy defined in the (whole) A^n is Minkowskian if and only if for some associated metric $e(x, y)$ the distances xy and $e(x, y)$ are proportional on each given line L. This will then be true for all associated metrics $e(x, y)$.*

This means there is a function $\varphi(L) > 0$ such that

(17.2) $$x\,y = \varphi(L)\cdot e(x, y) \qquad \text{for} \qquad x, y \in L.$$

The sufficiency is trivial. To show necessity, let xy be a Minkowski metric and L a given line in A^n. For a given associated euclidean metric $e(x, y)$ take on L two points a_0 and a_1 with $e(a_0, a_1) = 1$. Then put $a_0\,a_1 = \varphi > 0$ and it is to be shown that $xy = \varphi\, e(x, y)$ for all x, y on L. On L choose the euclidean arclength σ as parameter such that a_0, a_1 correspond to $\sigma = 0$ and $\sigma = 1$, and denote generally by a_σ the point corresponding to σ. Since a_i is the euclidean mid-point of a_{i-1} and a_{i+1}, it follows that $a_i\,a_{i+1} = \varphi$ if i is an integer, for $a_2\,a_1 = a_1\,a_0 = a_{-1}\,a_0 = \ldots$. Since a_0 is the mid-point of a_{-i} and a_i, it follows that $(a_{-i}\,a_0\,a_i)$ and then by (6.6) that $(a_k\,a_l\,a_m)$ if $k < l < m$ are integers, and hence $a_k\,a_l = \varphi\,|k - l|$.

Since $a_{k+1/2}$ is the euclidean and hence the Minkowski mid-point of a_k, a_{k+1} the same argument shows that $a_\rho\,a_\sigma = \varphi\,|\rho - \sigma|$ when ρ, σ have the form $k/2$, and similarly when of the form $k/2^n$, k integer.

As we assume xy and $e(x, y)$ are equivalent on L the rest easily follows.

(17.3) *A Minkowski metric in A^n is equivalent to the euclidean metrics associated with A^n.*

This will follow from:

(17.4) In a definite associated euclidean metric $e(x, y)$ the function $\varphi(L)$ defined in (17.2) depends continuously on L.

We show this by induction with respect to n; in A^1 there is nothing to prove. Assume that the statement holds in A^{n-1}, and let L_ν be any sequence of lines in A^n tending to the line L. On L take two distinct points a_1, a_2 and a hyperplane H_i through a_i that does not contain L. For large ν the line L_ν intersects H_i in a point a_i^ν and $e(a_i^\nu, a_i) \to 0$. The distance xy in A^n induces in H_i a distance with the same property (that $e(x, y)$ and xy are proportional on a line in H_i). By the inductive assumption $e(a_i^\nu, a_i) \to 0$ implies $a_i^\nu a_i \to 0$, hence by (2.1)

$$|\varphi(L_\nu)\, e(a_1^\nu, a_2^\nu) - \varphi(L)\, e(a_1, a_2)| = |a_1^\nu a_2^\nu - a_1 a_2| \leqslant a_1^\nu a_1 + a_2^\nu a_2 \to 0.$$

Since $e(a_1^\nu, a_2^\nu) \to e(a_1, a_2) \neq 0$, it follows that $\varphi(L_\nu) \to \varphi(L)$.

Next we show that

(17.5) $\qquad \varphi(L) = \varphi(L') \qquad$ *if* $\qquad L \,||\, L'$,

where $L \,||\, L'$ means that L and L' are parallel.

An equivalent statement is:

(17.5') $a\,b = a'\,b'$ if a, b and a', b' are opposite sides of a parallelogram.

For if $a, b \in L$ and $a', b' \in L'$ then $e(a, b) = e(a', b')$, hence $a\,b = a'\,b'$, if and only if $\varphi(L) = \varphi(L')$.

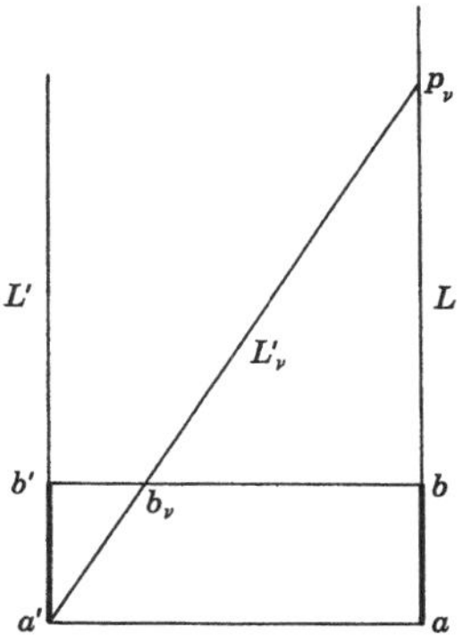

Fig. 12.

To prove (17.5') let $(a\,b\,p_\nu)$ and let the line L_ν' through p_ν and a' intersect $T(b, b')$ in b_ν. When $bp_\nu = \varphi(L)\, e(b, p_\nu) \to \infty$, then $b_\nu \to b'$; hence, by the preceding remark, $a'\,b_\nu = \varphi(L_\nu')\, e(a', b_\nu) \to \varphi(L')\, e(a', b') = a'\,b'$. But

$$\frac{a\,b}{a\,p_\nu} = \frac{e(a, b)}{e(a, p_\nu)} = \frac{e(a', b_\nu)}{e(a', p_\nu)} = \frac{a'\,b_\nu}{a'\,p_\nu},$$

hence

$$1 - \frac{a\,a'}{a\,p_\nu} = \frac{a p_\nu - a\,a'}{a\,p_\nu} \leqslant \frac{a'\,p_\nu}{a\,p_\nu} = \frac{a'\,b_\nu}{a\,b} \leqslant \frac{a'\,a + a\,p_\nu}{a\,p_\nu} = 1 + \frac{a\,a'}{a\,p_\nu}.$$

Since $a\,a'/a\,p_\nu \to 0$ it follows that $a'\,b' = \lim a'\,b_\nu = a\,b$.

This implies:

(17.6) *A Minkowski metric in A^n is invariant under the translations of A^n.*

The central reflection of A^n in p is the affinity $x_i' = -x_i + 2\,p_i$, which maps any point $x \neq p$ on the point x' on the line through p and x, for which p is the mid-point of x and x'. Any two points a, b not on a line through p go into points a', b' which form together with a, b a parallelogram, hence $a\,b = a'\,b'$. Conversely if a parallelogram is given then the central reflection in the intersection of the diagonals will interchange opposite sides. Therefore (17.6) yields:

(17.7) *A Minkowski metric in A^n is invariant under the central reflections of the A^n.*

The similitude

$$x_i' = b\,x_i + a_i, \qquad b \neq 0,$$

transforms any euclidean segment $T(x, y)$ into a segment $T(x', y')$ which is parallel to $T(x, y)$ and has length $e(x', y') = |b|\,e(x, y)$. It follows from (17.5) that $x'y' = |b|\,xy$.

(17.8) *If xy is a Minkowski metric in A^n with affine coordinates $x_1, \ldots, x_n$ then under the similitude $x_i' = bx_i + a_i$ all distances xy are multiplied by* $|b|$; that is, $x'y' = |b|\,x\,y$.

The Spheres

We now formulate several relevant facts in one theorem.

(17.9) *A Minkowski metric in A^n is finitely compact and M-convex. Prolongation is possible in the large, or $\rho(p) \equiv \infty$. The spheres $K(p, \rho)$ (the loci $p\,x = \rho > 0$) are convex and homothetic surfaces in A^n, and p is the affine center of $K(p, \rho)$. The space is a G-space and therefore straight, if and only if the spheres are strictly convex.*

Because of (17.5) $\varphi(L)$ takes (for a definite associated euclidean metric) all its values on the compact set of lines through a fixed point. Since $\varphi(L)$ is continuous, (17.4), it has a finite maximum β and a positive minimum α, so that

$$\alpha\, e(x, y) \leqslant xy \leqslant \beta\, e(x, y).$$

This implies that xy is finitely compact.

By (17.8) the similitude $x_i' = \sigma\,(x_i - p_i)/\rho + q_i$ maps $K(p, \rho)$ on $K(q, \sigma)$, therefore the spheres are homothetic. To prove the convexity of $K(p, \rho)$ let $pa = pb = \rho$. It is to be shown that $pc \leqslant \rho$ if $c = (1 - \tau)\, a + \tau\, b, 0 < \tau < 1$. Since the statement is trivial when p lies on $L(a, b)$, assume that $L(a, p)$ and $L(b, p)$ are different, and let the parallel to $L(p, a)$ through b intersect $L(p, c)$ in d. Then

$$\frac{bd}{\rho} = \frac{bd}{pa} = \frac{e(b, d)}{e(p, a)} = \frac{e(c, d)}{e(p, c)} = \frac{cd}{pc}$$

hence

$$\frac{\rho + bd}{\rho} = \frac{pb + bd}{\rho} = \frac{pc + cd}{pc} = \frac{pd}{pc}.$$

The triangle inequality $pb + bd \geqslant pd$ implies now $\rho \geqslant pc$.

It is seen at the same time that $pb + bd > pd$ implies $\rho > pc$, hence the strict convexity of $K(p, \rho)$. Since segments are unique in a straight space, it follows that the $K(p, \rho)$ are strictly convex, when xy defines a G-space.

Conversely, when the spheres are strictly convex, then xy defines a G-space. It is to be shown that prolongation is unique. Since prolongation of $T(p, b)$ is always possible along $L(p, b)$ it suffices to show that, if $pb+bd=pd$ for d not on $L(p, b)$, a sphere can be found which is not strictly convex. If we take a on the line through p parallel to $L(b, d)$ (on the proper side of p) such that $pa = pb$, then the preceding argument shows $pc = pa = pb$, so that $K(p, pa)$ is not strictly convex.

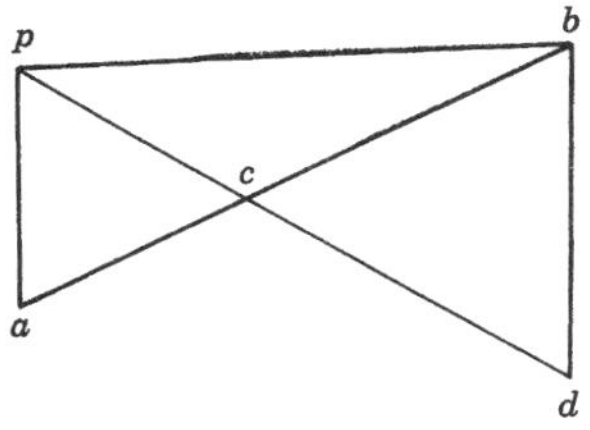

Fig. 13.

In view of all the properties which are necessary for a metric to be Minkowskian it is quite surprising how few are sufficient according to the original definition.

Different Minkowski metrics defined on the same model of the A^n are not necessarily intrinsically different, they may be isometric. The following theorem states that there are as many different Minkowski metrics as there are affinely not equivalent, closed convex surfaces with center.

(17.10) *Two Minkowski metrics xy and $m(x, y)$ in A^n are isometric if and only if an affinity exists which maps the sphere $K : px = 1$ on the sphere $K' : m(p, x) = 1$.*

Note: In view of (17.6) and (17.8) it would have amounted to the same to require that $px = \rho > 0$ can be mapped by an affinity on $qx = \sigma > 0$.

Assume an affinity Φ exists mapping K on K'. Since p is the affine center of both K and K' it stays fixed under Φ. If $e(x, y)$ is any euclidean metric associated with A^n denote the line functions corresponding to xy and $m(x, y)$ by $\varphi(L)$ and $\varphi'(L)$ respectively.

Let two distinct points y, z and their images $y' = y\Phi$, $z' = z\Phi$ be given. If L and L' are the parallels to $L(y, z)$ and $L(y', z')$ through p, then

$$yz = \varphi(L)\, e(y, z), \qquad y'z' = \varphi(L')\, e(y', z').$$

Because Φ leaves p fixed, $L' = L\Phi$. An intersection x of K and L goes under Φ into an intersection x' of L' and K', and

$$1 = px = \varphi(L)\, e(p, x) = m(p, x') = \varphi'(L')\, e(p, x').$$

As an affinity, Φ multiplies all euclidean distances on parallel lines by the same factor, hence

$$yz = e(y; z) : e(p, x) = e(y', z') : e(p, x') = m(y', z').$$

Conversely, assume the two metrics to be isometric, that is, a mapping Φ' of the A^n on itself exists such that $yz = m(y\Phi', z\Phi')$. If Φ'' is the translation sending $p' = p\Phi'$ into p, then $\Phi = \Phi'\Phi''$ also preserves distance.

$$yz = m(y\Phi, z\Phi) \qquad \text{and} \qquad p\Phi = p.$$

Hence Φ maps K on K'. The affine lines coincide with the geodesics only when the spheres are strictly convex. But they may be characterized intrinsically as those geodesics which go into themselves under any translation that carries a point of L into another point of L. Therefore Φ maps each affine line on an affine line. It is therefore induced by a projectivity and hence is an affinity.

The theorem implies that the euclidean metrics in A^n are those Minkowskian metrics whose spheres are affinely equivalent to a euclidean sphere, that is, ellipsoids:

(17.11) *A Minkowskian metric in A^n is euclidean if and only if its spheres are ellipsoids.*

Representation by convex functions

Next we discuss the representation of a Minkowski metric in A^n in terms of convex functions. The function $F(x) = F(x_1, \ldots, x_n)$ defined in a convex set C of A^n (which contains more than one point) is "*convex*", if for any two distinct points x, y in C and any τ, $0 < \tau < 1$.

(17.12) $$F\big((1-\tau)\, x + \tau y\big) \leqslant (1-\tau)\, F(x) + \tau\, F(y).$$

If the inequality always holds, then $F(x)$ is called "*strictly convex*".

At present those convex functions will interest us which are at the same time positive homogeneous: $F(x)$ defined in all of A^n is "*positive homogeneous*" (more specifically, of order 1) if,

(17.13) $$F(\mu\, x) = \mu\, F(x) \qquad \text{for} \qquad \mu > 0, \qquad F(0) = 0.$$

A positive homogeneous function is convex if and only if

(17.14) $$F(x + y) \leqslant F(x) + F(y) \quad \textit{for any } x, y.$$

For if $F(x)$ is convex, then

$$\frac{1}{2}\, F(x + y) = F\left(\frac{x}{2} + \frac{y}{2}\right) \leqslant \frac{1}{2}\, F(x) + \frac{1}{2}\, F(y).$$

And if (17.14) holds then for $0 < \tau < 1$

$$F\big((1-\tau)\, x + \tau\, y\big) \leqslant F\big((1-\tau)\, x\big) + F(\tau\, y) = (1-\tau)\, F(x) + \tau\, F(y).$$

A positive homogeneous function satisfies the relation

(17.15) $$F\big((1-\tau)\, x + \tau\, y\big) = (1-\tau)\, F(x) + \tau\, F(y), \quad \text{if} \quad y = \mu\, x \quad \text{with} \quad \mu > 0$$

and can therefore never be strictly convex. For such functions we will use the term "*strongly convex*" if it is convex and (17.15) represents for $x \neq y$ different from 0 the only case where the equality sign holds.

(17.16) *Let $F(x)$ be positive homogeneous and positive for $x \neq 0$. Then $F(x)$ is (strongly) convex if and only if the set $F(x) \leqslant 1$ is (strictly) convex.*

For let $F(x) \leqslant 1$ be (strictly) convex. In proving $F(x + y) \leqslant F(x) + F(y)$ we may because of the preceding remark assume that x and y are different from 0 and that not $y = \mu\, x$ with $\mu > 0$. Then $F\big(x/F(x)\big) = F\big(y/F(y)\big) = 1$ and the points $x/F(x)$ and $y/F(y)$ are distinct. Since $F(x) \leqslant 1$ is (strictly) convex

$$\frac{x + y}{F(x) + F(y)} = \frac{F(x)}{F(x) + F(y)}\, \frac{x}{F(x)} + \frac{F(y)}{F(x) + F(y)}\, \frac{y}{F(y)}$$

lies in (the interior of) $F(x) \leqslant 1$. Hence

$$\frac{F(x+y)}{F(x)+F(y)} = F\left(\frac{x+y}{F(x)+F(y)}\right) \leqslant 1,$$

with the inequality sign when $F(x) \leqslant 1$ is strictly convex.

Conversely let $F(x)$ be (strongly) convex and $F(x) \leqslant 1$, $F(y) \leqslant 1$. Since $F(x) \leqslant 1$ is starshaped with respect to the origin it suffices again to assume that x and y are distinct from 0 and that not $y = \mu x$, $\mu > 0$. Then

$$F((1-\tau)x + \tau y) \leqslant (1-\tau)F(x) + \tau F(y) \leqslant 1$$

with the inequality when $F(x)$ is strongly convex.

The relation of convex functions to Minkowskian metrics is then given by:

(17.17) *A Minkowski metric* xy *in* A^n *with given affine coordinates* $x_1, \ldots, x_n$ *has the form*

$$xy = F(x-y) \tag{17.18}$$

where

1) $F(x) > 0$ *for* $x \neq 0$,

2) $F(\mu x) = |\mu| F(x)$, *for any real* μ,

3) $F(x)$ *is convex.*

Conversely, if $F(x)$ *is defined in* A^n *and has these three properties, then* (17.18) *is a Minkowski metric. The metric defines a G-space if and only if* $F(x)$ *is strongly convex.*

Proof. Let xy be a Minkowski metric. If z denotes the origin we define

$$F(x) = zx.$$

Property 1) is then obvious, 2) follows from (17.8), and 3) from (17.9) and (17.16). Since the metric is invariant under translations, the distance of x and y equals the distance of $x-y$ to the origin z, which proves (17.18).

Conversely, if a function $F(x)$ with properties 1), 2), 3) is given, and xy is defined by (17.18), then xy is invariant under translations, and $|x| F(\pm x|x|^{-1}) = xz$ because of 2). Therefore xy is on each line proportional to $|x-y|$, hence will be Minkowskian, if we can establish that it is a distance. $xy > 0$ for $x \neq y$ follows from 1). The symmetry $xy = yx$ is equivalent to $F(x-y) = F(y-x)$ which is contained in 2). The triangle inequality reduces to

$$F(x-u) = F((x-y)+(y-u)) \leqslant F(x-y) + F(y-u)$$

which follows from the convexity of $F(x)$.

The statement regarding G-spaces is a consequence of (17.9) and (17.16).

This theorem contains as a special case the general expression for the associated euclidean metrics $e(x, y)$ in a given affine coordinate system: the sphere $e(z, x) = 1$ is an ellipsoid, see (17.11), with center z and has therefore an equation of the form

$$\sum g_{ik} x_i x_k = 0,$$

where the form $\Sigma g_{ik} x_i x_k$ is positive definite, because every line $x_i = ty_i$, $y \neq 0$ intersects $K(z, 1)$ for some $t \neq 0$, so that $\Sigma g_{ik} y_i y_k = 1/t^2 > 0$. Since $e(z, x)$ must satisfy 2) it follows that

$$e(x, y) = \left[\sum g_{ik} (x_i - y_i)(x_k - y_k)\right]^{1/2}$$

Motions

Denote by $E(x, y)$ the associated euclidean metric in which the unique ellipsoid E of minimal volume that contains $K(z, 1)$ and has centre z, is the unit sphere (compare (16.10)).

A motion of the Minkowski space is according to the second part of the proof of (17.10) an affinity

$$x_i' = \sum a_{ik} x_k + b_i.$$

Since translations are Minkowskian motions, $x_i' = \Sigma a_{ik} x_k$ is also a motion. It therefore carries $K(z, 1)$ and, as in the proof of (16.11), also E into itself. This means that it is a motion for the metric $E(x, y)$. Therefore the original affinity is a motion for $E(x, y)$. Thus we have proved:

(17.19) *For a given Minkowski metric in A^n there is a euclidean metric associated to A^n such that each motion of the Minkowski metric is also a motion of the euclidean metric.*

This implies together with (17.6), (17.7), and (16.11):

(17.20) *The group of motions of an n-dimensional Minkowski space is isomorphic to a subgroup of the motions of E^n which contains all translations and central reflections.*

If for given points p, x, y with $px = py > 0$ a motion exists which leaves p fixed and carries x into y, then the metric is euclidean.

At this point we break off the systematic development of Minkowskian geometry and discuss some isolated properties which will prove of importance later on. Since this book is concerned with G-spaces, we restrict ourselves to this case, although all properties hold with slight modifications for the general case.

Limit spheres

We introduce for a general metric space the notation $K(a, b) = K(a, ab)$ for the locus of points x with $ax = ab$. If $x(\tau)$ represents a straight line, it will be shown later that the sphere $K(x(\tau), x(0))$ converges for $\tau \to \infty$ to a limit, which we call a *limit sphere*.

In a Minkowskian G-space $x(\tau)$ is an affine line, all the spheres $K(x(\tau), x(0))$ are homothetic and have at $x(0)$ the same tangent cone T. It follows immediately that $\lim_{\tau \to \infty} K(x(\tau), x(0)) = T$. Hence, compare (16.6):

(17.21) *In a Minkowskian G-space the tangent cones of the spheres are the limit spheres. The limit spheres are hyperplanes if and only if the spheres are differentiable.*

Distance of point from line. Perpendiculars

If in a Minkowski G-space (axb) and $ax : xb = (1 - \tau') : \tau'$ then x is, because of (17.2), the point $(1 - \tau')\, a + \tau'\, b$, hence (17.8) yields:

(17.22) If in a Minkowskian G-space $(ga'a)$, $(gb'b)$ and

$$ga' : ga = gb' : gb = \tau \qquad \textit{then} \qquad a'b' : ab = \tau.$$

(17.23) *If $x(\tau)$ represents a geodesic in a Minkowskian G-space which does not pass through p, then $px(\tau)$ is a strictly convex function of τ.*

Given $\alpha < \beta$ and $0 < \tau < 1$, determine a' with $(pa'x(\alpha))$ and $pa' : px(\alpha) = 1 - \tau$. If $c = x((1 - \tau)\, \alpha + \tau\beta)$, then by (17.22) $a'c : px(\beta) = \tau$, hence

$$pc < pa' + a'c = (1 - \tau)\, px(\alpha) + \tau px(\beta).$$

Since $px(\tau) \to \infty$ for $|\tau| \to \infty$, and $px(\tau)$ is strictly convex, it reaches its minimum at exactly one point $x(\tau_0)$, which is the foot of p on the geodesic $x(\tau)$ according to the definition in section 1. Of course, convexity means much more than that the minimum is reached at one point, which is sufficient for the uniqueness of the foot. These questions will occupy us very much later on. It will be seen that uniqueness of the foot of a given point p on a given line L

is, in any straight space, equivalent to convexity of the spheres and implies that all points on the geodesic $\mathfrak{g}(p, f)$ have f as foot on L. This leads naturally to the definition:

The straight line L in a G-space is called a "*perpendicular*" to the set M at f, if $f \in L \cap M$ and every point of L has f as foot on M.

In Minkowski spaces the existence of perpendiculars can be verified immediately:

(17.24) *If p does not lie on the r-flat L_r of a Minkowskian G-space, then p has exactly one foot f on L_r and $L(p, f)$ is a perpendicular to L_r.*

The existence of at least one foot follows from (2.21). Then L_r cannot contain an interior point $K(p, f)$, and hence lies on a supporting hyperplane L_{n-1} of $K(p, f)$ at f. This hyperplane, and hence L_r, cannot contain a second point of $K(p, f)$ since $K(p, f)$ is strictly convex. The spheres $K(x, f)$ with x on $L(p, f)$ and on the same side of f as p are homothetic and therefore have also L_{n-1} as supporting plane. The fact that spheres have parallel supporting planes at diametrically opposite points shows the same for the x on $L(p, f)$ on the other side of f.

Asymmetry of perpendicularity

This theorem shows in particular that there is exactly one perpendicular G to a given line L through a given point p not on L. It is very important to realize that L need not be perpendicular to G, or, as we will briefly say, that perpendicularity between lines is in general not symmetric. In fact, according to a surprising theorem of Blaschke [1] it is a very exceptional occurrence:

(17.25) *If, in a Minkowskian G-space of dimension greater than two, perpendicularity between lines is symmetric, then the metric is euclidean.*

Proof. Consider a supporting hyperplane H of a sphere $K(p, 1)$ at a point g. The line $G = L(p, g)$ is perpendicular to all lines in H through g. If the hyperplane parallel to H through p intersects K in Z, then $L(p, z)$ is for $z \in Z$ parallel to some line L_z in H. Since G is perpendicular to L_z it is perpendicular to $L(p, z)$. By hypothesis $L(p, z)$ is also perpendicular to G, hence also to the parallel G_z to G through z. This means that G_z must be a supporting line of $K(p, 1)$ at z. Therefore Z is the locus of the points in which the supporting lines of $K(p, 1)$ parallel to G touch $K(p, 1)$. Thus $K(p, 1)$ satisfies the hypothesis of Blaschke's Theorem (16.14) and is therefore an ellipsoid. (17.11) shows that the metric is euclidean.

The assumption that the space has dimension greater than two is essential. This is not to say that perpendicularity is always symmetric in a Minkowskian plane, but that *there are curves in the plane other than the ellipses which lead to symmetric perpendicularity.*

In a Minkowskian G-plane the perpendicular G to a line L at a given point f of L is unique. For a second perpendicular at f would be perpendicular to L after being translated. It would then intersect G, and the point of intersection would have two feet on L, contrary to (17.24).

If perpendicularity is symmetric, then L is perpendicular to G and therefore unique. On the other hand every supporting line at f to a circle $K(p, f)$ with $p \in G$ has G as perpendicular. Hence the *circles are differentiable.* If the line G through p intersects $K(p, 1)$ at f_1 and f_2, and L_1, L_2 are the tangents of $K(p, 1)$ at these points, then the line L parallel to L_i through p must intersect $K(p, 1)$ in the points z_1, z_2 at which the tangents G_1, G_2 are parallel to G. Thus $K(p, 1)$ *shares with the ellipse certain properties of conjugate diameters.*

That this property cannot generally hold for differentiable circles is obvious: for should the tangent at z_i turn out to be parallel to G, the circle can obviously be modified in the vicinity of z_1 and z_2, so that the tangents are no longer parallel to G.

However, as stated before, there are other curves than ellipses with the properties of conjugate diameters derived above[1]. In polar co-ordinates r, φ take a differentiable strictly convex arc $r = f(\varphi)$, $0 \leqslant \varphi \leqslant \pi/2$, with $f(0) = 1$ and $r \cos\varphi = 1$ as tangent and $f(\pi/2) = 1$ and $r \sin\varphi = 1$ as tangent.

Let $r = g(\varphi)$ be the image of $r = f(\varphi)$ under the polar reciprocity in $r = 1$. This arc is also strictly convex and differentiable, passes through $(0, 1)$ and $(\pi/2, 1)$ and has the same tangents as $r = f(\varphi)$ at these points. Then the equations

$$r = f(\varphi), \quad 0 \leqslant \varphi \leqslant \pi/2, \qquad r = g(\varphi - \pi/2), \quad \pi/2 \leqslant \varphi \leqslant \pi,$$

$$r = f(\varphi - \pi), \quad \pi \leqslant \varphi \leqslant 3\pi/2, \qquad r = g(\varphi - 3\pi/2), \quad 3\pi/2 \leqslant \varphi \leqslant 2\pi$$

define a strictly convex differentiable curve with the origin as center.

The curves constructed in this way, taken as unit circles of Minkowskian geometries, yield *all different, that is non-isometric, two dimensional Minkowskian G-spaces with symmetric perpendicularity.* That these geometries enjoy symmetric perpendicularity is readily verified. That they exhaust all of them follows from (17.10). For any differentiable curve C' with center z or $r = 0$ can be transformed by exactly one affinity into a curve C with centre z which passes through $(0, 1)$ and $(\pi/2, 1)$ and has tangents $r \cos\varphi = 1$ and $r \sin\varphi = 1$ at these points.

18. Hilbert's Geometry

Plane Minkowskian geometry arises from the euclidean through replacing the ellipse as unit circle by a convex curve. In a somewhat similar way a geometry discovered by Hilbert[1] arises from Klein's Model of hyperbolic geometry through replacing the ellipse as absolute locus by a convex curve.

If the points a_1, a_2, d, c lie on a line in A^n and a_1, a_2 between c and d, then $a_i = (1-\tau_i)\,c + \tau_i d,\ \ 0 < \tau_i < 1$ and

$$(18.1)\qquad R(a_1, a_2, d, c) = \frac{1-\tau_1}{1-\tau_2}\,\frac{\tau_2}{\tau_1} > 1,\ if\ \tau_2 > \tau_1, \text{ where } R() \text{ denotes cross ratio.}$$

Therefore $R(a_1, a_2, d, c) = 1/R(a_1, a_2, c, d)$ implies that $|\log R(a_1, a_2, c, d)| = \log R(a_1, a_2, d, c)$ if a_2 lies between a_1 and d. Since in (18.1) $R(a_1, a_2, d, c) \to \infty$ if either $\tau_1 \to 0$ or $\tau_2 \to 1$ (or both), and obviously for $a_3 = (1-\tau_3)\,c + \tau_3 d$ with $\tau_3 \geqslant \tau_2$

$$(18.2)\qquad R(a_1, a_2, d, c)\, R(a_2, a_3, d, c) = R(a_1, a_3, d, c),$$

it follows that

$$(18.3)\qquad h(a_1, a_2) = |\log R(a_1, a_2, d, c)|$$

metrizes the open segment from c to d such that it becomes a straight line. In fact the metric is simply a hyperbolic metric in P^1.

If $x_\tau = (1-\tau)\,c + \tau\, d$, observe that for $0 < \tau_c < \tau_1 < \tau_2 < \tau_d < 1$:

$$(18.4)\qquad \begin{cases} R(x_{\tau_1}, x_{\tau_2}, x_{\tau_d}, c) = \dfrac{\tau_d-\tau_1}{\tau_d-\tau_2}\,\dfrac{\tau_2}{\tau_1} > \dfrac{1-\tau_1}{1-\tau_2}\,\dfrac{\tau_2}{\tau_1} = R(x_{\tau_1}, x_{\tau_2}, d, c), \\[2ex] R(x_{\tau_1}, x_{\tau_2}, d, x_{\tau_c}) = \dfrac{1-\tau_1}{1-\tau_2}\,\dfrac{\tau_2-\tau_c}{\tau_1-\tau_c} > \dfrac{1-\tau_1}{1-\tau_2}\,\dfrac{\tau_2}{\tau_1} = R(x_{\tau_1}, x_{\tau_2}, d, c), \\[2ex] R(x_{\tau_1}, x_{\tau_2}, x_{\tau_d}, x_{\tau_c}) > R(x_{\tau_1}, x_{\tau_2}, d, c). \end{cases}$$

Now let D be the interior of a closed convex surface C in A^n. Any two distinct points a_1, a_2 in D determine a line $L(a_1, a_2)$ which intersects C in two points c, d. Then (18.3) defines the *Hilbert metric in* D, if we complete the definition by $h(a, a) = 0$, which agrees with (18.1). We could add a factor k as in the definition of the non-euclidean metrics. Metrics in D corresponding to different factors are not isometric; however, we will only be concerned with a single Hilbert metric at a time.

With this distance the intersection of any line with D, if not empty, becomes a straight line. Therefore the space will be a finitely compact, M-convex metric space with $\rho(p) = \infty$ if it can be shown that

$$h(a, b) + h(b, c) \geqslant h(a, c) \textit{ if } a, b, c \textit{ are not collinear.}$$

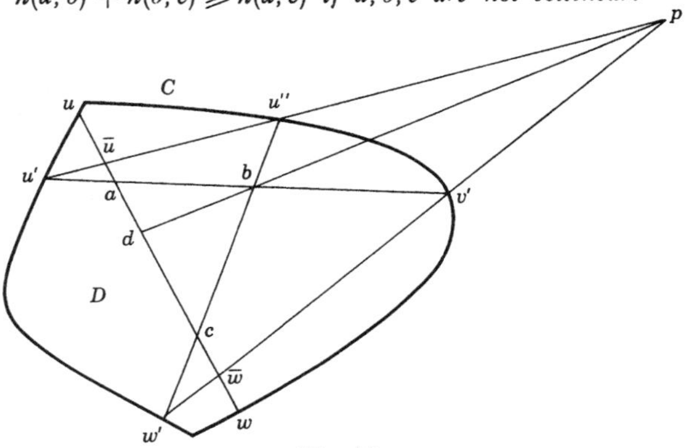

Fig. 14.

This follows with the notations of the figure (p may be at infinity) from

$$R(a, b, v', u') = R(a, d, \overline{w}, \overline{u}) \geqslant R(a, d, w, u)$$

$$R(b, c, w', u'') = R(d, c, \overline{w}, \overline{u}) \geqslant R(d, c, w, u),$$

see (18.4), hence by (18.2)

$$h(a, b) + h(b, c) \geqslant h(a, c)$$

with the equality only when $\overline{u} = u$ and $\overline{w} = w$. In that case C contains two proper non-collinear, coplanar segments. Therefore:

(18.5) *The Hilbert distance $h(a, b)$ defines in D a finitely compact M-convex metric space in which prolongation is possible in the large. $h(a, b)$ defines a straight Desarguesian space, if and only if, C does not contain two proper coplanar, but not collinear segments.*

The Hilbert metric will not be studied for its own sake. It will merely be used.

(1) to illustrate convexity questions,

(2) to construct other Desarguesian spaces,

(3) to provide examples for the theory of parallels.

In this section we discuss only points (1) and (2) and begin with the convexity proporties which for $n = 2$ are found in Pedersen [1].

Convexity of equidistant loci

(18.6) *Let p, f be two distinct points in D, let $L(p, f)$ intersect C in b and a, and let H_a, H_b be supporting planes of C at a and b.*

If H_p' and H_f' are the hyperplanes through p and f in the pencil determined by H_a and H_b, then

(18.7) $\quad h(p, f) \leqslant h(x, y) \quad$ *if* $\quad x \in H_p = H_p' \cap D, \quad y \in H_f = H_f' \cap D.$
If the space is straight then $h(p, f) = h(p, y)$ *only for* $y = f$. *If D is strictly convex then* $h(p, f) = h(x, y)$ *only for* $x = p$ *and* $y = f$.

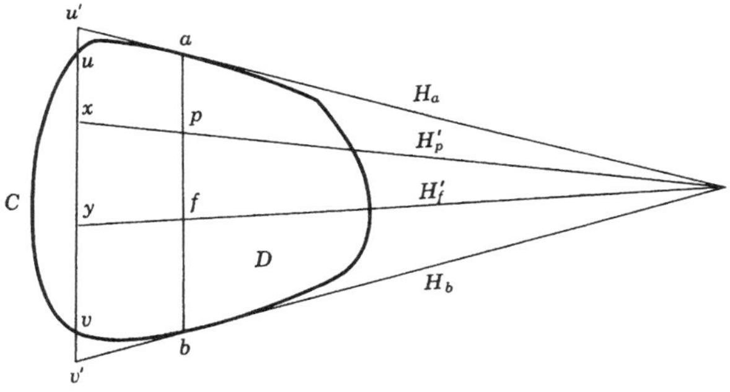

Fig. 15.

Proof. Let the notation be such that p lies between a and f. Put

$$u' = L(x, y) \cap H_a, \qquad v' = L(x, y) \cap H_b$$

and denote the intersections of $L(x, y)$ with C by u and v such that x lies between u and y. Since p, f, b, c and x, y, v', u' are the intersections of $L(p, f)$ and $L(x, y)$ with four planes in a pencil it follows from (18.4) that

$$R(x, y, v, u) \geqslant R(x, y, v', u') = R(p, f, b, a)$$

which is equivalent to (18.7).

For equality it is necessary that $u = u'$ and $v = v'$. If C is strictly convex this can happen only when $u = a$ and $v = b$, hence when $x = p$ and $y = f$.

If we only know that the space is straight, then it is possible that for instance b lies on a proper segment contained in C. If $u = a$ and v is a point of this segment, equality holds in (18.7). However if $x = p$ and $y \neq f$ then $u \neq a$ and $v \neq b$, and equality in (18.7) implies the presence of two non-collinear, coplanar segments on C, so that the space is not straight.

(18.6) has various applications. The relation $h(p, y) \geqslant h(p, f)$ for $y \in H_f$ means that H_f' is a supporting plane of $K(p, f)$ at f. Since p and f were arbitrary in D it follows from (16.3) that:

(18.8) *The spheres of a Hilbert geometry are (closed) convex surfaces (in A^n). They are strictly convex if the space is straight.*

This may be generalized as follows:

(18.9) *If Q is a convex set in D, then $S(Q, \rho)$ is convex. $S(Q, \rho)$ is strictly convex if D is strictly convex.*

If $\bar{Q}$ is the closure of Q in the sense of Hilbert geometry, then any point x satisfies $h(x, \bar{Q}) = h(x, Q)$. Therefore it may be assumed that Q is closed and $S(Q, \rho) \neq D$. If p is a boundary point of $S(Q, \rho)$ then $pQ = \rho > 0$. If f is a foot of p on Q, then Q and $S(p, \rho)$ have no common point. Hence $K(p, f)$ and Q have at f a common supporting hyperplane H_f'. If a and b denote again the intersections of $L(p, f)$ with C, then C has at a and b supporting planes H_a, H_b such that their pencil contains H_f'.

This is an obvious consequence of the preceding discussion if $K(p, f)$ is differentiable at f. Since the spheres $K(p, f)$ are differentiable for smooth C, the assertion follows in the general case by approximation of C with smooth surfaces[2].

If H_p' denotes the plane in this pencil through p, and $x \in H_p = H_p' \cap D$, then a segment connecting x to a foot of z of x on Q contains a point y of H_f'. It follows from (18.7) that

$$h(x, Q) = h(x, z), \qquad h(x, y) \geqslant h(p, f)$$

with inequality for $x \neq p$ and strictly convex D. Therefore H_f' is a supporting plane of $S(Q, \rho)$ at p, so that $S(Q, \rho)$ is convex, and strictly so if D is strictly convex. A particular case of (18.9) is of special interest.

(18.10) *If the intersection L_r of an r-flat L_r' in A^n with D is not empty, then the equidistant locus $h(x, L_r) = \rho > 0$ lies on a convex surface in A^n, which is strictly convex if D is strictly convex.*

We will see in Chapter V that (18.9) is in general spaces much stronger than convexity of spheres. On the other hand the convexity of $S(Q, \rho)$ for any convex Q, follows from the convexity of $S(T, \rho)$ for segments T, compare (36.21).

Peakless functions

The type of function which is needed to describe these facts analytically does not seem to have received a name in the literature. Since these functions will be met frequently, we venture to define:

A function $f(\tau)$ which is defined and continuous on a convex set of the real τ-axis is called "*peakless*" if

$$(18.11) \qquad f(\tau_2 \leqslant \max(f(\tau_1), f(\tau_3)) \qquad \text{for} \qquad \tau_1 < \tau_2 < \tau_3$$

and the equality sign implies $f(\tau_1) = f(\tau_3)$.

A function is peakless if and only if it belongs to one of the following types: $f(\tau)$

1) is constant 2) is strictly increasing or decreasing 3) takes its minimum at one point, decreases strictly to the left of the point and increases strictly to the right 4) takes its minimum at all points of an interval and decreases strictly to the left and increases strictly to the right of this interval.

We call a peakless function "*strictly peakless*" if it is not constant in any interval, thereby excluding cases 1) and 4).

Convex functions are peakless, but not conversely. Nevertheless peaklessness may be interpreted as a degenerate form of convexity: If $f(\tau)$ is convex and positive, then $f^{\alpha}(\tau)$ is convex for every $\alpha > 1$, hence the requirement that $f^{\alpha}(\tau)$ be convex becomes weaker as α increases. For $\alpha \to \infty$ it becomes the condition (18.11). For if $f^{\alpha}(\tau)$ is convex and positive, $\tau_2 = (1-\theta)\,\tau_1 + \theta\,\tau_3$, $0 < \theta < 1$, and for example $f(\tau_1) \geqslant f(\tau_3)$ then

$$f(\tau_2) \leqslant f(\tau_1)\left[1 - \theta + \theta\left(\frac{f(\tau_3)}{f(\tau_1)}\right)^{\alpha}\right]^{1/\alpha}$$

and the equality sign holds only when $f(\tau_2) = f(\tau_1) = f(\tau_3)$. For $\alpha \to \infty$ this becomes (18.11) with the condition for the equality sign.

We may now formulate (18.10) as follows.

(18.12) *In a Hilbert geometry, if $x(\tau)$ represents a geodesic G which lies on an affine line*[3] *then the distance $f(\tau) = x(\tau)\,Q$ is a peakless function for any convex set Q in D.*

If the space is straight, then $px(\tau)$ is strictly peakless. If D is strictly convex then all functions $f(\tau)$ are strictly peakless.

For if $f(\tau_1) = \rho = max\,(f(\tau_1), f(\tau_3))$ and $\tau_1 < \tau_2 < \tau_3$, then the convexity of the closure F of $S(Q, \rho)$ implies that $f(\tau_2) \leqslant \rho$. If $f(\tau_2) = f(\tau_1)$, then the convexity of F implies that the line $x(\tau)$ contains for $\tau \geqslant \tau_2$ no point of $S(Q, \rho)$, hence $f(\tau_3) \geqslant f(\tau_2) = f(\tau_1)$ and $f(\tau_3) = f(\tau_1)$.

In euclidean and hyperbolic geometry, if a' is the midpoint of a and p, and b' is the midpoint of b and p, and a, b, p are not collinear, then

$$2\,a'\,b' \leqslant a\,b. \tag{18.13}$$

Equality holds in the euclidean, and as we saw also in the Minkowskian geometry, and inequality in the hyperbolic.

It will be seen in Chapter V, compare (36.5, 13), that (18.13) implies the convexity of the functions $x(\tau)\,y(\tau)$, $x(\tau)\,H$, and $y(\tau)\,G$, if $x(\tau)$ and $y(\tau)$ represent geodesics G, H.

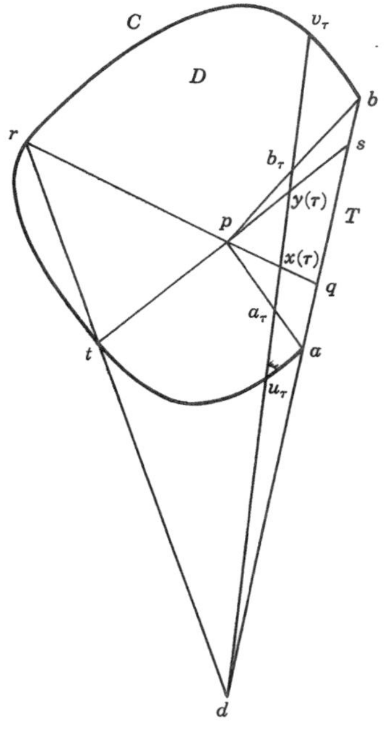

Fig. 16.

We will now convince ourselves that these functions are not convex in arbitrary Hilbert geometries:

In the affine plane consider a closed convex curve C which possesses exactly one supporting line that intersects C in a proper segment T. Let a, b be the endpoints and q, r distinct interior points of T such that q lies between a and s. In the Hilbert geometry defined in the interior D of C consider two lines G, H through a point p of D, and with endpoints q and s respectively. Choose representations $x(\tau)$ and $y(\tau)$ of G and H such that $x(0) = y(0) = p$ and $x(\tau) \to q$, $y(\tau)\ s \to$ for $\tau \to \infty$. Let the line $L(r, t)$ through the other endpoints r of G and t of H intersect $L(a, b)$ in d (possibly at infinity). The line L_τ through d and $x(\tau)$ intersects G in $y(\tau)$ because

$$R(p, x(\tau), q, r) = R(p, y(\tau), s, t).$$

Denote the intersections of L_τ with C by u_τ and v_τ, where $x(\tau)$ lies between $y(\tau)$ and u_τ. Finally, L_τ intersects the affine segments from p to a and b in the points a_τ and b_τ. Then

$$h(x(\tau), H) \leqslant h(x(\tau), y(\tau)) = \log R(x(\tau), y(\tau), v_\tau, u_\tau) < \log R(x(\tau), y(\tau), b_\tau, a_\tau) =$$
$$= \log R(q, s, b, a).$$

Thus the functions $h(x(\tau), H)$ and $h(x(\tau), y(\tau))$ vanish for $\tau = 0$ and are positive and bounded for $\tau > 0$, hence they cannot be convex.

This implies the existence of two numbers $\tau_2 > \tau_1 \geqslant 0$ such that

$$2\,h\left[x\left(\frac{1}{2}(\tau_1+\tau_2)\right), G\right] > h(x(\tau_1), G) + h(x(\tau_2), G),$$

and of two numbers $\tau'' > \tau' \geqslant 0$ such that

$$2\,h\left[x\left(\frac{1}{2}(\tau'+\tau'')\right), y\left(\frac{1}{2}(\tau'+\tau'')\right)\right] > h(x(\tau'), y(\tau')) + h(x(\tau''), y(\tau'')).$$

If, for sufficiently large τ_0, the arc of C from $x(\tau_0)$ to $y(\tau_0)$ which contains T is replaced by an arc through q and s which forms with the remaining arc of C a strictly convex curve, then none of the points in the cross ratios which must be calculated to find the six distances occurring in the last two inequalities will change. Therefore the functions $x(\tau)\, y(\tau)$ and $x(\tau)\, H$ also fail to be convex for general strictly convex C, although now $x(\tau) y(\tau) \to \infty$ and $x(\tau)\, H \to \infty$ for $|\tau| \to \infty$.

METRICS IN CONVEX SETS OF A^n

We now come to point 2) and prove:

(18.14) *Given a non-empty convex open set D in A^n, then D can be metrized as a straight space in which the intersections of D with the lines in A^n are the straight lines.*

First let D be a set that does not contain any complete line in A^n. If C is the surface bounding D, then any line $L(p, q)$ connecting two points p, q in D intersects C at least once, say at a. With b we denote the second intersection of $L(p, q)$ with C if it exists, otherwise the point of $L(p, q)$ at infinity. Then

$$h(p, q) = \log |R(p, q, a, b)|$$

defines a metric in D which is finitely compact and M-convex and with $\rho(p) = \infty$. The proof is the same as for (18.5), in fact the present case can be reduced to (18.5) by taking as new plane at infinity a finite plane which does not intersect $D \cup C$.

If C is strictly convex, then $h(p, q)$ defines a straight space, otherwise $L(p, q) + e(p, q)$, where $e(p, q)$ is any euclidean metric associated with A^n, will define a straight space in D with the intersections of the affine lines with D as geodesics. The proof is quite trivial and may be left to the reader.

If D does contain a whole straight line and is not the whole A^n, then a maximal r, $1 \leqslant r \leqslant n-1$ exists such that D contains an r-flat L_r'. An $(n-r)$-flat that intersects L_r' in a point, intersects D in a set D_{n-r} which does not contain a whole line. In D_{n-r} we introduce a Hilbert metric $h_{n-r}(p', q')$ as in the first part of the proof. If now p and q are any two points of D, let p' and q' be the intersection of the r-flats parallel to L_r' through p and q with D_{n-r} (these r-flats lie entirely in D). Then

$$pq = h_{n-r}(p', q') + e(p, q)$$

metrizes D such that it is a straight space whose geodesics are the intersections of the affine lines with D. The proof is again trivial.

There are of course many other ways of constructing metrics satisfying (18.14).

We give two simple examples which illustrate the degree of freedom in choosing the metric and also settle two points connected with convexity and parallels respectively.

It was shown in (6.9) that in any metric space $x, y \in S(p, \rho)$ implies $T(x, y) \subset S(p, 2\rho)$. To show that the factor 2 is the best possible we referred to the spherical geometry. This leaves the question, *whether* 2 *is the best factor in straight spaces. We are going to show that this is so,* even when the geodesics are the whole lines of A^2.

Let the function $\varphi(\tau)$ be defined by

$$\varphi(\tau) = 0 \quad \text{for} \quad \tau < 0, \quad \varphi(\tau) = \tau, \quad \text{for} \quad 0 \leqslant \tau \leqslant 1, \quad \varphi(\tau) = 1 \quad \text{for} \quad \tau > 1.$$

If x, y are affine coordinates in A^2, then for $p_i = (x_i, y_i)$, $i = 1, 2$,

$$p_1p_2 = \varepsilon[(x_1 - x_2)^2 + (y_1 - y_2)^2]^{1/2} + |\varphi(x_1) - \varphi(x_2)| + |\varphi(y_1) - \varphi(y_2)|$$

defines, for every $\varepsilon > 0$, a metric for which the affine lines are the geodesics.

The points $a = (2, 0)$, $b = (1, 1)$, $c = (0, 2)$ lie in this order on a segment, but with $z = (0, 0)$

$$za = zc = 2\varepsilon + 1, \qquad \text{and} \qquad zb = \sqrt{2} \cdot \varepsilon + 2,$$

hence $zb < \lambda\, za = \lambda\, zc$ will not be correct for a given $\lambda < 2$ if ε is small enough.

As second example we consider the set $D : x > 0$, $0 < y < 1$ of an affine (x, y)-plane. The function $\psi(\tau) = -\dfrac{1}{\tau} + \dfrac{1}{1-\tau}$ is monotone in $0 < \tau < 1$, hence for $p_i = (x_i, y_i)$, $i = 1, 2$,

$$p_1p_2 = [(x_1-x_2)^2 + (y_1-y_2)]^{1/2} + \left|\frac{1}{x_1} - \frac{1}{x_2}\right| + |\psi(y_1) - \psi(y_2)|$$

defines a straight space in which the intersections of the affine lines with D are the geodesics. In particular, if $0 < \alpha < \beta$ then $y = \alpha x$, $y = \beta x$, $0 < y < 1$ are geodesics G and H. Call G^+ and H^+ the orientations of G and H for which traversing G^+ or H^+ in the positive sense corresponds to decreasing x. Let $p(\tau)$ and $q(\tau)$ represent G and H such that ordinates of $p(0)$ and $q(0)$ equal $\frac{1}{2}$ and increasing τ corresponds to decreasing x.

The lines G^+ and H^+ have a common "positive end-point", and would be parallels in the language of hyperbolic geometry and *are asymptotes in our language* (Section 11 and next chapter), *the opposite orientations are not asymptotes. Nevertheless $p(\tau)H$ and $q(\tau)G$ are bounded for $\tau \leqslant 0$ and* $\lim p(\tau)H = \lim q(\tau)G = \infty$ *if* $\tau \to \infty$. For if p_y and q_y are the points $p(\tau)$ and $q(\tau)$ with ordinate y, then

$$p_y q_y = (\alpha^{-1} - \beta^{-1})\, y + (\beta - \alpha)\, y^{-1} < \alpha^{-1} - \beta^{-1} + 2\,(\beta - \alpha) \quad \text{if} \quad \tfrac{1}{2} \leqslant y < 1.$$

To prove the second assertion let $\tau_i \to \infty$ and $p(\tau_i) = (x_i, \alpha x_i)$, then $x_i \to 0$. For any numbers σ_i and $q(\sigma_i) = (\overline{x}_i, \beta\overline{x}_i)$

$$p(\tau_i)\, q(\sigma_i) > \left|\frac{1}{x_i} - \frac{1}{\overline{x}_i}\right| + \left| -\frac{1}{\alpha\, x_i} + \frac{1}{\beta\, \overline{x}_i} + \frac{1}{1-\alpha\, x_i} - \frac{1}{1-\beta\overline{x}_i}\right|$$

If the first term on the right is bounded, then $\overline{x}_i \to 0$ and the second tends to ∞ because $\beta > \alpha$.

CHAPTER III

PERPENDICULARS AND PARALLELS

19. Introduction

Perpendicular and parallel lines are basic concepts of elementary geometry. One would think that the extension of these concepts to more general spaces must have attracted the attention of mathematicians, but this is not the case[1], and so it happens that the present chapter is less related to the work of others than anything else in this book. Therefore this introduction is quite detailed.

A perpendicular to a straight line H is a straight line L that intersects H at a point f and such that every point x of L has f as its foot on H. In the discussion of Minkowskian geometry we saw that perpendicularity need not be symmetric. The line H touches the spheres $K(x, f)$ with center x through f at f and otherwise lies outside $K(x, f)$. We may say that H is a supporting line of $K(x, f)$ and expect that in straight spaces the existence of perpendiculars is related to convexity of spheres. In fact, these conditions are equivalent. In Section 20 we first discuss briefly local convexity of sets in a G-space, then, also locally, the relation of convexity of spheres to uniqueness of the foot of a point on a segment. Then we apply these results to straight spaces and find that other equivalent conditions to convexity of spheres are: the function $px(\tau)$ is peakless[2] when $x(\tau)$ represents a straight line; a sphere intersects a line at most twice.

The latter condition may be applied to spheres of radius less than λ in a space of the elliptic type whose geodesics have length 2λ. It is proved in Section 21 that for spaces of dimension greater than two *this condition alone makes a space of the elliptic type actually an elliptic space.* For two dimensions the problem offers peculiar difficulties and is open. One of the difficulties is that the Desarguesian character is for a two-dimensional space a much more complex notion than for a higher dimensional space.

The spheres are convex in any simply connected Riemann space with negative curvatures. Such spaces are, by Beltrami's Theorem, Desarguesian only when the curvature is constant. Therefore convexity of spheres is a comparatively weak condition, when the situation regarding parallels is as in hyperbolic space. We have just seen that it is extremely restrictive in spaces of the elliptic type, and expect therefore that *the boundary case, where parallels behave in the euclidean way, will prove very interesting.*

This question presupposes a *theory of parallels*. In view of the great fascination which parallels exerted for a long time it is most surprising that hardly any further work was done, after hyperbolic geometry had been discovered.

We develop a theory of parallels in Sections 22 and 23. First we deal with G-spaces. Every point of a non-compact G-space is origin of at least one ray, that is a set isometric to the non-negative real axis. If $\mathfrak{r}$ is any ray, q any point, then the limit $\mathfrak{s}$ of a convergent sequence of segments $T(q_\nu, z_\nu)$ with $q_\nu \to q$ and z_ν tending on $\mathfrak{r}$ to infinity, is called a *co-ray* from q to $\mathfrak{r}$. The co-ray from q to $\mathfrak{r}$ is in general not unique, but the co-ray to $\mathfrak{r}$ from any point $q_0 \neq q$ of $\mathfrak{s}$ is unique and is a subray of $\mathfrak{s}$. The proof of this remarkable fact is based on another characterization of co-rays. The spheres $K(z_\nu, p)$, $p \in \mathfrak{r}$, converge when z_ν tends on $\mathfrak{r}$ to infinity to a set $K_\infty(\mathfrak{r}, p)$, *the limit sphere through p with $\mathfrak{r}$ as central ray*. The intersection of a coray $\mathfrak{s}$ to $\mathfrak{r}$ with $K_\infty(\mathfrak{r}, p)$ is a foot of any point of $\mathfrak{s}$ on $K_\infty(\mathfrak{r}, p)$. Hence the co-rays to $\mathfrak{r}$ are the orthogonal trajectories of the limit sphere $K_\infty(\mathfrak{r}, p)$.

Applying this to straight spaces we find in Section 23 that if $\mathfrak{g}^+$ is an oriented straight line and q a point not on $\mathfrak{g}$, then as z tends on $\mathfrak{g}^+$ in the positive direction to infinity, the line $\mathfrak{g}^+(q, z)$, so oriented that z follows q converges to an oriented line $\mathfrak{a}^+$, which we call the *asymptote* through q to $\mathfrak{g}^+$. The asymptote to $\mathfrak{g}^+$ through any other point of $\mathfrak{a}$ coincides with $\mathfrak{a}^+$.

Strangely enough *the asymptote relation is in general neither symmetric nor transitive, even in two-dimensional straight spaces.* Hardly any of the statements which we might venture to make on the basis of our euclidean or hyperbolic experience is true in general. We saw, for instance, at the end of the last chapter that the distance from $\mathfrak{g}$ of a point x traversing $\mathfrak{a}^+$ in the positive direction may tend to infinity, even when $\mathfrak{g}^+$ is also an asymptote to $\mathfrak{a}^+$. A notable exception is that $\mathfrak{r}$ and $\mathfrak{s}$ must be co-rays to each other, when sequences $x_\nu \in \mathfrak{r}$ and $y_\nu \in \mathfrak{s}$ tending to infinity exist such that $x_\nu y_\nu \to 0$.

The absence of properties in general spaces leads to the task of ascertaining for special classes of spaces whether, and which of, their properties yield additional facts on parallels. There will be many instances later on where this is carried out.

Returning to our original problem we now formulate the *parallel axiom* in a straight space as follows: asymptotes have the symmetry property and, if $\mathfrak{a}^+$ is an asymptote to $\mathfrak{g}^+$, then the opposite orientation $\mathfrak{a}^-$ of $\mathfrak{a}^+$ is an asymptote to the opposite orientation $\mathfrak{g}^-$ of $\mathfrak{g}^+$. In the plane case this means exactly the same as the usual requirement: given a line $\mathfrak{g}$ and a point p not on $\mathfrak{g}$, there is exactly one line through p that does not intersect $\mathfrak{g}$.

We then prove in Section 24 that a straight space of dimension greater than two is *Minkowskian with differentiable spheres*, in the usual sense, *if it satisfies the parallel axiom and its spheres are convex and differentiable* in a sense which we define. The differentiability condition, as formulated here, is natural enough, but is a blemish, inasmuch as it is not known whether the theorem holds without it. We deduce this theorem from the fact that a higher dimensional straight space with flat limit spheres is Minkowskian with differentiable spheres. A third property characteristic for these spaces is that a set on which every point has a unique foot must be convex and closed.

We also obtain a *characterization of the higher dimensional euclidean geometry in terms of the parallel axiom and the property of symmetric perpendicularity.* Here it is not necessary to assume differentiability of spheres. We know from our discussion of plane Minkowskian geometry that these properties do not characterize the euclidean plane among two-dimensional straight spaces. Whether the characterizations of Minkowskian spaces which we gave, carry over to the plane is not known. We face here the same difficulty as in the elliptic case. As already mentioned, the convexity of the circles means that the functions $px(\tau)$ are peakless. The stronger condition, that for some fixed $\alpha \geqslant 1$ the functions $px(\tau)^{\alpha}$ are convex, implies in conjunction with the parallel axiom that the metric is Minkowskian. This is shown in Section 25. No differentiability hypothesis enters here, so that we have a characterization of all plane Minkowskian geometries.

20. Convexity of spheres and perpendicularity

Before discussing the implications of convexity of spheres we discuss briefly general convex sets in *G*-spaces.

Convex sets in *G*-spaces

The set M in a G-space R is *convex* if the segment $T(x, y)$ is unique for any two points x, y in the closure $\overline{M}$ of M and if $x, y \in M$ implies $T(x, y) \in M$. If M is convex, then $\overline{M}$ is convex because $x_\nu \to x$, $y_\nu \to y$, where $x_\nu, y_\nu \in M$ implies $T(x_\nu, y_\nu) \to T(x, y)$, since $T(x, y)$ is unique. The set M is *strictly convex* if $x, y \in \overline{M}$ implies that any point z with $(x\,z\,y)$ is an interior point of M, i. e., $S(z, \rho) \subset M$ for sufficiently small positive ρ. These definitions are the same as in affine space. We show that some of the elementary properties of convex sets in affine space extend to this general case:

(20.1) *Let M be convex. If x is an interior point of M and y is any point of $\overline{M}$ then every point z with $(x\, z\, y)$ is an interior point of $\overline{M}$.*

For there is a positive ρ such that $S(x, \rho) \in M$. Then $V = \cup_u T(y, u)$ where $u \in S(x, \rho)$, lies in M and $V - y$ is an open set. Hence no point of $V - y$ can lie on the boundary of M, so that $z \in V - y \subset M$.

A corollary of (20.1) is:

(20.2) The interior points of a convex set form a convex set. If x, y lie on the boundary of the convex set M, then either $T(x, y)$ lies on the boundary of M or every point z with $(x\, z\, y)$ is an interior point of M.

For if $T(x, y)$ does not lie entirely on the boundary of M, it contains an interior point z_0 of M. Then $T(x, y) = T(x, z_0) \cup T(z_0, y)$ and every point of $T(x, z_0) - x$ and $T(z_0, y) - y$ is by (20.1) an interior point of M.

In general spaces we cannot introduce supporting planes because planes will not exist. However, with some precautions supporting lines may be defined and proved to exist in sufficient numbers.

It follows from (8.5) that for any two points x, y in $S(p, \rho(p)/4)$ a segment $U(x, y)$ exists that has x as midpoint, passes through y and has length $\rho(p)$. If $\rho(p) = \infty$, then $U(x, y)$ means the entire geodesic through x and y.

If M is a convex set with interior points in $S(p, \rho(p)/4)$ and q a boundary point of M, then a *supporting line* of M at q is a line $U(q, x)$ through q that contains no interior point of M.

Denote by V_q the set of points formed by q and all points which lie on supporting lines of M at q. Let A_q consist of the points x in $S = S(p, \rho(p)/4)$ for which an interior point y of M with $(x\, y\, q)$ exists. Finally denote by B_q the set formed by the points x in S for which an interior point y of M with $(x\, q\, y)$ exists. It follows from (20.1) and (20.2) and the definition of supporting lines that the sets A_q, B_q are disjoint and

$$A_q \cup B_q \cup (V_q \cap S) = S.$$

If there is at least one supporting line of M at p, the point q need of course not be mentioned explicitly as element of V_q. There is certainly no supporting line, when the space is one-dimensional. That there are supporting lines in every other case, is contained in

(20.3) *V_q separates A_q from B_q in $S(p, \rho(p)/4)$.*

We may assume that the space has dimension greater than one. Let a curve $x(\tau)$, $\alpha \leqslant \tau \leqslant \beta$, in S be given with $x(\alpha) \in A_q$ and $x(\beta) \in B_q$ with $x(\tau) \neq q$ for all τ. It is to be shown that $x(\tau_0) \in V_q$ for at least one τ_0. Let τ_0 be the least upper bound of the τ for which an interior point y_τ of M with $(x(\tau)\ y_\tau\ q)$ exists. Then $\tau_0 > \alpha$. There can be no interior point y of M with $(x(\tau_0)\ y\ q)$ because this would imply the existence of y_τ with $(x(\tau)\ y_\tau\ q)$ for small $\tau - \tau_0 > 0$.

Also, no interior point y of M with $(y\ q\ x\ (\tau_0))$ can exist because then for small $\tau_0 - \tau > 0$ there would be a point $\bar{y}_\tau$ with $(\bar{y}_\tau\ q\ x(\tau))$. But by the definition of τ_0 there is a $\tau_1 < \tau_0$ arbitrarily close to τ_0 and an interior point y_{τ_1} of M with $(q\ y_{\tau_1}\ x(\tau_1))$. But then $(\bar{y}_{\tau_1}\ q\ y_{\tau_1})$ contradicts (20.1).

This proof yields a little more than was stated in the theorem. The point $x(\tau_0)$ is not an arbitrary point in S on an arbitrary supporting line of M at q, but would in euclidean space be a point of the supporting cone of M at q. For a later application we formulate this concept rigorously in the general case.

Denote as *ray* with origin $q \in S$ a segment with q as initial point and of length $\rho(p)/2$. If $\rho(p) = \infty$ we mean a half geodesic $x(\tau)$, $\tau \geqslant 0$ with $x(0) = q$. If the ray Q with origin q lies on a supporting line of M at q and is limit of rays (with origin q) that contain interior points of M, then we call R a *supporting ray* of M at q. Our proof contains:

(20.4) If $M \subset S(p, \rho(p)/4)$ is a convex set with interior points of an at least two-dimensional G-space, then the supporting rays of M at a given boundary point q form a set that separates A_q (which contains all interior points of M) from B_q.

CONVEXITY OF SPHERES AND UNIQUENESS OF FEET

We apply these concepts to spheres: For convenience we also call $K(p, \rho)$, $0 < \rho < \rho_1(p)$, (strictly) convex if $S(p, \rho)$ is (strictly) convex. Then $\bar{S}(p, \rho) = S(p, \rho) \cup K(p, \rho)$ is also (strictly) convex. We show first that convexity of all spheres with a fixed radius implies strict convexity of all spheres with a smaller radius.

(20.5) *Let* $\eta = \inf_{p \in F} \rho(p) > 0$, *and let* $K(p, \sigma_0)$ *be convex for* $p \in F$ *and a fixed positive* $\sigma_0 \leqslant \eta$. *Then each sphere* $K(q, \sigma)$ *with* $0 < \sigma < \sigma_0$ *and* $\bar{S}(q, \sigma_0 - \sigma) \subset F$ *is strictly convex.*

Proof. A not strictly convex $K(q, \sigma)$ would yield points x, y, z with $(x\ z\ y)$, $x q = y q \leqslant \sigma$ and $q z \geqslant \sigma$. Since $x, y \in S(q, \sigma_0)$ and $K(q, \sigma_0)$ is convex, $z q < \sigma_0$. Since $\sigma_0 \leqslant \eta \leqslant \rho(q)$ there is a point p with $(p\ q\ z)$ and $p z = \sigma_0$. Then $p q = p z - q z = \sigma_0 - \sigma$; hence $p \subset F$. Moreover $p x < p q + q x \leqslant p q + \sigma \leqslant \sigma_0$, similarly $p y < \sigma_0$. By hypothesis $S(p, \sigma_0)$ is convex, but $x, y \in S(p, \sigma_0)$ and z with $(x\ z\ y)$ does not lie in $S(p, \sigma_0)$.

We now turn to the connection of convexity of spheres with uniqueness of feet. The underlying idea is to make use of the fact that a point of contact q of a supporting line of $S(p, \rho)$ at a boundary point $q \in K(p, \rho)$ must be a foot of p on the supporting line. It will prove simpler however, not to use this approach explicitly. We begin with a simple, but very useful observation which holds without any further assumptions.

(20.6) *If f is a foot of p on the set M in a G-space and (pxf) then f is the unique foot of x on M.*

For if $y \in M$, then

$$xy \geqslant py - px \geqslant pf - px = xf,$$

which shows that f is a foot of x and that $xy = xf$ only for (pxy), but then $y = f$ by (8.1).

Next we show:

(20.7) *If the spheres $K(p, \sigma)$ are strictly convex for $0 < \sigma \leqslant \sigma_0 < \rho_1(p)$ then p has exactly one foot on any segment that contains points of $S(p, \sigma_0)$.*

Since segments are compact sets, p has at least one foot on any segment. Assume p has two different feet f_1, f_2 on a segment T which contains points of $S(p, \sigma_0)$. Then $p f_1 = p f_2 = \sigma < \sigma_0$ and $p x \geqslant p f_i$ for $(f_1 x f_2)$. This contradicts the strict convexity of $K(p, \sigma)$.

The following statement is a sort of converse of (20.7):

(20.8) *Let the subset F of R contain with x the sphere $S(x, \sigma(x))$, where $0 < \sigma(x) < \rho_1(x)$ and $|\sigma(x) - \sigma(y)| \leqslant xy$. If every point x of F has exactly one foot on any segment that contains points of $S(x, \sigma(x))$, then $K(x, \sigma)$ is strictly convex for $x \in F$ and $0 < \sigma < \sigma(x)$.*

If this were not true then a point q in F and points a, b, c with (abc), $0 < qa \leqslant \sigma < \sigma(q)$, $0 < qc \leqslant \sigma < \sigma(q)$, and $qb \geqslant \sigma$ would exist. Since the foot f_q of q on $T(a, c)$ is unique, it cannot be the point b. Assume f_q lies on $T(a, b)$ and let (quc). Then $u \in S(q, \sigma(q)) \subset F$ and by hypothesis $uc = qc - qu < \sigma(q) - qu \leqslant \sigma(u)$. Therefore u has exactly one foot on $T(a, c)$, which varies by (2.23) continuously with u, so that there is a u_0 with (qu_0c) whose foot on $T(a, c)$ is b. But

$$u_0 f_{u_0} = u_0 b \geqslant qb - qu_0 \geqslant qc - qu_0 = u_0 c$$

which contradicts the assumption that the foot of u_0 on $T(a, c)$ is unique.

The implications of these facts for straight spaces are important.

(20.9) ***Theorem:*** *In a straight space the following statements are equivalent:*

(a) *The spheres are convex.*

(b) *The spheres are strictly convex.*

(c) *A given point has exactly one foot on a given segment.*

(d) *A given point has exactly one foot on a given straight line.*

(e) *A straight line and a sphere intersect in at most two points.*

(f) *If $x(\tau)$ represents any straight line and p is any point, then $px(\tau)$ is strictly peakless.*[1]

(g) *The function $px(\tau)$ under (f) is peakless.*

Proof. We show $(a) \to (b) \to (c) \to (d) \to (e) \to (f) \to (g) \to (a)$. The implications $(a) \to (b) \to (c)$ are contained in (20.5) and (20.7). Two distinct feet f_1, f_2 of p on a straight line are also feet of p on $T(f_1, f_2)$, hence (d) follows from (c).

If (d) holds, then a sphere $K(q, \sigma)$ cannot contain three points a, b, c on a line G. For if (abc) then the foot f_q of q on G is different from a, b, c and we obtain the same contradiction as in the proof of (20.8).

Let (e) hold. With the notation of (f) the function $px(\tau)$ attains its minimum λ_0 for some τ_0. Since $px(\tau) \to \infty$ for $|\tau| \to \infty$, each value $\lambda > \lambda_0$ is taken at least once for $\tau > \tau_0$ and at least once for $\tau < \tau_0$. Since $K(p, \lambda)$ intersects $x(\tau)$ at most twice, the value λ is attained exactly once on either side of τ_0. The value λ_0 cannot be attained a second time at $\tau_1 > \tau_0$ say. For if $\tau_0 < \tau' < \tau_1$ then $\lambda' = px(\tau') > \lambda_0$, since $K(p, \lambda_0)$ intersects $x(\tau)$ at most twice. But then $px(\tau'') = px(\tau')$ for a suitable $\tau'' > \tau_1$ and λ' would be attained thrice. Therefore $px(\tau)$ reaches its minimum at τ_0, decreases for $\tau \leqslant \tau_0$, increases for $\tau \geqslant \tau_0$ and is therefore strictly peakless. Obviously $(f) \to (g)$.

Finally, (g) implies (a). For if $K(q, \sigma)$ is not convex, then a, b, c with (abc), $0 < aq \leqslant \sigma$, $0 < cq \leqslant \sigma$ and $bq > \sigma$ exist. If $x(\tau)$ represents the straight line through a and b then $qx(\tau)$ is not peakless.

Perpendiculars

According to the definition in Section 17 the straight line G is perpendicular to the straight line H at f if $f = G \cap H$ and every point x of G has f as foot on H. It follows from (20.6) that f is then the unique foot of x and that perpendiculars to H at different points of H cannot intersect. It is remarkable that this also suffices for the existence of perpendiculars to G.

(20.10) *If every point of the straight space R of dimension at least two, has exactly one foot on the straight line G, then there is exactly one perpendicular to G through a given point p not on G.*

The set formed by the perpendiculars to G at a given point f of G decomposes the space into two arcwise connected sets. If R is two-dimensional, then there is exactly one perpendicular to G through a given point f of G.

Proof. Let f be the foot of p on G and (pfy). To establish the first part it suffices to show that f is the foot of y on G. For by applying this result to points x with (yfx) we find that every point of $\mathfrak{g}(p, f)$ has f as foot on G.

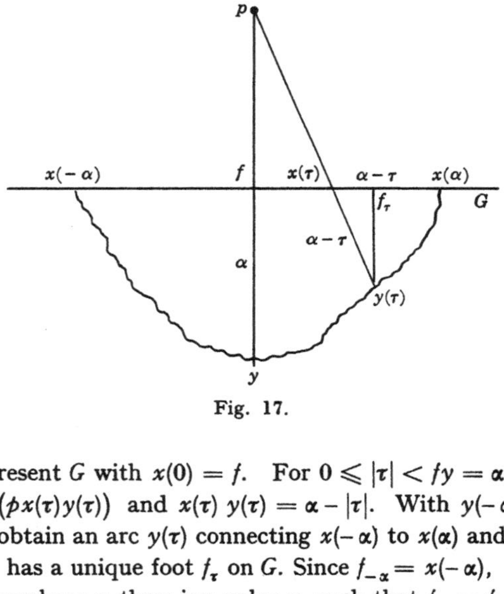

Fig. 17.

Let $x(\tau)$ represent G with $x(0) = f$. For $0 \leqslant |\tau| < fy = \alpha$ denote by $y(\tau)$ the point with $(px(\tau)y(\tau))$ and $x(\tau)\,y(\tau) = \alpha - |\tau|$. With $y(-\alpha) = x(-\alpha)$ and $y(\alpha) = x(\alpha)$ we obtain an arc $y(\tau)$ connecting $x(-\alpha)$ to $x(\alpha)$ and with $y(0) = y$. Each point $y(\tau)$ has a unique foot f_τ on G. Since $f_{-\alpha} = x(-\alpha)$, $f_\alpha = x(\alpha)$ and f depends continuously on τ, there is a value τ_0 such that $f_{\tau_0} = f$. We show that $\tau_0 = 0$. If $\tau \neq 0$ then

$$y(\tau)f_\tau + x(\tau)\,p \leqslant y(\tau)\,x(\tau) + x(\tau)\,p = y(\tau)\,p < y(\tau)\,f + fp \leqslant y(\tau)\,f + x(\tau)\,p$$

hence $y(\tau)f_\tau < y(\tau)\,f$, and $f_\tau \neq f$.

Let W be the set formed by all perpendiculars to G at the given point f of G. W separates the points whose feet on G have $\tau < 0$ from those whose feet have $\tau > 0$. For, any curve connecting a point of the first type to a point of the second must, by a now familiar argument, contain a point whose foot is f, and hence lies on W. If p and q are points whose feet are $x(\tau_1)$ and $x(\tau_2)$

with $\tau_i > 0$, then $T(p, x(\tau_1)) \cup T(x(\tau_1), x(\tau_2)) \cup T(x(\tau_2), q)$ is an arc connecting p and q and not intersecting W.

Finally, if R is two-dimensional, then at least one perpendicular to G at f must exist because W decomposes the plane. If there were two different perpendiculars H_1 and H_2 to G at f, then the segment $T(p_1, p_2)$ connecting points $p_i \in H_i$ on different sides of G would intersect G in a point $s \neq f$, and

$$p_1p_2 = p_1s + sp_2 < p_1f + fp_2$$

but $p_1f < p_1s$ and $p_2f < p_2s$.

If H is perpendicular to G at f, we call — as already mentioned — G *transversal to* H *at* f.

(20.11) If the spheres of a straight space are convex, then the transversals of a given line H at a given point f coincide with the supporting lines of $K(p, f)$ at f, where p is any point of H different from f.

Here — as in A^n — a supporting line $K(p, f)$ at f means a supporting line of $S(p, pf)$ or $\bar{S}(p, pf)$ at f.

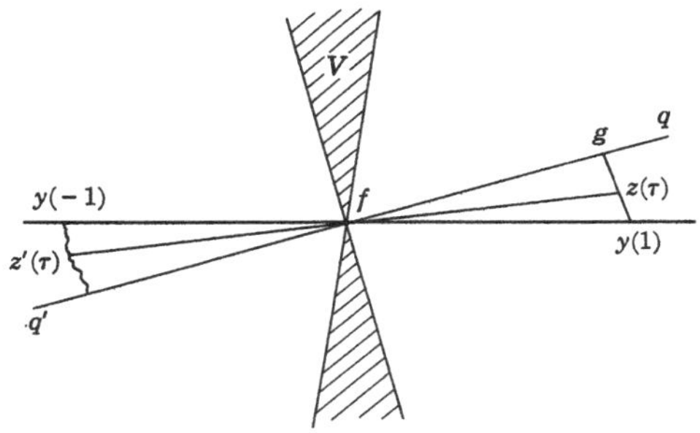

Fig. 18.

For if G is a supporting line of $K(p, f)$ at f, then f is the foot of p on G, hence H is by (20.10) and (20.9) perpendicular to G, or G is transversal to H.

Conversely, if G is transversal to H, then f is the foot of p on G, hence G contains no points of $S(p, pf)$, which is in the interior of $K(p, f)$.

(20.12) *The transversals to a given line H at a given point f in a straight space with convex spheres form a set V that decomposes the space into two arcwise connected sets.*

(20.3) and (20.11) show that V decomposes the space and, if $y(\tau)$ represents H with $y(0) = f$, then V separates, by the proof of (20.3), the points $y(\tau)$

with $\tau < 0$ from the points $y(\tau)$ with $\tau > 0$. We show that every point q not on V can be connected either to $y = y(1)$ or to $y(-1)$ by an arc which does not meet V.

The foot g of y on the geodesic $\mathfrak{g}(f, q)$ is different from f because q does not lie in V. If $z(\tau)$, $0 \leqslant \tau \leqslant gy$, represents $T(g, y)$, with $z(0) = g$, then $yz(\tau) \leqslant y(1)\, g < yf$, hence $z(\tau)$ does not lie on V. If now q lies on the same side of f on $\mathfrak{g}(f, q)$ as g, then $T(q, g) \cup T(g, y)$ is an arc connecting q to y which does not meet V.

If $(q'fg)$, let $z'(\tau)$ be the point for which f is the midpoint of $z(\tau)$ and $z'(\tau)$. We know already that $\mathfrak{g}(f, z(\tau))$ is not transversal to H, hence $z'(\tau)$ does not lie in V and $z'(\tau)$ is an arc connecting $z'(0)$ to $y(-1)$ that does not meet V. Then $T(q', z'(0))$ together with this arc connects q' to $y(-1)$ without meeting V.

21. Characterization of the higher-dimensional elliptic geometry.

Convexity of spheres was defined only for sets where the segment connecting two points is unique and does therefore not apply to an arbitrary sphere in a space of the elliptic type. However, since the geodesics behave in many ways like projective lines, we may expect that condition (*e*) in (20.9) will lead to a reasonable analogue to convexity of large spheres. In an elliptic space whose geodesics have length 2λ, the locus $K(p, \lambda)$ is a hyperplane and hence does not satisfy (*e*), but spheres $K(p, \sigma)$ with $\sigma < \lambda$ will satisfy (*e*). Imposing this as a condition on a space of the elliptic type has amazingly strong consequences:

(21.1) ***Theorem:*** *Let R be a space of the elliptic type of dimension greater than two whose geodesics have length 2λ. If a geodesic always intersects a sphere of radius $\sigma < \lambda$ in at most two points, then R is an elliptic space.*

It should be noted, that according to a later theorem, (31.2), which has already been mentioned several times, all geodesics in any space of the elliptic type have the same length, so that the above formulation does not imply an additional hypothesis.

For brevity put $K(p, \lambda) = K(p)$. The set $K(p)$ consists of the points conjugate to p on the different great circles through p. The relation $q \in K(p)$ means that $qp = \lambda$ and hence implies $p \in K(q)$. We prove first that the loci $K(p)$ are flat, which means in the present case that $K(p)$ contains with any two distinct points x, y the entire goedesic $\mathfrak{g}(x, y)$.

For let D_1 and D_2 be the two arcs of $\mathfrak{g}(x, y)$ with endpoints x and y. At least one of these arcs must lie on $K(p)$, because $pD_1 < \lambda$ and $pD_2 < \lambda$ would imply that a sphere $K(p, \sigma)$ with $\max pD_i < \sigma < \lambda$ intersects $\mathfrak{g}(x, y)$ at least four times.

Assume $D_1 \subset K(p)$ and $pD_2 < \lambda$. Let z be an interior point of D_1 and $(zp'p)$. Then $p'z < p'x$ and $p'z < p'y$ because $\lambda = pp' + p'z = pz = px < pp' + p'x$. If p' is sufficiently close to p, we would also have $p'D_2 < p'x$ and $p'D_2 < p'y$. But then a sphere $K(p', \sigma)$ with $\max (p'z, p'D_2) < \sigma < \min (p'x, p'y)$ would intersect $\mathfrak{g}(x, y)$ at least four times. Our theorem will therefore be contained in the following:

(21.2) ***Theorem:*** *A space of the elliptic type and dimension greater than two is elliptic, if the locus of the points conjugate to a given point is flat.*

In the proof we define a K-space to be any space of the elliptic type in which the loci $K(p)$ are flat. Any flat subset F of a K-space is — with the same metric — a K-space. For if $p \in F$, then $K(p) \cap F$ is flat and the locus of conjugate points to p in F. We prove first:

(21.3) A K-space R' is homeomorphic to P^m with a suitable m. The loci of conjugate points are homeomorphic to P^{m-1}.

For any point p_1 in R' construct the locus $K_1 = K'(p_1)$ of the conjugate points to p_1. If K_1 consists of one point p_2, then R' is a great circle and we are finished. Otherwise, let p_2 be any point in K_1. Then $K_2 = K'(p_2) \cap K_1$ is the conjugate locus to p_2 in K_1. If this is one point p_3 we stop. Otherwise let p_3 be any point in K_2 and construct $K_3 = K'(p_3) \cap K_2$. Proceeding in this way we obtain points $p_1, p_2, \ldots$ in R' each of which is conjugate to all others. Hence if $i \neq k$, $p_i p_k > \varkappa > 0$ ($p_i p_k = \lambda$ if we assume that all great circles have the same length, which is not necessary for the proof of (21.2)). Since the space is compact there is only a finite number m of p_i. Then K_{m-1} is a great circle, K_{m-2} is homeomorphic to P^2, K_{m-3} is homeomorphic to P^3 as union of the great circles through p_{m-2} and the points of K_{m-3}. Proceeding in the same way we see that $K_1 = K'(p_1)$ is homeomorphic to P^{m-1} and R' homeomorphic to P^m.

Next we observe:

(21.4) If p and q are different points in R and $r \in K(p) \cap K(q)$, $s \in \mathfrak{g}(p, q)$, then r and s are conjugate. For any $\bar{p} \neq q$ on $\mathfrak{g}(p, q)$ we have $K(\bar{p}) \cap K(q) = K(p) \cap K(q)$.

For $r \in K(p) \cap K(q)$ implies $p \in K(r)$, $q \in K(r)$, hence $s \in \mathfrak{g}(p, q) \subset K(r)$. Applying this result to a point $t \in K(\bar{p}) \cap K(q)$ yields because of $\mathfrak{g}(\bar{p}, q) = \mathfrak{g}(p, q)$ that $t \in K(p) \cap K(q)$, hence $K(\bar{p}) \cap K(q) \subset K(p) \cap K(q)$. In the same way we find $K(p) \cap K(q) \subset K(\bar{p}) \cap K(q)$.

It is now easy to show that:

(21.5) *A K-space R of dimension greater than two is Desarguesian.*

R is by (21.3) homeomorphic to a P^n with $n > 2$. Because of (14.5) and (21.3) it suffices to show that for any given n points $p_1, \ldots, p_n$ a point q exists such that $p_i \in K(q)$ for $i = 1, \ldots, n$.

Put $K_1 = K(p_1)$. By (21.3) $\dim K_1 = n - 1$. If $p_2 = p_1$ put $K_2 = K_1$. Otherwise let p_2' be the conjugate point to p_1 on $\mathfrak{g}(p_1 p_2)$. Then (21.4) yields $K_2 = K(p_1) \cap K(p_2) = K(p_1) \cap K(p_2')$. Since $p_2' \in K(p_1)$ the set K_2 is the conjugate locus to p_2' in K_1, hence has by (21.3) dimension $n - 2$. In either case $n_2 = \dim K_2 \geqslant n - 2$.

If $K(p_3)$ contains K_2 put $K_3 = K_2$. Otherwise the same method yields a point p_3' in K_2 such that $K_3 = K_2 \cap K(p_3') = K_2 \cap K(p_3)$, so that K_3 is by (21.3) homeomorphic to $P^{n_2 - 1}$.

In either case $K_3 = K(p_1) \cap K(p_2) \cap K(p_3)$ and $n_3 = \dim K_3 \geqslant n - 3$. Continuing in this manner we find

$$\dim K(p_1) \cap K(p_2) \cap \ldots \cap K(p_n) \geqslant 0.$$

Therefore this intersection contains at least one point q. But $q \in K(p_i)$ implies $p_i \in K(q)$ which proves the assertion.

We now identify our space with P^n and use projective and affine terminology. We show next:

(21.6) *In the affine space with $K(p)$ as hyperplane at infinity a sphere $K(p, \sigma)$, $0 < \sigma < \lambda$, is an ellipsoid whose affine center is p.*

Proof. Let $x \in K(p, \sigma)$, and x' be the conjugate point to x on $\mathfrak{g}(p, x)$. Since every point of $K(x')$ has distance λ from x', the point x is, in a trivial way, a foot of x' on $K(x')$. Because of (20.6) the point x is then the only foot of p on $K(x')$. Hence $K(x')$ is, in affine language, a supporting plane of $K(p, \sigma)$ at x. Therefore $K(p, \sigma)$ is a convex surface.

Let $z \in K(p)$. Then $K(z)$ contains p and intersects $K(p, \sigma)$ in a hyper plane set Z. If $x \in Z$ and x' is defined as before, then $\mathfrak{g}(z, x) \subset K(x')$ because $x \in K(x')$ and the relations $p \in K(z)$, $x \in K(z)$ imply $x' \in \mathfrak{g}(p, x) \subset K(z)$ so that also $z \in K(x')$. Therefore $\mathfrak{g}(z, x)$ is a supporting line of $K(p, \sigma)$. Since $z \in K(p)$ the lines $\mathfrak{g}(z, x)$, $x \in Z$ are all parallel. The supporting lines of $K(p, \sigma)$ parallel to $\mathfrak{g}(z, x)$ therefore touch $K(p, \sigma)$ in the plane set Z. Blaschke's Theorem (16.17) shows that $K(p, \sigma)$ is an ellipsoid. The affine center of this ellipsoid is p because the plane $K(z)$ containing Z passes through p.

It should be noticed that if we had any differentiability hypotheses like those for Finsler spaces in Section 15, the theorem would now be an immediate consequence of Beltrami's Theorem. For the finite spheres $K(p, \sigma)$ being ellipsoids (actually, homothetic ellipsoids) implies that an "infinitesimal sphere" about p, is an ellipsoid, so that the metric is Riemannian.

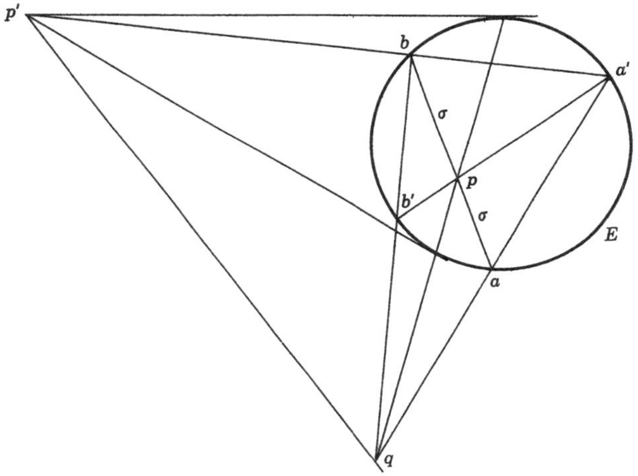

Fig. 19.

In absence of differentiability assumptions we proceed as follows: For an arbitrary point q let $p \in K(q)$ and let a, b be the intersection of $K(p, \sigma)$ with a line through p that neither passes through q nor lies in $K(q)$. The two-dimensional plane P through q, a, b intersects $K(q)$ in a line. If p' is the conjugate point of p on this line, then the intersection E of $K(p, \sigma)$ with P is by (21.6) an ellipse with affine center p in P with $\mathfrak{g}(q, p')$ as line at infinity. The proof of (21.6) contains the fact that the line connecting the points of contact of the tangents to E from p' passes through p and q. Therefore p, p', q is a self polar triangle with respect to E. Hence $a' = \mathfrak{g}(q, a) \cap \mathfrak{g}(p', b)$ and $b' = \mathfrak{g}(q, b) \cap \mathfrak{g}(p', a)$ are points of E collinear with p. It follows that $pa' = pa$, $ab = a'b'$.

If $\mathfrak{g}(a, a')$ intersects $\mathfrak{g}(p, p')$ or $K(q)$ in a_0, then q, a_0, a, a' is a harmonic quadruple. The mapping Φ of P^n on itself, which maps q and every point of $K(q)$ on itself and maps any point $a \neq q$ and not on $K(q)$ on the point a' on $\mathfrak{g}(q, a)$ for which the points $q, a_0 = \mathfrak{g}(q, a) \cap K(q)$, a, a' are harmonic, is a projectivity (more particularly in the terminology of projective geometry, a harmonic homology)[1]. As a projectivity it maps $\mathfrak{g}(a, p)$ on $\mathfrak{g}(p, a')$. The relations $ap = a'p$ and $ab = a'b'$ show that Φ preserves distance on $\mathfrak{g}(a, p)$. Since p is arbitrary in $K(q)$ the mapping Φ preserves distances in general, hence is a motion.

This shows that R possesses a reflection in the arbitrary hyperplane $K(q)$. It is well known, and will be proved later, that R must then be elliptic. This establishes (21.2) and (21.1).

The two main ideas of this proof were (1) using the flatness of $K(p)$ to establish the Desarguesian character of the space, (2) using Blaschke's Theorem to show that the spheres are ellipsoids. Neither of these methods works in two dimensions, and the problem of whether Theorem (20.1) holds for two-dimensional spaces is open. One would conjecture that the theorem does not hold without assuming the Theorem of Desargues, but will hold if this theorem is assumed.

The hypothesis that a sphere $K(p, \sigma)$ does not intersect a geodesic more than twice, implies that the space is Desarguesian and the metric elliptic. The Desarguesian character alone does not imply that the metric is elliptic. G. Hamel [1] has shown that compact Desarguesian spaces other than the elliptic spaces exist.

Spherelike spaces

A not necessarily Desarguesian metrization of P^2 as a G-space of the elliptic type leads by means of the standard 2 to 1 mapping of the sphere S^2 on P^2 to a metrization of S^2 (compare Section 28) in which all geodesics are great circles in the metric sense, and any geodesic passing through a point $\overline{a}$ of S^2 also passes through the antipodal point $\overline{a}'$ to $\overline{a}$. It is a wellknown conjecture of Blaschke that the only Riemannian metric on S^2 with this property is the spherical. The examples of Hamel just mentioned show that there are non-Riemannian metrics on S^2 in which the ordinary great circles on S^2 are the geodesics. If Blaschke's conjecture is correct, the proof must therefore use emphatically the Riemannian character of the metric, and not so much the configurational properties of the geodesics.

This connection with metrics on the sphere leads naturally to the question, whether our theorems (21.1) and (21.2) cannot be used to characterize the higher dimensional spherical metrics. To achieve this we define.

A G-space R is "*spherelike*" if all geodesics are great circles and each geodesic which passes through a given point $\bar{a}$ also passes through a second point $\bar{a}'$, the "*antipodal*" point to $\bar{a}$.

There cannot be more than one antipodal point to $\bar{a}$, because two distinct great circles cannot intersect more than twice. Since two geodesics through $\bar{a}$ intersect a second time, they have the same length, and $\bar{a}$, $\bar{a}'$ are conjugate to each other on both geodesics, see (9.12). Hence $\bar{a}'$ is also antipodal to $\bar{a}$, and the geodesic through two not antipodal points is unique. *Any two geodesics have the same length*, because there is a geodesic intersecting both.

(21.7) *The mapping of a spherelike space $\bar{R}$ on itself which mates antipodal points is a motion.*

We must show that $\bar{a}\,\bar{b} = \bar{a}'\,\bar{b}'$ for any two points $\bar{a}$, $\bar{b}$ in $\bar{R}$ and their antipodal points $\bar{a}'$, $\bar{b}'$. This is obvious for $\bar{a} = \bar{b}$ or $\bar{a} = \bar{b}'$. If neither is the case, then $\bar{a}$ and $\bar{b}$ are not antipodal, hence the geodesic $\bar{\mathfrak{g}}$ through $\bar{a}$ and $\bar{b}$ is unique and contains both $\bar{a}'$ and $\bar{b}'$. But $\bar{\mathfrak{g}}$ is a great circle, that is, congruent to an ordinary circle, therefore $\bar{a}\,\bar{b} = \bar{a}'\,\bar{b}'$.

We can now establish the connection with spaces of the elliptic type which leads to a characterization of spherical spaces:

(21.8) *The non-ordered pairs $a = (\bar{a}, \bar{a}')$ of antipodal points of a spherelike space $\bar{R}$ form with the metric*

$$a\,b = \min(\bar{a}\,\bar{b}, \bar{a}\,\bar{b}'), \tag{21.9}$$

a space R of the elliptic type in which the great circles have length $\lambda = \bar{a}\,\bar{b}'$.

The mapping Φ of $\bar{R}$ on itself which mates antipodal points has period two and forms therefore together with the identity a discrete group of motions of $\bar{R}$. It follows from a general theorem on covering spaces that R is with the metric (21.9) a G-space, compare (29.1).

In the present case a direct proof of this fact is simple: because of (21.7)

$$a\,b = \min(\bar{a}\,\bar{b}, \bar{a}\,\bar{b}') = \min(\bar{b}\,\bar{a}, \bar{b}\,\bar{a}') = b\,a.$$

Obviously $a\,b = 0$ if and only if $a = b$. The triangle inequality holds because (21.7) yields the inequalities

$$\bar{a}\,\bar{b} + \bar{b}\,\bar{c} \geqslant \bar{a}\,\bar{c}, \quad \bar{a}\,\bar{b}' + \bar{b}\,\bar{c} = \bar{a}'\,\bar{b} + \bar{b}\,\bar{c} \geqslant \bar{a}'\,\bar{c} = \bar{a}\,\bar{c}'$$

$$\bar{a}\,\bar{b} + \bar{b}\,\bar{c}' \geqslant \bar{a}\,\bar{c}', \quad \bar{a}\,\bar{b}' + \bar{b}\,\bar{c}' = \bar{a}'\,\bar{b} + \bar{b}\,\bar{c}' \geqslant \bar{a}'\,\bar{c}' = \bar{a}\,\bar{c},$$

which also show that $(a\,b\,c)$ only if $\bar{a}$, $\bar{b}$, $\bar{c}$ lie on a great circle in $\bar{R}$.

R is obviously finitely compact. The definition (21.9) applied to the points on one great circle $\bar{\mathfrak{g}}$ in $\bar{R}$ amounts to identifying antipodal or conjugate points on $\bar{\mathfrak{g}}$, and therefore yields in R a great circle $\mathfrak{g}$ with half the length of $\bar{\mathfrak{g}}$, that is λ. It is now clear that R is a G-space in which all geodesics are great circles of length λ. The great circle through two distinct points a, b of R is unique, because if $a = (\bar{a}, \bar{a}')$ and $b = (\bar{b}, \bar{b}')$ then $\bar{a}$ and $\bar{b}$ are not antipodal and the great circle through $\bar{a}$ and $\bar{b}$ is unique.

In $\bar{R}$ the spheres $K(\bar{p}, \sigma)$, $0 < \sigma < \lambda/2$, and $K(\bar{p}', \lambda - \sigma)$ coincide. $K(\bar{p}, \sigma)$ and $K(\bar{p}', \sigma)$ correspond to the sphere $K(p, \sigma)$ in R. Therefore, if the geodesic $\bar{\mathfrak{g}}$ in $\bar{R}$ intersects the sphere $K(\bar{p}, \sigma)$, $0 < \sigma < \lambda/2$, at most twice, the same holds for the corresponding geodesic $\mathfrak{g}$ in R and the sphere $K(p, \sigma)$.

If this is true for any geodesic $\bar{\mathfrak{g}}$ in $\bar{R}$ and any sphere $K(\bar{p}, \sigma)$, $0 < \sigma < \lambda/2$, and $\bar{R}$ is at least three-dimensional, then the metric in R is by (20.1) elliptic. We conclude that the metric in $\bar{R}$ is spherical. Thus we have the theorems:

(21.10) *Let R be a spherelike space of dimension greater than two, whose geodesics have length 4λ. If a geodesic intersects a sphere of radius $\sigma < \lambda$ always in at most two points, then R is a spherical space.*

(21.11) *An at least three-dimensional spherelike space whose geodesics have length 4λ is spherical if its spheres of radius λ are flat.*

22. Limit spheres and co-rays in G-spaces.

For the reason outlined in the introduction to this chapter we now turn to the theory of parallels for rays and straight lines, which we develop first for general non-compact G-spaces and then apply to straight spaces.

A *"ray"* with origin p is a half geodesic $x(\tau)$ $\tau \geqslant 0$ with $x(0) = p$ such that $x(\tau_1)\, x(\tau_2) = |\tau_1 - \tau_2|$ for $\tau_i \geqslant 0$. In other words, a ray is a set isometric to the ray $\tau \geqslant 0$ of the real axis. In a compact space the distances are bounded and rays cannot exist. However:

(22.1) *In a non-compact G-space every point is origin of at least one ray.*

For if q is a given point, then points q_ν with $q\, q_\nu \to \infty$ exist. Let $x_\nu(\tau)$ represent a geodesic for which the subarc $0 \leqslant \tau \leqslant q\, q_\nu$ of $x_\nu(\tau)$ represents a segment $T(q, q_\nu)$. Because of (8.14) $x_\mu(\tau)$ converges for a suitable subsequence $\{\mu\}$ of $\{\nu\}$ to a geodesic $x(\tau)$. Then $x(0) = q$, since $x_\mu(0) = q$. Moreover, if $\tau_1 \leqslant \tau_2$ are any non-negative numbers, then $q\, q_\mu > \tau_2$ from a certain subscript on. Then $x_\mu(\tau_1) x_\mu(\tau_2) = \tau_2 - \tau_1$ because $x_\mu(\tau)$ is a segment for $0 \leqslant \tau \leqslant qq_\mu$. It follows that $x(\tau_1)x(\tau_2) = \tau_2 - \tau_1$, hence $x(\tau)$, $\tau \geqslant 0$ represents a ray.

THE FUNCTION $\alpha(\mathfrak{r}, p)$

The theory of parallels is dominated by a simple function which we now define. If $z(\tau)$, $\tau \geqslant 0$, represents a ray $\mathfrak{r}$ with origin q and $0 \leqslant \tau_1 < \tau_2 < \tau_3$ then for any point p

$$(22.2)\quad z(\tau_1)\,p \geqslant \tau_3 - \tau_1 - pz(\tau_3) \left\} \begin{array}{ll} > \tau_2 - \tau_1 - pz(\tau_2) & \text{unless} \\ = \tau_2 - \tau_1 - pz(\tau_2) & \text{if} \end{array} \right\{ \begin{array}{l} (pz(\tau_2)\,z(\tau_3)) \\ \text{or } p = z(\tau_2). \end{array}$$

For

$$z(\tau_1)\,p \geqslant z(\tau_1)\,z(\tau_3) - pz(\tau_3) = \tau_2 - \tau_1 + z(\tau_2)\,z(\tau_3) - pz(\tau_3) \geqslant \tau_2 - \tau_1 - pz(\tau_2),$$

because $pz(\tau_2) \geqslant pz(\tau_3) - z(\tau_2)\,z(\tau_3)$. The equality sign holds only for $(pz(\tau_2)\,z(\tau_3))$ or $p = z(\tau_2)$.

Applying (22.2) with $\tau_1 = 0$ it is seen that $\tau - pz(\tau)$ is a bounded non-decreasing function of τ, therefore

$$(22.3)\qquad \alpha(\mathfrak{r}, p) = \lim_{\tau \to \infty} \big(p\,z(\tau) - \tau\big)$$

exists and is finite. Obviously

$$(22.4)\qquad \alpha(\mathfrak{r}, z(\tau_0)) = -\tau_0 \quad \text{for any} \quad \tau_0 \geqslant 0.$$

$$(22.5)\qquad \alpha(\mathfrak{r}, x) - \alpha(\mathfrak{r}, y) = \lim_{\tau \to \infty} \big(xz(\tau) - yz(\tau)\big) \leqslant xy;$$

hence

$$|\alpha(\mathfrak{r}, x) - \alpha(\mathfrak{r}, y)| \leqslant xy,$$

so that $\alpha(\mathfrak{r}, x)$ is a continuous function of x.

(22.6) If $\mathfrak{r}_0$ is the subray $z_0(\tau) = z(\tau + \tau_0)$, $\tau \geqslant 0$, $\tau_0 > 0$ of $\mathfrak{r}$ then

$$\alpha(\mathfrak{r}_0, p) = \alpha(\mathfrak{r}, p) + \tau_0.$$

This follows from $\alpha(\mathfrak{r}_0, p) = \lim \big(pz_0(\tau) - \tau\big) = \lim \,[pz(\tau) - (\tau - \tau_0)]$.

It is not possible to make statements about the behaviour of $\alpha(\mathfrak{r}, p)$ under general changes of $\mathfrak{r}$. There is however an important relation between the functions $\alpha(\mathfrak{r}, p)$ and $\alpha(\mathfrak{r}_1, p)$ and the *distance* $\delta_\infty(\mathfrak{r}, \mathfrak{r}_1)$ *of* $\mathfrak{r}$ *and* $\mathfrak{r}_1$ *at infinity*, which is defined as follows: if $r(\tau)$ and $r_1(\tau)$ represent $\mathfrak{r}$ and $\mathfrak{r}_1$ then

$$\delta_\infty(\mathfrak{r}, \mathfrak{r}_1) = \liminf_{\tau \to \infty,\ \tau_1 \to \infty} r(\tau)\, r_1(\tau_1).$$

For any two rays $\mathfrak{r}$, $\mathfrak{r}_1$ and any two points p, q

$$(22.7)\qquad |\alpha(\mathfrak{r}, p) + \alpha(\mathfrak{r}_1, q) - \alpha(\mathfrak{r}, q) - \alpha(\mathfrak{r}_1, p)| \leqslant 2\,\delta_\infty(\mathfrak{r}_2, \mathfrak{r}_1).$$

For by the definition of $\delta_\infty(\mathfrak{r}, \mathfrak{r}_1)$ there are sequences $\tau^\nu \to \infty$ and $\tau_1^\nu \to \infty$ such that $r(\tau^\nu)\, r(\tau_1^\nu) \to \delta_\infty(\mathfrak{r}, \mathfrak{r}_1)$, and

$$|pr(\tau^\nu) - pr_1(\tau_1^\nu) - [qr(\tau^\nu) - qr_1(\tau_1^\nu)]| \leqslant 2\, r(\tau^\nu)\, r_1(\tau_1^\nu).$$

The limit of the left side of this inequality equals by (22.5) the left side of (22.7). As a corollary of (22.7) we find

(22.8) *If* $\delta_\infty(\mathfrak{r}, \mathfrak{r}_1) = 0$, *then* $\alpha(\mathfrak{r}, x) - \alpha(\mathfrak{r}_1, x)$ *is constant.*

Limit spheres

The loci $\alpha(\mathfrak{r}, x) =$ constant are called the *"limit spheres" with central ray* $\mathfrak{r}$.

Since $\alpha(\mathfrak{r}, x)$ is defined for every x and single valued, there is exactly one limit sphere through a given point p with a given central ray $\mathfrak{r}$; we denote it by $K_\infty(\mathfrak{r}, p)$. Obviously

(22.9) $\alpha(\mathfrak{r}, x) = \alpha(\mathfrak{r}, p)$ *is the equation of* $K_\infty(\mathfrak{r}, p)$.

Since $\alpha(\mathfrak{r}, x)$ is a continuous function of x, limit spheres are closed sets. We conclude from (22.6) that

(22.10) $K_\infty(\mathfrak{r}, p) = K_\infty(\mathfrak{r}_0, p)$ if $\mathfrak{r}_0$ is a subray of $\mathfrak{r}$.

The name and notation for limit sphere derive from the fact that they are in straight spaces, as will be seen presently, always limits of spheres. In general we can only prove:

(22.11) *If* $z(\tau)$ *represents the ray* $\mathfrak{r}$, *and* $\tau_\nu \to \infty$, $p_\nu \to p$, *then*

$$\limsup K(z(\tau_\nu), p_\nu) \subset K_\infty(\mathfrak{r}, p).$$

For if $y_\lambda \in K(z(\tau_\lambda), p_\lambda)$ and $y_\lambda \to y$, where $\{\lambda\}$ is a subsequence of $\{\nu\}$, then by (22.5)

$$|\alpha(\mathfrak{r}, y) - \alpha(\mathfrak{r}, p)| = \lim |yz(\tau_\lambda) - pz(\tau_\lambda)| \leqslant \lim |y_\lambda z(\tau_\lambda) - p_\lambda z(\tau_\lambda)| + \\ + \lim p\, p_\lambda + \lim y y_\lambda = 0,$$

hence $y \in K_\infty(\mathfrak{r}, p)$.

The interior and exterior of $K(r, p)$ are by definition the sets of points x with $rx < rp$ and $rx > rp$, respectively. Correspondingly, we call the set $\alpha(\mathfrak{r}, x) < \alpha(\mathfrak{r}, p)$ the *"interior"* and the set $\alpha(\mathfrak{r}, x) > \alpha(\mathfrak{r}, p)$ the *"exterior of* $K_\infty(\mathfrak{r}, p)$".

(22.12) If $z(\tau)$ represents the ray $\mathfrak{r}$ and $0 \leqslant \tau_1 \leqslant \tau_2$ then $K(z(\tau_2), z(\tau_1))$ and its interior lie, except for $z(\tau_1)$, in the interior of $K_\infty(\mathfrak{r}, z(\tau_1))$.

For (22.2) and (22.5) show that for

(22.13)
$$\alpha(\mathfrak{r}, z(\tau_1)) - \alpha(\mathfrak{r}, x) = \lim_{\tau \to \infty} [z(\tau_1)\, z(\tau) - xz(\tau)] \geqslant$$
$$\geqslant \tau - \tau_1 - xz(\tau) \geqslant \tau_2 - \tau_1 - xz(\tau_2).$$

Put $z(\tau_i) = z_i$. If x is in the interior of $K(z_2, z_1)$ then $\tau_2 - \tau_1 > xz_2$ hence $\alpha(\mathfrak{r}, z_1) > \alpha(\mathfrak{r}, x)$ so that x lies in the interior of $K_\infty(\mathfrak{r}, z_1)$. If x lies on $K(z_2, z_1)$ then $\tau_2 - \tau_1 = xz_2$. Because of (22.2) the equality sign holds in (22.13) only if $(xz_2z(\tau))$. But then uniqueness of prolongation yields $x = z_1$.

We can now prove the following addition to (22.11):

(22.14) ***Theorem:*** *If $z(\tau)$ represents the ray $\mathfrak{r}$ and $\mathfrak{r}$ is a subray of a ray $\mathfrak{r}_0$ that intersects $K_\infty(\mathfrak{r}, p)$ at a point q then for any sequence $\tau_\nu \to \infty$*

$$\lim K(z(\tau_\nu), q) = K_\infty(\mathfrak{r}, p) = K_\infty(\mathfrak{r}, q).$$

Such a ray $\mathfrak{r}_0$ always exists when $\mathfrak{r}$ lies on a straight line.

Let $\mathfrak{r}$ be the subray $z(\tau) = z_0(\tau + \tau_0)$, $\tau_0 \geqslant 0$ of the ray $\mathfrak{r}_0$ represented by $z_0(\tau)$. By hypothesis there is a $\bar{\tau} \geqslant 0$ such that $z_0(\bar{\tau}) = q$. Because of (22.10) it suffices to show with $z_\nu = z(\tau_\nu) = z_0(\tau_\nu + \tau_0)$ that

$$\lim K(z_\nu, q) = K_\infty(\mathfrak{r}_0, q).$$

(22.11) reduces this equation to

$$K_\infty(\mathfrak{r}_0, q) \subset \liminf K(z_\nu, q).$$

If $x \in K_\infty(\mathfrak{r}_0, q)$ then (22.12) yields $xz_\nu \geqslant \tau_\nu + \tau_0 - \bar{\tau}$, when $\tau_\nu + \tau_0 > \bar{\tau}$. A segment $T(x, z_\nu)$ contains therefore a point x_ν with $x_\nu z_\nu = \tau_\nu + \tau_0 - \bar{\tau}$ or $x_\nu \in K(z_\nu, q)$. We have $\alpha(\mathfrak{r}_0, x) = \alpha(\mathfrak{r}_0, q)$ because $x \in K_\infty(\mathfrak{r}, q)$ and conclude from (22.4) that $xz_\nu - \tau_\nu - \tau_0 \to -\bar{\tau}$ or

$$xx_\nu = xz_\nu - x_\nu z_\nu = xz_\nu - \tau_\nu + \bar{\tau} - \tau_0 \to 0$$

which shows $x \in \liminf K(z_\nu, q)$.

If $\alpha_0 = \alpha(\mathfrak{r}, p) < 0$ then $K_\infty(\mathfrak{r}, p)$ intersects $\mathfrak{r}$ at the point $z(-\alpha_0)$, compare (22.4) and (22.9). If $\alpha_0 > 0$, then $K_\infty(\mathfrak{r}, p)$ will not intersect $\mathfrak{r}$. But if $\mathfrak{r}$ lies on a straight line, we can find a subray $\mathfrak{r}_0$ of this line which contains $\mathfrak{r}$ and such that the distance of the origins of $\mathfrak{r}$ and $\mathfrak{r}_0$ equals α_0. If $z_0(\tau)$ represents $\mathfrak{r}_0$, then $z(\tau) = z_0(\tau + \alpha_0)$. Because of (22.6) the point $z_0(0)$ lies on $K(\mathfrak{r}_0, p)$ and hence by (22.10) on $K(\mathfrak{r}, p)$.

This fact implies, of course, that in a straight space every limit sphere is a limit of spheres. In that case we show for completeness sake:

(22.15) Corollary: If $z(\tau)$ represents a ray $\mathfrak{r}$ in a straight space and $\tau_\nu \to \infty$, $p_\nu \to p$, then

$$\lim_{\nu\to\infty} K(z(\tau_\nu), p_\nu) = K_\infty(\mathfrak{r}, p).$$

For we may assume $K_\infty(\mathfrak{r}, p)$ intersects $\mathfrak{r}$ in a point q. If $z_\nu = z(\tau_\nu)$ then $0 = \alpha(\mathfrak{r}, p) - \alpha(\mathfrak{r}, q) = \lim (z_\nu p - z_\nu q) = \lim (z_\nu p_\nu - z_\nu q)$.

Because the space is straight there is for every point of one of the two spheres $K(z_\nu, p_\nu)$ and $K(z_\nu, q)$ a point (on the line through z) on the other sphere whose distance from the first point is $|z_\nu p_\nu - z_\nu q|$. This implies

$$\lim K(z_\nu, p_\nu) = \lim K(z_\nu, q) = K_\infty(\mathfrak{r}, q).$$

Co-rays

Let $z(\tau)$ represent the ray $\mathfrak{r}$. A ray $\mathfrak{s}$ with origin q is called a *co-ray* from q to $\mathfrak{r}$, if a sequence of segments $T(q_\nu, z(\tau_\nu))$ with $q_\nu \to q$ and $\tau_\nu \to \infty$ exists which tends to $\mathfrak{s}$. More precisely, if $w(\tau)$ represents $\mathfrak{s}$ and $w_\nu(\tau)$ represents the geodesic for which $w_\nu(0) = q_\nu$ and the subarc $0 \leqslant \tau \leqslant q_\nu z(\tau_\nu)$ is $T(q_\nu, z(\tau_\nu))$, then $w_\nu(\tau) \to w(\tau)$ for $\tau \geqslant 0$. The proof of (22.1) shows that every point q is origin of at least one co-ray to $\mathfrak{r}$.

The co-rays to $\mathfrak{r}$ play the role of parallels to $\mathfrak{r}$. Our aims are to characterize co-rays to $\mathfrak{r}$ (1) in terms of the function $\alpha(\mathfrak{r}, x)$, this is accomplished by (22.16) and its converse (22.20); (2) as orthogonal trajectories, as it were, of the limit spheres $K_\infty(\mathfrak{r}, x)$. This is contained in (22.18).

(22.16) *If $w(\tau)$ represents a co-ray to $\mathfrak{r}$ then*

$$\alpha(\mathfrak{r}, w(\tau'')) - \alpha(\mathfrak{r}, w(\tau')) = \tau' - \tau'' \quad \textit{for any} \quad \tau', \tau'' \geqslant 0.$$

Let $w(\tau) = \lim w_\nu(\tau)$, where $w_\nu(\tau)$ has the same properties as above. Assume $\tau' > \tau''$. For large ν we have $(q_\nu\, w_\nu(\tau')\, z(\tau_\nu))$ hence

$$w_\nu(\tau')\, w(\tau') + w_\nu(\tau'')\, w(\tau'') \geqslant$$

$$|w(\tau'')\, z(\tau_\nu) - w(\tau')\, z(\tau_\nu) - [w_\nu(\tau'')\, z(\tau_\nu) - w_\nu(\tau')\, z(\tau_\nu)]| =$$
$$= |w(\tau'')\, z(\tau_\nu) - w(\tau')\, z(\tau_\nu) - (\tau' - \tau'')|.$$

Since the left side tends to zero as $\nu \to \infty$, the assertion follows from (22.5).

For the converse we need the properties of the co-rays as orthogonal trajectories of the $K_\infty(\mathfrak{r}, x)$. A first step is:

(22.17) $\alpha(\mathfrak{r}, p) - \alpha(\mathfrak{r}, q) = pq > 0$ *if and only if q lies in the interior of $K_\infty(\mathfrak{r}, p)$ and is a foot of p on $K_\infty(\mathfrak{r}, q)$.*

If $\alpha(\mathfrak{r}, p) - \alpha(\mathfrak{r}, q) = pq > 0$ then q lies in the interior of $K_\infty(\mathfrak{r}, p)$ and it follows from (22.5) and (22.9) that for $x \in K_\infty(\mathfrak{r}, q)$

$$px \geqslant \alpha(\mathfrak{r}, p) - \alpha(\mathfrak{r}, x) = \alpha(\mathfrak{r}, p) - \alpha(\mathfrak{r}, q) = pq.$$

Therefore q is a foot of p on $K_\infty(\mathfrak{r}, q)$.

Conversely, let q be in the interior of $K_\infty(\mathfrak{r}, p)$ and a foot of p on $K_\infty(\mathfrak{r}, q)$. Then $\alpha(\mathfrak{r}, p) > \alpha(\mathfrak{r}, q)$. On a co-ray $p(\tau)$ from p to $\mathfrak{r}$ consider the points $p(\tau_0)$ with $\tau_0 = \alpha(\mathfrak{r}, p) - \alpha(\mathfrak{r}, q)$. By (22.16), $\alpha(\mathfrak{r}, p) - \alpha(\mathfrak{r}, p(\tau_0)) = \tau_0$ hence $\alpha(\mathfrak{r}, p(\tau_0)) = \alpha(\mathfrak{r}, q)$ and $p(\tau_0) \in K_\infty(\mathfrak{r}, q)$. Moreover $\tau_0 = pp(\tau_0) \geqslant pq$ because q is a foot of p. The definition of τ_0 and (22.5) show that $\tau_0 = pq$.

We can now prove:

(22.18) *If $\mathfrak{s}$ is a co-ray to $\mathfrak{r}$ with origin q then any point $q_0 \neq q$ of $\mathfrak{s}$ is a foot of q on $K_\infty(\mathfrak{r}, q_0)$ and is the only foot of any third point $q_1 \neq q$ of $\mathfrak{s}$ on $K_\infty(\mathfrak{r}, q_0)$. Also q is a foot of q_0 on $K_\infty(\mathfrak{r}, q)$.*

That q_0 is a foot of q is contained in (22.16) and (22.17). These two facts together with (20.6) show that q_0 is the only foot of q_1 if $(q\, q_1 q_0)$. If $(q\, q_0 q_1)$ take q_2 with $(q\, q_1 q_2)$ then also $(q_0 q_1 q_2)$. There is a foot f of q_1 on $K_\infty(\mathfrak{r}, q_0)$ because this set is closed. We conclude from (22.16) and (22.5) that

$$q_0 q_2 = \alpha(\mathfrak{r}, q_0) - \alpha(\mathfrak{r}, q_2) = \alpha(\mathfrak{r}, f) - \alpha(\mathfrak{r}, q_2) \leqslant f q_2$$
$$\leqslant f q_1 + q_1 q_2 \leqslant q_0 q_1 + q_1 q_2 = q_0 q_2.$$

Therefore $(f\, q_1 q_2)$ and $f q_1 = q_0 q_1$, and uniqueness of prolongation shows $q_0 = f$. Letting q_0 tend to q we deduce from the continuity of $\alpha(\mathfrak{r}, x)$ in x that q is a foot of q_0 on $K_\infty(\mathfrak{r}, q)$.

The co-ray from a given point to a given ray need not be unique. An example is obtained by considering the surface

$$\xi^2 + \eta^2 = 1 \quad \text{for} \quad \zeta \geqslant 0, \qquad \xi^2 + \eta^2 + \zeta^2 = 1 \quad \text{for} \quad \zeta < 0$$

in ordinary cartesian (ξ, η, ζ)-space with the length of the shortest join on the surface as distance. The geodesics through $p = (0, 0, -1)$ are the intersections of the surface with the planes through the ζ-axis. The point p decomposes each of these geodesics into two rays, and each of these rays is a co-ray to every other ray with origin p. However, if $\mathfrak{r}$ is one ray with origin p and $w_0(\tau)$ represents another, then for any $\eta > 0$ the subray $w(\tau) = w_0(\tau + \eta)$ of $w_0(\tau)$ is a co-ray to $\mathfrak{r}$ and the only one with origin $w(0) = w_0(\eta)$.

It is most remarkable that this is not a special property of the example, but holds in general.

(22.19) ***Theorem:*** *If $\mathfrak{s}$ is a co-ray from q to $\mathfrak{r}$ and q_0 is any point of $\mathfrak{s}$ different from q, then the co-ray from q_0 to $\mathfrak{r}$ is unique and coincides with the sub-ray of $\mathfrak{s}$ with origin q_0.*

For let $w(\tau)$ represent $\mathfrak{s}$ and $q_0 = w(\tau_0)$, $\tau_0 > 0$. If $q(\tau)$ represents any co-ray to $\mathfrak{r}$ with origin q_0, then (22.16) yields for $\tau' > \tau_0$

$$\alpha(\mathfrak{r}, w(\tau_0)) - \alpha(\mathfrak{r}, w(\tau')) = \tau' - \tau_0 = \alpha(\mathfrak{r}, q_0) - \alpha(\mathfrak{r}, q(\tau' - \tau_0)).$$

Therefore $q(\tau' - \tau_0) \in K_\infty(\mathfrak{r}, w(\tau'))$. By (22.18) the point $q(\tau' - \tau_0)$ is a foot and $w(\tau')$ is the only foot of $w(\tau_0) = q_0$ on $K_\infty(\mathfrak{r}, w(\tau'))$, hence $q(\tau' - \tau_0) = w(\tau')$, which shows that $q(\tau)$ is a sub-ray of $w(\tau)$.

We can now establish the converse of (22.16):

(22.20) *If $w(\tau)$ represents a geodesic and*

$$\alpha(\mathfrak{r}, w(\tau'')) - \alpha(\mathfrak{r}, w(\tau')) = \tau' - \tau'' \qquad \textit{for} \qquad \tau', \tau'' \geqslant 0$$

then $w(\tau)$ represents for $\tau \geqslant 0$ a co-ray to $\mathfrak{r}$.

For $w(\tau)$, $\tau \geqslant 0$, is a ray because by (7.3) and (22.5) for $\tau' > \tau''$

$$\tau' - \tau'' \geqslant w(\tau'')\, w(\tau') \geqslant \alpha(\mathfrak{r}, w(\tau'')) - \alpha(\mathfrak{r}, w(\tau')) = \tau' - \tau''.$$

It follows from (22.17) and (20.6) that for positive η and τ the point $w(\eta + \tau)$ is the only foot of $w(\eta)$ on $K_\infty(\mathfrak{r}, w(\eta + \tau))$, and from (22.18) that $w_\eta(\tau) = w(\eta + \tau)$, $\tau \geqslant 0$, is a co-ray from $w(\eta)$ to $\mathfrak{r}$. Letting η tend to zero we see from the definition of co-ray that $w(\tau)$, $\tau \geqslant 0$, also represents a co-ray to $\mathfrak{r}$.

It will be shown that even in a two-dimensional straight space the fact, that $\mathfrak{s}$ is a co-ray to $\mathfrak{r}$, does not imply that $\mathfrak{r}$ is a co-ray to $\mathfrak{s}$. It is therefore important to know conditions under which the relation of co-ray to ray is symmetric. Since each ray is a co-ray to itself the following statement contains a sufficient condition:

(22.21) *If $\alpha(\mathfrak{r}, x) - \alpha(\mathfrak{s}, x)$ is constant* (that is, independent of x) *then every co-ray to one of the rays $\mathfrak{r}$, $\mathfrak{s}$ is also a co-ray to the other.*

The hypothesis yields $\alpha(\mathfrak{r}, x) - \alpha(\mathfrak{r}, y) = \alpha(\mathfrak{s}, x) - \alpha(\mathfrak{s}, y)$ for any two points x, y. If $w(\tau)$ is a co-ray to $\mathfrak{r}$ then by (22.16)

$$\tau' - \tau'' = \alpha(\mathfrak{r}, w(\tau'')) - \alpha(\mathfrak{r}, w(\tau')) = \alpha(\mathfrak{s}, w(\tau'')) - \alpha(\mathfrak{s}, w(\tau')).$$

Now the converse (22.20) of (22.16) shows that $w(\tau)$ is a co-ray to $\mathfrak{s}$. This result yields together with (22.8) and (22.19) the following facts:

(22.22) ***Theorem:*** *If* $\delta_\infty(\mathfrak{r}, \mathfrak{s}) = 0$, *then every co-ray to one of the rays* $\mathfrak{r}, \mathfrak{s}$ *is also a co-ray to the other.*

(22.23) *If the origin of the ray* $\mathfrak{s}$ *is an interior point of the ray* $\mathfrak{r}$ *and* $\delta_\infty(\mathfrak{r}, \mathfrak{s}) = 0$, *then* $\mathfrak{s}$ *is a sub-ray of* $\mathfrak{r}$.

If $x(\tau)$ *and* $y(\tau)$ *represent two distinct straight lines with a common point then* $\liminf_{|\tau|\to\infty, |\sigma|\to\infty} x(\tau)\, y(\sigma) > 0.$

It should be noticed that the condition $\delta_\infty(\mathfrak{r}, \mathfrak{s}) = 0$ for the rays $\mathfrak{r} : z(\tau)$ and $\mathfrak{s} : w(\tau)$ does not imply that the distances $z(\tau)\mathfrak{s}$ or $w(\tau)\mathfrak{r}$ are bounded for $\tau \geqslant 0$.

An example is furnished by the surface of revolution

$$\xi^2 + \eta^2 = \frac{1}{\zeta} + e^\zeta \sin^2 \zeta, \qquad \zeta > 0,$$

in the ordinary (ξ, η, ζ)-space with the length of the shortest join on the surface as distance. The meridians of this surface are geodesics and straight lines in the sense of G-spaces. Let $\mathfrak{g}_1$ be the meridian $\eta = 0$, $\xi < 0$, $\mathfrak{g}_2$ the meridian $\eta = 0$, $\xi > 0$ and $\mathfrak{r}_k$ the sub-ray $\zeta \geqslant 1$ of $\mathfrak{g}_k$. Then $\mathfrak{r}_k$ contains the points $q_{k\nu} = \left((-1)^k (\nu\pi)^{-1},\ 0, \nu\pi\right)$, $k = 1, 2$ with $q_{1\nu}\, q_{2\nu} \leqslant \pi(\nu\pi)^{-1} = \nu^{-1}$ hence $\delta_\infty(\mathfrak{r}_1, \mathfrak{r}_2) = 0$. But $\mathfrak{r}_k$ also contains the points

$$p_{k\nu} = \left((-1)^k \left\{\frac{1}{\nu\pi + \pi/2} + e^{\nu\pi + \pi/2}\right\}, 0, \nu\pi + \pi/2\right)$$

with $p_{1\nu}\, \mathfrak{g}_2 = p_{2\nu}\, \mathfrak{g}_1 > e^{\nu\pi}$, so that $x_2\, \mathfrak{r}_1$ and $x_1\, \mathfrak{r}_2$ are not bounded for $x_i \in \mathfrak{r}_i$.

23. Asymptotes and parallels in straight spaces.

We now apply the results of the preceding section to straight spaces. In that case (22.19) may be extended as follows.

(23.1) *If in a straight space* $\mathfrak{s}$ *is a co-ray to* $\mathfrak{r}$, *then any ray that contains* $\mathfrak{s}$ *is a co-ray to* $\mathfrak{r}$. *The co-ray from any point to any ray is unique.*

A representation $w(\tau)$, $\tau \geqslant 0$, of $\mathfrak{s}$ may be extended to a representation $w(\tau)$, $-\infty < \tau < \infty$, of the straight line containing $\mathfrak{s}$. Let $z(\tau)$ represent $\mathfrak{r}$. Since $\mathfrak{s}$ is a co-ray to $\mathfrak{r}$ there is a sequence $q_\nu \to w(0)$ and a sequence $\tau_\nu \to \infty$ such that $w_\nu(\tau) \to w(\tau)$ for all τ, where $w_\nu(\tau)$ represents the straight line $\mathfrak{g}(q_\nu, z(\tau_\nu))$ with $w_\nu(0) = q_\nu$, $w_\nu(q_\nu\, z(\tau_\nu)) = z(\tau_\nu)$. This implies for $\tau' > \tau''$ and large ν, that $\tau' - \tau'' = w_\nu(\tau'')\, z(\tau_\nu) - w_\nu(\tau')\, z(\tau_\nu)$,

$$w(\tau')\, w_\nu(\tau') + w(\tau'')\, w_\nu(\tau'') \geqslant \left|\tau' - \tau'' - [w(\tau'')\, z(\tau_\nu) - w(\tau')\, z(\tau_\nu)]\right|;$$

hence by (22.5)

$$\alpha(\mathfrak{r}, w(\tau'')) - \alpha(\mathfrak{r}, w(\tau')) = \tau' - \tau''.$$

Since τ', τ'' are arbitrary it follows from (22.20) that the ray $w_0(\tau) = w(\eta + \tau)$ is for any real η a co-ray to $\mathfrak{r}$.

The uniqueness of the co-ray to $\mathfrak{r}$ from a given point is a consequence of this and (22.19).

Asymptotes

The usual terminology used for parallels in euclidean or hyperbolic space refers to oriented straight lines rather than to rays.

A geodesic $\mathfrak{g}$ represented by $x(\tau)$ in a G-space may be traversed in two different senses, corresponding to increasing or decreasing τ. We orient $\mathfrak{g}$ by distinguishing one of these senses as positive, the other as negative and use $\mathfrak{g}^+$ or $\mathfrak{g}^-$ as symbol for an oriented geodesic. $x(\tau)$ is a representation of $\mathfrak{g}^+$, if $x(\tau)$ represents $\mathfrak{g}$ and traversing $\mathfrak{g}^+$ in the positive sense corresponds to increasing τ. The representations $x(\tau + \beta)$ of $\mathfrak{g}$ and only these represent $\mathfrak{g}^+$. It is convenient to denote the other orientation of $\mathfrak{g}$ by $\mathfrak{g}^-$. If $x(\tau)$ represents $\mathfrak{g}^+(\mathfrak{g}^-)$ then $x(-\tau + \beta)$ represents $\mathfrak{g}^-(\mathfrak{g}^+)$.

If $x(\tau)$ represents the oriented straight line $\mathfrak{g}^+$ then a positive sub-ray of $\mathfrak{g}^+$ (or negative sub-ray of $\mathfrak{g}^-$) is a ray of the form $y(\tau) = x(\tau + \beta)$, $\tau \geqslant 0$, and a negative sub-ray of $\mathfrak{g}^+$ (or positive sub-ray of $\mathfrak{g}^-$) has the form $y(\tau) = x(-\tau + \beta)$, $\tau \geqslant 0$. The oriented straight line $\mathfrak{g}_\nu^+$ tends to the oriented straight line $\mathfrak{g}^+$ if $x_\nu(\tau) \to x(\tau)$ for suitable representations $x_\nu(\tau)$ of $\mathfrak{g}_\nu^+$ and $x(\tau)$ of $\mathfrak{g}^+$. If $\mathfrak{g}_\nu^+$ tends to $\mathfrak{g}^+$ or $\mathfrak{g}^-$ then $\mathfrak{g}_\nu^-$ tends to $\mathfrak{g}^-$ or $\mathfrak{g}^+$ respectively.

We may then express our results (23.1) and (22.18) as follows:

(23.2) ***Theorem:*** *In a straight space, if $x(\tau)$ represents the oriented line $\mathfrak{g}^+$, and $q_\nu \to q$, $\tau_\nu \to \infty$, then the line $\mathfrak{g}^+(q_\nu, x(\tau_\nu))$ so oriented that $x(\tau_\nu)$ follows q_ν tends to an oriented line $\mathfrak{a}^+$, which is independent of the choice of the sequences $\{q_\nu\}$ and $\{\tau_\nu\}$. The line $\mathfrak{a}^+$ is called the asymptote to $\mathfrak{g}^+$ through q. The asymptote to $\mathfrak{g}^+$ through any point q' on $\mathfrak{a}^+$ coincides with $\mathfrak{a}^+$.*

The feet of q on the limit spheres $K_\infty(r, x)$ are unique and $\mathfrak{a}^+$ is their locus. so that $\mathfrak{a}$ is perpendicular to all $K_\infty(\mathfrak{r}, x)$, where r is a positive sub-ray of $\mathfrak{g}^+$.

In any straight Desarguesian space defined in a convex subset C of A^n, the oriented line $\mathfrak{a}^+$ is therefore an asymptote to the oriented line $\mathfrak{g}^+$, if points traversing $\mathfrak{a}^+$ and $\mathfrak{g}^+$ in the positive sense tend to the same point on the boundary of C. Thus we see:

(23.3) *In a straight Desarguesian space asymptotes have the following two properties:*

Symmetry. If $\mathfrak{a}^+$ *is an asymptote to* $\mathfrak{g}^+$, *then* $\mathfrak{g}^+$ *is an asymptote to* $\mathfrak{a}^+$.

Transitivity. If $\mathfrak{a}^+$ *is an asymptote to* $\mathfrak{b}^+$ *and* $\mathfrak{b}^+$ *is an asymptote to* $\mathfrak{c}^+$, *then* $\mathfrak{a}^+$ *is an asymptote to* $\mathfrak{c}^+$.

We observe:

(23.4) *In any straight space the Transitivity Property implies the Symmetry Property.*

In a two-dimensional straight space the two properties are equivalent.

Proof. In an arbitrary straight space let $\mathfrak{a}^+$ be an asymptote to $\mathfrak{g}^+$ and p any point of $\mathfrak{g}$. If $\mathfrak{c}^+$ denotes the asymptote to $\mathfrak{a}^+$ through p, then transitivity yields that $\mathfrak{c}^+$ is an asymptote to $\mathfrak{g}^+$. Since $\mathfrak{g}^+$ is asymptote to itself it follows from (23.2) that $\mathfrak{c}^+ = \mathfrak{g}^+$. Whether the converse holds for higher dimensions is an open question.

To prove the converse in the plane case, let $\mathfrak{a}^+$ be an asymptote to $\mathfrak{b}^+$ and $\mathfrak{b}^+$ an asymptote to $\mathfrak{c}^+$. Since the assertion is otherwise trivial, we may assume that $\mathfrak{a}^+$, $\mathfrak{b}^+$, $\mathfrak{c}^+$ are distinct. The line $\mathfrak{a}$ does not intersect $\mathfrak{b}$ nor does $\mathfrak{b}$ intersect $\mathfrak{c}$. Moreover, $\mathfrak{a}$ cannot intersect $\mathfrak{c}$, because the Symmetry Property would imply that $\mathfrak{a}^+$ and $\mathfrak{c}^+$ are two different asymptotes to $\mathfrak{b}^+$ through their intersection. For an indirect proof we assume that the asymptote $\mathfrak{d}^+$ to $\mathfrak{c}^+$ through some point p of $\mathfrak{a}^+$ is different from $\mathfrak{a}^+$ and distinguish three cases:

1) $\mathfrak{b}^+$ lies between $\mathfrak{a}^+$ and $\mathfrak{c}^+$. Because $\mathfrak{a}^+$ is asymptote to $\mathfrak{b}^+$ the line $\mathfrak{d}$ must intersect $\mathfrak{b}$, hence $\mathfrak{b}^+$ is not an asymptote to $\mathfrak{c}^+$. 2) $\mathfrak{c}$ lies between $\mathfrak{a}$ and $\mathfrak{b}$. Since $\mathfrak{d}$ does not intersect $\mathfrak{c}$ it cannot intersect $\mathfrak{b}$, hence $\mathfrak{a}^+$ cannot be asymptote to $\mathfrak{b}^+$. 3) $\mathfrak{a}$ lies between $\mathfrak{b}$ and $\mathfrak{c}$. By symmetry $\mathfrak{c}^+$ is an asymptote to $\mathfrak{d}^+$. Now $\mathfrak{b}^+$, $\mathfrak{c}^+$, $\mathfrak{d}^+$ are in the same position as $\mathfrak{a}^+$, $\mathfrak{b}^+$, $\mathfrak{c}^+$ under case 2). Hence $\mathfrak{b}^+$ is an asymptote to $\mathfrak{d}^+$ but then $\mathfrak{a}^+$ and $\mathfrak{d}^+$ would, because of symmetry, be two different asymptotes through p to $\mathfrak{b}^+$.

In order to show that neither symmetry nor transitivity hold in general straight spaces, it suffices to give an example for the absence of symmetry. We show:

(23.5) *The following may occur in a two-dimensional straight space:*

(a) *There are oriented straight lines* $\mathfrak{a}^+$ *and* $\mathfrak{g}^+$ *such that* $\mathfrak{a}^+$ *is an asymptote to* $\mathfrak{g}^+$, *but neither is* $\mathfrak{g}^+$ *an asymptote to* $\mathfrak{a}^+$ *nor* $\mathfrak{g}^-$ *to* $\mathfrak{a}^-$.

(b) *There are oriented straight lines* $\mathfrak{a}^+$ *and* $\mathfrak{g}^+$ *such that* $\mathfrak{a}^+$ *is an asymptote to* $\mathfrak{g}^+$, *but* $\mathfrak{g}^+$ *is not an asymptote to* $\mathfrak{a}^+$, *although* $\mathfrak{g}^-$ *and* $\mathfrak{a}^-$ *are asymptotes to each other.*

We base the construction of such examples on the following observation:

Denote by H the branch $\xi < 0$ of the hyperbola $\xi\eta = -1$ in the cartesian (ξ, η)-plane. Given any slope $\mu > 0$ and any positive number η, there is exactly one chord of H that has slope μ and length η.

If Σ' denotes the totality of curves obtained from H by translations $\xi' = \xi + \alpha$, $\eta' = \eta + \beta$, we may therefore say: given any two distinct points $p_i = (\xi_i, \eta_i)$ such that $\mu = \dfrac{\eta_2 - \eta_1}{\xi_2 - \xi_1} > 0$, there is exactly one curve in Σ' that passes through p_1 and p_2. We denote by Σ the system of curves in the plane consisting of Σ' and all lines $\eta = \mu\,\xi + \beta$ with $\mu \leqslant 0$ and the lines $\xi = \beta$. Then Σ has the property that two distinct points of the plane lie on exactly one curve of Σ. It was shown in Theorem (11.2) that the plane can be metrized such that it becomes a straight space with the curves in Σ as straight lines.

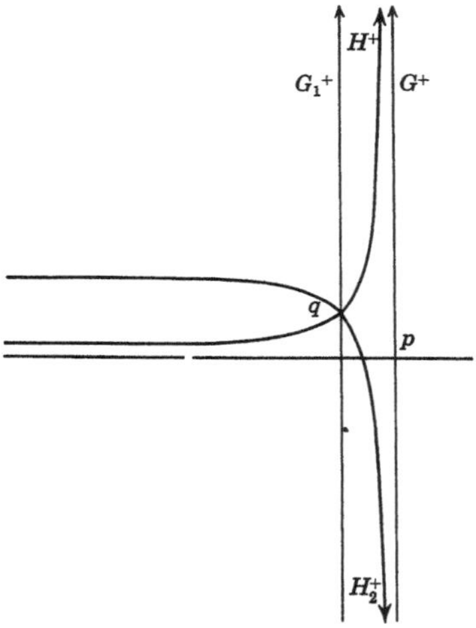

Fig. 20.

This furnishes an example for (*b*). For let H^+, G^+, G_1^+ be the orientations of H, the line $G : \xi = 0$, and the line $G_1 : \xi = -1$ for which increasing η corresponds to the positive sense. The curves H and G_1 intersect at $q = (-1, 1)$. The curve in Σ through $p = (0, 0)$ and $(-1, \nu)$ is an ordinary line which tends for $\nu \to \infty$ to G, therefore G^+ is the asymptote to G_1^+ through p. Since H^+ lies, from q on, between G_1 and G, the line G_1^+ cannot be the asymptote to G^+ through q. Actually, H^+ is an asymptote to G^+.

On the other hand, G_1^- is obviously an asymptote to G^-, and the hyperbola branches in Σ' through p and $(-1, -\nu)$ tend for $\nu \to \infty$ to G, hence G^- is also an asymptote to G_1^-.

To obtain an example for (a) we define Σ' as before, denote by H_1 the branch $\xi < 0$ of the hyperbola $\xi\,\eta = 1$, and by Σ'' the totality of all curves obtained from H_1 by translations. The system of curves Σ_1 is defined to consist of all curves in Σ', in Σ'', and of the lines $\xi = \beta$ and $\eta = \beta$. Again, any two distinct points of the plane lie on exactly one curve in Σ_1, and by the

Theorem (11.2) the plane can be metrized such that it becomes a two-dimensional straight space with the curves in Σ_1 as straight lines.

Let G^+, G_1^+, H^+, p and q be defined as before and denote by H_2^+ the curve in Σ_1 through q obtained from H_1 by the translation $\xi' = \xi + 2$, $\eta' = \eta$ and so oriented that decreasing η corresponds to traversing H_2^+ in the positive sense. This time G^+ is an asymptote to G_1^+ and G^- is an asymptote to G_1^-, but the asymptotes to G^+ and G^- through q are H^+ and H_2^+, hence different from G_1^+ and G_1^-.

If the straight lines $\mathfrak{g}$ and $\mathfrak{a}$ in a straight space have the property that, with proper orientations, $\mathfrak{a}^+$ is an asymptote to $\mathfrak{g}^+$ and also $\mathfrak{a}^-$ to $\mathfrak{g}^-$, then we call $\mathfrak{a}$ *parallel* to $\mathfrak{g}$. The two examples which we just constructed also show:

(23.6) *The following may occur in a two-dimensional straight space:*

$\mathfrak{a}$ *is parallel to* $\mathfrak{g}$, *but neither orientation of* $\mathfrak{g}$ *is an asymptote to either orientation of* $\mathfrak{a}$.

$\mathfrak{a}$ *is parallel to* $\mathfrak{g}$, *an orientation* $\mathfrak{g}^+$ *of* $\mathfrak{g}$ *is an asymptote to an orientation* $\mathfrak{a}^+$ *of* $\mathfrak{a}$, *but* $\mathfrak{g}^-$ *is not an asymptote to* $\mathfrak{a}^-$.

The Parallel Axioms

With these examples in mind we are led to the following conditions as adequate extensions to straight spaces of the properties of parallels as they are found in the foundations of euclidean and hyperbolic geometry.

Parallel Axiom: The asymptotes have the symmetry property[1] *and: If* $\mathfrak{a}^+$ *is an asymptote to* $\mathfrak{g}^+$, *then* $\mathfrak{a}^-$ *is an asymptote to* $\mathfrak{g}^-$.

Hyperbolic Axiom: The asymptotes have the symmetry property and: If $\mathfrak{a}^+$ *is an asymptote to* $\mathfrak{g}^+$ *and different from* $\mathfrak{g}^+$, *then* $\mathfrak{a}^-$ *is not an asymptote to* $\mathfrak{g}^-$.

Some remarks on these conditions are in order.

(23.7) *The Parallel Axiom holds in a straight two-dimensional space, if and only if for a given line* $\mathfrak{g}$ *and a given point* p *not on* $\mathfrak{g}$, *exactly one line through* p *exists which does not intersect* $\mathfrak{g}$.

Thus in the case of straight planes, the present Parallel Axiom is identical with the usual. The usual parallel axiom in space is only formulated when planes exist or the space is Desarguesian, and therefore implies transitivity.

To prove (23.7) observe that if in any straight plane the lines $\mathfrak{a}$, $\mathfrak{b}$ carrying the asymptotes $\mathfrak{a}^+$ and $\mathfrak{b}^+$ through p to the two orientations $\mathfrak{g}^+$ and $\mathfrak{g}^-$ are different, infinitely many lines through p exist which do not intersect $\mathfrak{g}$, and that $\mathfrak{a} = \mathfrak{b}$ implies that $\mathfrak{a}$ is the only line through p that does not intersect $\mathfrak{g}$.

Therefore the present parallel axiom implies the usual, and the usual axiom implies $\mathfrak{a} = \mathfrak{b}$ and clearly also symmetry.

The situation is somewhat different for the Hyperbolic Axiom. It is not hard to construct similar examples as above which show: symmetry does not follow from the assumption that, if $\mathfrak{a}^+$ is an asymptote to $\mathfrak{g}^+$ and different from $\mathfrak{g}^+$, the asymptote to $\mathfrak{g}^-$ is different from $\mathfrak{a}^-$. Scrutinizing the treatment of hyperbolic geometry found in the literature[2], we find that axioms of congruence or mobility are introduced before the hyperbolic parallel postulate, and that symmetry is derived with the help of these axioms.

If $a(\tau)$ represents an asymptote $\mathfrak{a}^+$ to $\mathfrak{g}^+ \neq \mathfrak{a}^+$, then in euclidean and hyperbolic geometries $a(\tau)\,\mathfrak{g}$ is bounded for $\tau \geqslant 0$. In hyperbolic geometry we have in addition $a(\tau)\,\mathfrak{g} \to \infty$ if $\tau \to -\infty$ and $a(\tau)\,\mathfrak{g} \to 0$ if $\tau \to \infty$. We are going to show with examples that no statements of this type are possible in general, even under strong additional assumptions:

(23.8) *Let $a(\tau)$ represent an asymptote $\mathfrak{a}^+$ to $\mathfrak{g}^+ \neq \mathfrak{a}^+$ in a two-dimensional Desarguesian space which satisfies the Hyperbolic Axiom. It may happen that*

(a) *$a(\tau)\,\mathfrak{g} \to \infty$ for $\tau \to \infty$ and $a(\tau)\,\mathfrak{g}$ is bounded for $\tau \leqslant 0$.*

(b) *The sets $S(Q, \rho)$ are convex for convex Q and $\alpha > a(\tau)\,\mathfrak{g} > \beta > 0$ for all real τ.*

(c) *The sets $S(Q, \rho)$ are convex for convex Q and $a(\tau)\,\mathfrak{g} \to 0$ for $\tau \to \infty$, whereas $a(\tau)\,\mathfrak{g}$ is bounded for $\tau < 0$.*

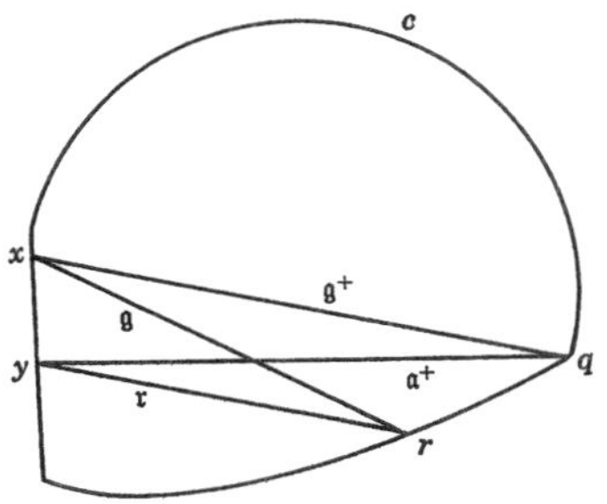

Fig. 21.

The example at the end of Section 18 proves (a). Our interest in (b) and (c) derives from the fact that slightly stronger convexity conditions will, without Desargues' Theorem, be shown to exclude these occurrences. To obtain examples for (b) and (c) we take a convex curve C in A^2 which contains exactly one proper segment T and is not differentiable at a point q not on T, and as metric the Hilbert distance in the interior D of C. If x and y are interior points of T, then the open segments $\mathfrak{g}^+$ from x to q and $\mathfrak{a}^+$ from y to q oriented towards q furnish an example for (b). If we replace q by a point r on $C - T$ where C is differentiable then the open segments $\mathfrak{g}$ from x to r and $\mathfrak{r}$ from y

to r furnish an example for (c). Elementary arguments about cross ratios confirm these statements[3].

(22.10) shows that for any two positive subrays $\mathfrak{r}_1$, $\mathfrak{r}_2$ of the oriented line $\mathfrak{g}^+$ the limit spheres $K_\infty(\mathfrak{r}_1, x)$ and $K_\infty(\mathfrak{r}_2, x)$ coincide. We may therefore denote these limit spheres, also, by $K_\infty(\mathfrak{g}^+, x)$. In euclidean geometry $K_\infty(\mathfrak{g}^+, x) = K_\infty(\mathfrak{g}^-, x)$. The Parallel Axiom does not generally imply this relation for (17.21) shows:

(23.9) *In a Minkowskian G-space in which the spheres are not differentiable, there are straight lines* $\mathfrak{g}$, *such that* $K_\infty(\mathfrak{g}^+, x) \neq K_\infty(\mathfrak{g}^-, x)$ *for every point* x. *Also there are lines* $\mathfrak{a}^+$ *and* $\mathfrak{g}^+ \neq \mathfrak{a}^+$ *which are asymptotes to each other such that* $K_\infty(\mathfrak{g}^+, x) \neq K_\infty(\mathfrak{a}^+, x)$ *for every* x.

It suffices to choose $\mathfrak{g}$ such that $K(p, y)$ is not differentiable at y for $p, y \in \mathfrak{g}$, and to choose as $\mathfrak{a}^+$ any asymptote to $\mathfrak{g}^+$ different from $\mathfrak{g}^+$.

By means of this example we conclude from (22.21):

(23.10) *The following is a sufficient but not necessary condition for the (Symmetry and) Transitivity Property of the asymptotes in a straight space*:

(23.11) $\alpha(\mathfrak{r}, x) - \alpha(\mathfrak{s}, x)$ *is independent of* x *when* $\mathfrak{s}$ *is a co-ray to* $\mathfrak{r}$.

If $\alpha(\mathfrak{r}, x) - \alpha(\mathfrak{s}, x)$ is independent of x for two rays $\mathfrak{r}, \mathfrak{s}$ of a G-space, then $K_\infty(\mathfrak{r}, p) = K_\infty(\mathfrak{s}, p)$ for every p.

For the equation of $K_\infty(\mathfrak{r}, p)$ is $\alpha(\mathfrak{r}, x) - \alpha(\mathfrak{r}, p) = 0$ and this implies $\alpha(\mathfrak{s}, x) - \alpha(\mathfrak{s}, p) = 0$. If the space is straight, then an asymptote $\mathfrak{a}^+$ to $\mathfrak{b}^+$ is by (23.2) the locus of the feet of any point q of $\mathfrak{a}^+$ on the limit spheres $K_\infty(\mathfrak{b}^+, x) = K_\infty(\mathfrak{a}^+, x)$. If $\mathfrak{b}^+$ is an asymptote to $\mathfrak{c}^+$, then $K_\infty(\mathfrak{c}^+, x) = K_\infty(\mathfrak{b}^+, x) = K_\infty(\mathfrak{a}^+, x)$, hence $\mathfrak{a}^+$ is the locus of the feet of q on $K_\infty(\mathfrak{c}^+, x)$, hence an asymptote to $\mathfrak{c}^+$. Thus the Transitivity Property holds.

That the condition (23.11) is not necessary follows from the fact that it implies $K_\infty(\mathfrak{a}^+, x) = K_\infty(\mathfrak{g}^+, x)$ when $\mathfrak{a}^+$ is an asymptote to $\mathfrak{g}^+$, which is not necessary because of (23.9).

In euclidean space, if $\mathfrak{a}^+$ is an asymptote to $\mathfrak{b}^+$ and $\mathfrak{r}$ is a positive subray of $\mathfrak{a}^+$, and $\mathfrak{s}$ is a negative subray of $\mathfrak{b}^+$, then it is easily verified that

(23.12) $\alpha(\mathfrak{r}, x) + \alpha(\mathfrak{s}, x)$ *is constant.*

This leads to the question whether (23.12) is in G-spaces, and more particularly in straight spaces, connected with the existence of parallels. We show:

(23.13) *If (23.12) holds for the two rays* $\mathfrak{r}$ *and* $\mathfrak{s}$ *of a G-space then any co-ray to* $\mathfrak{r}$ *(or* $\mathfrak{s}$*) is positive subray of a straight line* $\mathfrak{g}^+$, *and every negative subray of* $\mathfrak{g}^+$ *is a co-ray to* $\mathfrak{s}$ *(or* $\mathfrak{r}$*).*

The relation (23.12) implies $\alpha(\mathfrak{r}, x) - \alpha(\mathfrak{r}, p) = \alpha(\mathfrak{s}, p) - \alpha(\mathfrak{s}, x)$ for any p and x, hence $K_\infty(\mathfrak{r}, p) = K_\infty(\mathfrak{s}, p)$ for any p. Moreover, the interior of $K_\infty(\mathfrak{r}, p)$ coincides with the exterior of $K_\infty(\mathfrak{s}, p)$.

If $\mathfrak{p}_1$ and $\mathfrak{p}_2$ are co-rays from p to $\mathfrak{r}$ and $\mathfrak{s}$, they have no common point but p, because the interiors of $K_\infty(\mathfrak{r}, p)$ and $K_\infty(\mathfrak{s}, p)$ are disjoint. Any point a_1 of $\mathfrak{p}_1$ has by (22.18) p as foot on $K_\infty(\mathfrak{r}, p) = K_\infty(\mathfrak{s}, p)$. We deduce from (22.17) and (22.20) that p lies on a co-ray from a_1 to $\mathfrak{s}$, and then from (22.19) that $\mathfrak{p}_2$ is a subray of this co-ray. Therefore $\mathfrak{p}_1$ and $\mathfrak{p}_2$ lie on a straight line $\mathfrak{g}$ which, in the orientation $\mathfrak{g}^+$ for which $\mathfrak{p}_1$ is a positive subray has the property that every negative subray of $\mathfrak{g}^+$ is a co-ray to $\mathfrak{s}$.

For straight spaces we obtain from (22.13):

(23.14) *If* $\alpha(\mathfrak{r}, x) + \alpha(\mathfrak{s}, x)$ *is constant for two rays* $\mathfrak{r}, \mathfrak{s}$ *on the same line* $\mathfrak{g}$, *then every point of the space lies on a parallel to* $\mathfrak{g}$ *(or, if* $\mathfrak{a}^+$ *is an asymptote to* $\mathfrak{g}^+$, *then so is* $\mathfrak{a}^-$ *to* $\mathfrak{g}^-$*).*

For (23.13) implies that for any orientation $\mathfrak{g}^+$ of $\mathfrak{g}$ one of the rays $\mathfrak{r}, \mathfrak{s}$ is positive and the other negative. The rest follows from (23.13) and the definition of parallel. The converse of (23.14) is not true since we saw in the proof of (23.13) that if $\alpha(\mathfrak{r}, x) + \alpha(\mathfrak{s}, x)$ is independent of x, then $K_\infty(\mathfrak{g}^+, p) = K_\infty(\mathfrak{g}^-, p)$ and this need — by (23.9) — not be true even when the parallel axiom holds.

24. Characterizations of the higher dimensional Minkowskian geometry.

It was explained in the introduction to the chapter why it is reasonable to expect interesting results if a space has convex spheres and satisfies the parallel axiom.

The Desarguesian case.

We begin our investigation by proving:

(24.1) ***Theorem:*** *A Desarguesian space in which the parallel axiom holds and the spheres are convex is Minkowskian.*

The parallel axiom implies that we deal with a metric defined in an entire affine space A^n. It is to be shown that the affine mid-point of a segment $T(p_1, p_2)$ coincides with the mid-point in the sense of the given metric. By considering a (two-dimensional) plane through $\mathfrak{g}(p_1, p_2)$ we reduce the problem to showing that the metric in a given plane P is Minkowskian.

A circle $K(q, \sigma)$ in P is a convex curve. It suffices to see that $K(q, \sigma)$ possesses parallel supporting lines at the end-points p_1, p_2 of a given diameter $T(p_1, p_2)$ of $K(q, \sigma)$. For then q is by (16.7) the affine center of $K(q, \sigma)$, hence the affine midpoint of $T(p_1, p_2)$ as well as in the sense of the metric. Supporting lines of $K(q, \sigma)$ at p_1 and p_2 are transversals to $\mathfrak{g}(p_1, p_2)$. Therefore (24.1) is contained in the following fact which holds without assuming Desargues' Theorem.

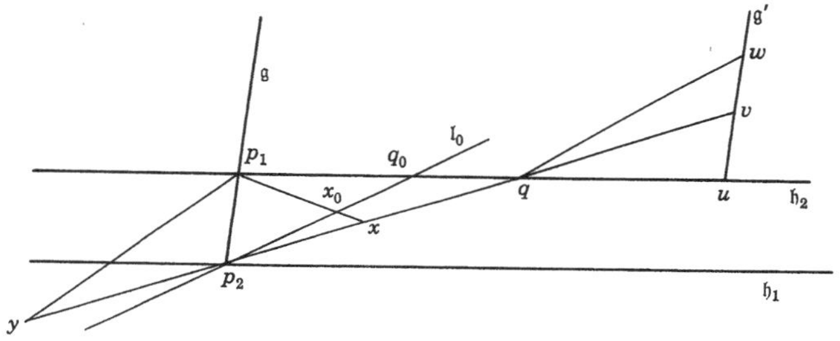

Fig. 22.

(24.2) *If a two-dimensional straight space satisfies the parallel axiom and has convex circles, then a line* $\mathfrak{g}$ *which is perpendicular to one of two parallel lines* $\mathfrak{h}_1, \mathfrak{h}_2$ *is also perpendicular to the other.*

Put $p_i = \mathfrak{g} \cap \mathfrak{h}_i$ and assume that $\mathfrak{g}$ is perpendicular to $\mathfrak{h}_1$. There is a transversal $\mathfrak{l}_0$ to $\mathfrak{g}$ at p_2 (see (20.12)). If $\mathfrak{l}_0 = \mathfrak{h}_2$ then nothing is to be proved — (23.9) shows that there may actually be a transversal $\mathfrak{l}_0 \neq \mathfrak{h}_2$ when the assumptions of (24.2) are satisfied.

If $\mathfrak{l}_0 \neq \mathfrak{h}_2$ let $q_0 = \mathfrak{l}_0 \cap \mathfrak{h}_1$. We show that for any q with $(p_1 q_0 q)$ the point p_2 is the foot of p_1 on $\mathfrak{l}$, the line joining p_2, q. It follows then for $p_1 q \to \infty$ from the parallel axiom that p_2 must also be the foot of p_1 on $\mathfrak{h}_2$.

If x lies on $\mathfrak{l}$ on the same side of p_2 as q, then $T(p_1, x)$ intersects $\mathfrak{l}_0$ in a point x_0 with $p_1 x > p_1 x_0 > p_1 p_2$, hence x is not the foot of p_1 on $\mathfrak{l}$. Let y be any point satisfying $(q\, p_2 y)$; through a point u with $(p_1 q u)$ draw the parallel $\mathfrak{g}'$ to $\mathfrak{g}$. Since $\mathfrak{g}'$ is the only line through u which does not intersect $\mathfrak{g}$ it must be the perpendicular to $\mathfrak{h}_1$ at u. The line $\mathfrak{g}'$ intersects $\mathfrak{l}$ at some point v and the parallel to $\mathfrak{g}(p_1, y)$ through q in a point w with (wvu), hence $wv < wu < wq$. Therefore $\mathfrak{g}(p_1, y)$ cannot be perpendicular to $\mathfrak{l}$. For if it were, then $\mathfrak{g}(q, w)$ would be perpendicular to $\mathfrak{l}$ as the only line through w which does not intersect $\mathfrak{g}(p_1, y)$. Hence $\mathfrak{g}$ is perpendicular to $\mathfrak{l}$, and (24.2) is proved.

The general case

Theorem (24.1) shows that the main problem is to find conditions under which a straight space with the parallel axiom and convex spheres is Desarguesian. It may be that no additional conditions are necessary but this seems hard to prove. In higher dimensions we will see that it suffices to add a condition, which in smooth spaces amounts to differentiability of the spheres. We need the following:

Lemma. Let G_2 be distinct from, and parallel to, G_1 and let L intersect G_i at p_i. Then an arc A with the following properties exists: (1) A connects points q, q' on L with (qp_2q') and contains no other point of L. (2) A intersects G_2 in a point r (3) for every other point y of A the line $\mathfrak{g}(p_2, y)$ intersects G_1, in a point x.

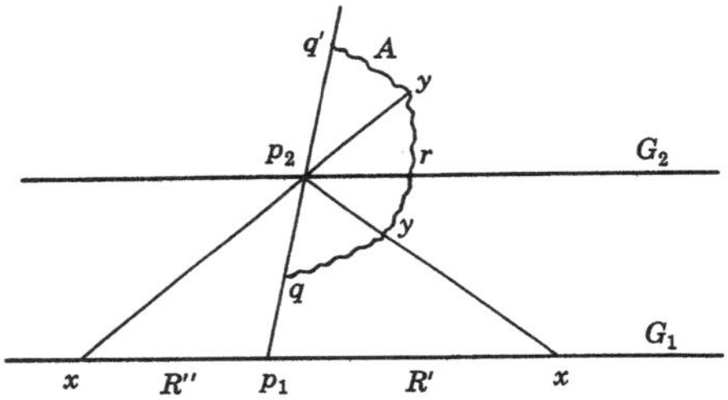

Fig. 23.

Let $p_2G_1 = 2\eta$ (> 0). The point p_1 divides G_1 into two rays R' and R''. For $x \in R'$ define y by (p_2yx) and $p_2y = \eta$. For $x \in R''$ define y by (xp_2y) and $p_2y = \eta$. If p_1 is considered as point of R' the point y is the point q with (p_2qp_1) and $p_2q = \eta$, and if p_1 is a point of R'' the point y is the point q' with (p_1p_2q') with $p_2q' = \eta$. If $p_1x \to \infty$ either for $x \in R'$ or $x \in R''$ the point y tends, because of the parallel axiom, to the same point r on G_2.

We apply this lemma to prove:

(24.3) *If G_1, G_2 are parallel and intersect L at p_1, p_2, and G_1 is perpendicular to L, then G_2 is perpendicular to L.*

For if A is constructed as in the lemma, then for reasons of continuity at least one point of A has p_2 as foot on L, but all lines $\mathfrak{g}(p_2, y)$ with $y \neq r$ on A intersect G_1, hence r must have p_2 as foot.

A second application of the lemma is:

(24.4) If G_1 intersects L at p_1 and is not perpendicular to L then a given point p_2 of L is foot of at least one point x_2 of G_1.

Let G_2 be the parallel to G_1 through $p_2 \neq p_1$.

We construct A according to the lemma. At least one point y_2 of A must have p_2 as foot on L. This time it cannot be r, because then G_2 would be perpendicular to L, and this would by (24.3) imply that G_1 is perpendicular to L. Since $y_2 \neq r$ the line $\mathfrak{g}(p_2, y_2)$ intersects G_1 in a point x_2.

This has the important corollary.

(24.5) *If a, b are distinct points of $K_\infty(\mathfrak{g}^+, q)$ then the parallel to $\mathfrak{g}$ through a (or b) is perpendicular to $\mathfrak{g}(a, b)$.*

For if the parallel G to $\mathfrak{g}$ through a were not perpendicular to $\mathfrak{g}(a, b)$ then G would by (24.4) contain a point x whose foot on $\mathfrak{g}(a, b)$ is b, then $xb < xa$, but a is by (23.2) the foot of x on $K_\infty(\mathfrak{g}^+, q)$.

We now define differentiability of spheres:

If the sphere $K(q, \sigma)$ in a straight space is convex, then it is called "*differentiable at $x \in K(q, \sigma)$*" if no proper subset of the set W, formed by the points on the supporting lines of $K(q, \sigma)$ at x, decomposes the space.

This implies because of (20.4) that W must coincide with the set of all points on the supporting rays of $K(q, \sigma)$ at x, and therefore in those spaces, in which we can speak of differentiability in the usual sense, our definition of differentiability is equivalent to the ordinary one. If $K(q, \sigma)$ is differentiable everywhere, we call it simply differentiable. As mentioned above, the following theorem then holds:

(24.6) ***Theorem:*** *A straight space of dimension greater than two in which the spheres are convex and differentiable and the parallel axiom holds, is a Minkowski space with differentiable spheres.*

To prepare the proof, consider any limit sphere $K = K_\infty(\mathfrak{g}^+, q)$. If a is a fixed point of K then $\mathfrak{g}(a, b)$ is for all $b \neq a$ on K transversal to the parallel L to $\mathfrak{g}$ through a, see (24.5). The transversals to L at a are just the supporting lines of $K(p, a)$ at a, where p is any point of L different from a, compare (20.11). Therefore K lies in the set W of all points on the supporting lines of $K(p, a)$ at a. The limit sphere K decomposes the space, but no subset of K does, because a parallel to $\mathfrak{g}$ intersects K in exactly one point and contains points on both sides of K. Since $K(p, a)$ is differentiable at a, it follows that $\mathfrak{g}(a, b) \subset W = K$.

Thus K contains with any two distinct points a, b the entire straight line $\mathfrak{g}(a, b)$. Since K is closed it is, in our previous terminology, flat.

SPACES WITH FLAT LIMIT SPHERES

The theorem (24.6) is therefore a consequence of the following:

(24.7) ***Theorem:*** *A straight space of dimension greater than two is a Minkowski space with differentiable spheres, if and only if the limit spheres are flat.*

The necessity follows immediately from (17.21). The sufficiency is more difficult. We decompose the rather long proof into several steps.

(a) If the line L intersects $K = K_\infty(\mathfrak{g}^+, p)$ at a, but does not lie in K, then the two rays into which a decomposes L, lie on different sides of K. We express this briefly by saying that L crosses K (at a).

For otherwise points p, q on L with $(p\ a\ q)$ would lie on the same side of K. There is a sequence a_ν (for instance on the parallel to $\mathfrak{g}$ through a) which approaches a and lies on the other side of K. Then $\mathfrak{g}(p, a_\nu) \to \mathfrak{g}(p, a)$, so that $\mathfrak{g}(p, a_\nu)$ would for large ν contain points close to q, on the same side of K as q, consequently $\mathfrak{g}(p, a_\nu)$ would intersect K twice without being contained in K.

(b) $K_\infty(\mathfrak{g}^+, p) = K_\infty(\mathfrak{g}^-, p)$. After this has been proved we will simply write $K_\infty(\mathfrak{g}, p)$ when convenient.

First let $p \in \mathfrak{g}$. Since $S(u, up) \cap S(v, vp) = 0$ if u, v are points of $\mathfrak{g}$ with $(u\ p\ v)$, the limit sphere $K_\infty(\mathfrak{g}^-, p)$ lies in the exterior of or on $K = K_\infty(\mathfrak{g}^+, p)$. A line $\mathfrak{g}(p, x)$ with $x \in K_\infty(\mathfrak{g}^-, p)$, $x \neq p$, lies in $K_\infty(\mathfrak{g}^-, p)$ since the limit spheres are flat, and does therefore not cross K, and so lies by (a) in K. It follows that $K(\mathfrak{g}^-, p) \subset K$. Similarly $K \subset K(\mathfrak{g}^-, p)$.

If p does not lie on $\mathfrak{g}$, there are by (22.14) points q_1, q_2 on $\mathfrak{g}$ such that $p \in K_\infty(\mathfrak{g}^+, q_1)$, $p \in K_\infty(\mathfrak{g}^-, q_2)$. Since p lies in $K_\infty(\mathfrak{g}^+, q_1) = K_\infty(\mathfrak{g}^-, q_1)$ and also in $K_\infty(\mathfrak{g}^-, q_2)$ it follows that $q_1 = q_2$, hence

$$K_\infty(\mathfrak{g}^+, p) = K_\infty(\mathfrak{g}^+, q_1) = K_\infty(\mathfrak{g}^-, q_1) = K_\infty(\mathfrak{g}^-, p).$$

(c) *If $\mathfrak{a}^+$ is an asymptote to $\mathfrak{g}^+$, then $\mathfrak{a}^-$ is an asymptote to $\mathfrak{g}^-$.*

This follows from (b) because $\mathfrak{a}$ is the locus, as x varies, of the feet of any of its points on the limit sphere $K_\infty(\mathfrak{g}^+, x)$, see (23.2), and then by (b) also the locus of feet of the same points on $K_\infty(\mathfrak{g}^-, x)$, hence an asymptote to $\mathfrak{g}^-$.

(d) *The spheres are convex.*

For if four points a, b, c, p with $(a\ c\ b)$ and $pc \geqslant \max(pa, pb) = \rho$ existed and $\mathfrak{r}$ denotes the ray with origin c containing p, then $K_\infty(\mathfrak{r}, c)$ would by (22.12) contain a and b in its interior. Then $\mathfrak{g}(a, b)$ would not cross $K_\infty(\mathfrak{r}, c)$ at c in contradiction to (a).

Since in Desarguesian spaces the asymptotes always have the transitivity property, it follows from (c), (d), (14.1) and (24.1) that Theorem (24.7) holds when we can show that the space is finite dimensional and that any three non-collinear points lie on a plane (= two-flat).

We begin with two observations on more general spaces.

(24.8) *A straight space is homeomorphic to the topological product of any limit sphere* $K = K_\infty(\mathfrak{g}^+, q)$ *and a straight line.*

For an asymptote to $\mathfrak{g}^+$ intersects K in exactly one point x, depends continuously on x, and two different asymptotes intersect K in different points.

(24.9) If R', a subspace of R, is with the metric of the G-space R itself a G-space and $\mathfrak{r}$ is a ray in R', q a point in R', then $K' = K_\infty(\mathfrak{r}, q) \cap R'$ is the limit sphere with central ray $\mathfrak{r}$ through q in R'.

This follows from the fact that K' and $K_\infty(\mathfrak{r}, q)$ are defined by the equation $\alpha(\mathfrak{r}, x) = \alpha(\mathfrak{r}, q)$ and distances for points in R' coincide with the distances in R.

Returning to the assumption that the space is straight and has flat limit spheres we show

(e) *The space has finite dimension.*

The proof will imply that it is homeomorphic to E^n with a suitable $n \geqslant 3$. Take any limit sphere $K_1 = K_\infty(\mathfrak{g}_1, q)$, $q \in \mathfrak{g}_1$ and on $\mathfrak{g}_1$ a point p_1 with $p_1q = 1$. Then $p_1K_1 = 1$.

In K_1 take any line $\mathfrak{g}_2$ through q and put $K_2 = K_1 \cap K_\infty(\mathfrak{g}_2, q)$ and let $p_2 \in \mathfrak{g}_2$ with $p_2q = 1$. Since K_2 is by (24.9) a limit sphere in K_1 we have $p_2K_2 = 1$. If dim $K_2 = 1$ we stop, otherwise we proceed in the same way obtaining points $p_1, p_2, p_3, \ldots$ and sets $K_1, K_2, \ldots$ such that $p_i q = 1$, $p_i K_i = 1$, $p_{i+1} \in K_i \subset K_{i-1}$. Hence $p_i p_k \geqslant 1$ for $k \neq i$.

Because the space is finitely compact, there is only a finite number of p_i. If p_{n-1} is the last, then K_{n-1} is a line, K_{n-2} by (24.8) the product of K_{n-1} with a line, hence homeomorphic to E^2. Continuing this process we find that K_1 is homeomorphic to E^{n-1} and the whole space to E^n.

We precede the proof that three given points lie on a plane by a remark on more general spaces.

(24.9) *If in a straight space* $p \in \mathfrak{g}$, *then a line* L *through* p *that contains a point* z *in the interior of* $K_\infty(\mathfrak{g}^+, p)$ *is not transversal to* $\mathfrak{g}$.

For if $x(\tau)$ represents the positive subray $\mathfrak{r}$ of $\mathfrak{g}$ with origin p, then, by the definition of the interior of a limit sphere, $\alpha(\mathfrak{r}, z) < \alpha(\mathfrak{r}, p)$. Hence (22.5) yields that $zx(\tau) - px(\tau) < 0$ for large τ, so that p cannot be a foot of $x(\tau)$ on L.

We apply this to show that in spaces with flat limit spheres

(f) $K = K_\infty(\mathfrak{g}, p)$, $p \in \mathfrak{g}$, *is the union of all transversals to* $\mathfrak{g}$ *at* p.

Since for $x \in \mathfrak{g}$, $x \neq p$, the point p is the foot of x on K, every line in K through p is transversal to $\mathfrak{g}$. Any other line through p contains, because of (a), points in the interior of K and is therefore by (24.9) not transversal to $\mathfrak{g}$.

If L_ν is transversal to G_ν and $L_\nu \to L$, $G_\nu \to G$, then L is transversal to G. This follows immediately from the continuity properties of feet. Therefore we conclude from (f) that $K_\infty(\mathfrak{g}, p)$, $p \in \mathfrak{g}$, varies continuously with $\mathfrak{g}$ and p. We can now prove

(g) *Any three non-collinear points a, b, c lie in a plane.*

We prove this by induction with respect to the dimension n of the space. So far we have not used that the dimension is at least three, we can therefore start the induction at $n = 2$ where nothing is to be proved. Assume (g) to be true for $n - 1 \geqslant 2$.

Put $\mathfrak{g} = \mathfrak{g}(a, b)$ and $K = K_\infty(\mathfrak{g}(a, b), c)$. Then K is by (24.8) an $(n - 1)$-dimensional flat subspace. Because of (24.9) the limit spheres in K are flat, hence K contains, by the inductive assumption, a plane P through c and $f = K \cap \mathfrak{g}$. The circle C in P with center f through c is homeomorphic to an ordinary circle (this is true in any straight two-dimensional space because of Pasch's Axiom).

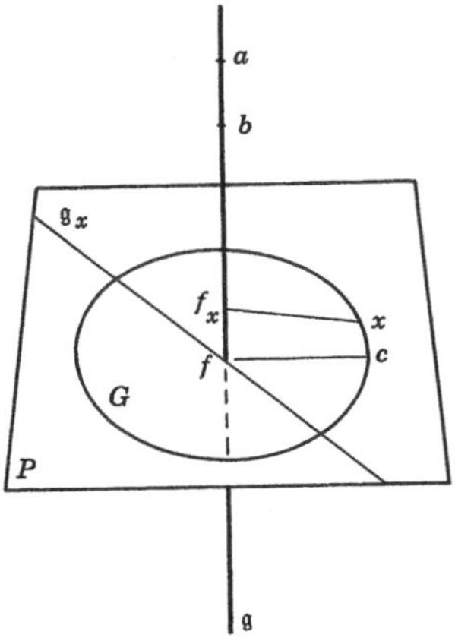

Fig. 24.

For any point $x \in C$ let f_x be the foot on the geodesic $\mathfrak{g}$ and put $K_x = K_\infty(\mathfrak{g}(x, f_x), f_x)$. Because of (f) the set K_x contains $\mathfrak{g}$. The line $\mathfrak{g}(x, f)$ crosses K_x by (a) and lies in P. Hence P contains points on both sides of K_x. Since K_x separates such points, $K_x \cap P$ separates P and is therefore a straight line $\mathfrak{g}_x$. The line $\mathfrak{g}_x$ does not contain x and varies continuously with x since K_x does. It intersects C in two diametrically opposite points. Therefore a position x_0 of x on C exists for which $\mathfrak{g}_{x_0}$ passes through c. Since $\mathfrak{g}_{x_0} = \mathfrak{g}(c, f) \subset K_{x_0}$, this limit sphere contains c and $\mathfrak{g}$, hence a, b, c. By the induction hypothesis there is a plane in K_{x_0} which passes through a, b, c. This completes the proof of (24.7), compare (24.1) and (14.1).

EUCLIDEAN SPACES

Remembering that symmetric perpendicularity in a Minkowski space of dimension greater than two implies that the space is euclidean, see (17.25), we are led to conjecture:

(24.10) ***Theorem:*** *A straight space of dimension greater than two is euclidean, if it satisfies the parallel axiom and the following condition: if p is not on the straight line $\mathfrak{h}$ and f is a foot of p on $\mathfrak{h}$, then f is the foot on $\mathfrak{g}(p, f)$ of any point x on $\mathfrak{h}$.*

Let $(f\,p\,y)$. Applying the hypothesis to x and $\mathfrak{g}(p, f)$ (instead of p and $\mathfrak{h}$), we see that f is a foot of y on $\mathfrak{h}$, hence f is by (20.6) the only foot of p on $\mathfrak{h}$. It follows from Theorem (20.9) that the spheres are convex; moreover, $\mathfrak{g}(p, f)$ is perpendicular to $\mathfrak{h}$ and $\mathfrak{h}$ to $\mathfrak{g}(p, f)$. Theorem (24.10) will therefore follow from (24.6) and (17.25), if we can show that the spheres are differentiable.

The supporting lines of $K(q, p)$ at p are the transversals to $\mathfrak{g} = \mathfrak{g}(q, p)$ at p. Call the set of all points on these transversals W.

It is to be shown that for any $a \in W$ the set $W - a$ does not separate the space. This is obvious for $a = p$. Let $a \neq p$, and $\mathfrak{g}'$ the parallel to $\mathfrak{g}$ through a. The line $\mathfrak{g}'$ does not contain a second point b of W. For then $\mathfrak{g}(p, a)$ and $\mathfrak{g}(p, b)$ would by (24.3) be transversals to $\mathfrak{g}'$, hence also perpendiculars to $\mathfrak{g}'$, but they intersect at p.

If u, v are on $\mathfrak{g}$ with $(u\,p\,v)$, then W separates $S(u, u\,p)$ from $S(v, v\,p)$. On the other hand $\mathfrak{g}'$ contains for large $p\,u$ and $p\,v$ points of $S(u, u\,p)$ and $S(v, v\,p)$. Since W decomposes the space into exactly two sets, and these sets can be connected by a segment on $\mathfrak{g}'$, the set $W - a$ does not decompose the space.

We know from the discussion at the end of Section 17, that (24.10) does not hold if the restriction on the dimension is omitted.

A CONDITION EQUIVALENT TO FLATNESS OF LIMIT SPHERES

We finally come to a perhaps less important, but very surprising characterization of Minkowski geometry. In euclidean spaces the following theorem holds:

(24.11) *A (non-empty) set F in E^n is closed and convex if, and only if, every point of the space has exactly one foot on F* [1].

That convex sets in E^n have this property is obvious. In fact we have:

(24.12) *A given point of a straight space has exactly one foot on a given closed convex set if, and only if, the spheres are convex.*

If a given point has exactly one foot on a given closed convex set F, then this holds in particular for segments, hence the spheres are by (20.9) convex.

Conversely, if F is any closed convex set, then two different feet f_1, f_2 of p on F are feet of p on $T(f_1, f_2) \subset F$, hence the spheres are not convex.

Thus the necessity part of (24.11) is common to a wide class of spaces. The sufficiency part is remarkably restrictive.

(24.13) ***Theorem:*** *In a straight space the closed convex sets are the only sets on which every point has exactly one foot if, and only if, the limit spheres are flat.*

This yields a conjunction with (24.7).

(24.14) ***Theorem:*** *Among the straight spaces of dimension greater than two, the Minkowski spaces with differentiable spheres are characterized by the property that a non-empty set on which every point has exactly one foot is convex and closed.*

The necessity of the condition in (24.13) is simple: If $K = K_\infty(\mathfrak{g}^+, p)$ is a limit sphere in a straight space then a given point q lies on exactly one asymptote $\mathfrak{a}$ to $\mathfrak{g}^+$. The point where $\mathfrak{a}$ intersects K is the only foot of q on K. Since K is closed it must by hypothesis be a convex set, hence K contains with any two distinct points x, y the segment $T(x, y)$. We have to show that this implies that $\mathfrak{g}(x, y) \subset K$.

Both the interior and the exterior of K form with K convex sets. If, for example, a is an interior point of K and b is in the interior of, or on K, then $T(a, b)$ cannot contain a point c in the exterior of K because $T(a, c)$ would intersect K in a point a' and $T(a', b)$ would not be entirely on K.

If now x, y are two distinct points of K and $(x\,y\,z)$, then z cannot lie either in the interior or in the exterior of K because every point of $T(z, x)$ except x would then by (20.1) also lie in the interior or exterior of K respectively, whereas $T(x, y)$ lies in neither. Hence $z \in K$ and $\mathfrak{g}(x, y) \subset K$.

It is less simple to show that the flatness of the limit spheres implies that a closed set on which every point has exactly one foot, is convex. We adapt the proof given by Jessen [1] for the euclidean case.

Let M be a non-empty set on which every point has exactly one foot. M must be closed since an accumulation point of M which does not belong to M has no foot on M.

Assume for an indirect proof that M is not convex. Then there are two points a, b in M and a point c with (acb) not in M. Since M is closed there is a positive η such that $S = \bar{S}(c, \eta)$ does not intersect M. Denote generally by p' the unique foot of p on M, and put $S_p = \bar{S}(p, pp')$.

The set F formed by the points p for which $S_p \supset S$, is closed because $pp' = pM$ depends continuously on p. The flatness of the limit spheres implies that F is bounded. For if F contained a sequence $\{z_\nu\}$ with $z_\nu c \to \infty$, then for a suitable subsequence $\{\mu\}$ of $\{\nu\}$ the segments $T(c, z_\mu)$ tend to a ray $\mathfrak{r}$. In a Minkowski space with differentiable spheres, and also for a two-dimensional straight space in which the limit circles are straight lines, it is easily seen that the spheres $K(z_\lambda, z'_\lambda)$ tend for a suitable subsequence $\{\lambda\}$ of $\{\mu\}$ to a limit sphere $K_\infty(\mathfrak{r}, d)$ with $\mathfrak{r}$ as central ray. But $K_\infty(\mathfrak{r}, d)$ cannot be flat because the interior of $K(z_\lambda, z'_\lambda)$ contains S but neither a nor b.

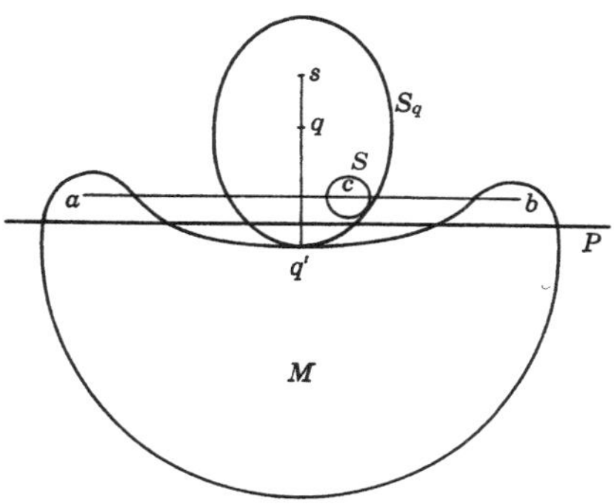

Fig. 25.

Since F is bounded and closed the function $pp' = pM$ reaches on F a maximum at some point q. Choose a plane P parallel to the tangent plane of $K(q, q')$ at q' such that it separates q' from S. The intersection A of S_q and the closed halfspace bounded by P and containing S is disjoint from M because q' is the only foot of q on M. For a point s with (sqq') sufficiently close to q we can find a sphere $\overline{S}(s, \delta)$ that contains A and is disjoint from M. Then $\delta > qq'$, because there is a point y on $P \cap K(q, q')$ whose foot on $\mathfrak{g}(q, q')$ falls on the same side of q as q'. Since the distance of y to a variable point of $\mathfrak{g}(q, q')$ is peakless, $ys > yq = qq'$, hence $ss' = sM \geqslant ys > qq'$. This contradicts the maximum property of q because $S_s \supset \overline{S}(s, \delta) \supset A \supset S$.

25. Characterization of the Minkowski plane

It is an open question whether Theorems (24.6) and (24.7) remain true when the restriction that the dimension be greater than two, is dropped. The difficulty derives, of course, from the complicated nature of Desargues' Theorem as compared to the existence of planes. But we will establish a theorem analogous to (24.6) under a stronger condition than the convexity of the circles, without assuming that the circles are differentiable.

In these considerations some facts on convex curves are needed. The proofs are so similar to the usual proofs in euclidean geometry that we might leave them to the reader. However, these facts will occur again and again, and therefore it may be well to discuss them briefly once for all.

CONVEX CURVES IN TWO-DIMENSIONAL G-SPACES

As we saw previously in Section **13**, every point of a two-dimensional G-space has a neighborhood D homeomorphic to E^2 with the following properties: for any two points x, y the segment $T(x, y)$ is unique, lies in D, and is contained in a unique segment whose endpoints lie on the boundary of D. The latter segment under omission of its endpoints we denote as the *line* $G(x, y)$ in D. If the space is straight we let D coincide with the space and $G(x, y) = \mathfrak{g}(x, y)$.

A convex curve in D [1] is a component of the boundary of a proper convex subset of D. A convex curve is always the complete boundary of a convex subset of D. An arc is convex if it lies on a convex curve.

A supporting line of a curve C is a line that contains at least one point of C, but does not separate any two points of C. The existence of a supporting line at a given point p of a convex curve C can be deduced from (20.3) but may also be seen as follows: Orient C, and let x approach p on C from the right. The line $G(p, x)$ revolves (because of the convexity) monotonically about p, but does not pass through a position where it contains a point of C to the left of p. Hence $G(p, x)$ tends to a limit G_r which is called the "*right tangent*" of C at p and is a supporting line of C.

If G_k is a supporting line of C at p_k, and p_k approaches p from the right, then $\lim_{k \to \infty} G_k = G_r$. This follows from the fact that any line which can be approached by a subsequence of $\{G_k\}$ is a supporting line of C at p.

If the right and left tangents of C at p coincide, then no other supporting line at p exists and we call C "*differentiable*" at p and the unique supporting line a "*tangent*". This definition applied to a convex circle $K(q, \rho)$ of a two-dimensional straight space agrees with the definition of the last section: any supporting line G of $K(q, \rho)$ at one of its points p decomposes the space, and no subset of G does. Hence $K(q, \rho)$ is differentiable at p if, and only if, the supporting line at p is unique.

The set of points where a convex curve is not differentiable is at most denumerable.

The standard proof of this fact in E^2 uses parallels and is therefore applicable only with a modification.

Let G_r and G_l be the right and left tangents of C at p. Let v be on the left, u on the right of p on C. If G_u and G_v are supporting lines of C at u and v, then $G_v \to G_l$ and $G_u \to G_r$ as $u \to p$ and $v \to p$. Since G_r and G_l intersect at p, the lines G_u and G_v intersect when u and v are sufficiently close to p.

About every point p we construct an arc $\widehat{uv}$ for which G_u and G_v intersect at a point w. Since a countable number of such arcs cover C it suffices to see that arcs of this type contain an at most denumerable number of points, where C is not differentiable.

Let q_1 and q_2 be two points on the arc where C is not differentiable, q_1 to the right of q_2. Traversing $T(u, w)$ from u towards w we meet in this order the intersections x_1', x_1'', x_2', x_2'' of $T(u, w)$ with the right tangent at q_1, the left tangent at q_1, the right tangent at q_2 the left tangent at q_2. Therefore $x_1'' \neq x_1'$, $x_2'' \neq x_2'$, but x_2' may coincide with x_1''. Thus $T(x_1', x_1'')$ and $T(x_2', x_2'')$ have at most one endpoint in common and are non-degenerate. There can be only a countable number of such intervals on $T(u, w)$.

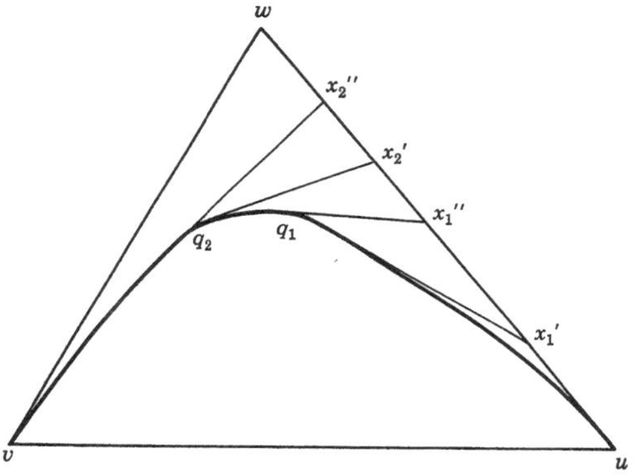

Fig. 26.

We formulate our results as:

(25.1) *A convex curve in a convex domain of a two-dimensional G-space possesses at every point p right and left tangents. They are supporting lines and coincide if and only if the supporting line at p is unique. The set of points where they do not coincide*, that is, where the curve is not differentiable, *is empty or countable.*

Just as in euclidean geometry the existence of supporting lines is characteristic for convexity:

(25.2) *A closed Jordan curve C in D which has a supporting line at each point, is convex.*

The proof runs exactly as usual: If a segment $T(a, b)$ with a, b inside or on C existed, that contains a point c outside of C, let a' and b' be points on $T(c, a) \cap C$ and $T(c, b) \cap C$ respectively, and let p be a point inside C but not on the line $G(a, b)$. Then $G(p, c)$ intersects C in a point q with (cqp) and in a point r with (qpr). Obviously any line through q separates two of the three points a', b', r on C, contrary to our assumption.

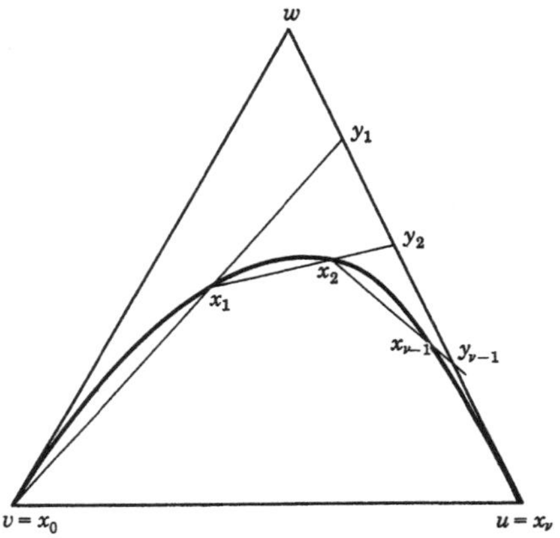

Fig. 27.

Finally we show:

(25.3) *A convex arc A has finite length.*

Since A is subset of a convex curve it can be covered by a finite number of arcs of the type constructed in the proof of (25.1). It suffices, therefore, to show that such an arc has finite length. If $x_0 = v, \; x_1, \ldots, \; x_{\nu-1}, \; x_\nu = u$ follow each other from left to right on A, denote by y_k the intersection of $G(x_{k-1}, x_k)$ with $T(u, w)$. Then $w, y_1, \ldots, y_{\nu-1}, u$ lie in this order on $T(w, u)$ and

$$vw + wy_1 \geqslant vy_1 = x_0x_1 + x_1y_1$$

$$x_1y_1 + y_1y_2 \geqslant x_1y_2 = x_1x_2 + x_2y_2$$

$$x_{\nu-2}y_{\nu-2} + y_{\nu-2}y_{\nu-1} \geqslant x_{\nu-2}y_{\nu-1} = x_{\nu-2}\,x_{\nu-1} + x_{\nu-1}\,y_{\nu-1}$$

$$x_{\nu-1}\,y_{\nu-1} + y_{\nu-1}\,u \geqslant x_{\nu-1}\,u = x_{\nu-1}\,x_\nu.$$

By addition of these relations we find

$$vw + wu \geqslant x_0\,x_1 + x_1\,x_2 + \ldots + x_{\nu-1}\,x_\nu.$$

Therefore the length of A cannot surpass $vw + wu$.

It is remarkable that an exact proof of (25.2) in E^2 was already given by Archimedes. In fact, his proof is essentially the same as the present one.

Imbedding of straight spaces

We are going to establish our characterization of plane Minkowskian geometry by imbedding the plane in a three-dimensional space. We define the imbedding first. If R is a straight space, define a metric space R^* with points x_τ, $x \in R$, τ real, by means of the distance

(25.4) $$x_\tau\,y_\sigma = (xy^\alpha + |\tau - \sigma|^\alpha)^{1/\alpha}, \qquad \alpha > 1.$$

Then R^* is straight. The subsets $\tau = \text{const.}$ of R^* are isometric to R. This is the special case of (8.15) where R' is the real τ-axis with the absolute value of the difference as distance.

We add the observation:

(25.5) *If $z(\xi)$ represents a straight line in R, $z^*(\zeta)$ represents a straight line of R^* which lies over $z(\xi)$, and $pz(\xi)^\alpha$ is a convex function of ξ, then $[p_\tau z^*(\zeta)]^\alpha$ is for any τ, a convex function of ζ.*

For if $z^*(\zeta_i) = z_{\tau_i}(\xi_i)$, $i = 1, 2$, then

$$z_\theta^* = z^*((1-\theta)\zeta_1 + \theta\,\zeta_2) = z_{(1-\theta)\tau_1+\theta\tau_2}((1-\theta)\,\xi_1 + \theta\,\xi_2)$$

because the metric in the plane over $z(\xi)$ is Minkowskian, compare Section 17.

If $0 \leqslant \theta \leqslant 1$ the convexity of $pz(\xi)^\alpha$ and $|\sigma|^\alpha$ yield

$$\begin{aligned} p_\tau z_\theta^{*\alpha} &= (pz((1-\theta)\,\xi_1 + \theta\,\xi_2))^\alpha + |\tau - (1-\theta)\,\xi_1 - \theta\,\xi_2|^\alpha \\ &\leqslant (1-\theta)\,pz(\xi_1)^\alpha + \theta\,pz(\xi_2)^\alpha + (1-\theta)\,|\tau - \xi_1|^\alpha + \theta\,|\tau - \xi_2|^\alpha \\ &= (1-\theta)\,p_\tau z^*(\zeta_1)^\alpha + \theta\,p_\tau\,z^*(\zeta_2)^\alpha. \end{aligned}$$

Characterization of the Minkowski plane

The convexity of circles in a straight space was seen (Theorem (20.9)) to be equivalent to the statement that $px(\tau)$ is peakless when $x(\tau)$ represents a line. We also saw in Section 18, that the convexity of $px(\tau)^\alpha$ degenerates for $\alpha \to \infty$ into peaklessness. This elucidates the bearing of the principal theorem of the present section:

(25.6) ***Theorem:*** *A two-dimensional straight space is Minkowskian if, and only if, it satisfies the parallel axiom and an $\alpha \geqslant 1$ exists such that $py^\alpha \leqslant \frac{1}{2}(px^\alpha + pz^\alpha)$ whenever y is the midpoint of x and z.*

The necessity follows from (17.23) and the fact that $px(\tau)^\alpha$ is convex if $px(\tau)$ is convex.

If $x(\tau)$ represents a geodesic, then the condition in (25.6) amounts to $px\left(\frac{\tau_1+\tau_2}{2}\right)^\alpha \leqslant (\frac{1}{2})\,px(\tau_1)^\alpha + (\frac{1}{2})\,px(\tau_2)^\alpha$. Since $px(\tau)$ is a continuous function of τ this implies that $px(\tau)^\alpha$ is convex. Because convexity of $px(\tau)$ implies convexity of $px(\tau)^\alpha$ for $\alpha > 1$ it suffices to consider the case $\alpha > 1$.

Denote the given space by R. Since $px(\tau)^\alpha$ is convex $px(\tau)$ is peakless hence the circles in R are convex, see (20.9).

Let a be a point on the circle $K(p, 1)$, where this circle is differentiable, and let H be the tangent or unique supporting line at a. Let $\mathfrak{g}^+$ be the orientation of $\mathfrak{g}(a, p)$ for which p follows a. Then H is the unique transversal to $\mathfrak{g}$ at a and also supporting line of the limit circles $K_\infty(\mathfrak{g}^+, a)$ and $K_\infty(\mathfrak{g}^-, a)$. It follows that $H = K_\infty(\mathfrak{g}^+, a) = K_\infty(\mathfrak{g}^-, a)$. For, since $K_\infty(\mathfrak{g}^+, a)$ decomposes the plane, it suffices to show that $H \supset K_\infty(\mathfrak{g}^+, a)$. If b is any point of $K_\infty(\mathfrak{g}^+, a)$ different from a, then $\mathfrak{g}(a, b)$ is by (24.5) transversal to $\mathfrak{g}$, hence $\mathfrak{g}(a, b) = H$ and $b \in H$.

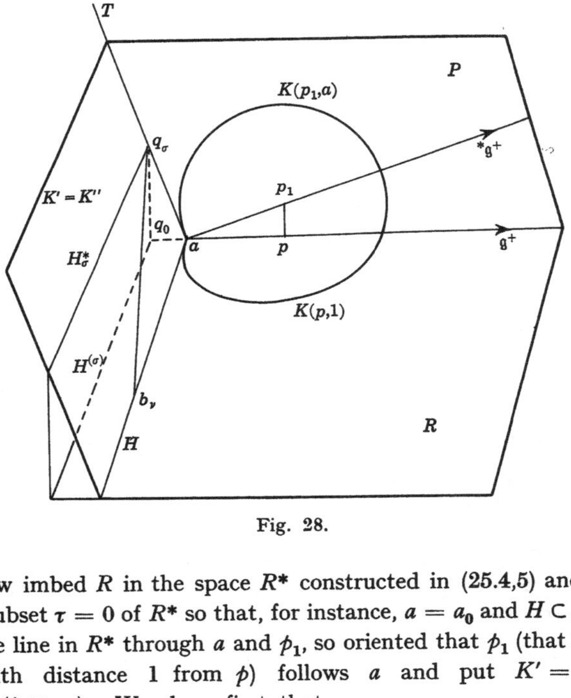

Fig. 28.

We now imbed R in the space R^* constructed in (25.4,5) and identify R with the subset $\tau = 0$ of R^* so that, for instance, $a = a_0$ and $H \subset R^*$. Denote by $^*\mathfrak{g}^+$ the line in R^* through a and p_1, so oriented that p_1 (that is, the point over p with distance 1 from p) follows a and put $K' = K_\infty(^*\mathfrak{g}^+, a)$, $K'' = K_\infty(^*\mathfrak{g}^-, a)$. We show first that

$$H \subset K' \cap K'' = K_\infty(^*\mathfrak{g}^+, a) \cap K_\infty(^*\mathfrak{g}^-, a).$$

To show $H \subset K'$ consider a sequence of points $x^\nu_{\tau_\nu}$ on $^*\mathfrak{g}^+$ which tend in the positive direction to infinity. Then $\tau_\nu \to \infty$. Since the points in R^* over $\mathfrak{g}$ form a Minkowski plane P and $p_0 a = p p_1 = 1$, we have

$$x^\nu{}_0\, a = x^\nu{}_0\, x^\nu_{\tau_\nu} = \tau_\nu$$

and if b lies on the limit circle $H = K_\infty(\mathfrak{g}^+, a)$ in R

$$0 \leqslant \varepsilon_\nu = x^\nu{}_0 b - x^\nu{}_0 a \to 0$$

see (22.5). Therefore

$$x^\nu_{\tau_\nu} b - x^\nu_{\tau_\nu} a = (x^\nu{}_0 b^\alpha + \tau^\alpha_\nu)^{1/\alpha} - (x^\nu{}_0 a^\alpha + \tau^\alpha_\nu)^{1/\alpha} = [(\varepsilon_\nu + \tau_\nu)^\alpha + \tau^\alpha_\nu]^{1/\alpha} - 2^{1/\alpha}\tau_\nu$$
$$\leqslant (\tau^\alpha_\nu + 2\alpha\varepsilon_\nu\tau_\nu^{\alpha-1} + \tau^\alpha_\nu)^{1/\alpha} - 2^{1/\alpha}\tau_\nu, \text{ for large } \nu.$$

Hence

$$x^\nu_{\tau_\nu} b - x^\nu_{\tau_\nu} a \leqslant 2\tau_\nu\,[(1 + \alpha\varepsilon_\nu/\tau_\nu)^{1/\alpha} - 1] \leqslant 2\tau_\nu\left(1 + 2\alpha\frac{1}{\alpha}\frac{\varepsilon_\nu}{\tau_\nu} - 1\right) = 4\varepsilon_\nu$$

for large ν, so that $x^\nu_{\tau_\nu} b - x^\nu_{\tau_\nu} a \to 0$ which means, because of (22.5) that $b \in K'$, hence $H \subset K'$, and in the same way $H \subset K''$.

We want to show that $K' = K''$ and that this set is flat. Since the metric in P is Minkowskian with differentiable circles, the tangent T to the circle $P \cap K(p_1, a)$ at a lies in K' and K''. It follows from (25.5) that the spheres in R^* are convex. Therefore the interiors of K' and K'' form together with K' and K'' convex sets. The intersection of these sets is convex and, since the interiors of K' and K'' are disjoint, this intersection equals $K' \cap K''$. Hence $K' \cap K''$ is convex.

Since $K' \cap K''$ is also closed it contains the closure F of the union of all segments $T(x, y)$ with $x \in H$, $y \in T$.

Let q_σ be any point of T different from a. Orient H. If b^ν tends on H^+ in the positive direction to infinity, then $T(q_\sigma, b^\nu)$ tends by definition to a co-ray $\mathfrak{r}^*$ to H. The segment $T(q, b^\nu)$ lies over $T(q_\sigma, b^\nu)$ which tends to a co-ray to H^+ in R. Because the parallel axiom holds in R, the latter co-ray lies on the parallel $H^{(\sigma)}$ to H through q_σ. Hence $\mathfrak{r}^*$ lies over $H^{(\sigma)}$. Similarly the co-ray to H^- through q_σ lies over $H^{(\sigma)}$. The two co-rays lie therefore on a parallel H^*_σ to H through q_σ. If $x(\xi)$ represents $H^{(\sigma)}$, then $x_\sigma(\xi)$ must represent H^*_σ, because H^*_σ lies in the Minkowski plane formed by the points over $H^{(\sigma)}$, and cannot intersect $H^{(\sigma)}$ because both $H^{(\sigma)}$ and H^*_σ are parallel to H. The line $x_\sigma(\xi)$ is the only line in the Minkowski plane that does not intersect H.

The line H^*_σ lies in F because $T(q_\sigma, b^\nu) \subset F$ implies $\mathfrak{r}^* \subset F$. Putting $H^*_0 = H^{(0)} = H$ we see that

$$\bigcup_{-\infty < \tau < \infty} H^*_\tau \subset F \subset K' \cap K''.$$

The set on the left decomposes R, and since no subset of K' or K'' decomposes R we find $K' = K'' = \cup\, H^*_\tau$. We know now that $K' = K' \cap K''$ is a convex set and conclude as in the first part of the proof of (24.14) that K' is flat.

The lines $H^{(\tau)}$ are parallel to H in R. Consider any three of these lines $H^{(\tau_1)}$, $H^{(\tau_2)}$, $H^{(\tau_3)}$ and their intersections c^1, c^2, c^3 and d^1, d^2, d^3 with two other lines L_c and L_d in R. The Minkowski plane over L_c intersects K' in three points $c^{\,1}_{\tau_1}, c^{\,2}_{\tau_2}, c^{\,3}_{\tau_3}$ which must lie on a line because K' is flat. Since the geometry in the plane over L_c is Minkowskian

$$\frac{c^{\,1}_{\tau_1}\, c^{\,3}_{\tau_3}}{c^{\,1}_{\tau_1}\, c^{\,2}_{\tau_2}} = \frac{|\tau_1 - \tau_3|}{|\tau_1 - \tau_2|} = \frac{c^1 c^3}{c^1 c^2}.$$

Applying the same arguments to d^1, d^2, d^3 we obtain

$$\text{(25.7)} \qquad \frac{d^1 d^3}{d^1 d^2} = \frac{|\tau_1 - \tau_3|}{|\tau_1 - \tau_2|} = \frac{c^1 c^3}{c^1 c^2}.$$

We now remember that H was the supporting line of $K(p, 1)$ at any point a where this circle is differentiable. Take any second supporting line N of $K(p, 1)$, not parallel to H, at another point where $K(p, 1)$ is differentiable.

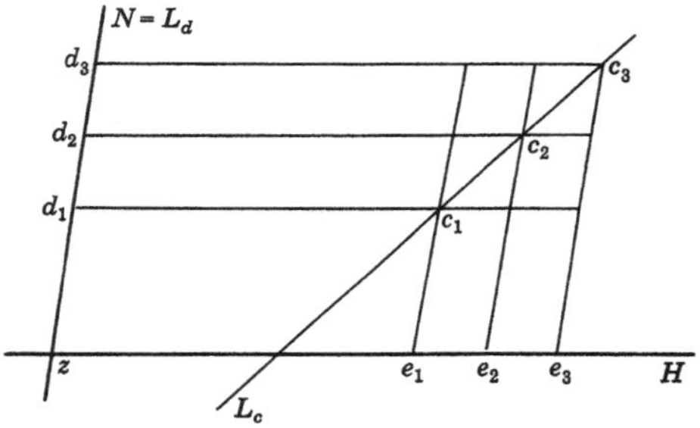

Fig. 29.

If c^1, c^2, c^3 are points on a line L_c which is parallel neither to H nor N ($= L_d$) and the parallel through c^i to H and N intersects N and H at d^i and e^i then the relation (25.7) shows that

$$\text{(25.8)} \qquad \frac{d^1 d^3}{d^1 d^2} = \frac{c^1 c^3}{c^1 c^2} = \frac{e^1 e^3}{e^1 e^2}.$$

Let $z = H \cap N$ and orient H and N. For any point c we define coordinates ξ, η as follows: If the parallel through c to N (or H) intersects H (or N) at e (or d) we put $\xi = \pm\, z\,e$ (or $\eta = \pm\, z\,d$) according to whether e (or d) follows or precedes z on H (or N). The equation (25.8) then states for $c_i = (\xi_i, \eta_i)$ that

$$\frac{\eta_1 - \eta_3}{\eta_1 - \eta_2} = \frac{\xi_1 - \xi_3}{\xi_1 - \xi_2}$$

and may be interpreted to mean that the variable point c_3 on the line L_c through c_1 and c_2 satisfies a linear equation.

This means, of course, that the lines in R are the ordinary lines or that Desargues' Theorem holds. The theorem now follows from (24.1).

Remark on ellipses and hyperbolas

In a two-dimensional straight space R we may define the *ellipse* E with foci p and q and eccentricity ε as the locus:

$$(25.9) \qquad px + qx = pq/\varepsilon, \qquad \varepsilon < 1.$$

It is easily seen that E is homeomorphic to a circle. The same reasoning as was used for circles, shows that the ellipses with foci p and q are convex if and only if for any straight line $x(\tau)$ the function $px(\tau) + qx(\tau)$ is peakless. Since peaklessness of $px(\tau)$ and $qx(\tau)$ does not imply that their sum is peakless, the convexity of the circles does not imply the convexity of the ellipses.

However, if $px(\tau)$ is convex for every line and point, then $px(\tau) + qx(\tau)$ is convex, because the sum of two convex functions is convex. Therefore *ellipses are convex in many geometries*, for instance, in a Minkowski plane or on any simply connected surfaces with negative curvature in E^3.

One might therefore expect that the two branches

$$(25.10) \qquad px - qx = pq/\varepsilon, \qquad qx - px = pq/\varepsilon, \qquad \varepsilon > 1$$

of the hyperbola with foci p, q and eccentricity ε can be convex curves in many different geometries. *This is not so.*

For it is readily seen that if the two branches of a hyperbola are convex, the convex set bounded by either branch does, as in E^2, not contain the other branch. If p and q are given and the branches (25.10) are convex for every $\varepsilon > 1$, we see for $\varepsilon \to \infty$ that the locus $px - qx = 0$ must bound two convex domains and is therefore a straight line. It will appear later that the only two-dimensional straight spaces in which the loci $px = qx$ are straight lines, are the euclidean and the hyperbolic planes, see Theorem (47.4).

Relations to the calculus of variations

We conclude this chapter by stating briefly how the present assumptions and terminology are related to those customary in the calculus of variations. No proofs will be given: the results will not be used, but they offer no difficulty to those familiar with the calculus of variations. On the other hand, proofs understandable to the non-expert would require repeating a long list of standard definitions, notations and results. All concepts used here are found, for instance, in Bliss [1].

We assume that our G-space is a Finsler space R satisfying the conditions (a) (b) (c) (d) (i) (j) (k) of Section 15. It was mentioned before that:

(25.11) *R is straight if and only if it is simply connected and no geodesic contains a pair of mutually conjugate points.*

The "*Non-conjugacy hypothesis*", see for instance Morse-Hedlund [2], means that the universal covering space (see Section 28) is straight.

In the calculus of variations the curve $p(\alpha)$ is called transversal to the curve $q(\beta)$ at the common point $r = p(\alpha_0) = q(\beta_0)$ if

$$\sum_{i=1}^{n} F_{\xi_i}(\alpha, dq/d\beta)\, dp_i/d\alpha = 0 \qquad \text{for} \qquad \alpha = \alpha_0, \qquad \beta = \beta_0.$$

This means in terms of the local Minkowskian geometry at r, that the line containing the line element parallel to dp through the point where the line G through r with direction dq intersects $F(x, \xi) = 1$, lies in the tangent plane of $F(x, \xi) = 1$. Therefore the line through r parallel to dp in the local Minkowski geometry is transversal in our sense to the line G.

If $x(\tau)$ and $y(\sigma)$ represent geodesics intersecting at $r = x(0) = y(0)$, and p is a foot of $x(\tau_1)$, $\tau_1 \neq 0$ (where $x(\tau)$ represents for $0 \leqslant \tau \leqslant \tau_1$ a segment S_{τ_1}) on a segment T_δ: $y(\sigma)$, $|\sigma| \leqslant \delta$, $\delta > 0$, then $y(\sigma)$ is transversal to $x(\tau)$ at p:

$$\text{(25.12)} \qquad \sum_{i=1}^{n} F_{\xi_i}(x, dx/d\tau)\, dy_i/d\sigma = 0 \qquad \textit{for} \qquad \tau = \sigma = 0.$$

(25.13) *If (25.12) is satisfied and $x(\tau)$ represents a segment S_{τ_1} for $0 \leqslant \tau \leqslant \tau_1$, then p is the unique foot $x(\tau_1)$ on T_δ if S_{τ_1} does not contain a focal point of T_δ.*

The strong satement that p is the *unique* foot of $x(\tau_1)$ is not needed for our purposes, it can be avoided by using (20.6).

There is a positive number $\alpha(p) \leqslant \rho(p)$ which depends continuously on p, such that for any segment $T_{\alpha(p)}$ represented by a geodesic $y(\sigma)$ with $y(0) = p$ for $|\sigma| \leqslant \alpha(p)$ and any geodesic $x(\tau)$, $x(0) = p$ intersected transversally by $y(\sigma)$ at p the segment $S_{\alpha(p)}$ represented by $x(\tau)$ for $0 \leqslant \tau \leqslant \alpha(p)$ contains no focal point of $T_{\alpha(p)}$.

Therefore (20.8) implies that for every point p of R the spheres $K(p, \rho)$ are strictly convex for sufficiently small positive ρ. This result is due to Whitehead [1].

If the space is straight we may formulate the following *Non-focality-hypothesis*: If the geodesic $y(\sigma)$ intersects the geodesic $x(\tau)$ at $p = x(0) = y(0)$ transversally then a positive δ exists such that no point $x(\tau)$, $\tau \neq 0$ is a focal point of $T_\delta : y(\sigma)$, $|\sigma| \leqslant \delta$.

The Non-focality hypothesis implies (and is implied by) convexity of spheres (see L. W. Green [1]): If, with the previous notation, the point $x(\tau_1)$, $\tau_1 \neq 0$ has two feet $p = y(0)$ and $y(\sigma_1)$, $\sigma_1 > 0$ on $y(\sigma)$, then $x(\tau_1)\, y(\sigma)$ would either be constant for $0 \leqslant \sigma < \sigma_1$, or it would reach a maximum for some value σ', $0 < \sigma' < \sigma_1$. The first case is impossible because then p would not be the unique foot of $x(\tau_1)$ on any T_δ.

In the second case $y(\sigma)$ would intersect the geodesic through $x(\tau_1)$ and $y(\sigma')$ transversally at $y(\sigma')$ which contradicts (25.13). Thus we have found:

(25.14) *A simply connected Finsler space is straight and has convex spheres if and only if it satisfies the Non-conjugacy and Non-focality hypotheses.*

It should be mentioned that by defining focal points in a more general way than here, the Non-focality hypothesis may be, and often is, formulated such as to imply the Non-conjugacy hypothesis.

Finally we mention the following consequence of our results:

(25.15) *A straight Riemann space which has convex spheres and satisfies the parallel axiom, is euclidean.*

Since perpendicularity is symmetric in a Riemann space, the theorem follows for spaces of dimension greater than two from (24.10). If the dimension equals two, then the arguments in the proof of (24.6) show that the limit circles are geodesics. Since limit circles with the same central ray are equidistant, the assertion follows immediately from the Gauss-Bonnet Theorem.

L. W. Green [1] contains for two-dimensions the result: If in a straight Riemannian plane with convex circles two non-intersecting lines H, G exist such that xH and yG are bounded for $x \in G$ and $y \in H$, then the metric in the strip bounded by H and G is euclidean, if the Gauss curvature is bounded from below. Our examples (23.8 b, c) show that this theorem has no direct analogue in Finsler spaces.

CHAPTER IV

COVERING SPACES

26. Introduction

Covering spaces, have for a long time, played an important part in the theory of functions of a complex variable and in topology. Very satisfactory accounts of the general topological theory are found in the literature, in particular in Section 46 of Pontrjagin's Topological Groups, and in Chapters 7, 8 of Seifert and Threlfall's Lehrbuch der Topologie. In view of this it would not be necessary to treat covering spaces here, were it not that for G-spaces questions of metric nature arise, which have not interested the topologists. Although covering spaces have been used in differential geometry, no book treats the general theory from the point of view of this field.

The G-space $\overline{R}$ *is a covering space of the G-space* R *if a locally isometric mapping* Ω *of* $\overline{R}$ *on* R *exists*; this means that there is for every point $\overline{p}$ of $\overline{R}$ a positive $\eta_{\overline{p}}$ such that Ω maps $S(\overline{p}, \eta_{\overline{p}})$ isometrically on $S(p, \eta_{\overline{p}})$, where $p = \overline{p}\Omega$. We establish in Section 27 the elementary properties of locally isometric mappings and find, in particular, that Ω maps $S(\overline{p}, \rho(p)/2)$ isometrically on $S(p, \rho(p)/2)$. In contrast to the topological theory, the number $\rho(p)/2$ is not only independent of the choice of $\overline{p}$ in $p\Omega^{-1}$, but also of the choice of Ω. This has as a consequence, that a locally isometric mapping of a compact G-space on itself is one-to-one and hence a motion, a theorem which has no analogue in topology. In the proofs we refer to the above two accounts of the topological theory, whenever their methods may be applied without change.

Then (Section 28) we construct the *universal covering space* $\overline{R}$ of a given space R very much as in the topological theory; new questions are, whether $\overline{R}$ is again finitely compact and whether $\overline{R}$ is unique in the metric sense, that is up to isometries. The answer to both questions is affirmative. Next we show that the *fundamental group* $\mathfrak{F}$ *of* R *may be realized as the group of those motions* Ψ *of* $\overline{R}$, *which lie over the identity motion of* R, i. e. satisfy $\Psi\Omega = \Omega$. With the topology of Section 4 the group $\mathfrak{F}$ is discrete and no motion of $\mathfrak{F}$ has a fixed point.

If $\mathfrak{G}$ is a discrete group of motions without fixed points of the G-space R', then identification of the points $x\Phi$, $\Phi \in \mathfrak{G}$, in R' yields a new G-space R, and R' is locally isometric to R (Section 29). To visualize the space R obtained in this way we employ, as usual, *fundamental sets*. The construction of such sets shows that a compact G-space has finite connectivity.

Section 30 deals with G-spaces in which every point has a neighborhood $S(p, \rho_p)$ isometric to a sphere in a Minkowskian, hyperbolic or spherical space. It is shown that their universal covering spaces are the Minkowskian, hyperbolic or spherical spaces.

Using a previous result, (17.19), we reduce the study of locally Minkowskian spaces to that of locally euclidean spaces and find, following Killing, that all locally Minkowskian spaces have finite connectivity. We discuss all two-dimensional locally euclidean spaces, and give some examples of two-dimensional locally hyperbolic spaces.

The theory of covering spaces is used in Section 31 to establish the result already mentioned on several occasions, that *in a G-space in which the geodesic through two points is unique, either all geodesics are straight lines or all are great circles of the same length.* If, in the latter case, the dimension is greater than one, the space has a spherelike space (see Section 21) as two sheeted universal covering space. The proof is long but is quite simple under the assumptions of classical differential geometry. This is not due to the differentiability assumptions as such, but to the hypothesis that the space is a topological manifold, without which differentiability cannot be formulated.

Shortest curves in a given (non-trivial) class of freely homotopic curves of a G-space R are *closed geodesics*, and all closed geodesics are obtained in this way when the universal covering space $\overline{R}$ of R is straight, Section 32. In that case the closed geodesics in R correspond to the lines in $\overline{R}$ which go into themselves under motions in the fundamental group of R, considered as a group of motions of $\overline{R}$. This connection leads to many results of which we mention the following: In a compact space without conjugate points and with an abelian fundamental group no geodesic has multiple points and the geodesics in a given free homotopy class cover the space simply.

In the remainder of the chapter these ideas are applied to the study of two-dimensional compact manifolds from which they originally evolved.

The interest in the *torus* derives from the remarkable theorem of E. Hopf [1] that a Riemannian metrization of the torus without conjugate points is euclidean. Section 33 characterizes all systems of curves on the torus which can serve as geodesics without conjugate points in terms of systems of curves in the plane as the universal covering space of the torus. The most recondite property is the *parallel axiom.* Since we also find that Desargues' Theorem need not hold, we see that *E. Hopf's result belongs to the few Riemannian theorems which have no analogue in Finsler spaces.*

The surfaces of *higher genus* have evoked considerable interest through various phenomena such as *transitive geodesics*, i. e. geodesics which approach any geodesic curve of finite length arbitrarily closely. The existence of such geodesics was first shown by Nielsen [1] for compact orientable surfaces with constant negative curvature, then by Koebe [1] for more general topological types. The condition of constant negative curvature was greatly reduced by Morse-Hedlund [2], and Green [1] implies a further substantial reduction. The behavior of the geodesics became the object of a detailed study in the theory of symbolic dynamics, see Morse-Hedlund [1].

A significant section of this theory can be established by our methods for non-Riemannian metrics and without differentiability hypotheses, in fact, the material is much too abundant to be fully discussed here. In Section **34** we therefore merely extend Nielsen's original results to compact orientable manifolds whose universal covering space is straight, contains no pair of parallel lines and has the further property that the distance $x(\tau)\mathfrak{s}$ of a point $x(\tau)$ traversing a ray $x(\tau)$, $\tau \geqslant 0$ from another ray $\mathfrak{s}$ with the same origin $x(0)$ tends to infinity. It will be seen in the next chapter that the conditions are satisfied when the given surface has negative curvature in the general sense defined there.

27. Locally isometric spaces

The mapping Ω of the G-space $\overline{R}$ on the G-space R is called *"locally isometric"* if for every point $\overline{p}$ of $\overline{R}$ a positive number $\eta_{\overline{p}}$ exists such that Ω maps $S(\overline{p}, \eta_{\overline{p}})$ isometrically on $S(\overline{p}\Omega, \eta_{\overline{p}})$. Obviously Ω is continuous.

If Ω maps the set $\overline{A}$ in $\overline{R}$ on the set A in R we say that $\overline{A}$ *lies over* A. In particular, any point in $p\Omega^{-1}$ *lies over* p. In contrast to the purely topological theory of locally homeomorphic spaces we do not require that inf $\eta_{\overline{p}} > 0$ for $\overline{p} \in p\Omega^{-1}$ because an even stronger statement than this can be proved, compare (27.10) below. We also say that the curve $\overline{x}(\tau)$ in $\overline{R}$ lies over the curve $\overline{x}(\tau)\,\Omega$ in R and so forth. If $\overline{x}$ is any point in $\overline{R}$ we denote $\overline{x}\Omega$ by x.

Elementary properties

We list a few immediate consequences of the definition:

(27.1) *Let* $\eta(\overline{p}) = \sup \eta$, *where* η *traverses the numbers for which* Ω *maps* $S(\overline{p}, \eta)$ *isometrically on* $S(p, \eta)$. *Then* Ω *maps* $S(\overline{p}, \eta(\overline{p}))$ *isometrically on* $S(p, \eta(\overline{p}))$. *Moreover, either* $\eta(\overline{p}) \equiv \infty$ *and* Ω *is an isometry, or* $0 < \eta(\overline{p}) < \infty$ *and* $|\eta(\overline{p}) - \eta(\overline{q})| \leqslant \overline{p}\,\overline{q}$.

For a proof we cannot refer to (7.5) since two spaces are involved. However, the arguments run as in the proof of (7.5): If $\overline{x}, \overline{y} \in S(\overline{p}, \eta(\overline{p}))$, then also $\overline{x}, \overline{y} \in S(\overline{p}, \eta)$ with $\eta < \eta(\overline{p})$ hence $\overline{x}\,\overline{y} = xy$. If $x, y \in S(p, \eta(\overline{p}))$ are given then again $\eta < \eta(\overline{p})$ exists such that $x, y \in S(p, \eta)$, hence Ω maps $S(\overline{p}, \eta(\overline{p}))$ isometrically on all of $S(p, \eta(\overline{p}))$. That Ω is an isometry if $\eta(\overline{p}) = \infty$, and that then $\eta(\overline{q}) = \infty$ for any $\overline{q}$, is obvious.

If $\overline{q}\,\overline{p} < \eta(\overline{q}) < \infty$, then $S(\overline{q}, \eta(\overline{p}) - \overline{q}\,\overline{p}) \subset S(\overline{p}, \eta(\overline{p}))$ and $S(q, \eta(\overline{p}) - pq) \subset S(p, \eta(\overline{p}))$ moreover $\overline{p}\,\overline{q} = pq$. The isometry of $S(\overline{p}, \eta(\overline{p}))$ on $S(p, \eta(\overline{p}))$ induces therefore an isometry of $S(\overline{q}, \eta(\overline{p}) - \overline{p}\,\overline{q})$ on $S(q, \eta(\overline{p}) - \overline{p}\,\overline{q})$. This shows $\eta(\overline{q}) \geqslant \eta(\overline{p}) - \overline{p}\,\overline{q}$ for any $\overline{q}$. Similarly $\eta(\overline{p}) \geqslant \eta(\overline{q}) - \overline{p}\,\overline{q}$.

Although most of the following statements are trivial when $\eta(\overline{p}) = \infty$, we do not exclude this possibility, because there are interesting cases, where $\eta(\overline{p}) = \infty$ can be deduced from properties of R.

(27.2) Ω *preserves length*: $\lambda(\overline{x}) = \lambda(x)$ *if* $\overline{x}(\tau)$, $\alpha \leqslant \tau \leqslant \beta$, *is a curve in* $\overline{R}$.

For if Δ is any partition with $\|\Delta\| \leqslant \min \eta(\overline{x}(\tau))$ then $\lambda(\overline{x}, \Delta) = \lambda(x, \Delta)$, and (27.2) follows from (5.2).

(27.3) $\quad \overline{x}\,\overline{y} \geqslant x\,y$.

For if $\overline{z}(\tau)$ represents a segment from $\overline{x}$ to $\overline{y}$, then $\overline{x}\,\overline{y} = \lambda(\overline{z}) = \lambda(z) \geqslant xy$. We have as a corollary

(27.4) *If* $\overline{x}(\tau)$, $\alpha \leqslant \tau \leqslant \beta$ *is a curve in* $\overline{R}$ *and* $\overline{x}(\tau)\,\Omega = x(\tau)$ *represents a segment, then* $\overline{x}(\tau)$ *represents a segment and* $\overline{x}(\alpha)\,\overline{x}(\beta) = x(\alpha)\,x(\beta)$.

For $x(\alpha)\,x(\beta) = \lambda(x) = \lambda(\overline{x}) \geqslant \overline{x}(\alpha)\,\overline{x}(\beta) \geqslant x(\alpha)\,x(\beta)$

(27.5) *If* $\overline{x}(\tau)$ *represents a geodesic in* $\overline{R}$ *then* $x(\tau) = \overline{x}(\tau)\,\Omega$ *represents a geodesic in* R.

If $x(\tau)$ *represents a geodesic in* R *and a point* $\overline{x}$ *over* $x(\alpha)$ (α arbitrary) *is given, then exactly one representation* $\overline{x}(\tau)$ *of a geodesic in* $\overline{R}$ *exists such that* $\overline{x}(\tau)\,\Omega = x(\tau)$ *and* $\overline{x}(\alpha) = \overline{x}$.

The first part is obvious since $x(\tau)$ will for a given τ_0 represent a segment when $|\tau - \tau_0| \leqslant \beta(\tau_0) = \min [\rho(x(\tau_0)).\eta(\overline{x}(\tau_0))/2]$.

To prove the second part denote by $\Omega(\overline{p})$ the isometric mapping of $S(\overline{p}, \eta(\overline{p}))$ on $S(p, \eta(\overline{p}))$ induced by Ω, and notice that $\Omega(\overline{p})$ has an inverse $\Omega^{-1}(\overline{p})$.

Let $x(\tau)$ be given in R and $\overline{x}$ over $x(\alpha)$ in $\overline{R}$. Then $x(\tau)$ represents a segment for $|\tau - \alpha| \leqslant \beta(\alpha)$ and $\Omega^{-1}(\overline{x})$ maps this segment on a segment $x(\tau)\Omega^{-1}(\overline{x}) = \overline{x}(\tau)$, $|\tau - \alpha| \leqslant \beta(\alpha)$. There is exactly one representation $\overline{x}'(\tau)$ of a geodesic in $\overline{R}$ which coincides with $\overline{x}(\tau)$ for $|\tau - \alpha| \leqslant \beta(\alpha)$. By the first part of this proof $\overline{x}'(\tau)\,\Omega = x'(\tau)$ represents a geodesic in R. Since $x'(\tau) = x(\tau)$ for $|\tau - \alpha| \leqslant \beta(\alpha)$ we have $x'(\tau) = x(\tau)$ for all τ, therefore $\overline{x}'(\tau)$ lies over $x(\tau)$.

Let $\overline{x}(\tau)$ be any geodesic in $\overline{R}$ with $\overline{x}(\tau)\,\Omega = x(\tau)$ and $\overline{x}(\alpha) = \overline{x}$. Then $\overline{x}(\tau) = x(\tau)\,\Omega^{-1}(\overline{x}) = \overline{x}'(\tau)$ for $|\tau - \alpha| \leqslant \beta(\alpha)$ hence $\overline{x}(\tau) = \overline{x}'(\tau)$ for all τ.

In more geometric terms we may formulate the second part of (27.5) as follows:

(27.6) *Given a geodesic $\mathfrak{g}$ in R and a segment $\overline{\mathfrak{s}}$ which lies over a proper segment $\mathfrak{s}$ on $\mathfrak{g}$. Then exactly one geodesic $\overline{\mathfrak{g}}$ in $\overline{R}$ over $\mathfrak{g}$ containing $\overline{\mathfrak{s}}$ exists.*

It is, in general, not true that there is only one geodesic $\overline{\mathfrak{g}}$ in $\overline{R}$ over a given geodesic $\mathfrak{g}$ in R and through a given point $\overline{p}$ over a point p of $\mathfrak{g}$. But it follows from (27.5, 6) that

(27.7) *The sum of the multiplicities at $\overline{x}$ of the geodesics in $\overline{R}$ through $\overline{x}$, which lie over a given geodesic $\mathfrak{g}$ in R through x, equals the multiplicity of $\mathfrak{g}$ at x.*

In particular, if x is a simple point of $\mathfrak{g}$ and $\overline{x}$ lies over x, then the geodesic through $\overline{x}$ over $\mathfrak{g}$ is unique and has a simple point at $\overline{x}$.

For the definition of multiplicity compare Section 9.

These results yield the following lemma:

(27.8) *If the segment $T(a, b) = T$ in R is unique and $\overline{a}$ is a given point over a then there is exactly one segment $\overline{T}$ in $\overline{R}$ which begins at $\overline{a}$, ends at a point over b, and has length ab. Also $\overline{T}\,\Omega = T$.*

Here and in the following considerations it will prove convenient to use the term "*geodesic curve*" for a geometric curve $\mathfrak{c}$ which has a parametrization, $x(\tau)$, $\alpha \leqslant \tau \leqslant \beta$, which is part of a representation of a geodesic. We call such a parametrization a *representation* of $\mathfrak{c}$.

Let $x(\tau)$, $\alpha \leqslant \tau \leqslant \beta$, represent T with $x(\alpha) = a$, $x(\beta) = b$. By (27.5) there is a geodesic curve $\overline{x}(\tau)$ with $\overline{x}(\tau)\,\Omega = x(\tau)$ and $\overline{x}(\alpha) = \overline{a}$. It follows from (27.4) that $\bar{x}(\tau)$ represents a segment $\bar{T}$ of length $\bar{a}\,\bar{b} = a\,b$, where $\bar{b} = \bar{x}(\beta)$ lies over b.

Let $\bar{y}(\tau)$, $\alpha \leqslant \tau \leqslant \beta$ represent any segment of length $a\,b$ in $\bar{R}$ with $y(\alpha) = \bar{a}$ and $\bar{y}(\beta)\,\Omega = b$. Then $\bar{y}(\tau)\,\Omega = y(\tau)$ is by (27.5) a geodesic curve which has length $a\,b$, see (27.2), and is therefore a segment $T(a, b)$. Since $T(a, b)$ is unique, $y(\tau) = x(\tau)$ in $[\alpha, \beta]$, and (27.5) yields $\bar{y}(\tau) = \bar{x}(\tau)$ in $[\alpha, \beta]$.

The main results

Denote by $U(p)$ *the set of points* x *in* R *(including* p*) for which the segment* $T(p, x)$ *is unique.* If $x \in U(p)$ then $T(p, x) \subset U(p)$. The closure of $U(p)$ is the whole space. For if $y \in R$ then $T(p, x)$ is unique for any point x with $(p\,x\,y)$ and y can be approached by such points x. The set $U(p)$ enters in statements on the extent to which Ω is topological.

(27.9) ***Theorem:*** *A given point* $\bar{p}$ *over* p *determines uniquely a set* $\bar{U}(\bar{p}) \supset \bar{p}$ *through the requirement that* Ω *maps* $\bar{U}(\bar{p})$ *topologically on* $U(p)$. *The set* $\bar{U}(\bar{p})$ *contains* $S(\bar{p}, \rho(p))$. *If* $\bar{p}'$ *is a second point over* p *then* $\bar{U}(\bar{p}') \cap \bar{U}(\bar{p}) = 0$.

Let $x \in U(p)$. By (27.8) there is exactly one point $\bar{x}$ over x which is endpoint of a segment $T(\bar{p}, \bar{x})$ of length px. Putting $\bar{x} = x\Psi$ and $\bar{U}(\bar{p}) = U(p)\Psi$ we want to show that Ψ is topological. Since Ψ^{-1} coincides on $\bar{U}(\bar{p})$ with Ω it suffices to see that Ψ is continuous.

This is obvious: if $x_\nu \to x$ with $x_\nu, x \in U(p)$, then $\bar{p}\,x_\nu\Psi = \bar{p}\,\bar{x}_\nu = p x_\nu$ implies that $\{\bar{x}_\nu\}$ is bounded. If $\bar{y}$ is an accumulation point of $\{\bar{x}_\nu\}$, then for a suitable subsequence a segment $T(\bar{p}, \bar{x}_\theta)$ tends to segment $T(\bar{p}, \bar{y})$ of length $\lim \bar{p}\,\bar{x}_\theta = \lim p x_\theta = px$, hence $\bar{y} = \bar{x}$ by (27.8).

Let $\bar{V}$ be any set containing $\bar{p}$ and mapped by Ω topologically on $U(p)$. If $\bar{y} \in \bar{V}$ then $\bar{V}$ contains a curve $\bar{x}(\tau)$ lying over the segment from p to y. It follows again from (27.8) that $\bar{y} = y\Psi$, hence $\bar{V} = \bar{U}(\bar{p})$.

If for a point $\bar{p}' \neq \bar{p}$ over p the sets $\bar{U}(\bar{p}')$ and $\bar{U}(\bar{p})$ had a common point $\bar{y}$, then $T(\bar{p}, \bar{y})$ and $T(\bar{p}', \bar{y})$ would be different segments over $T(p, y)$ contradicting (27.8).

Finally let $\bar{y} \in S(\bar{p}, \rho(p))$. If $\bar{x}(\tau)$, $0 \leqslant \tau \leqslant \bar{p}\,\bar{y}$ represents a segment $T^+(\bar{p}, \bar{y})$, then $x(\tau) = \bar{x}(\tau)\,\Omega$ represents a geodesic curve from x to y of length $\bar{p}\,\bar{y} < \rho(p)$, so that $x(\tau)$ represents by (8.5) a segment $T(p, y)$ and $T(p, y)$ is the only segment from p to y. Therefore $y \in U(p)$ and $\bar{y} = y\Psi \in \bar{U}(\bar{p})$.

This theorem has a very important consequence which sharply distinguishes the metric theory developed here from the topological theory. To formulate it adequately we denote the number $\eta(\bar{p})$, which depends on Ω, by $\eta(\bar{p}, \Omega)$, and have then:

(27.10) ***Theorem:*** *For any locally isometric mapping Ω of a G-space $\bar{R}$ on the G-space R the number $\eta(\bar{p}, \Omega)$ satisfies the inequality $\eta(\bar{p}, \Omega) \geqslant \rho(p)/2$.*
If $\bar{p}_1\Omega = \bar{p}_2\Omega = p$, $\bar{p}_1 \neq \bar{p}_2$ then $\bar{p}_1\bar{p}_2 \geqslant 2\rho(p)$.

This means that Ω maps $S(\bar{p}, \rho(p)/2)$ isometrically on $S(p, \rho(p)/2)$ and is seen as follows: If $\bar{x}, \bar{y} \in S(\bar{p}, \rho(p)/2)$ then $x, y \in S(p, \rho(p)/2)$. The segment $T(x, y)$ is unique and lies, because of (6.9), in $S(p, \rho(p))$. Under the mapping Ψ of the preceding proof the segment $T(x, y)$ goes into a curve from $\bar{x} = x\Psi$ to $\bar{y} = y\Psi$ which is by (27.8) a segment $T(\bar{x}, \bar{y})$ with $\bar{x}\,\bar{y} = x\,y$. The assertion $\bar{p}_1\bar{p}_2 \geqslant 2\rho(p)$ follows from $\bar{U}(\bar{p}_1) \cap \bar{U}(\bar{p}_2) = 0$ and $S(\bar{p}_i, \rho(p)) \subset \bar{U}(\bar{p}_i)$.

Thus, $\inf \eta(\bar{p}, \Omega) > 0$ and $\inf \bar{p}_1\bar{p}_2 > 0$ not only for the points $\bar{p}$ which lie over the same point p under a fixed mapping Ω, but simultaneously for all mappings Ω.

No such statement is possible in the topological theory: the mapping $\bar{p} = e^{i\tau} \to e^{i\nu\tau} = p$, of the unit circle of the complex plane on itself is locally homeomorphic for every positive integer ν, but there is evidently no positive lower bound for the distances of distinct points $\bar{p}_1$, $\bar{p}_2$ corresponding to the same p and all $\nu = 1, 2, \ldots$.

A G-space $\bar{R}$ which possesses a locally isometric mapping on the G-space R is called a "*covering space*" of R. Since a G-space is obviously arcwise and locally arcwise connected (by segments), and $\inf \eta(\bar{p}) > 0$ the topological theory of covering spaces applies. Many of the facts are intuitively quite obvious. They have, in fact, for a long time been used without exact proofs; such proofs, like those of many elementary topological facts, often require cumbersome details. Fortunately, excellent accounts are found in the literature, so that there is no need to repeat the details here. We will refer in particular to § **46** in Pontrjagin's Topological Groups and to Chapters 7, 8 in Seifert and Threlfall's Lehrbuch der Topologie; we quote these sources as P and ST respectively.

We use the old notation: Ω is a locally isometric mapping of the G-space $\bar{R}$ on the G-space R.

(27.11) *For a given curve $x(\tau)$, $\alpha \leqslant \tau \leqslant \beta$, in R and a given point $\bar{a}$ over $x(\alpha)$, there is exactly one curve $\bar{x}(\tau)$ in $\bar{R}$ over $x(\tau)$, i. e. $\bar{x}(\tau)\,\Omega = x(\tau)$ with $\bar{x}(\alpha) = \bar{a}$.* ($ST$ p. 184).

Homotopy

Two curves H and K from a to b in a G-space R are *homotopic*, $H \sim K$, if in R the curve H can be continuously deformed into K keeping a and b fixed. The exact definition and elementary properties, such as symmetry and transitivity, are found in P.

(27.12) *If $\overline{H}$ and $\overline{K}$ lie over H and K and $\overline{H} \sim \overline{K}$ then $H \sim K$. If $H \sim K$ and $\overline{H}$ and $\overline{K}$ lie over H and K and begin at the same point, then they end at the same point and $\overline{H} \sim \overline{K}$.*

The first part is obvious. For proof of the second part see ST pp. 186, 187.

(27.13) *The number of points of $\overline{R}$ which lie over a given point of R is countable and is the same for different points of R.*

The relation $\overline{p}_1\overline{p}_2 \geqslant 2\,\rho(p)$ in (27.10) and the finite-compactness of $\overline{R}$ show that the number of points over p is countable. Let p and q be any two different points of R and K any curve from p go q. If $\overline{p}_1, \overline{p}_2$ are different points over p, then the curves $\overline{K}_i$ which lie over K and begin at $\overline{p}_i$ end at different points $\overline{q}_1, \overline{q}_2$ over q because $\overline{q}_1 = \overline{q}_2$ and $K^{-1} \sim K^{-1}$ would, because of (27.12), imply $\overline{K}_1^{-1} \sim \overline{K}_2^{-1}$ and $\overline{p}_1 = \overline{p}_2$.

We now prove a theorem which clearly exhibits the difference between the topological and the metric theory of covering spaces:

(27.14) ***Theorem:*** *A locally isometric mapping of a compact G-space on itself is a motion.*

For a proof we first observe:

(27.15) A one-to-one locally isometric mapping Ω of a G-space $\overline{R}$ on a G-space R is an isometry.

For then Ω^{-1} is a locally isometric mapping of R on $\overline{R}$ and it follows from (27.3) that the mapping is isometric. This reduces (27.14) to proving: if Φ is a locally isometric mapping of the G-space R on itself which is not one-to-one then R is not compact.

Let p_1, p_2 be two different points with $p_i\Phi = p$. We may consider R as its own covering space. Because of (27.13), since there are two distinct points over p, there are two distinct points p_{i1} and p_{i2} over p_i. Moreover $p_{1i} \neq p_{2j}$ because $p_1 \neq p_2$. By construction $p_{ij}\Phi^2 = p$. Similarly there are points p_{ijk}, $i, j, k = 1, 2$, which are all different with $p_{ij1}\Phi = p_{ij2}\Phi = p_{ij}$. Moreover $p_{ijk}\Phi^3 = p$. Induction yields 2^n different points $p^1, \ldots, p^{2^n}$ with $p^\nu\Phi^n = p$.

Since Φ^n is a locally isometric mapping of R on itself it follows from (27.10) that $p^ip^j \geqslant 2\,\rho(p)$ for $i \neq j$, and n being arbitrary, the space R is not compact.

An example

The assumption, in (27.14), that the space be compact is essential. An instructive example for a non-compact space which possesses locally isometric, but not isometric mappings on itself may be obtained as follows:

Let $\overline{R}$ be the hyperbolic plane referred to limit circle coordinates ξ, η. This means that $\eta =$ const. are asymptotes to each other, $\xi =$ const. are the limit circles with $\eta =$ const. as orthogonal trajectories, ξ is arc length on the asymptotes, and η is arc length on $\xi = 0$. The line element then takes the form

$$ds^2 = d\xi^2 + e^{-2\xi/k} d\eta^2, \quad \text{where } k \text{ is the space constant}^1.$$

Clearly, ds^2 remains invariant under the transformations

$$\xi' = \xi + \gamma, \qquad \eta' = \eta\, e^{\gamma/k} + \beta,$$

which are therefore motions of $\overline{R}$.

Let $\beta > 0$ and denote by Φ the motion $\xi' = \xi$, $\eta' = \eta + \beta$, and let R' be the strip $0 \leqslant \eta < \beta$. By identifying (ξ, β) with $(\xi, 0)$ we obtain, topologically speaking, a cylinder R, which can be provided with a hyperbolic metric because Φ is a motion. An exact formulation of this process, corresponding to the general procedure to be given in Section 29, is as follows:

For any point $\overline{p}$ in $\overline{R}$ denote by p the set of points $\overline{p}\,\Phi^\nu$, $\nu = 0, \pm 1, \pm 2, \ldots$. Every p has exactly one element p' in R'. The classes p are the points of R. If $\delta(\overline{p}, \overline{q})$ is the distance in $\overline{R}$ we define the distance in R by

$$\delta(p, q) = \min_{\nu,\mu} \overline{p}\,\Phi^\mu\, \overline{q}\,\Phi^\nu = \min_{\nu} \overline{p}\,\overline{q}\,\Phi^\nu.$$

It is easily seen that

$$\delta(p, q) = \min\{\delta(p', q'), \delta(p', q'\Phi), \delta(p', q'\Phi^{-1})\}$$

and this may be considered as the exact definition of the identification of $(\xi, 0)$ with (ξ, β).

The mapping $\overline{p} \to p$ of $\overline{R}$ is locally isometric. If we use that a chord of length γ subtends on a limit circle an arc of length $2\,k \sinh(\gamma/2\,k)$ [2] we find $\eta(p) = \beta/2\,k$ Area sinh $\xi/2\,k$, if ξ is the (common) abscissa of the points of $\overline{R}$ in the class p.

For any positive integer ν the motion ($\gamma = -k \log \nu$, $\beta = 0$)

$$\xi' = \xi - k \log \nu, \qquad \eta' = \eta/\nu,$$

of $\overline{R}$ maps the strip $0 \leqslant \eta < \nu\,\beta$ on the strip R' and defines therefore a locally isometric ν-to-1 mapping of R on itself. Thus, (27.14) cannot be extended to general non-compact G-spaces.

This does not exclude the possibility that (27.14) holds for special non-compact spaces: We have for instance:

(27.16) *If the fundamental group of R is not isomorphic to a proper subgroup of itself, then a locally isometric mapping of R on itself is a motion.*

For the concept of fundamental group the reader is referred to P and ST. The hypothesis is satisfied for any simply connected space, for any metrization of P^n as a G-space and many others. (27.16) follows from:

(27.17) *A locally isometric mapping Ω of a G-space $\overline{R}$ on a G-space R is an isometry when the fundamental group of $\overline{R}$ is not isomorphic to a proper subgroup of the fundamental group of R.*

To prove (27.17) let $\overline{p}\,\Omega = p$. Two curves in $\overline{R}$ which begin and end at $\overline{p}$ are by (27.12) homotopic if and only if their images in R are homotopic. Therefore Ω associates with a class of homotopic curves beginning and ending at $\overline{p}$, a class of homotopic curves beginning and ending at p, and thus yields an isomorphic mapping of the fundamental group of $\overline{R}$ in the fundamental group of R, see ST p. 188. If there are two different points of $\overline{R}$ over p, the fundamental group of $\overline{R}$ is mapped on a proper subgroup of R. Under the hypothesis of (27.17) there can then be only one point $\overline{p}$ over p and (27.17) follows from (22.15).

28. The universal covering space

Among all covering spaces of a given G-space R there is exactly one (up to isometries) which is simply connected. It is called the universal covering space of R. This is the main result of the present section.

Auxiliary results

(28.1) *If $H : x(\tau)$ and $K : y(\tau)$ are curves defined in $[\alpha, \beta]$ with $x(\alpha) = y(\alpha)$, $x(\beta) = y(\beta)$ and $T[x(\tau), y(\tau)]$ is unique, then $K \sim H$. A sufficient condition for the uniqueness of $T[x(\tau), y(\tau)]$ is $x(\tau)\, y(\tau) < \rho_1(x(\tau))$.*

Since $T[x(\tau), y(\tau)]$ is unique it depends continuously on τ. If $x_\sigma(\tau)$, $0 \leqslant \sigma \leqslant 1$, denotes the point on $T[x(\tau), y(\tau)]$ with $y_\sigma(\tau)y(\tau) = \sigma\, x(\tau)y(\tau)$ then $y_0(\tau) = y(\tau)$, $y_1(\tau) = x(\tau)$ and $y_\sigma(\tau)$ yields a continuous deformation of K into H.

(28.2) *Among all curves from a to b in R which are homotopic to a given curve K from a to b there is at least one rectifiable shortest one. Each such shortest curve is a geodesic curve from a to b, and degenerates into a if, and only if, $a = b$ and $K \sim 0$.*

(28.1) implies the existence of polygons $\bigcup_{k=1}^{n-1} T(a_k, a_{k+1})$ with $a_1 = a$, $a_n = b$ which are homotopic to K, hence $\delta = \inf \lambda(\mathfrak{c})$, $\mathfrak{c} \sim K$, is finite. If $\delta = 0$ then $a = b$ and $K \sim 0$ because (28.1).

If $\delta > 0$ choose $\mathfrak{c}_\nu \sim K$ with $\lambda(\mathfrak{c}_\nu) \to \delta$. Because of (5.16) there is a subsequence $\{\mathfrak{c}_\mu\}$ of $\{\mathfrak{c}_\nu\}$ which tends uniformly to a curve $\mathfrak{c}$ from a to b and $\lambda(\mathfrak{c}) \leqslant \liminf \lambda(\mathfrak{c}_\mu) = \delta$. For large μ we deduce from (28.1) that $\mathfrak{c}_\mu \sim \mathfrak{c}$, hence also $\mathfrak{c} \sim K$. The definition of δ then implies $\lambda(\mathfrak{c}) = \delta$.

Let $y(\sigma)$, $0 \leqslant \sigma \leqslant \delta$, be the standard representation of $\mathfrak{c}$ (in terms of arc length, see Section 5). Then $y(\sigma_1)\, y(\sigma_2) \leqslant |\sigma_1 - \sigma_2|$, see (5.12). If $0 < \sigma_2 - \sigma_1 < \rho_1(y(\sigma_1))/2$, then $y(\sigma_1)y(\sigma_2) = \sigma_2 - \sigma_1$ so that $y(\sigma)$ represents a segment for $\sigma_1 \leqslant \sigma \leqslant \sigma_2$. Otherwise (28.1) would permit replacing this subarc by a segment, thereby decreasing the length of $\mathfrak{c}$ without destroying the relation $\mathfrak{c} \sim K$. This shows that $y(\sigma)$ is part of a representation of a geodesic or $\mathfrak{c}$ is a geodesic curve. If $a = b$ but K is not homotopic to 0 then (28.1) shows that $\mathfrak{c}$ must contain points outside of $S(a, \rho_1(p))$, hence $\lambda(\mathfrak{c}) \geqslant 2\rho_1(p)$. This proves the necessity of $K \sim 0$ for $\lambda(\mathfrak{c}) = 0$ or $\mathfrak{c} = a$.

The construction of the universal covering space

Fix a point p in R. With every point x in R and every class K of curves homotopic to a given curve K from p to x we associate a point $\bar{x}_K$. The totality of these points forms, with

$$\bar{x}_K \bar{y}_L = \inf \lambda(\mathfrak{c}), \qquad \mathfrak{c} \sim K^{-1} L \tag{28.3}$$

as distance, a space $\bar{R}$, of which we want to show that it is a G-space and that $\bar{x}_K \to x$ is a locally isometric mapping of $\bar{R}$ on R.

It follows from (28.2) that $\bar{x}_K \bar{y}_L$ is finite and vanishes only for $x = y$ and $K \sim L$. Since $\mathfrak{c} \sim K^{-1} L$ implies $\mathfrak{c}^{-1} \sim L^{-1} K$ and $\lambda(\mathfrak{c}^{-1}) = \lambda(\mathfrak{c})$ it follows that $\bar{x}_K \bar{y}_L = \bar{y}_L \bar{x}_K$. Because of (28.2) there is a curve $\mathfrak{c}$ such that

$$\lambda(\mathfrak{c}) = \bar{x}_K \bar{y}_L \qquad \text{and} \qquad \mathfrak{c} \sim K^{-1} L. \tag{28.4}$$

If $\bar{x}_K$, $\bar{y}_L$, $\bar{z}_M$ are given choose $\mathfrak{c}_1 \sim K^{-1} L$ and $\mathfrak{c}_2 \sim L^{-1} M$ such that $\lambda(\mathfrak{c}_1) = \bar{x}_K \bar{y}_L$ and $\lambda(\mathfrak{c}_2) = \bar{y}_L \bar{z}_M$. Then $\mathfrak{c}_1 \mathfrak{c}_2 \sim K^{-1} L L^{-1} M \sim K^{-1} M$, hence

$$\bar{x}_K \bar{z}_M \leqslant \lambda(\mathfrak{c}_1) + \lambda(\mathfrak{c}_2) = \bar{x}_K \bar{y}_L + \bar{y}_L \bar{z}_M.$$

Thus $\bar{R}$ is a metric space.

We show next that $\bar{R}$ is M-convex. Let $\bar{x}_K \neq \bar{y}_L$. With the notation of (28.4) let z be any point of $\mathfrak{c}$ different from x and y. Denote by $\mathfrak{c}_1$ and $\mathfrak{c}_2$ subcurves of $\mathfrak{c}$ such that $\mathfrak{c}_1\mathfrak{c}_2 = \mathfrak{c}$ and $\lambda(\mathfrak{c}_1) + \lambda(\mathfrak{c}_2) = \lambda(\mathfrak{c})$: Put $K\ \mathfrak{c}_1 = M$. Then $\mathfrak{c}_1 \sim K^{-1}M$ and $\mathfrak{c}_2 \sim \mathfrak{c}_1^{-1}\,\mathfrak{c} \sim M^{-1}KK^{-1}L = M^{-1}L$, hence

$$\lambda(\mathfrak{c}) = \lambda(\mathfrak{c}_1) + \lambda(\mathfrak{c}_2) \geqslant \overline{x}_K\,\overline{z}_M + \overline{z}_M\,\overline{y}_L \geqslant \overline{x}_K\,\overline{y}_L = \lambda(\mathfrak{c}) \quad \text{so that} \quad (\overline{x}_K\,\overline{z}_M\,\overline{y}_L).$$

To see the finite compactness of $\overline{R}$ let $\overline{p}_0\overline{x}_\nu < \beta$ where $\overline{x}_\nu = (\overline{x}_\nu)_{K_\nu}$. There is a geodesic curve $\mathfrak{c}_\nu \sim K_\nu$ with $\lambda(\mathfrak{c}_\nu) = \overline{p}_0\overline{x}_\nu$. Denote by $x_\nu(\tau)$ the representation of a geodesic which represents $\mathfrak{c}_\nu$ for $0 \leqslant \tau \leqslant \overline{p}_0\overline{x}_\nu$. Because of (8.14) there is a subsequence $\{x_\mu(\tau)\}$ of $\{x_\nu(\tau)\}$ which converges to a representation $x(\tau)$ of a geodesic, and uniformly so for $0 \leqslant \tau \leqslant \beta + 1$. We may also assume that $\lim \overline{p}_0\overline{x}_\mu = \lambda \leqslant \beta$ exists. Then $x(\tau)$ represents for $0 \leqslant \tau \leqslant \lambda$ a geodesic curve K, $\lim \lambda(\mathfrak{c}_\mu) = \lambda$, and $x_\mu = x_\mu(\overline{p}_0\overline{x}_\mu) \to x(\lambda)$. Let $\mathfrak{s}_\mu$ be the (oriented) segment from x to x_μ. Since $\mathfrak{c}_\mu$ tends uniformly to K it follows from (28.1) that $\mathfrak{c}_\mu\mathfrak{s}_\mu^{-1} \sim K$ for large μ or $\mathfrak{s} \sim K^{-1}K_\mu$ so that $\overline{x}_K\overline{x}_\mu \leqslant \lambda(\mathfrak{s}_\mu) = x_\mu\, x(\lambda) \to 0$. Hence $\overline{x}_K$ is an accumulation point of $\{\overline{x}_\nu\}$.

With $\sigma = \rho_1(p)/2$ we show that $\overline{x}_K \to x$ maps $S(\overline{q}_M, \sigma)$ isometrically on $S(q, \sigma)$. Let $\overline{x}_K, \overline{y}_L \in S(\overline{q}_M, \sigma)$. There is a curve $\mathfrak{c} \sim K^{-1}L$ such that $\lambda(\mathfrak{c}) = \overline{x}_K\overline{y}_L < 2\sigma = \rho_1(p)$. It follows from (28.1) that $\mathfrak{s}(x, y) \sim \mathfrak{c}$, therefore $\mathfrak{s}(x, y) \sim K^{-1}L$ and $\overline{x}_K\overline{y}_L \leqslant \lambda(\mathfrak{s}(x, y)) = xy$. But $\overline{x}_K\overline{y}_L \geqslant xy$ follows from the definition (28.3), so that $\overline{x}_K\overline{y}_L = xy$. Applying this to $\overline{y}_L = \overline{q}_M$ we see that $\overline{x}_K \to x$ maps $S(\overline{q}_M, \sigma)$ isometrically in $S(q, \sigma)$. If $x \in S(q, \sigma)$ is given, put $K = M\,\mathfrak{s}(q, x)$, then $M^{-1}K \sim \mathfrak{s}(q, x)$ and $\overline{q}_M\overline{x}_K \leqslant \lambda(\mathfrak{s}(q, x)) < \sigma$, so that $\overline{x}_K \to x$ maps $S(\overline{q}_M, \sigma)$ on $S(q, \sigma)$.

Finally, since $\rho(p) \leqslant \rho_1(p)$, the sphere $S(\overline{q}_M, \rho(p)/2)$ is isometric to $S(q, \rho(p)/2)$. The space $\overline{R}$ satisfies the axiom of local prolongability and the hypothesis of (8.2), and is therefore a G-space.

The space $\overline{R}$ is simply connected, see P. Theorem 59.

The uniqueness and universality of $\overline{R}$ will follow from

(28.5) *Let Ω be a locally isometric mapping of the simply connected G-space $\overline{R}$ on the G-space R, and Φ a locally isometric mapping of the G-space R' on R. If $\overline{p} \in \overline{R}$ and $p' \in R'$ are given points over the same point p of R (that is $\overline{p}\Omega = p'\Phi = p$), then a locally isometric mapping Ψ of $\overline{R}$ on R' exists such that $\overline{p}\Psi = p'$ and $\Psi\Phi = \Omega$.*

For a curve $\overline{K}$ in $\overline{R}$ from $\overline{p}$ to the arbitrary point $\overline{x}$ of $\overline{R}$ lies over a curve K from p to x in R. There is exactly one curve K' in R' over K beginning at p', see (27.11). Define $x' = \overline{x}\Psi$ as the end point of K'. This definition does not depend on the choice of $\overline{K}$. For if $\overline{L}$ is another curve in $\overline{R}$ from $\overline{p}$ to $\overline{x}$, then $\overline{K} \sim \overline{L}$ because $\overline{R}$ is simply connected. If $\overline{L}$ lies over the curve L in R and L'

is the curve in R' over L beginning at p', then (27.12) shows $K \sim L$ and $K' \sim L'$ and that K' and L' end at the same point x'.

The relation $\overline{x}\Psi\Phi = x = \overline{x}\Omega$ shows $\Psi\Phi = \Omega$. We know that Φ induces an isometry of $S(x', \rho(x)/2)$ on $S(x, \rho(x)/2)$. If $\Phi^{-1}(x')$ denotes the inverse of this induced mapping, then $\Psi = \Omega\Phi^{-1}(x')$ shows that Ψ maps $S(\overline{x}, \rho(x)/2)$ isometrically on $S(x', \rho(x)/2)$. It is obvious that Ψ maps $\overline{R}$ on all of R'; for if $x' \in R'$ is given, connect p' to x' by a curve K', put $K'\Phi = K$ and let $\overline{K}$ be the curve in $\overline{R}$ over K which begins at $\overline{p}$, its end point $\overline{x}$ has the property $\overline{x}\Psi = x'$.

If R' is also simply connected then it follows from (27.17) that Ψ is an isometry. Thus we have established the important fact:

(28.6) ***Theorem:*** *A given G-space R possesses a simply connected covering space* $\overline{R}$. *Any two simply connected covering spaces of R are isometric.* $\overline{R}$ *is covering space of any covering space R′ of R.*

Because $\overline{R}$ is unique up to isometries and covers every covering space of R, a simply connected covering space of R is called *"the universal"* covering space of R.

Motions of the universal covering space

If Ψ is a locally isometric mapping of R' on R then the motion Φ' of R' is said to *lie over the motion* Φ of R if $x'\Phi$ lies for every $x' \in R'$ over $x\Phi$. This means that $x'\Phi'\Psi = x\Phi = x'\Psi\Phi$, or that $\Phi'\Psi = \Psi\Phi$.

In general there is no motion of R' over a given motion of R, and even more rarely will a given motion of R' lie over a motion of R. However, the universal covering space of R, for which we reserve from now on the symbol $\overline{R}$, denoting other covering spaces of R by R', possesses greater mobility than the other covering spaces. It always has the first of the above two properties:

(28.7) *There is a motion* $\overline{\Phi}$ *of the universal covering space* $\overline{R}$ *(related to R by a definite locally isometric mapping* Ω*) over a given motion* Φ *of R.*

The proof runs on similar lines to that for (28.5): Select a point p in R, a point $\overline{p}$ in $\overline{R}$ over p and a point $\overline{q}$ over $q = p\Phi$. For any point $\overline{z} \in \overline{R}$ define the image $\overline{z}\overline{\Phi}$ as follows: let $\overline{K}$ be a curve from $\overline{p}$ to $\overline{z}$. It lies over a curve K from p to z, and $K\Phi$ is a curve from $q = p\Phi$ to $z\Phi$. Let $\overline{z}\overline{\Phi}$ be the end point of the curve over $K\Phi$ that begins at $\overline{q}$.

The point $\overline{z}\overline{\Phi}$ does not depend on the choice of $\overline{K}$: for if $\overline{L}$ also connects $\overline{p}$ to $\overline{x}$ in $\overline{R}$, then $\overline{K} \sim \overline{L}$, hence $L = \overline{L}\Omega \sim K$, $L\Phi \sim K\Phi$ because Φ is a motion, hence the curve over $L\Phi$ beginning at $\overline{q}$ also ends at $\overline{x}$.

$S(\bar{z}, \rho(z)/2)$ is mapped by Ω isometrically on $S(z, \rho(z)/2)$. Since Φ is a motion, $\rho(z) = \rho(z\Phi)$, moreover Φ maps $S(z, \rho(z)/2)$ isometrically on $S(z\Phi, \rho(z\Phi)/2)$. The step from $z\Phi$ to $\overline{z\Phi}$ coincides with $\Omega^{-1}(\overline{z\Phi})$, hence $\bar{\Phi}$ maps $S(\bar{z}, \rho(z)/2)$ isometrically on $S(\bar{z}\bar{\Phi}, \rho(z)/2)$. That $\bar{\Phi}$ maps $\bar{R}$ on itself is seen as in the proof of (28.5). Thus, $\bar{\Phi}$ is a locally isometric mapping of $\bar{R}$ on itself and must by (27.16) be a motion. $\bar{\Phi}$ lies over Φ by the construction.

We now turn to the important fact that the fundamental group of a G-space may be realized as the group of motions of the universal covering space. In the discussion we need the lemma

(28.8) *A motion Φ of a G-space which leaves all points of a sphere $S(p, \sigma)$, $\sigma > 0$, fixed is the identity.*

It is to be shown that $x\Phi = x$ if $x\,p \geqslant \sigma$. A segment $T(p, x)$ contains a point y with $(p\,y\,x)$ and $p\,y < \sigma$. Then $(p\Phi y\Phi x\Phi)$ or $(p\,y\,x\Phi)$ and $y\,x = y\Phi x\Phi = y\,x\Phi$, hence (8.1) yields $x\Phi = x$.

Before proceeding we remind the reader of the following definition:

A collection Σ of transformations of the set M on itself is called "*transitive*" on the subset N of M, if for any two points x, y in N a transformation in Σ exists carrying x into y. If this transformation in Σ is unique then Σ is said to be "*simply transitive*" on N.

The fact to which we alluded above is:

(28.9) **Theorem:** *The motions of the universal covering space $\bar{R}$ of the G-space R, which lie over the identity motion E of R, form a group $\mathfrak{F}$, which is isomorphic to the fundamental group of R. If $\bar{R}$ is compact then $\mathfrak{F}$ is finite.*

$\mathfrak{F}$ is (with the topology (4.7)) *discrete and simply transitive on the points which lie over a given point x of R.*

It is obvious that the motions of $\bar{R}$ over E form a group $\mathfrak{F}$. If $\bar{x}_1$ and $\bar{x}_2$ are two points of $\bar{R}$ over the same point x of R, then (28.5) yields the existence of a locally isometric mapping $\bar{\Phi}$ of $\bar{R}$ on itself which carries $\bar{x}_1$ into $\bar{x}_2$ and with $\bar{\Phi}\Omega = \Omega = \Omega E$. Because of (27.16) $\bar{\Phi}$ is a motion of $\bar{R}$ and it lies over E.

Therefore $\mathfrak{F}$ is transitive on the points over x. If $\bar{\Phi}_1$ and $\bar{\Phi}_2$ are elements of $\mathfrak{F}$ which carry $\bar{x}_1$ into $\bar{x}_2$, then $\bar{\Phi}_1\bar{\Phi}_2^{-1}$ leaves $\bar{x}_1$ fixed and lies over E, or $\bar{\Phi}_1\bar{\Phi}_2^{-1}\Omega = \Omega$. Applying $\Omega^{-1}(\bar{x}_1)$ to $S(x, \frac{1}{2}\rho(x_1))$ and both sides of the last equation we find that $\bar{\Phi}_1\bar{\Phi}_2^{-1}$ leaves every point of $S(\bar{x}_1, \frac{1}{2}\rho(x_1))$ fixed, hence is by the preceding lemma the identity. This shows that $\mathfrak{F}$ is simply transitive on $x\Omega^{-1}$.

Discreteness of $\mathfrak{F}$ under the metric (4.7) means that for every $\overline{\Phi} \in \mathfrak{F}$, a sphere $S(\overline{\Phi}, \rho)$, $\rho > 0$, (or $\delta_{\overline{p}}(\overline{\Phi}, \overline{\Psi}) < \rho$) exists which contains no other element of $\mathfrak{F}$. This follows immediately from $\overline{x}_1\overline{x}_2 \geqslant 2\,\rho(x)$ for two distinct points of $x\Omega^{-1}$ (compare (27.10)). That $\mathfrak{F}$ is isomorphic to the fundamental group of R follows now from standard arguments, see ST p. 197. The finiteness of $\mathfrak{F}$ for compact $\overline{R}$ follows from (4.15).

When we speak of the fundamental group of R, we shall always mean this group $\mathfrak{F}$ defined by means of a definite locally isometric mapping Ω of $\overline{R}$ on R, so that the symbol $\overline{R}$ itself actually stands for the pair $(\overline{R}, \Omega)$. Because of (27.13) and the simple transitivity of $\mathfrak{F}$ on $x\Omega^{-1}$ the number of elements in $\mathfrak{F}$ is countable. We denote them by $\Xi = \Xi_1, \Xi_2, \ldots$, where Ξ is the identity.

From (28.7) and (28.9) we deduce the following further facts on motions of $\overline{R}$ which lie over motions of R.

(28.10) ***Theorem:*** *The motion $\overline{\Phi}$ of $\overline{R}$ lies over a motion of R if and only if $\overline{\Phi}\mathfrak{F} = \mathfrak{F}\overline{\Phi}$.*

The motions of $\overline{R}$ which lie over the motions of a group Γ of motions of R form a group $\overline{\Gamma}$ and $\overline{\Gamma}/\mathfrak{F} \cong \Gamma$.

If Γ is closed, then $\overline{\Gamma}$ is closed. If Γ is (simply) transitive on R then $\overline{\Gamma}$ is (simply) transitive on $\overline{R}$.

Let $\overline{\Phi}_i$ lie over Φ_i, $i = 1, 2$. Then $\overline{\Phi}_1\,\overline{\Phi}_2\Omega = \overline{\Phi}_1\Omega\Phi_2 = \Omega\Phi_1\Phi_2$ and $\overline{\Phi}_1^{-1}\Omega = \overline{\Phi}_1^{-1}\Omega\Phi_1\Phi_1^{-1} = \overline{\Phi}_1^{-1}\overline{\Phi}_1\Omega\Phi_1^{-1} = \Omega\Phi_1^{-1}$. Hence the motions over elements of Γ form a group $\overline{\Gamma}$, and mapping $\overline{\Phi}$ on the motion over which it lies produces a homomorphism of $\overline{\Gamma}$ on Γ. The kernel of this homomorphism consists of the motions in $\overline{\Gamma}$ which are mapped on the identity motion of R, and is therefore $\mathfrak{F}$.

This applies in particular to the group Γ_R of all motions that R possesses. Then $\overline{\Gamma}_R$ consists of all motions $\overline{\Phi}$ of $\overline{R}$ which lie over motions Φ of R. As kernel of the homomorphism $\overline{\Gamma}_R \sim \Gamma_R$ the group $\mathfrak{F}$ is a normal subgroup of $\overline{\Gamma}_R$, hence $\overline{\Phi}\mathfrak{F} = \mathfrak{F}\overline{\Phi}$. This means the existence of relations

$$\Xi_\varkappa\overline{\Phi} = \overline{\Phi}\,\Xi_{\gamma(\varkappa)} \quad \text{and} \quad \overline{\Phi}\,\Xi_\varkappa = \Xi_{\mu(\varkappa)}\overline{\Phi}, \quad \varkappa = 1, 2, \ldots$$

We must show that these relations are also sufficient for $\overline{\Phi}$ to lie over a motion Φ of R. To construct Φ let $\overline{x}$ be any point over x and put $x\Phi = \overline{x}\overline{\Phi}\Omega$. The point $x\Phi$ does not depend on the choice of $\overline{x}$; for if $\overline{x}_1$ is another point over x, then a $\Xi_\varkappa$ with $\overline{x}\,\Xi_\varkappa = \overline{x}_1$ exists and

$$\overline{x}_1\overline{\Phi}\Omega = \overline{x}\,\Xi_\varkappa\overline{\Phi}\,\Omega = \overline{x}\overline{\Phi}\,\Xi_{\gamma(\varkappa)}\Omega = \overline{x}\overline{\Phi}\Omega = x\Phi.$$

The definition $\bar{x}\bar{\Phi}\Omega = x\Phi = \bar{x}\Omega\Phi$ implies $\bar{\Phi}\Omega = \Omega\Phi$, hence it remains to show that Φ is a motion or $x\Phi y\Phi = xy$.

We conclude from (27.8) that there are points $\bar{x}, \bar{y}$ over x, y such that $\bar{x}\bar{y} = xy$. Then (27.3) yields

$$xy = \bar{x}\bar{y} = \bar{x}\bar{\Phi}\bar{y}\bar{\Phi} \geqslant \bar{x}\bar{\Phi}\Omega\,\bar{y}\bar{\Phi}\Omega = x\Phi y\Phi.$$

Next choose by (27.8) a point $\bar{z}$ over y such that $\bar{x}\bar{\Phi}\bar{z} = x\Phi y\Phi$. Since $\bar{z}$ and $\bar{y}\bar{\Phi}$ both lie over y there is a $\Xi_\varkappa$ such that $\bar{y}\bar{\Phi}\Xi_\varkappa = \bar{z}$, then by (27.3)

$$x\Phi y\Phi = \bar{x}\bar{\Phi}\bar{z} = \bar{x}\bar{\Phi}\bar{y}\bar{\Phi}\Xi_\varkappa = \bar{x}\bar{\Phi}\bar{y}\Xi_{\mu(\varkappa)}\bar{\Phi} = \bar{x}\bar{y}\Xi_{\mu(\varkappa)} \geqslant xy.$$

This proves that Φ is a motion.

Let Γ be a group of motions in R which is closed in the topology of Section 4 and $\bar{\Gamma}$ the group of motions over Γ. We want to show that $\bar{\Gamma}$ is a closed subset of the group of all motions of $\bar{R}$. Let — with the notations of Section 4 — $\delta_{\bar{p}}(\bar{\Phi}_\nu, \bar{\Phi}) \to 0$, where $\bar{\Phi}_\nu \in \bar{\Gamma}$ and $\bar{\Phi}$ is any motion of $\bar{R}$. Then $\bar{x}\bar{\Phi}_\nu \to \bar{x}\bar{\Phi}$ for every $\bar{x}$ in $\bar{R}$, see (4.13); hence $\bar{x}\bar{\Phi}_\nu\,\Omega = \bar{x}\Omega\Phi_\nu = x\Phi_\nu \to \bar{x}\bar{\Phi}\Omega$. Put $\bar{x}\bar{\Phi}\Omega = x\Phi$. Then $x\Phi_\nu \to x\Phi$ and $x\Phi_\nu\, y\Phi_\nu = xy$ shows $x\Phi y\Phi = xy$. Moreover, $x\Phi$ must traverse (with x) all of R because $\bar{x}\bar{\Phi}$ traverses $\bar{R}$. Therefore Φ is a motion of R and it follows from (4.14) that $\delta_p(\Phi_\nu, \Phi) \to 0$, hence $\Phi \in \Gamma$ and $\bar{\Phi} \in \bar{\Gamma}$.

Let Γ be transitive on R. If two points $\bar{x}, \bar{y}$ on $\bar{R}$ are given, a motion Φ in Γ exists which takes x into y. Any motion $\bar{\Phi}_1$ over Φ carries $\bar{x}$ into a point $\bar{y}_1$ over y. There is an element $\Xi_\varkappa \in \mathfrak{F}$ carrying $\bar{y}_1$ into $\bar{y}$. Then $\bar{\Phi} = \bar{\Phi}_1\Xi_\varkappa$ will take $\bar{x}$ into $\bar{y}$ and $\bar{\Phi} \in \bar{\Gamma}$ because $\mathfrak{F} \subset \bar{\Gamma}$.

If Φ is the only motion in Γ taking x into y, then any two motions $\bar{\Phi}$ and $\bar{\Phi}_2$ in $\bar{\Gamma}$ which take $\bar{x}$ into $\bar{y}$ must lie over Φ, hence $\bar{\Phi}\bar{\Phi}_2^{-1}$ lies over the identity and is therefore an element of $\mathfrak{F}$ which leaves $\bar{x}$ fixed. Hence $\bar{\Phi}\,\bar{\Phi}_2^{-1} = \Xi$ or $\bar{\Phi} = \bar{\Phi}_2$ because $\mathfrak{F}$ is simply transitive on $x\Omega^{-1}$. This establishes (28.10).

Since the fundamental group need not be abelian it will, clearly, in general not be true that $\bar{\Phi}\Xi_\varkappa = \Xi_\varkappa\bar{\Phi}$ for every $\varkappa$. There is an important case, discovered by Killing, where these relations can be proved, namely when $\bar{\Phi}$ belongs to a one-parameter subgroup of motions which lie over motions of R:

(28.11) *Let Γ_R denote the group of all motions which R possesses. The relations $\bar{\Phi}\Xi_\varkappa = \Xi_\varkappa\bar{\Phi}$, $\varkappa = 1, 2, \ldots$ hold for a motion $\bar{\Phi}$ in $\bar{\Gamma}_R$, if a sequence $\{\bar{\Phi}_\nu\}$ of motions in $\bar{\Gamma}_R$ and integers λ_ν exist such that $\bar{\Phi}_\nu \to \Xi$ and $\bar{\Phi}_\nu^{\lambda_\nu} = \bar{\Phi}$.*

There is a $\Xi_{\sigma(\nu)}$ such that

$$\Xi_{\sigma(\nu)} \overline{\Phi}_\nu \, \Xi_\varkappa = \Xi_\varkappa \overline{\Phi}_\nu$$

because $\overline{\Phi}_\nu \, \Xi_\varkappa$ and $\Xi_\varkappa \overline{\Phi}_\nu$ lie over the same motion of R. By (4.11, 12)

$$\delta_{\overline{p}} \, (\Xi, \Xi_{\sigma(\nu)}) = \delta_{\overline{p}} \, (\overline{\Phi}_\nu \Xi_\varkappa, \Xi_{\sigma(\nu)} \overline{\Phi}_\nu \, \Xi_\varkappa) = \delta_{\overline{p}} \, (\overline{\Phi}_\nu \, \Xi_\varkappa, \Xi_\varkappa \overline{\Phi}_\nu) \leqslant$$
$$\leqslant \delta_{\overline{p}} \, (\overline{\Phi}_\nu \, \Xi_\varkappa, \Xi_\varkappa) + \delta_{\overline{p}} \, (\Xi_\varkappa, \Xi_\varkappa \, \overline{\Phi}_\nu) = \delta_{\overline{p}} \, (\overline{\Phi}_\nu, \Xi) + \delta_{\overline{q}} \, (\Xi, \overline{\Phi}_\nu),$$

where $\overline{q} = \overline{p}\Xi_\varkappa$.

The hypothesis $\overline{\Phi}_\nu \to \Xi$ then yields $\delta_{\overline{p}} \, (\Xi, \Xi_{\sigma(\nu)}) \to 0$. But $\mathfrak{F}$ is discrete, hence $\Xi_{\sigma(\nu)} = \Xi$ and $\overline{\Phi}_\nu \, \Xi_\varkappa = \Xi_\varkappa \overline{\Phi}_\nu$ from a certain ν on. The last relation yields

$$\overline{\Phi}_\nu^{\,2} \, \Xi_\varkappa = \overline{\Phi}_\nu \, \Xi_\varkappa \overline{\Phi}_\nu = \Xi_\varkappa \, \overline{\Phi}_\nu^{\,2}$$

and generally $\overline{\Phi}_\nu^\lambda \, \Xi_\varkappa = \Xi_\varkappa \, \overline{\Phi}_\nu^\lambda$, in particular the assertion for $\lambda = \lambda_\nu$.

29. Fundamental sets

Covered spaces

We now turn to the converse problem to the construction of the universal covering space: given a simply connected G-space $\overline{R}$ to find all G-spaces R covered by $\overline{R}$. It is clear that the solution consists in identifying the points of $\overline{R}$ which lie over the same point of the space that is to be constructed. These points of $\overline{R}$ are the points of the form $\overline{x}\,\Xi_\varkappa$. To formulate this adequately we denote as "*orbit*" $0(p, \Sigma)$ of p under the transformations in the collection Σ the set of points $p\Phi$ with $\Phi \in \Sigma$. The points of $\overline{R}$ over x form the orbit $0(\overline{x}, \mathfrak{F})$ of any point $\overline{x}$ over x.

The group $\mathfrak{F}$ is a discrete group of motions of $\overline{R}$ which is simply transitive on each orbit $0(\overline{x}, \mathfrak{F})$. The latter property implies that no motion in $\mathfrak{F}$ except the identity has fixed points. These properties are characteristic for fundamental groups.

(29.1) ***Theorem:*** *Let $\mathfrak{G}$ be a discrete group of motions without fixed points (except for the identity E') of the G-space R'. Then the orbits $x = 0(x', \mathfrak{G})$ form with the distance*

$$xy = 0(x', \mathfrak{G}) \, 0(y', \mathfrak{G})$$

a G-space R and $x' \to x$ is a locally isometric mapping of R' on R. Hence R' is a covering space of R.

Since $\mathfrak{G}$ is discrete each orbit is a discrete, hence closed set. Clearly

$$\text{either } 0(x', \mathfrak{G}) \cap 0(y', \mathfrak{G}) = 0 \qquad \text{or} \qquad 0(x', \mathfrak{G}) = 0(y', \mathfrak{G});$$

$$0(x', \mathfrak{G})\, 0(y', \mathfrak{G}) = \inf_{\Phi, \Psi \in G} x'\Phi y'\Psi = \inf x'y'\Psi\Phi^{-1} = x'0(y', \mathfrak{G}) = y'0(x', \mathfrak{G}).$$

Since $0(y', \mathfrak{G})$ is closed, x' has a foot on it; this means that a point $y'' \in 0(y', \mathfrak{G})$ exists such that

$$xy = 0(x', \mathfrak{G})\, 0(y', \mathfrak{G}) = x'y''. \tag{29.2}$$

This implies $xy = yx > 0$ unless $x = y$.

To prove the triangle inequality $xy + yz \geqslant xz$ choose x'' in $0(x', \mathfrak{G})$ and z'' in $0(z', \mathfrak{G})$ such that $x''y' = xy$ and $y'z'' = yz$. Then

$$xy + yz = x''y' + y'z'' \geqslant x''z'' \geqslant 0(x', \mathfrak{G})\, 0(z', \mathfrak{G}) = xz.$$

To show finite compactness let $x_\nu x < \beta$. If $x_\nu = 0(x_\nu', \mathfrak{G})$, $x = 0(x', \mathfrak{G})$, choose $x_\nu'' \in 0(x_\nu', \mathfrak{G})$ such that $x_\nu''x' = x_\nu x < \beta$. Because R' is finitely compact $\{x_\nu''\}$ contains a subsequence $\{x_\mu''\}$ which converges to a point y'. Then

$$x_\mu''y' \geqslant 0(x_\mu'', \mathfrak{G})\, 0(y', \mathfrak{G}) = x\, y$$

yields $x_\mu \to y$.

Let two distinct points x, y of R be given. Choose by (29.2) the points x', y' in the orbits x, y such that $x'y' = xy$. Then $x' \neq y'$. Since R' is M-convex a point z' with $(x'z'y')$ exists. Then

$$xy \leqslant xz + zy \leqslant x'z' + z'y' = x'y' = xy$$

proves (xzy), hence R is M-convex.

So far we have not used that the motions in $\mathfrak{G} - E'$ have no fixed points. This enters in the proof that, for any given x', a $\sigma(x') > 0$ exists such that $y'y'\Phi' > \sigma(x')$ for any $\Phi' \neq E'$ in $\mathfrak{G}$ and any $y' \in S(x', 1)$. If this were not correct a sequence y_ν' in $S(x', 1)$ and a sequence $\Phi_\nu' \neq E'$ in $\mathfrak{G}$ would exist such that $y_\nu'y_\nu'\Phi_\nu' \to 0$. Then

$$x'x'\Phi_\nu' \leqslant x'y_\nu' + y_\nu'y_\nu'\Phi_\nu' + y_\nu'\Phi_\nu'x'\Phi_\nu' = 2\, x'y_\nu' + y_\nu'y_\nu'\Phi_\nu'$$

would show that $x'x'\Phi_\nu'$, hence $\delta_{x'}(E', \Phi_\nu')$, is bounded (compare (4.8)), so that Φ_ν' would by (4.15) contain a subsequence $\{\Phi_\mu'\}$ converging to a motion Φ' in $\mathfrak{G}$. Since $\mathfrak{G}$ is discrete $\Phi_\mu' = \Phi' \neq E'$ for $\mu \geqslant \mu_0$ say. But then $y_\mu'y_\mu'\Phi' \to 0$ would imply that Φ' leaves every accumulation point of $\{y_\mu'\}$ fixed.

With $\eta = \min(1, \sigma(x')/3)$ we are going to show that $x' \to x$ maps $S(x', \eta)$ isometrically on $S(x, \eta)$. Let $y', z' \in S(x', \eta)$. Choose z'' by (29.2) such that $z'' \in 0(z', \mathfrak{G})$ and $z''y' = zy$, then $z'' = z'$. For otherwise, $z'' = z'\Phi'$ with $\Phi' \neq E'$, $\Phi' \in \mathfrak{G}$, and owing to the choice of η and the definition of $\sigma(x')$, $y'z'' \geqslant z'z'' - z'y' \geqslant 2\,\sigma(x')/3 > y'z' \geqslant yz$. This proves that $x' \to x$ maps $S(x', \eta)$ isometrically in $S(x, \eta)$. But if $y \in S(x, \eta)$ is given then y' in the orbit y exists such that $y'x' = yx$ and $y' \in S(x', \eta)$.

Since R' satisfies the axiom of local prolongability and the hypothesis of (8.3) it follows that the neighborhood $S[x, \min(1, \sigma(x')/3, \rho(x')]$ satisfies these two requirements. Hence R is a G-space.

This proves the theorem, we deduce from it:

(29.3) ***Theorem:*** *The fundamental group $\mathfrak{G}$ of any covering space R' of R is isomorphic to a subgroup of the fundamental group $\mathfrak{F}$ of R.*

For a given subgroup $\mathfrak{G}$ of $\mathfrak{F}$ there is a covering space R' of R with $\mathfrak{G}$ as fundamental group.

The first part is contained in the proof (27.17).

A subgroup $\mathfrak{G}$ of $\mathfrak{F}$ is discrete and none of its motions has fixed points. By (29.1) the orbits $x' = 0(\bar{x}, \mathfrak{G})$ of $\mathfrak{G}$ in the universal covering space $\bar{R}$ of R form with the distance $x'y' = 0(\bar{x}, \mathfrak{G})\, 0(\bar{y}, \mathfrak{G})$ a G-space R' and $\Psi : \bar{x} \to x'$ is a locally isometric mapping of $\bar{R}$ on R'. Since $\bar{R}$ is simply connected, $\mathfrak{G}$ is the fundamental group of R'. Also, $x' = 0(\bar{x}, \mathfrak{G}) \to x = 0(\bar{x}, \mathfrak{F})$ is a locally isometric mapping of R' on R, hence R' is a covering space of R.

Fundamental sets

The abstract method of identification of points used in (29.1) does not provide an intuitive picture of the space covered by a simply connected space $\bar{R}$. A well known procedure which frequently leads to a clear grasp of the geometric situation, is the construction of fundamental sets. Various methods have been used in special cases. One of these, called "méthode de rayonnement" in the French literature (presumably because of property (b) below) is particularly well suited to G-spaces.

Let $\bar{R}, R, \Omega, \mathfrak{F} = \{\Xi_1 = \Xi, \Xi_2, \ldots, \Xi_\varkappa, \ldots\}$ have the same meaning as before and assume that $\mathfrak{F}$ has at least two distinct elements. We then show

(29.4) ***Theorem:*** *For a given point $\bar{p}_1$ of $\bar{R}$ let the "fundamental set $H(\bar{p}_1)$" be defined to consist of the points $\bar{x}$ which satisfy the inequalities $\bar{x}\bar{p}_1 < \bar{x}\bar{p}_1\Xi_\varkappa$, $\varkappa = 2, 3, \ldots$.*

Then $H(\overline{p}_1)$ has the following properties:

(a) *$H(\overline{p}_1)$ is open and contains with $\overline{x}$ every segment $T(\overline{p}_1, \overline{x})$.*

(b) *A segment $T(\overline{p}_1, \overline{q})$ contains at most one point on the boundary of $H(\overline{p}_1)$.*

(c) *$H(\overline{p}_1 \Xi_\varkappa) = H(\overline{p}_1) \Xi_\varkappa$ and $H(\overline{p}_1) \Xi_\varkappa \cap H(\overline{p}_1) \Xi_\lambda = 0$ for $\varkappa \neq \lambda$.*

(d) *$H(\overline{p}_1) \Omega = H(\overline{p}_1 \Xi_\varkappa) \Omega \supset U(p)$* (defined in Section 27).

(e) *If $F(\overline{p}_1)$ is the closure of $H(\overline{p}_1)$ then $F(\overline{p}_1 \Xi_\varkappa) = F(\overline{p}_1) \Xi_\varkappa$ and $\cup_\varkappa F(\overline{p}_1) \Xi_\varkappa = \overline{R}$.*

(f) *If H^* is an open set containing $F(\overline{p}_1)$ then $H^* \cap H^* \Xi_\varkappa \neq 0$ for a suitable $\varkappa > 1$.*

(g) *If $\delta(M)$ denotes the diameter of the set M, then $\delta(F(\overline{p}_1)) \leqslant 2\,\delta(R)$, $\delta(F(\overline{p}_1)) = \infty$ if $\delta(R) = \infty$.*

(h) *A sphere $S(\overline{q}, \sigma)$, $0 < \sigma < \infty$ intersects only a finite number of $F(\overline{p}_1) \Xi_\varkappa$.*

(i) *$\mathfrak{F}$ can be generated by positive powers of those $\Xi_\varkappa$ for which a point $\overline{x}_\varkappa$ with $\overline{p}_1\overline{x}_\varkappa = \overline{p}_1\Xi_\varkappa\overline{x}_\varkappa$ exists.*

For a proof of all these facts put $\overline{p}_\varkappa = \overline{p}_1\Xi_\varkappa$ and $\overline{Q}_\varkappa = \bigcup_{\lambda \neq \varkappa} \overline{p}_\lambda$. Because $\overline{p}_\varkappa\overline{p}_\lambda \geqslant 2\,\rho(p)$ for $\varkappa \neq \lambda$, the set $\overline{Q}_\varkappa$ is closed. $H(\overline{p}_1)$ is the set of $\overline{x}$ with $\overline{p}_1\overline{x} < \overline{x}\overline{Q}_1$ and is obviously open, see (2.3). Let $\overline{x} \in H(\overline{p}_1)$ and $\overline{y} \in T(\overline{p}_1, \overline{x})$. Then $\overline{p}_\varkappa\,\overline{y} \geqslant \overline{p}_\varkappa\,\overline{x} - \overline{y}\,\overline{x} > \overline{p}_1\overline{x} - \overline{y}\overline{x} = \overline{p}_1\overline{y}$ for $\varkappa > 1$. This proves (a).

The assertion (b) is contained in the statement: if $\overline{x}$ lies on the boundary of $H(\overline{p}_1)$ and $(\overline{p}_1\overline{y}\overline{x})$ then $\overline{y} \in H(\overline{p}_1)$. If the hypothesis is satisfied then $\overline{p}_1\overline{x} \leqslant \overline{p}_\varkappa\,\overline{x}$ for $\varkappa > 1$, hence

$$\overline{p}_1\overline{y} = \overline{p}_1\overline{x} - \overline{x}\overline{y} \leqslant \overline{p}_\varkappa\overline{x} - \overline{x}\overline{y} \leqslant \overline{p}_\varkappa\overline{y}.$$

The equality sign would imply $\overline{p}_1\overline{y} = \overline{p}_\varkappa\overline{y}$ and $(\overline{p}_\varkappa\overline{y}\,\overline{x})$ hence, by uniqueness of prolongation, $\overline{p}_\varkappa = \overline{p}_1$. Thus $\overline{p}_1\overline{y} < \overline{p}_\varkappa\overline{y}$ for $\varkappa > 1$ or $\overline{y} \in H(\overline{p}_1)$.

Since $\mathfrak{F}$ is simply transitive on the set $\{\overline{p}_\nu\}$ we find $\overline{Q}_1\Xi_\varkappa = \overline{Q}_\varkappa$. Since moreover $\overline{p}_1\Xi_\varkappa = \overline{p}_\varkappa$ and $\Xi_\varkappa$ is a motion, it transforms the set $\overline{x}\overline{p}_1 < \overline{x}\overline{Q}_1$ into the set $\overline{x}\overline{p}_\varkappa < \overline{x}\overline{Q}_\varkappa$ which proves $H(\overline{p}_\varkappa) = H(\overline{p}_1)\,\Xi_\varkappa$. Also $H(\overline{p}_\varkappa) \cap H(\overline{p}_\lambda) = 0$ for $\varkappa \neq \lambda$, because $\overline{x} \in H(\overline{p}_\varkappa)$ implies

$$\overline{x}\overline{Q}_\lambda \leqslant \overline{x}\overline{p}_\varkappa < \overline{x}\overline{Q}_\varkappa \leqslant \overline{x}\overline{p}_\lambda,$$

so that $\overline{x}$ cannot lie in $H(\overline{p}_\lambda)$, hence (c).

We now turn to (d). Let $x \in U(p)$, then $T(p, x)$ is, by the definition of $U(p)$, unique. There is exactly one segment $T(\overline{p}_1, \overline{x})$ over $T(p, x)$ beginning at $\overline{p}_1$ and of length $\overline{p}_1\overline{x} = px$. If $\overline{x}$ did not lie in $H(\overline{p}_1)$, then $\overline{x}\overline{Q}_1 \leqslant \overline{x}\overline{p}_1$. Hence a $\overline{p}_\varkappa \neq \overline{p}_1$ would exist such that $\overline{x}\overline{p}_\varkappa \leqslant \overline{x}\overline{p}_1$. Then Ω would send a segment $T(\overline{p}_\varkappa, \overline{x})$ into a geodesic curve from p to x of length $\overline{x}\overline{p}_\varkappa \leqslant px$. Since $T(p, x)$

is the only segment from p to x this geodesic curve must coincide with $T(p, x)$. But then $T(\overline{x}, \overline{p}_\varkappa)$ and $T(\overline{x}, \overline{p}_1)$ would be two different segments over $T(x, p)$ beginning at $\overline{x}$ and ending at a point over p, which contradicts (27.8).

The relation $F(\overline{p}_\varkappa) = F(\overline{p}_1)\, \Xi_\varkappa$ follows from (c). Let $\overline{x}$ be an arbitrary point in $\overline{R}$. Connect $x = \overline{x}\Omega$ to p by a segment T and let $\overline{T}$ be the segment over T which begins at $\overline{x}$. It ends at a point $\overline{p}_\varkappa$ over p. The segment $T(p, y)$ is unique for any $y \neq x$ on T. Since $H(\overline{p}_\varkappa)\, \Omega \supset U(p)$ it follows that the point $\overline{y}$ of $\overline{T}$ over y lies in $H(\overline{p}_\varkappa)$; for $\overline{y} \to \overline{x}$ we find $\overline{x} \in F(\overline{p}_\varkappa)$, hence $\cup\, F(\overline{p}_\varkappa) = \overline{R}$.

For (f) let H^* be an open set containing $F(\overline{p}_1)$. Since $\mathfrak{F}$ has at least two elements it follows from (c) that $F(\overline{p}_1)$ is not the whole space. Therefore H^* contains a point $\overline{q}$ not in $F(\overline{p}_1)$ and a sphere $S(\overline{q}, \rho) \subset H^* - F(\overline{p}_1)$, $\rho > 0$, exists. Because of (e) the point $\overline{q}$ lies in an $F(\overline{p}_\varkappa)$, $\varkappa > 1$. By (b) all points on a segment $T(\overline{p}_\varkappa, \overline{q})$, except possibly $\overline{q}$, lie in $H(\overline{p}_\varkappa)$. Therefore $H(\overline{p}_\varkappa) \cap S(\overline{q}, \rho) \neq 0$ and it follows from $H^*\Xi_\varkappa \supset H(\overline{p}_\varkappa)$ that $H^* \cap H^*\Xi_\varkappa \neq 0$.

Let $\delta(R) < \infty$. If $\overline{T}$ is a segment from $\overline{p}_1$ to $\overline{x} \in F(\overline{p}_1)$, then $\overline{T}\Omega$ is a geodesic curve from p to x, which is a segment. Otherwise a segment $\overline{T}_1$ over a segment $T(x, p)$ beginning at $\overline{x}$ would end at a point $\overline{p}_\varkappa$ with $\overline{p}_\varkappa\overline{x} < \overline{p}_1\overline{x}$. But then $\overline{x}$ could not lie in $F(\overline{p}_1)$. Hence $\overline{p}_1\overline{x} \leqslant \delta(R)$ and $\overline{y}\,\overline{z} \leqslant 2\,\delta(R)$ for any two points in $F(\overline{p}_1)$.

If $\delta(R) = \infty$, then p is by (22.1) origin of a ray $\mathfrak{r}$, and the set $U(p)$ contains this ray; there is a ray $\overline{\mathfrak{r}}$ beginning at $\overline{p}_1$ over $\mathfrak{r}$, and $\overline{\mathfrak{r}}$ lies by (c) in $H(\overline{p}_1)$, consequently $\delta(F(\overline{p}_1) = \infty$.

To prove (h) choose $\varkappa$ such that $F(\overline{p}_\varkappa) \supset \overline{q}$. If $S(\overline{q}, \sigma) \cap F(\overline{p}_\lambda)$ contains a point $\overline{r}$ then $\overline{r}\overline{p}_\lambda \leqslant \overline{r}\overline{p}_\varkappa$, hence

$$\overline{p}_\lambda \overline{p}_\varkappa \leqslant \overline{p}_\lambda \overline{r} + \overline{r}\overline{q} + \overline{q}\overline{p}_\varkappa \leqslant \overline{p}_\varkappa \overline{r} + \overline{r}\overline{q} + \overline{q}\overline{p}_\varkappa$$
$$\leqslant 2\,\overline{q}\overline{p}_\varkappa + 2\,\overline{r}\overline{q} < 2\,\overline{q}\overline{p}_\varkappa + 2\,\sigma.$$

and there is only a finite number of $\overline{p}_\lambda$ which satisfy this inequality.

Finally we establish (i). Let $\Xi_\eta \neq \Xi_1$ be given. Connect $\overline{p}_1$ to $\overline{p}_\eta$ by an oriented segment $\overline{\mathfrak{s}}$. Denote the last point on $\overline{\mathfrak{s}}$ in $F(\overline{p}_1)$ by $\overline{q}_1$. There is a set $F(\overline{p}_{\varkappa_2})$ that contains $\overline{q}_1$ and points of $\overline{\mathfrak{s}}$ that follow $\overline{q}_1$. For the subsegment $\overline{\mathfrak{s}}_1$ of $\overline{\mathfrak{s}}$ from $\overline{q}_1$ to $\overline{p}_\eta$ has common points with only a finite number of $F(\overline{p}_\varkappa)$ because of (h). If none of these contained $\overline{q}_1$ and other points of $\overline{\mathfrak{s}}_1$, then the half open segment $\overline{\mathfrak{s}}_1 - \overline{q}_1$ would be the union of a finite number of sets which are closed in $\overline{\mathfrak{s}}$, whereas $\overline{\mathfrak{s}}_1 - \overline{q}_1$ is open in $\overline{\mathfrak{s}}$.

If $F(\overline{p}_{\varkappa_2})$ contains $\overline{p}_\eta$, then $F(\overline{p}_{\varkappa_2}) = F(\overline{p}_\eta)$ and we stop. Otherwise, let $\overline{q}_2$ be that last point of $F(\overline{p}_{\varkappa_2})$ on $\overline{\mathfrak{s}}$. There is then by the same argument a $F(\overline{p}_{\varkappa_3})$ that contains $\overline{q}_2$ and points of $\overline{\mathfrak{s}}$ following $\overline{q}_2$. Again, if $F(\overline{p}_{\varkappa_3})$ contains $\overline{p}_\eta$, then $F(\overline{p}_{\varkappa_2}) = F(\overline{p}_\eta)$ and we stop, otherwise we continue. It follows from (h) that the process must end after a finite number of steps, say $F(\overline{p}_{\varkappa_m}) = F(\overline{p}_\eta)$. The definition of $H(\overline{p}_{\varkappa_i})$ implies

$$\overline{p}_{\varkappa_i}\overline{q}_i = \overline{p}_{\varkappa_{i+1}}\overline{q}_i, \quad i = 1, \ldots, m-1, \ \varkappa_1 = 1.$$

Therefore $\Xi_{\varkappa_2} = \Xi_{\lambda_1}$ satisfies (i) and

$$F(\overline{p}_1) = F(\overline{p}_{\varkappa_2})\,\Xi_{\lambda_1}^{-1}$$

Since $F(\overline{p}_{\varkappa_2})$ and $F(\overline{p}_{\varkappa_3})$ have $\overline{q}_2$ as common boundary point, $\overline{q}_2\Xi_{\lambda_1}^{-1}$ is a common boundary point of $F(\overline{p}_1)$ and $F(\overline{p}_{\varkappa_3})\,\Xi_{\lambda_1}^{-1}$, and there is a Ξ_{λ_2} of the type required in (i) such that

$$F(\overline{p}_1)\,\Xi_{\lambda_2} = F(\overline{p}_{\varkappa_3})\,\Xi_{\lambda_1}^{-1} \qquad \text{or} \qquad F(\overline{p}_{\varkappa_3}) = F(\overline{p}_1)\,\Xi_{\lambda_2}\Xi_1.$$

Continuation of this process yields $\Xi_{\lambda_1}, \ldots, \Xi_{\lambda_m}$ such that

$$F(\overline{p}_\eta) = F(\overline{p}_{\varkappa_m}) = F(\overline{p}_1)\,\Xi_{\lambda_m}\ldots\Xi_{\lambda_1} \quad \text{or} \quad \Xi_\eta = \Xi_{\lambda_m}\ldots\Xi_{\lambda_1}$$

and the Ξ_{λ_j} satisfy (i). This establishes the theorem.

We add the following remark:

(29.5) *If the space $\overline{R}$ is straight then $H(\overline{p}_1)\,\Omega = U(p)$.*

We know from (d) that $H(\overline{p}_1)\,\Omega \supset U(p)$. Let $y \in H(\overline{p}_1)\,\Omega$ and $\overline{y}$ an original of y in $H(\overline{p}_1)$. Because $H(\overline{p}_1)$ is open and $\overline{R}$ is straight, $H(\overline{p}_1)$ contains by (a) a point $\overline{z}$ with $(\overline{p}_1\overline{y}\,\overline{z})$ and $T(\overline{p}_1, \overline{z}) \subset H(\overline{p}_1)$. Then $T(\overline{p}, \overline{z})\,\Omega = A$ is a geodesic curve in $H(\overline{p}_1)\,\Omega$ from p to z that contains y. If A were not a segment, let T be a segment from p to z and take in $\overline{R}$ the segment over T which ends at $\overline{z}$. It begins at a point $\overline{p}_\varkappa \neq \overline{p}_1$ because $T \neq A$ and segments in $\overline{R}$ are unique. Since $\overline{p}_1\overline{z}$ is the length of A and A is not a segment, $\overline{p}_\varkappa\overline{z} = pz < \overline{p}_1\overline{z}$, which contradicts the definition of $H(\overline{p}_1)$. Therefore A is a segment, (pyz), hence $T(p, y)$ is unique, $y \subset U(p)$ and $H(\overline{p}_1)\,\Omega \subset U(p)$.

(29.4) has the following corollary

(29.6) *If $F(\overline{p}_1)$ is bounded, then R has finite connectivity, that is, its fundamental group can be generated by a finite number of elements. A compact G-space has finite connectivity.*

For if R is compact then its diameter δ is finite, hence by (g), $\delta(F(\overline{p}_1) \leqslant 2\,\delta$, so that $F(\overline{p}_1)$ is bounded. Assume $\delta(F(\overline{p}_1)) \leqslant 2\,\delta$, whether R is compact or not. Any point $\overline{p}_\varkappa$ for which an $\overline{x}_\varkappa$ with $\overline{p}_1\overline{x}_\varkappa = \overline{p}\,\overline{x}_\varkappa$ exists has at most distance $4\,\delta$ from $\overline{p}_1$. The sphere $S(\overline{p}_1, 5\,\delta)$ intersects by (h) only a finite number of $F(\overline{p}_\varkappa)$, hence, the number of $\Xi_\varkappa$ for which an $\overline{x}_\varkappa$ with $\overline{p}_1\overline{x}_\varkappa = \overline{p}_1\,\Xi_\varkappa\,\overline{x}_\varkappa$ exists, is finite.

EXAMPLE

The point p of R or the set $\{\overline{p}_\varkappa\}$ in $\overline{R}$ was arbitrary in (29.4). The shape of $H(\overline{p}_1)$ may vary greatly with the choice of $\{\overline{p}_\varkappa\}$. As an example we consider the *Moebius strip* with a euclidean metric.

Its universal covering space is the euclidean plane E^2, the fundamental group $\mathfrak{F}$ consists with a proper choice of rectangular coordinates x, y of the motions $\Phi^\varkappa$, $\varkappa = 0, \pm 1, \pm 2, \ldots$ where

$$\Phi : x' = x + a, \qquad y' = -y, \qquad a > 0.$$

In this, as in many other special cases, it is not convenient to put the motions in $\mathfrak{F}$ in form of a sequence $\Xi_1, \Xi_2, \ldots$. If $\overline{p}_0 = (0, 0)$, then $\{\overline{p}_0\Phi^\varkappa\}$ is the set $\{\varkappa a, 0\}$. Then $H(\overline{p}_0)$ becomes the set $\overline{p}_0\overline{x} < \overline{p}_\varkappa\overline{x}$, $\varkappa \neq 0$, where $\overline{x} = (x, y)$, which is the strip $-a/2 < x < a/2$. The motion Φ carries $x = -a/2$

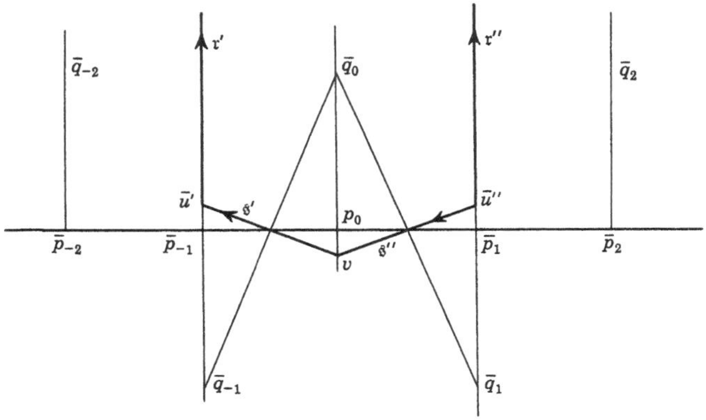

Fig. 30.

oriented in the sense of increasing y into $x = a/2$ oriented in the sense of decreasing y. To obtain R we therefore have to "twist" the strip $-a/2 < x < a/2$ through 180 degrees and then paste the boundaries together, hence R is obviously a Moebius strip.

If instead of $\overline{p}_0$ we choose $\overline{q}_0 = (0, b)$, $b > 0$ as starting point, then $\overline{q}_\varkappa = \overline{q}_0\Phi^\varkappa = (\varkappa a, (-1)^\varkappa b)$. The set $\overline{q}_0\overline{x} < \overline{q}_\varkappa\overline{x}$, $\varkappa \neq 0$, then is the intersection of the four open half planes containing $\overline{q}_0$ and bounded by the perpendicular bisectors of the segments $T(\overline{q}_0, \overline{q}_{-2})$, $T(\overline{q}_0, \overline{q}_2)$, $T(\overline{q}_0, \overline{q}_{-1})$, $T(\overline{q}_0, \overline{q}_1)$, hence the boundary of $H(\overline{q}_0)$ consists of the rays $\overline{\mathfrak{r}}'$, $\overline{\mathfrak{r}}''$: $x = -a$, $x = a$, $y \geqslant a^2/4b$

with origins $\bar{u}' = (-a, a^2/4b)$ and $\bar{u}'' = (a, a^2/4b)$ and the segments $\bar{\mathfrak{s}}', \bar{\mathfrak{s}}''$ connecting $\bar{v} = (0, -a^2/4b)$ to $\bar{u}'$ and $\bar{u}''$.

The motion Φ^2 in $\mathfrak{F}$ carries $\bar{\mathfrak{r}}'$ into $\bar{\mathfrak{r}}''$, these are therefore to be identified, and Φ carries $\bar{\mathfrak{s}}'^{-1}$ into $\bar{\mathfrak{s}}''$, and in particular $\bar{u}'$ into $\bar{v}$ and $\bar{v}$ into $\bar{u}''$. To obtain a picture we first identify $\bar{u}'$ with $\bar{v}$ and $\bar{v}$ with $\bar{u}''$ and then the two closed curves thus obtained with due regard to their orientation. The result must again be a Moebius strip but that is harder to visualize in the present case.

This example also shows that the elements generating $\mathfrak{F}$ according to part (i) in (29.4) depend on the choice of $\bar{p}_1$: For the first choice $\bar{p}_1 = (0, 0) = \bar{p}_0$ the motions Φ and Φ^{-1} are the only elements obtained, for the second choice $\bar{p}_1 = (0, b) = \bar{q}_0$ we obtain Φ^2 and Φ^{-2} as well. In both cases we find that with any $\Xi_\varkappa$ also $\Xi_\varkappa^{-1}$ belongs to the Ξ_λ determined by (i). This is, of course, a general feature of the procedure.

(29.7) *The set of elements in $\mathfrak{F}$ determined by (i) in (29.4) contains with any element its inverse.*

For if a point $\bar{x}_\varkappa$ with $\bar{p}_1\bar{x}_\varkappa = \bar{p}_\varkappa\bar{x}_\varkappa$ exists and $\Xi_\lambda = \Xi_\varkappa^{-1}$, then with $\bar{x}_\lambda = \bar{x}_\varkappa\Xi_\varkappa^{-1}$

$$\bar{p}_\lambda\bar{x}_\lambda = \bar{p}_1\Xi_\varkappa^{-1}\bar{x}_\varkappa\Xi_\varkappa^{-1} = \bar{p}_\varkappa\Xi_\varkappa^{-1}\bar{x}_\varkappa\Xi_\varkappa^{-1} = \bar{p}_1\bar{x}_\lambda$$

hence Ξ_λ is also one of the motions in (29.4 i).

30. Locally Minkowskian, hyperbolic or spherical spaces

If every point p of a G-space R has a neighborhood $S(p, \sigma_p)$, $\sigma_p > 0$ which is isometric to a sphere $S(p', \sigma_p)$ in a Minkowskian, hyperbolic or spherical space, then we call R *locally Minkowskian, hyperbolic or spherical* respectively.

We did not postulate, because it follows, that $S(p', \sigma_p)$ may for all p be considered as lying in the same Minkowski space, the same hyperbolic space, or the same spherical space. For if p, q are any two points of R, then $T(p, q)$ is contained in a finite number of spheres $S(r_i, \sigma_{r_i})$, $r_i \in T(p, q)$ such that two successive spheres overlap. The metric in the common part of two spheres determines the Minkowskian, spherical or hyperbolic metric in the entire union of the two spheres.

Simply connected spaces

We know that the Minkowski spaces, the hyperbolic spaces and the spherical spaces (the latter only if the dimension is greater than one) are simply connected. The question arises whether they are the only simply connected spaces which are locally Minkowskian, hyperbolic or spherical. The following theorem, and (28.6) answer this question affirmatively.

(30.1) ***Theorem:*** *The universal covering space of a locally Minkowskian, hyperbolic or spherical space (of dimension $\geqslant 2$) is a Minkowskian, hyperbolic or spherical space.*

The proof is essentially the same in all three cases; the spherical case offers a slight additional difficulty. We treat here only the latter, since a proof for the other two cases, in addition to being simpler, is also contained in the construction of the universal covering space of a space with non-positive curvature given in Section 38.

Let the sphere $S(p, \sigma_p)$, $\sigma_p > 0$ of the G-space R be isometric to a sphere in a spherical space $\overline{R}$ with radius β (or length of great circles $2\pi\beta$). It is clear that any sphere $S(q, \rho)$ contained in $S(p, \sigma_p)$ is then also isometric to a sphere in $\overline{R}$. By a standard argument, compare (7.5), we see that if $\sigma(p)$ is the least upper bound of those σ_p for which $S(p, \sigma_p)$ is isometric to a sphere in $\overline{R}$, then $\sigma(p)$ satisfies $|\sigma(p) - \sigma(q)| \leqslant pq$. If $\sigma(p) > \pi\beta$ for some p, then we have nothing to prove.

Choose an arbitrary point $\overline{p}$ in $\overline{R}$ and an arbitrary point p in R, denote by Ω an isometry of $S(\overline{p}, \sigma(p))$ on $S(p, \sigma(p))$. If $\sigma_0 = \sigma(p)/2$, then Ω induces an isometry $\overline{q} \to q$ of $K(\overline{q}, \sigma_0)$ on $K = K(q, \sigma_0)$. Denote by $\overline{x}_q(\tau)$, $x_q(\tau)$, $\tau \geqslant 0$, the representations of the half geodesics in $\overline{R}$ and R for which $\overline{x}_q(0) = \overline{p}$, $\overline{x}_q(\sigma_0) = \overline{q}$, $x_q(\sigma_0) = q$. Then $\overline{x}_q(\tau)\,\Omega = x_q(\tau)$ for $0 \leqslant \tau \leqslant \sigma(p)$. We extend Ω to the association $\overline{x}_q(\tau) \to x_q(\tau)$ for $\tau \geqslant 0$.

To show that Ω is a locally isometric mapping of $\overline{R}$ on R it must be ascertained first of all that it is single valued, or that the points $x_q(\nu\pi\beta)$ coincide for all odd ν and all $q \in K$, as well as for all even ν and all $q \in K$. Let $\nu_0 > 0$ be given and $\beta' > \nu_0\pi\beta$. Put $\sigma = 1/3 \min \sigma(x)$ for $px \leqslant \beta'$. Then $\sigma > 0$ because $\sigma(x)$ is continuous.

It follows from (8.11, 12) that there is an $\varepsilon > 0$ such that $qr < \varepsilon$, $q, r \in K$, implies $x_q(\tau')x_r(\tau'') < \sigma$ and $\overline{x}_q(\tau')\overline{x}_r(\tau'') < \sigma$ for $0 \leqslant \tau', \tau'' \leqslant \beta$. Then

$$\text{(30.2)} \quad \overline{x}_q(\tau')\,\overline{x}_r(\tau'') = x_q(\tau')\,x_r(\tau'') \quad \text{for} \quad qr < \varepsilon, \quad 0 \leqslant \tau', \tau'' \leqslant \beta', \text{ and}$$

$$|\tau' - \tau''| \leqslant \sigma.$$

For choose a partition $\tau_0 = 0 < \tau_1 < \ldots < \tau_m = \beta'$ of $[0, \beta']$ with $\max(\tau_{i+1} - \tau_i) < \sigma$. Then the set Σ_i of the six points $x_q(\tau_{i-1})$, $x_q(\tau_i)$, $x_q(\tau_{i+1})$, $x_r(\tau_{i-1})$, $x_r(\tau_i)$, $x_r(\tau_{i+1})$ lies in $S(x_q(\tau_i), \sigma)$. For $i = 1, 2$, Σ_i is isometric to the sextuple $\overline{\Sigma}_i$: $\overline{x}_q(\tau_i)$, $\overline{x}_q(\tau_i)$, $\overline{x}_q(\tau_{i+1})$, $\overline{x}_r(\tau_{i-1})$, $\overline{x}_r(\tau)$, $\overline{x}_r(\tau_{i+1})$. By induction we see that Σ_i and $\overline{\Sigma}_i$ are isometric for all i: For if Σ_{i-1} and $\overline{\Sigma}_{i-1}$ are isometric, then the quadruples $x_q(\tau_{i-1})$, $x_q(\tau_i)$, $x_r(\tau_{i-1})$, $x_r(\tau_i)$ and $\bar{x}_q(\tau_{i-1})$, $\bar{x}_q(\tau_i)$, $\bar{x}_r(\tau_{i-1})$, $\bar{x}_r(\tau_i)$ are isometric, the latter lies on a two-sphere in S^n, and determines together with the values of the τ_j the distances in $\bar{\Sigma}_i$. Since $\Sigma_i \subset S(x_q(\tau_i), \sigma)$ and $S(x_q(\tau_i), \sigma)$ is isometric to $S(\bar{x}_q(\tau_i), \sigma) \supset \bar{\Sigma}_i$, the distances in the first quadruple determine the distances in Σ_i in the same way, hence Σ_i and $\bar{\Sigma}_i$ are also isometric.

By first choosing the partition such that the points $\nu\pi\beta$, $0 \leqslant \nu \leqslant \nu_0$, occur as points in it, we see from $\bar{x}_q(\nu\pi\beta)\,\bar{x}_r(\nu\pi\beta) = 0$ that $x_q(\nu\pi\beta)\,x_r(\nu\pi\beta) = 0$, hence $x_q(\nu\pi\beta) = x_r(\nu\pi\beta)$ for fixed ν and $q, r < \varepsilon$. It follows that $x_q(\nu\pi\beta) = x_r(\nu\pi\beta)$ for fixed ν and any $q, r \in K$.

The same argument proves (30.2) for $\tau' = \tau''$. If $\tau' < \tau''$ we choose the partition such that τ' and τ'' occur as successive points, say $\tau_{i-1} = \tau'$, $\tau_{i+1} = \tau''$. Then the isometry of $\bar{\Sigma}_i$ and Σ_i proves (30.2).

This result implies that Ω defines a locally isometric mapping of $S(\bar{p}, \pi\beta)$ on the set $x_q(\tau)$, $q \in K$, $0 \leqslant \tau < \pi\beta$ in R. Moreover, the antipodal point $\bar{p}' = \bar{x}_q(\pi\beta)$ (q arbitrary on K) to $\bar{p}$ goes into the single point $p' = x_q(\pi\beta)$ in R. It follows from (30.2) and the isometry of $S(\bar{p}', \sigma(p'))$ and $S(p', \sigma(p'))$ that Ω defines a continuous mapping of $\bar{R}$ on the set $x_q(\tau)$, $0 \leqslant \tau \leqslant \pi\beta$, in R, and an isometric mapping of the set $\bar{x}_q(\tau)$, $q \in K$, $\pi\beta \leqslant \tau < \pi\beta + \sigma$ on the $x_q(\tau)$ with the same q and τ.

Hence $x_q(\pi\beta - \tau) = x_{q'}(\pi\beta + \tau)$ for $|\tau| < \sigma$ if q' is the antipodal point to q on the sphere K (which is isometric to an ordinary sphere). Since $x_q(\tau)$ and $x_{q'}(\tau)$ are the parts $\tau \geqslant 0$ of representations of geodesics defined for all τ, the relation $x_q(\pi\beta - \tau) = x_{q'}(\pi\beta + \tau)$ extends to all τ. This shows that Ω maps $\bar{R}$ on all of R and is locally isometric. Thus (30.1) is proved.

Locally spherical spaces of even dimensions

The number of topologically different types of locally Minkowskian, hyperbolic and spherical spaces increases with the dimension and a general discussion here of what is known is out of question. There is one exception

(30.3) *An even-dimensional locally spherical space is a sphere or an elliptic space.*

It means no restriction to assume that the radius equals 1. We represent the n-dimensional spherical space S^n in the form

$$S^n : x_0^2 + x_1^2 + \ldots + x_n^2 = 1$$

with rectangular coordinates x_i in E^{n+1}. The motions of S^n are

$$\Phi : x_i' = \sum_{\varkappa=0}^{n} a_{i\varkappa}\, x_\varkappa, \quad \sum_{\varkappa=0}^{n} a^2{}_{i\varkappa} = 1, \quad \sum_{\varkappa=0}^{n} a_{i\varkappa}\, a_{j\varkappa} = 0 \text{ for } j \neq i$$

The equation

$$\begin{vmatrix} a_{00}-\lambda & a_{01} & \dots & a_{0n} \\ a_{11} & a_{11}-\lambda & \dots & a_{1n} \\ \cdot & \cdot & \dots & \cdot \\ a_{n0} & a_{n1} & \dots & a_{nn}-\lambda \end{vmatrix} = 0$$

is of odd degree in λ and has therefore at least one real root which is not 0, because $|a_{ik}| \neq 0$. Therefore a point y on S^n exists such that

$$\lambda\, y_i = \sum a_{i\varkappa}\, y_\varkappa.$$

But the expression Σx_i^2 is invariant under Φ, i. e., $\Sigma x_i'^2 = \Sigma x_i^2$, hence

$$\sum \lambda^2 y_i^2 = \sum y_i^2 = 1, \quad \text{or} \quad \lambda = \pm 1.$$

Now let S^n be covering space of $R \neq S^n$. Then the fundamental group $\mathfrak{F}$ of R contains at least one motion Ξ_2 of S^n besides the identity $\Xi = \Xi_1$. Applying the above results to $\Phi = \Xi_2$ we find that $\lambda = -1$, because Ξ_2 would leave y fixed if $\lambda = 1$. Hence Ξ_2 interchanges y and $-y$; and Ξ_2^2 leaves y fixed. Since $\Xi_2^2 \subset \mathfrak{F}$, it must be the identity. It follows that $x\,\Xi_2 = -x$ for any x. For if $x_0\Xi_2 = x_0' \neq -x_0$, then the midpoint z of x_0 and x_0' would be unique. Since $x_0'\Xi_2 = x_0\Xi_2^2 = x_0$ the motion Ξ_2 would interchange x_0 and x_0' and therefore leave z fixed. Thus $\mathfrak{F}$ consists of the identity and $x_i' = -x_i$, hence R is an elliptic space.

Locally Minkowskian spaces

We next consider a locally Minkowskian G-space R. Its universal covering space $\overline{R}$ is a Minkowskian straight space. By (17.19) the space $\overline{R}$ may be remetrized as a euclidean space $\overline{R}_E$ with distance $E(\overline{x}, \overline{y})$ such that any motion of $\overline{R}$ is also a motion of $\overline{R}_E$ and the straight lines are the same in both spaces (if $\overline{x}(\tau)$ represents a geodesic in $\overline{R}$ then for a suitable $\varphi > 0$ depending on the geodesic, $\overline{x}(\varphi\tau)$ represents a geodesic in $\overline{R}_E$).

In particular, the fundamental group $\mathfrak{F}$ of R is also a group of motions for $\overline{R}_E$. The orbits $0(\overline{x}, \mathfrak{F})$ yield with the distance $E[0(\overline{x}, \mathfrak{F}), 0(\overline{y}, \mathfrak{F})]$ a remetrization R_E of R as locally euclidean space:

(30.4) *A locally Minkowskian space R may be remetrized as a locally euclidean space R_E such that every motion of R is also a motion of R_E and $x(\tau)$ represents a geodesic in R if, and only if for a suitable $\varphi > 0$ depending on the geodesic, $x(\varphi\tau)$ represents a geodesic in R_E.*

The study of locally Minkowskian spaces can therefore be reduced to a study of locally euclidean spaces. The advantage of the reduction lies in the simple form which the fundamental sets take in the euclidean case.

(30.5) *If $\overline{R}$ is euclidean or hyperbolic, then* $F(\overline{p}_1)$ (unless it is the whole space) *is the intersection of a countable number of closed halfspaces.* (The notations are those of (29.4).)

For the locus $\overline{x}\,\overline{p}_1 \leqslant \overline{x}\,\overline{p}_\varkappa$ is for each $\varkappa > 1$ a closed halfspace since the locus $\overline{x}\,\overline{p}_1 = \overline{x}\,\overline{p}_\varkappa$ is in both the euclidean and the hyperbolic geometries a hyperplane; $F(\overline{p}_1)$ is the intersection of these half spaces for $\varkappa \geqslant 2$. As we saw in the example of the Moebius strip, it will in general not be necessary to use all these half spaces in forming the intersection. We will show that in the euclidean case a finite number will always suffice. Thus we obtain the following theorem which exhibits the applicability of (30.4) and is due to Killing in the euclidean case:

(30.6) **Theorem:** *A locally Minkowskian space has finite connectivity.*

Because of (30.4) we may assume that the given space R is locally euclidean. Then $\overline{R}$ is the E^n with a suitable n. If $F(\overline{p}_1)$ is bounded, then (30.6) follows from (29.6). If $F(\overline{p}_1)$ is not bounded, then it contains at least one ray $\overline{\mathfrak{r}}$ with origin $\overline{p}_1$ since it is by (30.5) a convex set. Let $\overline{P}$ be the hyperplane perpendicular to $\overline{\mathfrak{r}}$ at $\overline{p}_1$. No point $\overline{p}_\varkappa$ lies on the same side of $\overline{P}$ as $\overline{\mathfrak{r}}$ because $\overline{P} = K_\infty(\overline{\mathfrak{r}}, \overline{p}_1)^1$ implies $\alpha(\overline{\mathfrak{r}}, \overline{x}) - \alpha(\overline{\mathfrak{r}}, \overline{p}_1) < 0$ for a point $\overline{x}$ in the interior of $K_\infty(\overline{\mathfrak{r}}, \overline{p}_1)$ hence by (22.5) $\overline{z}\,\overline{x} - \overline{z}\,\overline{p}_1 < 0$ for $\overline{z} \in \overline{\mathfrak{r}}$ and large $\overline{p}_1\overline{z}$ so that $\overline{z}$ can, by the definition of $H(\overline{p}_1)$, not be a point $\overline{p}_\varkappa$, $\varkappa > 1$.

Now let L_m be a flat of maximal dimension m through $\overline{p}_1$ contained in $F(\overline{p}_1)$, (of course, m may be zero). Then the preceding argument applied to all rays $\overline{\mathfrak{r}}$ with origin $\overline{p}_1$ in L_m shows that all $\overline{p}_\varkappa$ must lie in the $(n-m)$-flat L_{n-m} through $\overline{p}_1$ perpendicular to L_m.

The totality of the rays with origin $\overline{p}_1$ in $L_{n-m} \cap F(\overline{p}_1)$ is a convex cone C. The above argument shows that all $\overline{p}_\varkappa$, $\varkappa \geqslant 2$, must lie in the closed complementary cone C' to C. The intersection $C' \cap F(\overline{p}_1)$ is bounded because C contains all rays with origin $\overline{p}_1$ in $F(p_1) \cap L_{n-m}$, hence C' none. If $C' \cap F(\overline{p}_1) \subset S(\overline{p}_1, \rho)$, then all $\overline{p}_\varkappa$ for which a point $\overline{x}_\varkappa$ with $\overline{x}_\varkappa\overline{p}_\varkappa = \overline{x}_\varkappa\overline{p}_1$

exists must lie in $S(\bar{p}_1, 2\rho)$, and there is only a finite number of such $\bar{p}_x$. It follows from (29.4 i) that $\mathfrak{F}$ can be generated by a finite number of its elements, which means, by definition, that $\mathfrak{F}$ has finite connectivity.

A general property of all spaces with non-positive curvature is that their fundamental groups contain no finite sub-groups, except the group formed by identity (see 39.3). Therefore the fundamental groups of locally euclidean or hyperbolic spaces have no finite sub-groups. It was already mentioned that there are too many locally euclidean or hyperbolic spaces to attempt an enumeration. We will however briefly discuss the two-dimensional locally euclidean spaces:

Locally euclidean surfaces

The only motions of E^2 without fixed points are the translations and the translations followed by a reflection in a line parallel to the direction of the translation[2].

In case the fundamental group $\mathfrak{F}$ consists of the powers $\Phi^k : x' = x + ka$, $y' = y$ of the translation $\Phi : x' = x + a,\ y' = y$, then for $\bar{p}_1 = (0, 0)$ the fundamental set $H(\bar{p}_1)$ is the strip $-a/2 < x < a/2$, and the surface with $\mathfrak{F}$ as fundamental group is a *cylinder*.

If $\mathfrak{F}$ consists of the powers $\Phi^k : x' = x + ka,\ y' = (-1)^k y$ of $\Phi: x' = x + a$, $a > 0,\ y' = -y$ we obtain the Moebius strip R already discussed in the last section. The group $\mathfrak{F}$ contains as sub-group $\mathfrak{G}$ the group of translations $\Phi^{2k} : x' = x + 2ka,\ y' = y$. The surface R' belonging to $\mathfrak{G}$ as fundamental group is a cylinder, and R' is by (29.3) a two-sheeted[3] covering space of R.

We discuss briefly the *geodesics on the Moebius strip* R: If $F(\bar{p}_1)$ is constructed for $\bar{p}_1 = (0, 0)$ as in the preceding section, then its boundary is formed by the lines $x = \pm a/2$ which are to be identified such that the point $(-a/2, y)$ goes into $(a/2, -y)$.

The segment $-a/2 \leqslant x \leqslant a/2,\ y = 0$, yields on the surface a great circle of length a. If $b > 0$ then the two segments $-a/2 \leqslant x \leqslant a/2,\ y = b$ and $-a/2 \leqslant x \leqslant a/2,\ y = -b$ form together on R a simple closed geodesic of length $2a$. This is not a great circle, when $0 < b < a/2$, because the arcs of the geodesic connecting $(0, b)$ to $(0, -b)$ have length a, and cannot be segments, because the segment in E^2 from $(0, b)$ to $(0, -b)$ yields in R a geodesic connection of length $2b < a$. For $b \geqslant a/2$ the geodesics are great circles.

The lines $x = c$ yield simple geodesics on R. The parts $y \geqslant 0$ and $y \leqslant 0$ yield rays, but the whole curve $x = c$ is not a straight line, for, as we just saw, points (c, b) and $(c, -b)$ with $b > a/2$ have a shorter connection than the arc $-b \leqslant y \leqslant b$ of $x = c$.

To see the shape of the other geodesics on R, it suffices to consider those through $\overline{p}_1 = (0, 0)$, because the translations $x' = x + c$ of E^2 lie over motions of R. If a line through $\overline{p}_1$ different from the x-axis and the y-axis intersects

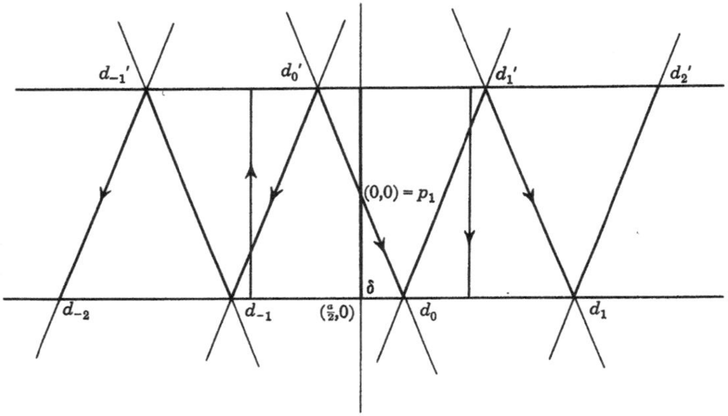

Fig. 31.

$x = a/2$ at $d_0 = (a/2, \delta)$ with $\delta > 0$ say, then it intersects $x = -a/2$ at $d_0' = (-a/2, -\delta)$. Putting $d_k = (a/2, (4k+1)\delta)$, $d_k' = (-a/2, (4k-1)\delta)$, $k = 0, \pm 1, \pm 2, \ldots$, we see that a half geodesic on R is obtained by traversing successively the following segments in E^2 : $T(d_0', d_0)$, $T(d_0', d_{-1})$, $T(d_1', d_1)$, $T(d_{-1}', d_{-2})$, $T(d_2', d_2)$, ... The other half geodesic on the same geodesic in R may be visualized similarly. We find that all the points d_i and d'_i are double points, that is, have multiplicity two.

We now turn to the case where $\mathfrak{F}$ *is not cyclic*. If $\mathfrak{F}$ consists of translations, we know from the theory of elliptic functions that in order to be discrete it must have the form

$$x' = x + m a, \qquad y' = y + n b, \quad m, n \text{ integers}, \qquad a > 0, \quad b > 0$$

in suitable affine coordinates x, y in E^2, or the points $\overline{p}_k$ form a lattice. If the x-axis and y-axis are perpendicular, then $F(\overline{p}_1)$ becomes a rectangle and the surface R is clearly a *torus*. If the x-axis and y-axis are not perpendicular, then $F(\overline{p}_1)$ will be a hexagon and it requires a little discussion to see that R

is a torus. However, we may use (30.4) and remetrize the plane with another euclidean metric in which the x-axis and y-axis are perpendicular, without changing the geodesics. Let $\bar{p}_1 = (0, 0)$. Then $F(\bar{p}_1)$ is the rectangle $-a/2 \leqslant x \leqslant a/2$, $-b/2 \leqslant y \leqslant b/2$. The lines $x =$ const. and $y =$ const.

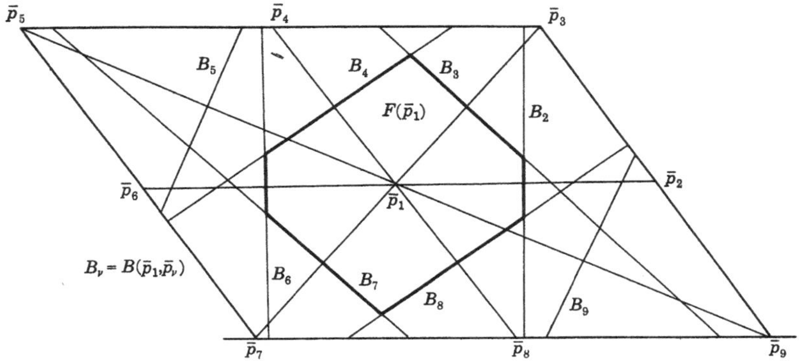

Fig. 32.

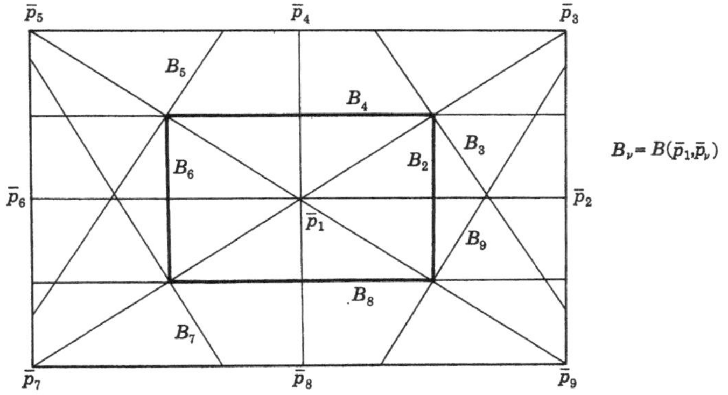

Fig. 33.

furnish great circles on R. The lines $y = m\,x + c$ furnish *simple closed geodesics*, if m is a rational multiple of b/a, and furnish *everywhere dense simple geodesics on* R, if $m\,a/b$ is irrational. For a given $\varepsilon > 0$ there are closed geodesics which enter every circular disk $S(p, \varepsilon)$ on R.

Finally, we have to consider the case where $\mathfrak{F}$ contains motions which are products of reflections and translations. Since, in any affine coordinates, the product of

$$x' = x + a, \quad y' = -y, \quad a \neq 0 \quad \text{and} \quad x' = -x, \quad y' = y + b, \quad b \neq 0,$$

namely $x'' = -x - a$, $y'' = -y + b$, has the fixed point $(-a/2, b/2)$ the only case left is, where the elements of $\mathfrak{F}$ have in suitable affine coordinates x, y the form

$$x' = x + m\,a, \quad y' = (-1)^m y + n\,b, \quad a > 0, b > 0, \quad m, n \text{ integers}.$$

If we interpret x, y again as rectangular coordinates in a suitable new euclidean metric and put $\bar{p}_1 = (0, 0)$, then $F(\bar{p}_1)$ is the same rectangle as above, only this time the segment from $(-a/2, -b/2)$ to $(-a/2, b/2)$ must be identified with the segment from $(a/2, b/2)$ to $(a/2, -b/2)$ in this order. The surface thus obtained is the *Klein bottle or the one-sided torus*. The motions

$$x' = x + 2\,m\,a, \qquad y' = y + n\,b$$

form a sub-group of $\mathfrak{F}$, hence the Klein bottle has an ordinary torus as two-sheeted covering space.

Since translations are motions of the Minkowskian geometry, the fundamental groups of the cylinder and torus can be realized as groups of motions in a given Minkowski metric. The fundamental groups of the Moebius strip and the Klein bottle can be realized with a given Minkowski metric only if there is a line in which the Minkowski plane possesses a reflection, which will in general not be the case.

(30.7) *The cylinder and the torus can be metrized with a given Minkowski metric, the Moebius strip and Klein bottle only with Minkowski metrics which possess a reflection in some line.*

Locally hyperbolic surfaces

The sphere, projective plane, torus, cylinder, Moebius strip, and Klein bottle can be provided with locally spherical or euclidean metrics and no other surfaces can. It is a fact that all other surfaces of finite connectivity, and many with infinite connectivity, can be provided with a locally hyperbolic metric. We will restrict ourselves to a few examples.

If *an orientable closed surface of genus* γ is made simply connected by the standard $2\,\gamma$ retrosections $A_1, B_1, \ldots, A_\gamma, B_\gamma$ through a point q, then a polygon of 4γ sides is obtained of the type indicated in the figures for $\gamma = 1, 2$, compare Seifert and Threlfall [1, p. 139]. The torus can be provided with a euclidean metric because the euclidean plane can be parqueted with rectangles. The hyperbolic plane can be parqueted with regular 4γ-gons for any $\gamma > 1$.

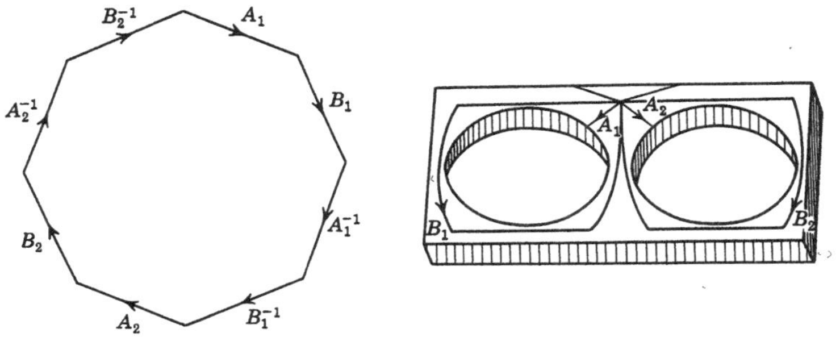

Fig. 34.

For if $\bar{p}_1$ is any point in the hyperbolic plane H, and $\bar{\mathfrak{r}}_1, \ldots, \bar{\mathfrak{r}}_{4\gamma}$ are rays with origin $\bar{p}_1$ such that $\bar{\mathfrak{r}}_i$ and $\bar{\mathfrak{r}}_{i+1}$ form the angle $2\pi/4\gamma = \pi/2\gamma$ ($\bar{\mathfrak{r}}_{4\gamma+1} = \bar{\mathfrak{r}}_1$), and $\bar{q}_i^\rho$ is the point on $\bar{\mathfrak{r}}_i$ with distance ρ from $\bar{p}_1$, then for small ρ the angle at $\bar{q}_i^\rho$ in the 4γ-gon $\bar{q}_1^\rho, \ldots, \bar{q}_{4\gamma}^\rho$ is close to the angle $(4\gamma - 2)\pi/4\gamma = (2\gamma - 1)\pi/2$ of the euclidean regular 4γ-gon. For $\rho \to \infty$ the angle tends monotonically to zero, hence there is exactly one value ρ_0 such that with $\bar{q}_i^{\rho_0} = \bar{q}_i$ the corresponding 4γ-gon $Q : \bar{q}_1, \ldots, \bar{q}_{4\gamma}$ has the angle $\pi/2\gamma$ at $\bar{q}_i$.

We can then parquet the hyperbolic plane H with 4γ-gons $Q_1 = Q, Q_2, Q_3, \ldots$ congruent to Q such that each vertex belongs to exactly 4γ polygons Q_i. After the sides of Q_1 have been labeled $A_1, B_1, A_1^{-1}, B_1^{-1}, \ldots, A_\gamma, B_\gamma, A_\gamma^{-1}, B_\gamma^{-1}$ in this order, the labeling of the sides of the other Q_i by the same symbols is determined by the requirement, that a common side of two Q_i must be labeled by inverse symbols.

This set of Q_i determines a discrete group $\mathfrak{F}$ of motions without fixed points of H: there is exactly one motion of H which carries Q_i into Q_j such that any side of Q_i goes into the side of Q_j with the same label. If $\bar{p}_i$ is the center of Q_i, then $H(\bar{p}_i)$ is the interior of Q_i.

The space R with H as universal covering space and $\mathfrak{F}$ as fundamental group is an orientable closed surface of genus γ with a locally hyperbolic metric.

As a last example we discuss the *torus with a hole.* Since our spaces are finitely compact we have to think of the hole as *an infinitely long tube.* We take the retrosections on the torus from a point in the hole. This situation

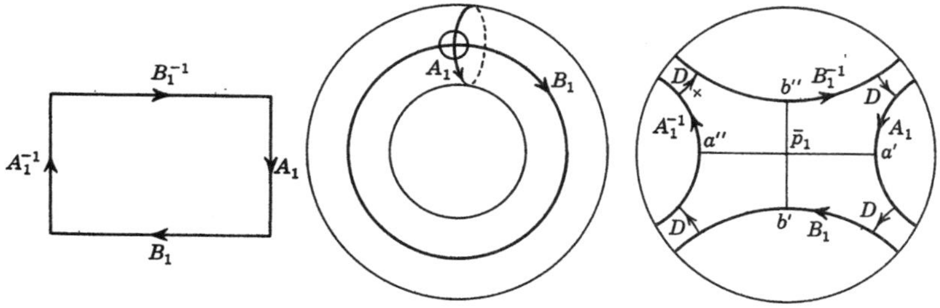

Fig. 35.

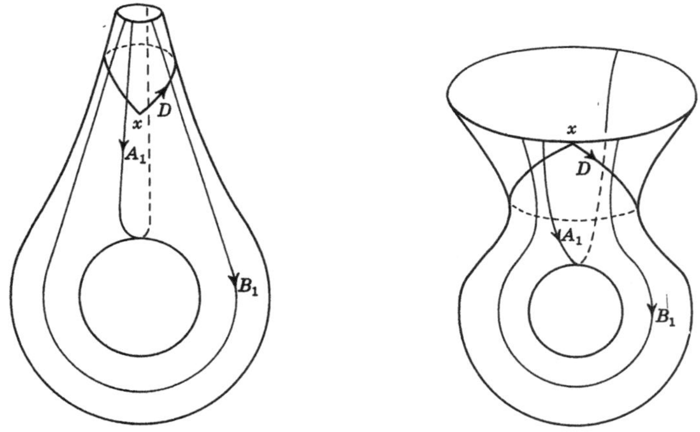

Fig. 36.

can be realized in the hyperbolic plane as follows: Take any two perpendicular lines G_1, G_2 through a point $\bar{p}_1$, then lines perpendicular to G_1 at points a', a'' which have equal distance from $\bar{p}_1$, and lines perpendicular to G_2 at points b', b'' with equal distance from $\bar{p}_1$ such that the latter lines do not intersect the former. Labeling the 4 perpendiculars through a', b', a'', b'' by $A_1, B_1, A_1^{-1}, B_1^{-1}$ respectively we can parquet H with sets $Q_1, Q_2, \ldots$ congruent to the set Q bounded by the four perpendiculars, and such that a common side of two sets has inverse labels. For instance, the translation along G_1 which carries a' into a'' produces such a set.

If the lines labeled A_1, B_1 (and hence A_1, B_1^{-1} etc.) are asymptotes then *the tube T corresponding to the hole on the torus contracts*: for a point x which tends on T to infinity, there is a geodesic monogon with vertex x whose length tends to zero. The image in Q of such a monogon consists of 4 different pieces as indicated in the figure. If A_1 and B_1 are not asymptotes, then *the tube T expands*, the length of a geodesic monogon D with vertex x tends to ∞ as x tends on T to infinity. We produce a surface of genus γ with a hole by replacing in the above construction of the 4γ-gon the sides $T(\bar{q}_i, \bar{q}_{i+1})$ by 4γ non-intersecting lines with equal distance from $\bar{p}_1$ and such that consecutive perpendiculars from $\bar{p}_1$ to the lines enclose equal angles (i.e., $\pi/2\gamma$). We obtain thus a surface of genus γ with a contracting tube, if consecutive lines are asymptotes and with an expanding tube, if they are not.

31. Spaces in which two points determine a geodesic.

By using covering spaces, it can now be proved that a G-space in which the geodesic through two points is unique is either straight or of the elliptic type and that all geodesics have the same length.

A THEOREM ON SPACES WHICH ARE NOT SIMPLY CONNECTED

We show first

(31.1) ***Theorem:*** *If the G-space R is not simply connected and has dimension at least two, if moreover each geodesic of R contains with any two points a segment connecting them, then R is of the elliptic type, and R has a spherelike space as two-sheeted universal covering space. Hence the geodesics in R all have the same length* (see Section 21).

We know from Theorem (9.6) that each geodesic of R is a straight line or a great circle. Since R is not simply connected its universal covering space $\bar{R}$ has at least two sheets, that is, at least two points of $\bar{R}$ lie over a given point of R.

The geodesics of $\bar{R}$ also have the property that they contain with any two points a segment connecting them. For let $\bar{\mathfrak{g}}$ be a geodesic in $\bar{R}$ and first $\bar{p}$, $\bar{q}$ two points of $\bar{\mathfrak{g}}$ such that $p = \bar{p}\Omega$, $q = \bar{q}\Omega$ are neither identical nor conjugate on $\mathfrak{g} = \mathfrak{g}(p, q) = \bar{\mathfrak{g}}\Omega$. If $\bar{\mathfrak{g}}$ did not contain a segment $T(\bar{p}, \bar{q})$, there would be a geodesic $\bar{\mathfrak{g}}_1 \neq \bar{\mathfrak{g}}$ containing such a segment. Because of (27.7) there is only one geodesic over $\mathfrak{g}$ through $\bar{p}$, hence $\bar{\mathfrak{g}}_1\Omega = \mathfrak{g}_1 \neq \mathfrak{g}$. But this contradicts (9.12), because p and q are not conjugate on $\mathfrak{g}$.

Since the point set carrying a geodesic in R is closed, it follows from (27.3) and (27.5, 6) that the point set carrying a geodesic $\bar{\mathfrak{g}}$ in $\bar{R}$ is closed. If now $\bar{p}, \bar{q}$ are arbitrary points of $\bar{\mathfrak{g}}$, we may choose $\bar{r}$ on $\bar{\mathfrak{g}}$ arbitrarily close to $\bar{q}$ such that p and r are neither identical nor conjugate on $\mathfrak{g}$. As we just proved, $\bar{\mathfrak{g}}$ contains a segment $T(\bar{p}, \bar{r})$, and since the point set carrying $\bar{\mathfrak{g}}$ is closed, $\bar{\mathfrak{g}}$ also contains a segment $T(\bar{p}, \bar{q})$. Thus, every geodesic in $\bar{R}$ is a straight line or a great circle.

We show next that there is exactly one geodesic $\bar{\mathfrak{g}}$ of $\bar{R}$ over a given geodesic $\mathfrak{g}$ of R. If both $\bar{\mathfrak{g}}_1$ and $\bar{\mathfrak{g}}_2$ lie over $\mathfrak{g}$ let p_1, p_2 be distinct points of $\mathfrak{g}$ which are not conjugate, and choose $\bar{p}_i \in \bar{\mathfrak{g}}_i$ over p_i, $i = 1, 2$. Let $\bar{\mathfrak{h}}$ be a geodesic in $\bar{R}$ through $\bar{p}_1$ and $\bar{p}_2$. Since p_1 is different from, and not conjugate to p_2 on $\mathfrak{g}$, it follows from (9.12) that $\mathfrak{g} = \mathfrak{h}$ and from (27.7) that $\bar{\mathfrak{g}}_i = \bar{\mathfrak{h}}$ because $\bar{\mathfrak{g}}_i$ and $\bar{\mathfrak{h}}$ have the common point $\bar{p}_i$ over p_i. Hence $\bar{\mathfrak{g}}_1 = \bar{\mathfrak{g}}_2$.

Since there is only one geodesic $\bar{\mathfrak{g}}$ over $\mathfrak{g}$, the geodesic $\bar{\mathfrak{g}}$ must, because of (27.6), contain all points which lie over a given point p of $\mathfrak{g}$. Therefore two different geodesics through a point $\bar{p}$ over p have all points over p in common. We noticed that there are at least two such points and it follows from (9.12) that there cannot be more than two, and that all geodesics through $\bar{p}$ are great circles (of the same length), which pass through the second point over p. Hence $\bar{R}$ is spherelike, and each geodesic in R has length equal to the distance of antipodal points in $\bar{R}$.

This theorem implies the theorem mentioned in the beginning of this section for a space R in which the geodesic through two points is unique, if R is a manifold of the type usually studied in differential geometry. For we know that each geodesic of R is a great circle or a straight line, see (9.9). Now if p is a point of R, and p' a point in an elliptic space E of the same dimension $n \geqslant 2$ as R whose geodesics have length at least $2\,\rho(p)$, we may associate the line elements of R at p linearly (or affinely) with the line elements of E at p', and the geodesic $\mathfrak{g}$ in R through p with the elliptic line $\mathfrak{g}'$ of E through p' which contains the line element corresponding to that of $\mathfrak{g}$ at p. By the method used in the proof of (10.7) we then map $\mathfrak{g}$ in $\mathfrak{g}'$, and produce in this way a topological mapping of R on a (proper or improper) subset R' of E. If $\mathfrak{g}$ is a great circle, then $\mathfrak{g}'$ is an entire elliptic line. The latter cannot be contracted to a point in E, therefore still less in R'. Hence $\mathfrak{g}$ cannot be contracted in R, so that R is not simply connected and therefore satisfies the hypotheses of the last theorem. Thus we have proved under certain differentiability assumptions:

(31.2) ***Theorem:*** *If the geodesic through two points of a G-space is unique, then the space is either straight or it is of the elliptic type and all geodesics have the same length. In the latter case, if the dimension is greater than one, the space has a spherelike space as two-sheeted universal covering space.*

The proof in the general case

The theorem also holds without differentiability hypotheses and we are going to outline a proof. Since it seems hard to see directly that the space is not simply connected when it contains a great circle, we prove this indirectly by constructing a two-sheeted covering space.

Assume then that R contains the great circle $\mathfrak{g}$, and let a be a point of $\mathfrak{g}$. If $\mathfrak{g}(a, d)$ is a great circle, $d \neq a$, then $\mathfrak{g}(a, x)$ is a great circle for $dx < \varepsilon$, if $\varepsilon > 0$ is properly chosen. This follows from (9.11). Call β the locus of the conjugate points to a on the different great circles through a. Put $S = S(a, \rho(a)/2)$, $K = K(a, \rho(a)/2)$. We define a mapping $x \rightarrow x^*$ of $R - \beta$ on S as follows: $a^* = a$. If $x \in R - \beta$ is different from a, then $T(a, x)$ is unique, and we define x^* as the point on $T(a, x)$ for which

$$a\,x^* = \frac{\rho(a)}{2} \cdot \frac{a\,x}{1 + a\,x} \cdot \frac{2 + \lambda(\mathfrak{g}(a, x))}{\lambda(\mathfrak{g}(a, x))} \quad \text{if} \quad \lambda(\mathfrak{g}(a, x)) < \infty,$$

$$= \frac{\rho(a)}{2} \cdot \frac{a\,x}{1 + a\,x} \quad \text{if} \quad \lambda(\mathfrak{g}(a, x)) = \infty.$$

Because of (9.11) the mapping $x \rightarrow x^*$ is one-to-one and continuous. Let K^* be the subset of those points y^* of K for which $\mathfrak{g}(a, y^*)$ is closed. We map the point y^* of K^* on the point y conjugate to a on $\mathfrak{g}(a, y^*)$. We then obtain a continuous mapping Ψ of $S^* = S + K^*$ on all of R. Each point x of β has two originals in S^* which are diametrically opposite points of K.

We now assign to each point of S^* a neighborhood $U(x^*)$ as follows: if x is not on β, let $S(x, \sigma_x) \cap \beta = 0$, $0 < \sigma_x < \rho(x)$ and define $U(x^*) = S(x, \sigma_x)\, \Psi^{-1}$. If $x \in \beta$ and x_1^*, x_2^* are the two originals of x, choose $\eta < \rho(a)/4$ such that $\mathfrak{g}(a, z)$ is a great circle for $z \in S(x_i^*, \eta)$. Choose y with $0 < \sigma_x < \rho(y)$ such that $\mathfrak{g}(a, y)$ intersects $S(x_i^*, \eta)$ for $y \in S(x, \sigma_x)$. This is possible because of (9.10). Let U_i be the set of the points y in $S(x, \sigma_x)$ for which a point $z_i \in S(x_i^*, \eta)$ with (yz_ia) exists. Then $U_1 \cup U_2 = S(x, \eta)$ and $U_1 \cap U_2 = \beta \cap S(x, \eta)$. For if (yz_ia) and $z_i \in S(x_i^*, \eta)$, $i = 1, 2$, then

$$z_1 z_2 > x_1^* x_2^* - 2\eta > \rho(a) - \rho(a)/2 > 2\eta.$$

If y were not conjugate to a, then $T(a, y)$ would be unique and contain both z_1 and z_2. But then

$$z_1z_2 = |az_1 - az_2| < \rho(a)/2 + \eta - (\rho(a)/2 - \eta) = 2\eta.$$

As neighborhood $U(x_i^*)$ of x_i^* we define $U_i\Psi^{-1}$.

Now let S^{**} be a second copy of S^* and Φ an isometry of S^* on S^{**}. Let x_1^* and x_2^* be diametrically opposite points of K^*. We then identify x_1^* with x_2^{**} and x_2^* with x_1^{**} [1]. The union of S^* and S^{**} with this identification will be denoted by $\bar{R}$. The mappings Ψ and $\Phi\Psi$ define a continuous mapping Ω of $\bar{R}$ on R. Under Ω^{-1} every point x of R has exactly two images x^* and x^{**} in $\bar{R}$. We say, as usual, that x^* and x^{**} lie over x.

We also associate with every point of $\bar{R}$ a neighborhood. If x does not lie in β we take the previously defined $U(x^*)$ as neighborhood of x^* and $U(x^{**}) = U(x^*)\,\Phi$ as neighborhood of x^{**}. If $x \in \beta$ we take $U(x_1^*)\,\Phi \cup U(x_2^*)$ as neighborhood of $x_1^{**} = x_2^*$ and $U(x_1^*) \cup U(x_2^*)\,\Phi$ as neighborhood of $x_1^* = x_2^{**}$. The neighborhoods $U(x^*)$ and $U(x^{**})$ of the two points x^* and x^{**} over x are then disjoint, and Ω maps both $U(x^*)$ and $U(x^{**})$ topologically on the neighborhood $S(x, \sigma_x)$ of x.

Now $\bar{R}$ is metrized in the following manner: For a continuous curve $\bar{c}$: $\bar{p}(\tau)$, $\alpha \leqslant \tau \leqslant \beta$, in $\bar{R}$ we define $\lambda(\bar{c})$ as the length of the curve $p(\tau) = \bar{p}(\tau)\,\Omega$ in R and define the distance of two arbitrary points $\bar{x}, \bar{y}$ in $\bar{R}$ by

$$\overline{xy} = \inf \lambda(\bar{c})$$

where $\bar{c}$ traverses all curves from $\bar{x}$ to $\bar{y}$ in $\bar{R}$. The number $\overline{xy}$ is always defined and finite. For any point x^* of S^* (or x^{**} of S^{**}) can be connected to a^* (or a^{**}) by a path whose image is one of the possibly two segments from x to a and the points a^* and a^{**} can be connected by a curve whose image is a given great circle through a in R.

The relations $\overline{xx} = 0$, $\overline{xy} = \overline{yx}$, $\overline{xy} + \overline{yx} \geqslant \overline{xz}$ are obvious, and so is $\overline{xy} > 0$ when $\bar{x}\Omega = x \neq y = \bar{y}\Omega$. If $\bar{x} \neq \bar{y}$ but $x = y$, then $\bar{x}$ and $\bar{y}$ have the form x^* and x^{**} and $\overline{xy} > 0$ follows from $U(x^*) \cap U(x^{**}) = 0$. We show next that Ω is a locally isometric mapping. The definition of $\overline{xy}$ implies $\overline{xy} \geqslant xy$. Under Ψ or $\Phi\Psi$ the neighborhood $U(\bar{p})$ of $\bar{p}$ goes into $S(p, \sigma_p)$. Let $U'(\bar{p})$ be the subset of $U(\bar{p})$ mapped on $S(p, \sigma_p/2)$. Let $\bar{x}, \bar{y} \in U'(\bar{p})$. The segment $T(x, y)$ lies entirely in $S(p, \sigma_p)$, hence we have for the image $\bar{c}$ of $T(x, y)$ in $U(\bar{p})$ the relation $xy = \lambda(\bar{c}) \geqslant \overline{xy}$. This shows that Ω maps $U'(\bar{p})$ isometrically on $S(p, \sigma_p/2)$.

The finite compactness of $\bar{R}$ is nearly obvious: if $\bar{p}\bar{x}_\nu < \delta$ then $p x_\nu < \delta$. Hence $\{x_\nu\}$ contains a subsequence $\{x_\mu\}$ which tends to a point x of R. The local isometry of R and $\bar{R}$ implies that, with proper notations, $\{x_\mu{}^*\}$ and $\{x_\mu{}^{**}\}$ tend to x^* and x^{**}. The sequence $\{\bar{x}_\mu\}$ contains infinitely many elements of at least one of the sequences $\{x_\mu{}^*\}$ or $\{x_\mu{}^{**}\}$.

The M-convexity of $\bar{R}$ follows from (5.18), the additivity of length and the definition of $\overline{xy}$. Thus we see that R possesses a two-sheeted covering space and is therefore not simply connected.

This consideration shows how much is implied by the usual assumptions of differential geometry.

A THEOREM ON TWO-DIMENSIONAL SPACES

The hermitian elliptic spaces of dimension greater than two, see Section 53, provide examples which show, that the straight spaces, the spherelike spaces, and those of the elliptic type do, for higher dimensions, or at least for even higher dimensions, not exhaust the spaces in which each geodesic has the property of containing with any two points a segment joining them. For two dimensions the situation is different.

(31.3) *Let R be a two-dimensional G-space in which every geodesic contains with any two points a segment connecting them. Then R is either homeomorphic to the plane and straight, or homeomorphic to the projective plane and of the elliptic type, or homeomorphic to the sphere and spherelike.*

Theorems (31.2) and (10.7) show that one of the first two cases enters, when the geodesic through two points is unique. Assume therefore that there are two different geodesics $\mathfrak{g}$, $\mathfrak{h}$ which have two distinct common points a and a'. Because of (9.12) $\mathfrak{g}$ and $\mathfrak{h}$ have no further common point, and a, a' are conjugate on both $\mathfrak{g}$ and $\mathfrak{h}$. Call T^1_g, T^2_g and T^1_h, T^2_h the two segments $T(a, a')$ on $\mathfrak{g}$ and $\mathfrak{h}$ respectively. We know, see Section 10, that $S(a, \rho(a)/2)$, is homeomorphic to a circular disc. Let u_i and v_j be the point on T^i_g and T^j_h respectively for which $au_i = av_j = \rho(a)/32$. If $x \in T(u_i, v_j)$ then $ax < \rho(a)/16$, moreover $u_iv_j < \rho(a)/16$, hence $xu_i < \rho(a)/16$, $xv_j < \rho(a)/16$. Since $\rho(x) > \rho(a) - xa > 15\,\rho(a)/16$, it follows that a, u_i, v_j lie in $S(x, \rho(x)/8)$.

Let $\mathfrak{g}_x$ be the geodesic which carries a segment $T(a', x)$ where x is an interior point of $T(u_i, v_j)$. By the local Axiom of Pasch (10.4), the geodesic $\mathfrak{g}_x$ intersects $T(u_i, a) \cup T(a, v_j)$. The point a is the only possible intersection, because each intersection of $\mathfrak{g}_x$ with $\mathfrak{g}$ or $\mathfrak{h}$ other than a' is conjugate to a' on $\mathfrak{g}_x$ and $\mathfrak{g}$ or $\mathfrak{h}$.

Therefore the unique segment $T(a, x)$ lies on $\mathfrak{g}_x$ (since $\mathfrak{g}_x$ contains a segment $T(a, x)$). The set $\cup\, T(a, x)$ with $x \in \cup\, T(u_i, v_j)$ covers a neighborhood of a, hence all geodesics through a are great circles that pass through a'. The geodesic $\mathfrak{g}(a, x)$ is unique because x is not conjugate to a and varies continuously with x. Therefore the space is homeomorphic to a sphere. The space being compact all geodesics must be great circles.

Consider any point $b \neq a, a'$. There is a geodesic $\mathfrak{k}$ through a and b. Let $\mathfrak{k}'$ be a second geodesic through b. Since $\mathfrak{k}'$ crosses $\mathfrak{k}$ at b and is a great circle, it must cross $\mathfrak{k}$ at some other point b'. The previous argument shows that each geodesic through b' also passes through b. Hence R is spherelike.

32. Free homotopy and closed geodesics

Let $\overline{R}$ be the universal covering space of R, Ω a locally isometric mapping of $\overline{R}$ on R, and $\{\Xi_\varkappa\} = \mathfrak{F}$ the fundamental group of R. If p is a point of R, and $\overline{p} = \overline{p}_1$ is a point over p, then $\{\overline{p}_\varkappa = \overline{p}_1\Xi_\varkappa\}$ is the set of all points over p. Let $\overline{\mathfrak{c}}$ be a curve in $\overline{R}$ from $\overline{p}_1$ to $\overline{p}_\varkappa$, $\overline{p}(\tau)$, $\alpha \leqslant \tau \leqslant \beta$, a parametrization of $\overline{\mathfrak{c}}$. Then $\mathfrak{c} : p(\tau) = \overline{p}(\tau)\,\Omega$ is a curve in R which begins and ends at p. If $\mathfrak{c}_0 : q(\tau)$ is any curve beginning and ending at p and homotopic to $\mathfrak{c}$, then the curve $\overline{\mathfrak{c}}_0$ over $\mathfrak{c}_0$ which begins at $\overline{p}_1$ also ends at $\overline{p}_\varkappa$, see (27.12). If $\mathfrak{c}_0$ begins and ends at p and is not homotopic to $\mathfrak{c}$, then the curve $\overline{\mathfrak{c}}_0$ over $\mathfrak{c}_0$ beginning at $\overline{p}_1$ does not end at $\overline{p}_\varkappa$, also because of (27.12). Thus the elements $\Xi_\varkappa$ of $\mathfrak{F}$ correspond in a one-to-one way to the classes of curves on R homotopic to each other and beginning and ending at the same point p. If $\Xi_\varkappa$ corresponds to the curve K and Ξ_λ to L, then $\Xi_\varkappa\,\Xi_\lambda$ correspond to the curve KL. This is the way in which the fundamental group is usually defined in topology.

Free homotopy

In this definition the point p is always kept fixed. If this requirement is dropped we are led to the concept of free homotopy:

The closed curve $\mathfrak{c} : q(\tau)$, $\alpha \leqslant \tau \leqslant \beta$, $q(\alpha) = q(\beta)$ is *freely homotopic* to the closed curve $\mathfrak{c}' : q'(\tau)$, $\alpha \leqslant \tau \leqslant \beta$, $q'(\alpha) = q'(\beta)$, if a family of curves $q_\sigma(\tau)$, $\alpha \leqslant \tau \leqslant \beta$, $q_\sigma(\alpha) = q_\sigma(\beta)$, $0 \leqslant \sigma \leqslant 1$, exists such that $q_\sigma(\tau)$ depends continuously on the two variables σ and τ and $q_0(\tau) = q(\tau)$, $q_1(\tau) = q'(\tau)$ for $\alpha \leqslant \tau \leqslant \beta$. For an exact formulation of this definition and the following remarks see Seifert and Threlfall [1, § 49].

For any closed curve $\mathfrak{c}$ and a given point p there is a curve $\mathfrak{c}'$ freely homotopic to $\mathfrak{c}$ for which $q'(\alpha) = q'(\beta) = p$. For if $\mathfrak{k}$ is any curve connecting p to $q(\alpha)$ then $\mathfrak{k}\,\mathfrak{c}\,\mathfrak{k}^{-1}$ is freely homotopic to $\mathfrak{c}$.

To find all classes of freely homotopic curves it suffices therefore to consider curves through a fixed point p. If, with the above notations, $\mathfrak{c}$ is freely homotopic to $\mathfrak{c}'$ and $q(\alpha) = q(\beta) = q'(\alpha) = q'(\beta) = p$ then under the deformation $q_\sigma(\tau)$ the point $q_\sigma(\alpha) = q_\sigma(\beta)$ traverses a closed curve $\mathfrak{k}$ which begins and ends at p. It is intuitively clear that $\mathfrak{c}'$ is homotopic to $\mathfrak{k}\mathfrak{c}\mathfrak{k}^{-1}$ in the ordinary sense. Conversely, if $\mathfrak{k}$ is any curve which begins and ends at p then $\mathfrak{k}\mathfrak{c}\mathfrak{k}^{-1}$ is freely homotopic to $\mathfrak{c}$. Thus the classes of freely homotopic curves in R correspond in a one-to-one way to the classes of conjugate elements in the fundamental group of R. In particular, the curves which can be freely contracted to a point, can also be deformed into a given point of the curve, leaving this point fixed during the process.

The identity in $\mathfrak{F}$ belongs to all curves in R which can be contracted to a point.

(32.1) *If a class of freely homotopic curves of a G-space R which are not contractable to a point, contains a shortest curve* $\mathfrak{c}$, *then* $\mathfrak{c}$ *is a closed geodesic.*

The statement that $\mathfrak{c}$ is a "*closed geodesic*" means this: there is a parametrization $x(\tau)$, $\alpha \leqslant \tau \leqslant \beta$, of $\mathfrak{c}$ and a representation $y(\tau)$ of a geodesic such that $y(\tau + \beta - \alpha) \equiv y(\tau)$ and $y(\tau) = x(\tau)$ for $\alpha \leqslant \tau \leqslant \beta$. Observe that $\beta - \alpha$ is the length of $\mathfrak{c}$.

The assertion (32.1) is obvious: if $p(\tau)$ is a parametrization of $\mathfrak{c}$ and $p(\tau_0)\, p(\tau) < \rho_1(p(\tau_0))$ for $|\tau_0 - \tau| \leqslant \eta$, $\eta > 0$, then the subcurve $p(\tau)$: $\tau_0 - \eta \leqslant \tau \leqslant \tau_0 + \eta$ of $\mathfrak{c}$ must be a segment, otherwise it could by (28.1) be deformed into a segment, thereby decreasing the length of $\mathfrak{c}$. It is clear that this argument also applies to the neighborhood of the (coinciding) initial and terminal points of $\mathfrak{c}$, since we no longer insist on keeping this point fixed.

Note: That a given free homotopy class need not contain a shortest curve is clear: the cylinder with the hyperbolic metric constructed in Section 27 provides an example. The homotopy class determined by the closed curve $x = \alpha$, $0 \leqslant y \leqslant \beta$, contains arbitrarily short curves, hence no curve of minimal length.

On the other hand on the ordinary cylinder (8.1) all circles $\zeta = \text{const.}$ traversed once are shortest curves which are freely homotopic to each other. A hyperboloid of one sheet in ordinary space with its intrinsic metric furnishes a third metrization of the cylinder (in the topological sense). Here the line of striction, traversed once, is the only shortest curve in its free homotopy class.

Existence of closed geodesics

It is intuitively clear that on closed or compact surfaces in ordinary space a free homotopy class will contain a shortest curve. This is true for general compact G-spaces.

(32.2) *A given class of freely homotopic curves in a compact G-space contains a shortest curve.*

If R is not compact and K is a given closed curve, then there is a shortest curve freely homotopic to K, if for a fixed point p the length of any curve homotopic to K, which contains points outside $S(p, \nu)$, tends with ν to ∞.

The latter may, for instance, be used to assert the existence of a shortest curve homotopic to a meridian K of a torus with an attached tube extending to infinity as at the end of Section 30 (but not necessarily with a hyperbolic metric). A curve homotopic to K always contains for a given point p and a sufficiently large, but fixed, σ points of $S(p, \sigma)$. Any curve homotopic to K containing points outside of $S(p, \nu)$ has then at least length $2(\nu - \sigma)$.

The proof of (32.2) is practically the same as that of (28.2): it follows from (28.1) that any closed curve is homotopic to a geodesic polygon, hence a given free homotopy class contains curves of finite length. Put $\delta = \inf \lambda(\mathfrak{c})$, where $\mathfrak{c}$ traverses all curves freely homotopic to K, which may, of course, be assumed to be not contractable to a point. It is to be shown that a $\mathfrak{c}_0$ exists in the class of K whose length equals δ.

In any event, there is a sequence $\{\mathfrak{c}_\mu\}$ of curves freely homotopic to K such that $\lambda(\mathfrak{c}_\mu) \to \delta$. Whether R is compact, or K satisfies the hypothesis of the second part of (32.2), we know that $\mathfrak{c}_\mu \subset S(p, \nu_0)$ for a suitable ν_0. Because of (5.16) a subsequence $\{\mathfrak{c}_\theta\}$ of $\{\mathfrak{c}_\mu\}$ tends uniformly to a curve $\mathfrak{c}_0$ with $\lambda(\mathfrak{c}_0) \leqslant \lim \lambda(\mathfrak{c}_\theta) = \delta$. But by (28.1) $\mathfrak{c}_0$ and $\mathfrak{c}_\theta$ are freely homotopic for large θ, hence $\mathfrak{c}_0$ is freely homotopic to K, and the definition of δ yields $\lambda(\mathfrak{c}_0) = \delta$.

Axial motions of straight spaces

There is a very interesting connection between certain types of motions and closed geodesics of R, when the universal covering space of R is straight. To find this connection we prove a few simple, but very important facts:

(32.3) **Theorem:** *If for a motion Φ of a G-space R a point z with $0 < zz\Phi = \inf_{x \in R} xx\Phi < \rho(z)/2$ exists, then $(zz\Phi z\Phi^2)$.*

If R is straight, then the converse holds: $(zz\Phi z\Phi^2)$ implies $zz\Phi = \inf_{x \in R} xx\Phi$.

Put $z_i = z\Phi^i$, $i = 0, 1, 2, \ldots$. To prove the first part denote by m the unique midpoint of $z = z_0$ and z_1. Then

$$\gamma = zz_1 \leqslant mm\Phi \leqslant mz_1 + z_1 m\Phi = zm + mz_1 = zz_1$$

hence $(mz_1 m\Phi)$. Since $(z_1 m\Phi z_2)$ the geodesic $x(\tau)$ with $x(0) = z$ and representing $T(z, z_1)$ for $0 \leqslant \tau \leqslant \gamma$, represents $T(z_1, z_2)$ for $\gamma \leqslant \tau \leqslant 2\gamma$ and a segment for $0 \leqslant \tau \leqslant \rho(z)$. The assertion (zz_1z_2) now follows from $\rho(z) > 2\gamma$.

In a straight space let (zz_1z_2). Then $(z_{i-1}z_iz_{i+1})$ for all i and for any point x:

$$nzz_1 = zz_n \leqslant zx + \sum_{i=1}^{n} x\Phi^{i-1}\, x\Phi^i + x\Phi^n z_n = 2\, zx + n\, xx\Phi.$$

Dividing the inequality by n and letting $n \to \infty$ yields the assertion $zz\Phi \leqslant xx\Phi$.

If for a motion Φ of a straight space a point z with $(zz\Phi z\Phi^2)$ exists we call Φ "*axial*" and the line carrying the points $z\Phi^i$ an "*axis of* Φ". The orientation of this line in which $z\Phi$ follows z is an *oriented axis* of Φ. In the opposite orientation it is then an axis of Φ^{-1}. Clearly, with Φ the motions Φ^i, $i \neq 0$, are also axial and have (because of $zz\Phi^i = \inf_{x \in R} xx\Phi^i > 0$) no fixed points. Φ may have more than one axis, but $uu\Phi$ has the same value $\gamma(\Phi) = \inf xx\Phi$ for a point u on any axis. Moreover,

(32.4) ***Theorem:*** *Two axes of the same motion of a straight space are parallel to each other.*

Let $\mathfrak{g}_1$ and $\mathfrak{g}_2$ be two different axes of the motion Φ, and let $\mathfrak{h}$ be the asymptote to an orientation $\mathfrak{g}_2^+$ of $\mathfrak{g}_2$ through a point q of $\mathfrak{g}_1$. The intersection f of $\mathfrak{h}$ with the limitsphere $K = K_\infty(\mathfrak{g}_2^+, q\Phi) = K_\infty(\mathfrak{g}_2^+, q)\Phi$ is the unique foot of q on K, compare Theorem (23.2). If $K_\infty(\mathfrak{g}_2^+, q)$ intersects $\mathfrak{g}_2$ at b then $K \cap \mathfrak{g}_2 = b\Phi$. Then for a positive sub-ray $\mathfrak{r}$ of $\mathfrak{g}_2^+$, see (22.17, 18), we have

$$qf = \alpha(\mathfrak{r}, q) - \alpha(\mathfrak{r}, f) = \alpha(\mathfrak{r}, b) - \alpha(\mathfrak{r}, b\Phi) = bb\Phi = \gamma(\Phi) = qq\Phi.$$

Hence $q\Phi$ is also a foot of q on K so that $q\Phi = f$ and $\mathfrak{h} = \mathfrak{g}_1$.

Theorem (32.3) leads naturally to the question how points z with $zz\Phi = \sup xx\Phi$ behave. We prove:

(32.5) ***Theorem:*** *If for a motion Φ of a G-space R a point z with $0 < zz\Phi = \sup_{x \in R} xx\Phi < \rho(z)/2$ exists, then $(zz\Phi z\Phi^2)$.*

If, in addition, R is straight, then $xx\Phi$ is constant (hence $(xx\Phi x\Phi^2)$ for all x).

For since $zz\Phi < \rho(z)/2$ a point u exists such that $z\Phi$ is the midpoint of z and u. Then $z\Phi^2$ is a midpoint of $z\Phi$ and $u\Phi$ and the only one because $\rho(z\Phi) = \rho(z)$. The relation

$$zz\Phi \geqslant uu\Phi \geqslant z\Phi u\Phi - z\Phi u = zu - z\Phi u = zz\Phi$$

shows that u is a midpoint of $z\Phi$ and $u\Phi$, hence $u = z\Phi^2$ which proves $(zz\Phi z\Phi^2)$.

If the space is straight we conclude from (32.3) that also $zz\Phi = \inf xx\Phi$, hence $xx\Phi$ is constant.

Axial motions and closed geodesics

The connection between homotopy classes and motions may now be formulated as follows:

(32.6) **Theorem.** *Let the universal covering space* $\overline{R}$ *of* R *be straight and let* $\Xi_k \neq \Xi_1$ *belong to the class of conjugate elements in* $\mathfrak{F}$ *determined by the class of curves freely homotopic to* K. *There is a closed geodesic (freely) homotopic to* K *if and only if* Ξ_k *is axial. The closed geodesics homotopic to* K *are the images* (under Ω) *of the axes of* Ξ_k *and have length* $\gamma(\Xi_k)$.

First let $\mathfrak{g}$ be a closed geodesic homotopic to K. Then a representation $x(\tau)$ of $\mathfrak{g}$ exists such that for a suitable $\alpha > 0$ (because $\Xi_k \neq \Xi_1$), $x(\tau + \alpha) \equiv x(\tau)$, and $x(\tau)$ is for $0 \leqslant \tau \leqslant \alpha$, a curve homotopic to K. If the point $\overline{x}$ over $x(0)$ is suitably chosen then $\overline{x}\Xi_k$ is the endpoint of the curve $\overline{x}(\tau), 0 \leqslant \tau \leqslant \alpha$, over $\overline{x}(\tau)$ beginning at $\overline{x}$, and $\overline{x}(\tau)$ is part of a representation of a straight line $\overline{\mathfrak{g}}$ in $\overline{R}$. Denote by L the lineal element (see the beginning of Section 9) represented by $x(\tau)$ for $|\tau| \leqslant \sigma = \sigma(x(0))/2$ and by $\overline{L}$ the segment (on $\overline{\mathfrak{g}}$) with center $\overline{x}$ over L. Because Ξ_k lies over the identity of R it carries $\overline{L}$ into a segment $\overline{L}'$ over L, and $\overline{\mathfrak{g}}$ into the straight line $\overline{\mathfrak{g}}'$ containing $\overline{L}'$. Since $\overline{\mathfrak{g}}$ lies over $\mathfrak{g}$ it contains the segment $\overline{L}''$ with center $\overline{x}(\alpha) = \overline{x}(0)\,\Xi_k$ that lies over the lineal element L'' represented by $x(\tau)$ for $|\tau - \alpha| \leqslant \sigma$. But $L'' = L$ because $\mathfrak{g}$ is closed, hence $\overline{L}'' = \overline{L}'$, or $\overline{\mathfrak{g}} = \overline{\mathfrak{g}}' = \overline{\mathfrak{g}}\,\Xi_k$. This shows that Ξ_k is axial and that the length α of the closed geodesic equals $\overline{x}(0)\,\overline{x}(\alpha) = \gamma(\Xi_k)$.

Now let Ξ_k be axial and let $\overline{x}$ lie on an axis $\overline{\mathfrak{g}}$ of Ξ_k. Let $\overline{x}(\tau)$ represent $\overline{\mathfrak{g}}$ such that $\overline{x}(0) = \overline{x}$ and $\overline{x}(\alpha) = \overline{x}\,\Xi_k$, $\alpha > 0$. We retrace the steps of the first part of this proof: let $\mathfrak{g}$ be the geodesic $x(\tau) = \overline{x}(\tau)\,\Omega$ and L the lineal element represented by $x(\tau)$ for $|\tau| \leqslant \sigma \leqslant \sigma(x(0))/2$. Since $\overline{x}(\alpha) = \overline{x}(0)\,\Xi_k$ the point $x(\alpha)$ coincides with $x(0)$. Denote by L'' the lineal element given by $x(\tau)$ for $|\tau - \alpha| \leqslant \sigma$ and by $\overline{L}''$ the segment over L'' with center $\overline{x}(\alpha)$, finally by $\overline{L}'$ the segment on $\overline{\mathfrak{g}}$ with center $\overline{x}(\alpha)$ and length 2σ. Then $\overline{\mathfrak{g}}\,\Xi_k = \overline{\mathfrak{g}}$ implies $\overline{L}' = \overline{L}''$, hence $L'' = L$ or $x(\tau + \alpha) = x(\tau)$ for $|\tau| \leqslant \sigma$. This implies $x(\tau + \alpha) \equiv x(\tau)$, hence $\mathfrak{g}$ is closed and has length $\alpha = \gamma(\Xi_k)$.

Since $\overline{\mathfrak{g}}$ was an arbitrary axis of Ξ_k we see that Ω maps the axes of Ξ_k on the closed geodesics homotopic to K.

If $\overline{x}'$ is any point over $x(0)$ and Ξ_j takes $\overline{x}'$ into $\overline{x}$, then $\overline{x}(\tau)\,\Xi_j^{-1}$ is the curve over $x(\tau)$ which begins at $\overline{x}'$. It ends at

$$\overline{x}(\alpha)\,\Xi_j^{-1} = \overline{x}\,\Xi_k\,\Xi_j^{-1} = \overline{x}'\,\Xi_j\,\Xi_k\,\Xi_j^{-1}$$

If $\overline{\mathfrak{g}}$ is an axis of Ξ_k then $\overline{\mathfrak{g}}\,\Xi_j^{-1}$ is an axis of $\Xi_j\,\Xi_k\,\Xi_j^{-1}$. This exhibits the correspondence between classes of conjugate elements in $\mathfrak{F}$ and the sets of freely homotopic closed geodesics in R when $\overline{R}$ is straight.

In this theorem we spoke of closed geodesics instead of shortest curves, but the hypothesis that R is straight makes this distinction unnecessary.

(32.7) *If the universal covering space $\overline{R}$ of R is straight, then the shortest curves in a (non-trivial) free homotopy class coincide with its closed geodesics. If R is compact then every motion in the fundamental group of R is axial.*

We saw in (32.1) that a shortest curve is a closed geodesic without the assumption that $\overline{R}$ is straight. It is easy to see that the assumption must be essential for the converse: a torus shaped surface in E^3 whose meridians are circles in planes through the z-axis with centers on $x^2 + y^2 = 1$, $z = 0$, but with varying radii provides an example. For not only the meridians of minimal, but also the meridians of maximal radius are closed geodesics.

To prove the converse, let $\mathfrak{g}$ be a closed geodesic and let $p(\tau)$, $0 \leqslant \tau \leqslant \beta$, $p(0) = p(\beta)$ be a curve freely homotopic to $\mathfrak{g}$. Let $\overline{p}$ be a point over $p(0)$, and $\overline{p}(\tau)$ the curve over $p(\tau)$ beginning at $\overline{p}$. Then $\overline{p}(\tau)$ ends at a point $\overline{p}(\beta)$ over $p(0)$ different from $\overline{p}$, because the considered homotopy class is non-trivial. Hence $\overline{p}(\beta) = \overline{p}\,\Xi_k$ for a suitable $k \neq 1$. Since the free homotopy class contains $\mathfrak{g}$, the motion Ξ_k leaves by (32.6) a line over $\mathfrak{g}$ invariant and $\mathfrak{g}$ has length $\gamma(\Xi_k)$. It follows from (27.2) and (32.3) that

$$\gamma(\Xi_k) \leqslant \overline{p}(0)\,\overline{p}(\beta) \leqslant \lambda(\overline{p}) = \lambda(p),$$

hence $\mathfrak{g}$ is a shortest curve in its homotopy class.

The second part of (32.7) is a corollary of the first part, (32.2) and (32.6).

Ordinary homotopy for straight $\overline{R}$

This theorem does not exclude the possibility that R might contain a closed geodesic contractable to a point. Actually there are not even geodesic monogons contractable to a point, when $\overline{R}$ is straight. A geodesic monogon is a geodesic curve $x(\tau)$, $\alpha \leqslant \tau \leqslant \beta$ with $x(\alpha) = x(\beta)$, (we do not require $x(\tau + \beta - \alpha) \equiv x(\tau)$). Our statement will follow from the following fact on ordinary (not free) homotopy.

(32.8) *If the universal covering space $\overline{R}$ of R is straight, then a given class of homotopic curves from p to q contains exactly one geodesic curve.*

For if $x(\tau)$, $y(\tau)$ are homotopic geodesic curves in R from p to q and $\overline{p}$ lies over p, let $\overline{x}(\tau)$, $\overline{y}(\tau)$ be the curves over $x(\tau)$, $y(\tau)$ beginning at $\overline{p}$. Because of (27.12) they end at the same point $\overline{q}$, and they are geodesic curves in $\overline{R}$. But since $\overline{R}$ is straight, there is only one geodesic curve from $\overline{p}$ to $\overline{q}$, hence the geodesic curves represented by $x(\tau)$ and $y(\tau)$ must coincide.

If we apply this to the case $p = q$ and the class of curves homotopic to 0, we obtain:

(32.9) *If the universal covering space $\overline{R}$ of R is straight, then R possesses no geodesic monogons or closed geodesics which are homotopic to zero or (freely) contractable to a point.*

The addition about contractable curves follows from the fact that a curve $\mathfrak{c}$ which can be freely contracted to a point a, can also be contracted to a point leaving a point of $\mathfrak{c}$ fixed.

For later application we generalize (32.8) slightly: In a G-space R let $p(\tau)$ and $q(\tau)$ be defined on the same connected set V of the τ-axis and continuous. If $\tau_1 \leqslant \tau_2$, $\tau_i \in V$, denote with $p(\tau_1, \tau_2)$ the curve defined by $p(\tau)$ for $\tau_1 \leqslant \tau \leqslant \tau_2$, and by $p(\tau_2, \tau_1)$ the curve $p(\tau_1, \tau_2)$ with the opposite orientation. Define $q(\tau_1, \tau_2)$ and $q(\tau_2, \tau_1)$ similarly. If the curve $\mathfrak{c}_i$, $i = 1, 2$ connects $p(\tau_i)$ to $q(\tau_i)$ then $\mathfrak{c}_1$ is called "*homotopic to $\mathfrak{c}_2$ along (p, q)*" if

$$\mathfrak{c}_1\, q(\tau_1, \tau_2)\, \mathfrak{c}_2^{-1}\, p(\tau_2, \tau_1) \sim 0.$$

Since then also

$$\mathfrak{c}_2\, q(\tau_2, \tau_1)\, \mathfrak{c}_1^{-1}\, p(\tau_1, \tau_2) \sim 0$$

the concept is symmetric and transitive (and of course reflexive).

(32.10) *If the universal covering space $\overline{R}$ of R is straight, and $\mathfrak{c}$ is a curve connecting $p(\tau_0)$ to $q(\tau_0)$, then for every $\tau \in V$ exactly one geodesic curve $\mathfrak{g}_\tau$ from $p(\tau)$ to $q(\tau)$ and homotopic to $\mathfrak{c}$ along (p, q) exists.*

For let an arbitrary curve over $\mathfrak{c}$ begin at $\overline{p}$ and end at $\overline{q}$, and let $\overline{p}(\tau)$, $\overline{q}(\tau)$ be the curves over $p(\tau)$, $q(\tau)$ beginning at $\overline{p}$ and $\overline{q}$ respectively. Because $\overline{R}$ is simply connected the segment $\overline{\mathfrak{s}}_\tau = \mathfrak{s}(\overline{p}(\tau), \overline{q}(\tau))$ satisfies $\overline{\mathfrak{s}}_{\tau_0} \sim \overline{\mathfrak{c}}$ and for any τ_1, τ_2,

$$\overline{\mathfrak{s}}_{\tau_1}\, \overline{q}\,(\tau_1, \tau_2)\, \overline{\mathfrak{s}}_{\tau_2}^{-1}\, \overline{p}\,(\tau_2, \tau_1) \sim 0.$$

Therefore $\overline{\mathfrak{s}}_\tau \Omega = \mathfrak{g}_\tau$ is a geodesic curve in R with $\mathfrak{g}_{\tau_0} \sim \mathfrak{c}$ and $\mathfrak{g}_\tau \sim \mathfrak{c}$ along (p, q), see (27.12).

For a given value $\tau_1 \in V$ let $\mathfrak{g}_1$ be any geodesic curve from $p(\tau_1)$ to $q(\tau_2)$ and homotopic to $\mathfrak{g}_{\tau_1}$ along (p, q). Then $\mathfrak{g}_1 \sim \mathfrak{g}_{\tau_1}$. Therefore the geodesic curve $\overline{\mathfrak{g}}_1$ over $\mathfrak{g}_1$ which begins at $\overline{p}(\tau_1)$ ends at $\overline{q}(\tau_1)$, and must coincide with $\overline{\mathfrak{s}}_{\tau_1}$ since this is the only geodesic curve in $\overline{R}$ from $\overline{p}(\tau_1)$ to $\overline{q}(\tau_1)$.

Spaces with abelian fundamental groups

Many of the preceding results were applications of (32.3). We now show that its counterpart (32.5) also has unexpectedly strong implications.

(32.11) ***Theorem.*** *If a compact G-space R has an abelian fundamental group and a straight universal covering space $\overline{R}$ then no geodesic has multiple points and the closed geodesics within a given free homotopy class have the same length and cover R simply.*

Proof. For an arbitrary point $\overline{p}_1$ in $\overline{R}$ define the set $F(\overline{p}_1)$ as in (29.4 e). Because of (29.4 g) the set $F(\overline{p}_1)$ is compact. If Ξ' is any non-trivial motion in the fundamental group $\mathfrak{F}$ of R there is a point $\overline{z} \in F(\overline{p}_1)$ such that

$$\overline{z}\,\overline{z}\,\Xi' = \max_{\overline{x} \in F(\overline{p}_1)} \overline{x}\,\overline{x}\,\Xi'.$$

If $\overline{y}$ is an arbitrary point of $\overline{R}$, a motion Ξ_k in $\mathfrak{F}$ exists such that $\overline{y}\,\Xi_k \in F(\overline{p}_1)$, see (29.4 e). Then, since $\mathfrak{F}$ is abelian,

$$\overline{y}\,\overline{y}\,\Xi' = \overline{y}\,\Xi_k\,\overline{y}\,\Xi'\,\Xi_k = \overline{y}\,\Xi_k\,\overline{y}\,\Xi_k\,\Xi' \leqslant \overline{z}\,\overline{z}\,\Xi' \quad \text{and} \quad \overline{z}\,\overline{z}\,\Xi' = \sup_{\overline{x} \in \overline{R}} \overline{x}\,\overline{x}\,\Xi'.$$

We now conclude from (32.5) that $\overline{x}\,\overline{x}\,\Xi'$ is constant and that every point of R lies on an axis of Ξ'. The assertion follows now immediately from (32.6) for the homotopy class determined by the (arbitrary) Ξ'.

Since a geodesic monogon lies in some free (non-trivial) homotopy class, no geodesic can have multiple points.

This theorem generalizes, of course, the well known situation on a torus of arbitrary dimension with a euclidean or Minkowskian metric. For a two-dimensional torus a theorem of E. Hopf [1] states that the only Riemannian metrization without conjugate points is the euclidean. In the next section we will see, however, that this theorem has no analogue in non-Riemannian spaces, so that (32.11) contains, even in two dimensions, statements about non-trivial metrics.

Other applications of (32.5) will be found in Section 52.

Translations

The results on axial motions can be carried further in the two-dimensional case. For brevity two-dimensional G-spaces will be called *"G-surfaces"* and the term *"straight plane"* will be used for a straight G-surface. An axial, orientation preserving motion of a straight plane R with axis $\mathfrak{g}$ will be called a *"translation of R with axis, or along,* $\mathfrak{g}$". The square of any axial motion of R is a translation.

The stronger statements which can be made on translations as compared to axial motions of general straight spaces, all derive from the following fact:

(32.12) **Theorem.** *Let* $\mathfrak{g}$ *be an axis of the translation* Φ *of the straight plane* R. *If* p *is any point not on* $\mathfrak{g}$, *then*

$$\bigcup_{i=-\infty}^{\infty} T(p\Phi^i, p\Phi^{i+1})$$

bounds together with $\mathfrak{g}$ *a convex subset of* R.

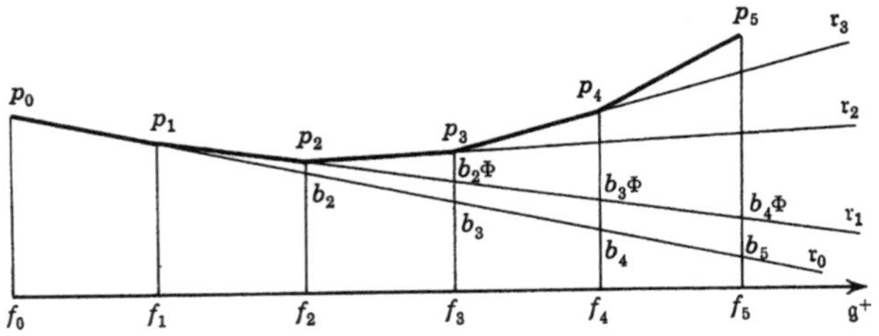

Fig. 37.

Put $p_i = p\Phi^i$. If $(p_{i-1}\, p_i\, p_{i+1})$ for one i then this relation holds for all i. The points p_i are then collinear and lie on an axis of Φ, see (32.3), and (32.12) is trivial. Orient the plane R such that p lies on the left of the oriented axis $\mathfrak{g}^+$ of Φ. If p_{i+2} lies for one i on the right (left) of $\mathfrak{g}^+(p_i, p_{i+1})$ the same holds for all i. To prove the theorem it must be shown that p_{i+2} lies on the right of $\mathfrak{g}^+(p_i, p_{i+1})$. Assume for an indirect proof that p_{i+2} lies on the left of $\mathfrak{g}^+(p_i, p_{i+1})$.

One of the two rays with origin p_i on $\mathfrak{g}^+(p_i, p_{i+1})$ does not intersect $\mathfrak{g}$, assume the ray $\mathfrak{r}_i$ containing p_{i+1} has this property (otherwise we work with the ray with origin p_{i+1} containing p_i and Φ^{-1} instead of Φ).

If f_0 is a foot of $p_0 = p$ on g then $f_i = f_0 \Phi^i$ is a foot of p_i on $\mathfrak{g}$ and $p_0 f_0 = p_i f_i$. Since p_{i+2} is on the left of $\mathfrak{g}^+(p_i, p_{i+1})$, the ray $\mathfrak{r}_i$ intersects $T(p_k, f_k)$ for all $k > i + 1$. The intersection b_k of $\mathfrak{r}_0$ with $T(p_k, f_k)$, $k > 1$, goes under Φ into the intersection $b_k\Phi$ of $\mathfrak{r}_1$ with $T(p_{k+1}, f_{k+1})$, and $(b_k\Phi\, b_{k+1} f_{k+1})$ because $\mathfrak{r}_1$ lies on the left of $\mathfrak{r}_0$. Moreover,

$$b_k f_k = b_k\Phi f_k\Phi = b_k\Phi f_{k+1} > b_{k+1} f_{k+1},$$

so that the sequence $\{b_k f_k\}$ has a limit, and $b_k\Phi f_{k+1}$ has the same limit. It follows that $b_k\Phi b_{k+1} \to 0$, which contradicts (22.23).

A first corollary is:

(32.13) *If a translation Φ has two different axes $\mathfrak{g}_1$ and $\mathfrak{g}_2$, then any point between $\mathfrak{g}_1$ and $\mathfrak{g}_2$ lies on an axis. All axes of Φ are parallel and any parallel to an axis is an axis.*

For if p is any point between $\mathfrak{g}_1$ and $\mathfrak{g}_2$ then $\bigcup_{i=-\infty}^{\infty} T(p\Phi^i, p\Phi^{i+1})$ bounds by (32.12) a convex set together with $\mathfrak{g}_1$ as well as $\mathfrak{g}_2$, hence must be a straight line, which is by (32.3) an axis. That any two axes are parallel follows from (32.4). The limit line of axes is also an axis. Unless the axes of Φ cover the plane, there is a point q not on an axis, and an axis $\mathfrak{h}$ closest to q. That q does not lie on a parallel to $\mathfrak{g}$ follows then from:

(32.14) *The asymptotes through a point q not on an axis of Φ to the two orientations of the axis $\mathfrak{h}$ of Φ closest to q are distinct and have distance 0 from $\mathfrak{h}$.*

Let $\mathfrak{a}$ be the asymptote through q to the oriented axis $\mathfrak{h}^+$ of Φ. The line $\mathfrak{a}\Phi^{-\nu}$ is also an asymptote to $\mathfrak{h}^+$ and lies for positive ν between $\mathfrak{a}$ and $\mathfrak{h}$ (since q does not lie on an axis). Therefore $\mathfrak{a}\Phi^{-\nu}$ converges for $\nu \to \infty$ to a line $\mathfrak{k}$. This line is invariant under Φ and is therefore an axis of Φ. By construction $\mathfrak{k}$ either separates $\mathfrak{h}$ from q or coincides with $\mathfrak{h}$. The first case is impossible since $\mathfrak{h}$ was the axis of Φ closest to $\mathfrak{g}$. Thus $\mathfrak{k} = \mathfrak{h}$. But $\mathfrak{a}\Phi^{-\nu} \to \mathfrak{k} = \mathfrak{h}$ and $\mathfrak{a}\Phi^{-\nu}\,\mathfrak{h} = \mathfrak{a}\mathfrak{h}$ implies $\mathfrak{a}\mathfrak{h} = 0$.

It follows more precisely, that there are representations $a(\tau)$ of $\mathfrak{a}^+$ and $y(\tau)$ of $\mathfrak{h}^+$ such that

(32.15) $$\lim_{\tau \to \infty} a(\tau)\, y(\tau) = 0.$$

Since $\mathfrak{h}^{-1}$ is an axis of Φ^{-1} we obtain the same result for the asymptote $\mathfrak{b}$ through q to $\mathfrak{h}^{-}$. The lines $\mathfrak{a}$ and $\mathfrak{b}$ cannot coincide because $\mathfrak{a}\Phi^{-1}$ lies between $\mathfrak{a}$ and $\mathfrak{h}$, whereas $\mathfrak{b}$ lies between $\mathfrak{b}\Phi^{-1}$ and $\mathfrak{h}$.

The following theorem constitutes a very interesting addition to the theory of parallels developed in Section 23, because it shows that the existence of a translation along $\mathfrak{g}$ excludes for the asymptotes to $\mathfrak{g}$ many of the strange phenomena occurring in perfectly general straight planes.

(32.16) **Theorem.** *The oriented asymptotes to an oriented axis of a translation Φ are all asymptotes to each other. $\delta_\infty(\mathfrak{r}_1, \mathfrak{r}_2)$ is finite for positive subrays $\mathfrak{r}_1, \mathfrak{r}_2$ of any two of these asymptotes and $\delta_\infty(\mathfrak{r}_1, \mathfrak{r}_2) = 0$ if Φ has only one axis.*

There are three cases to consider. (1) The axes of Φ cover the plane. (2) The axes of Φ form a closed half plane. Let this half plane be bounded by the oriented axis $\mathfrak{h}^+$ of Φ and denote the remaining open half plane by P. (3) The axes of Φ form a closed strip S bounded by two parallel lines. Then let P and Q be the open half planes remaining after removing S. (3) includes as special case the possibility (3′), that S consists of only one line, which is then the unique axis of Φ.

Case (1) is settled by (32.13). That $\delta_\infty(\mathfrak{r}_1, \mathfrak{r}_2)$ is finite for two positive subrays $\mathfrak{r}_1, \mathfrak{r}_2$ of two oriented axes $\mathfrak{g}_1^+$ and $\mathfrak{g}_2^+$ of Φ is obvious:

If $x_i(\tau)$ represents $\mathfrak{g}_i^+$ then $x_1(\tau)x_2(\tau)$ is periodic with period $\gamma(\Phi)$, hence

$$x_1(\tau)x_2(\tau) \leqslant \beta = \max_{0 \leqslant \tau \leqslant \gamma(\Phi)} x_1(\tau)x_2(\tau) \quad \text{for all } \tau. \tag{32.17}$$

In case (2) let $\mathfrak{a}$ and $\mathfrak{b}$ be two distinct oriented asymptotes to $\mathfrak{h}^+$, and let $\mathfrak{a}^+$ be between $\mathfrak{b}^+$ and $\mathfrak{h}^+$. If $\mathfrak{r}_a, \mathfrak{r}_b, \mathfrak{r}_h$ are positive subrays of $\mathfrak{a}^+, \mathfrak{b}^+, \mathfrak{h}^+$ respectively, then $\delta_\infty(\mathfrak{r}_b, \mathfrak{r}_h) = 0$ implies $\delta_\infty(\mathfrak{r}_a, \mathfrak{r}_b) = 0$ hence the lines $\mathfrak{a}^+, \mathfrak{b}^+, \mathfrak{h}^+$ are asymptotes to each other. It also follows from (22.22) that each oriented parallel $\mathfrak{g}^+$ to $\mathfrak{h}^+$ is an asymptote to $\mathfrak{b}^+$ and conversely. The finiteness of $\delta_\infty(\mathfrak{r}_b, \mathfrak{r}_g)$ for positive subrays of $\mathfrak{b}^+$ and $\mathfrak{g}^+$ follows from (32.17), applied to $\mathfrak{g}^+$ and $\mathfrak{h}^+$, and $\delta_\infty(\mathfrak{r}_b, \mathfrak{r}_h) = 0$.

In case (3) application of (22.22) shows immediately that all asymptotes to an oriented axis of Φ are asymptotes to each other, and (32.15) and (32.17) yield the rest, also the additional statement that in case (3′) any two asymptotes to the single axis have distance 0.

33. Metrics without conjugate points on the torus

We consider a two-dimensional torus R whose universal covering space is a straight plane P. Why this case is particularly interesting is explained in the introduction, Section 26, to this chapter. We add here the following remarks: Morse and Hedlund [2] had proved the result of E. Hopf on the torus under the stronger hypothesis that there are no focal points, which means in our language (see Section 25) that the circles in P are convex. This result is contained in our theorems (25.15) and (33.1). The question whether Morse and Hedlund's theorem can be extended to Finsler spaces by establishing that the metric is Minkowskian leads to the problem which we left unsolved in Section 25.

Since we operate exclusively in P we denote its points by $p, q, \ldots$ (instead of $\bar{p}, \bar{q}, \ldots$). The motions of the fundamental group of R may be assumed to be given by the translations

$$\Psi(m, n) : x' = x + m, \qquad y' = y + n, \qquad m, n \text{ integers}$$

in properly chosen cartesian coordinates in P. The following facts are contained in the proof of Theorem (32.5):

Each cartesian translation $\Psi = \Psi(m, n) \neq \Xi$ is also a translation of P as straight plane. Every point of P lies on an axis of Ψ and $pp\Psi$ is constant. All axes of Ψ are parallel to each other. If a line L in P contains q and $q\Psi(m, n)$ it contains all points $q\Psi(\nu m, \nu n)$, $\nu = \pm 1, \pm 2, \ldots$.

A very important and not at all trivial property of the geodesics in P is:

(33.1) ***Theorem.*** *If the universal covering plane P of a torus is straight, then P satisfies the parallel axiom.*

We assume again that the translations $\Psi(m, n)$ are the elements of the fundamental group. It will prove suggestive to call the line L rational if it contains two points of the form q and $q\Psi(m, n)$, $(m, n) \neq (0, 0)$. The rational lines are the axes of the various motions ($\neq \Xi$) of the fundamental group. For any point a not on L the axis of $\Psi(m, n)$ through a is by (32.5) the only geodesic in P through a that does not intersect L. Thus the parallel axiom holds for all rational lines.

By means of a topological transformation of the unit square $0 \leqslant x \leqslant 1$, $0 \leqslant y \leqslant 1$, on itself and doubly periodic extension of this transformation to the whole plane we can reach that the equations $x = \text{const.}$, $y = \text{const.}$

represent geodesics. Because the parallel axiom holds for these lines, every other line has a representation of the form

$$(1) \qquad y = f(x), \qquad -\infty < x < \infty,$$

with $f(x)$ either strictly increasing or strictly decreasing and $|f(x)| \to \infty$ for $|x| \to \infty$. It will be shown that for every such line the "*slope*"

$$(2) \qquad \lim_{x \to \pm\infty} \frac{f(x)}{x}$$

exists and is different from 0 *and* ∞.

Consider first the case where L is a rational line through the origin z and the point $(m, n) = z\Psi(m, n)$, $m \neq 0$, $n \neq 0$. Then $f(\nu m) = \nu f(m)$ and if $\nu m \leqslant x < (\nu + 1) m$, because $f(x)$ is monotone,

$$|f(x) - f(\nu m)| < |f[(\nu + 1) m] - f(\nu m)|,$$

hence with $|\theta_i| < 1$:

$$\frac{f(m)}{m} = \lim \frac{f(\nu m)}{\nu m} = \lim \frac{f(\nu m) + \theta_1 f(m)}{\nu m + \theta_2 m} = \lim \frac{f(x)}{x}$$

A rational line L_k obtained from L by the translation $\Psi(0, k)$ has the equation $y = f(x) + k$ so that L_k has the same slope as L. If L' is any line parallel to L, with the equation $y = f'(x)$, then L' lies for a suitable k between L and L_k so that L' also has this slope.

Now let $y = f(x)$ represent an arbitrary line. If it did not have a slope, then m and n different from 0 would exist such that

$$\liminf \frac{f(x)}{x} < \frac{n}{m} < \limsup \frac{f(x)}{x}.$$

The rational line containing the points $(0, f(0))\ \Psi(\nu m, \nu n)$ would then intersect L more than once, without coinciding with L.

The definition (2) of the slope implies

(3) *Lines with different slopes intersect.*

(4) *There is a line with a given slope* $\mu \neq 0$, *through a given point* $p = (x_0, y_0)$.

If $\mu = n/m$ then the line containing the points $p\Psi(\nu m, \nu n)$ satisfies (4). If μ is irrational, choose a sequence of rational numbers ρ_ν, $\nu = 1, 2, \ldots$, which increase and tend to μ and a rational number $\rho_0 > \mu$. Denote by L_i

a rational line through p with slope ρ_i. Then L_{i+1} lies between L_i, $i \geqslant 1$, and L_0, hence L_i tends to a limit line L through p. If $y = f_i(x)$ and $y = f(x)$ represent L_i and L respectively, then $f_0(x) > f(x) > f_i(x)$ for $x > x_0$ and $i \geqslant 1$; hence the slope of L is at least μ, and at most ρ_0. Since ρ_0 was arbitrary, L has slope μ.

Statements (3) and (4) show that the parallel axiom follows from:

(5) *There is at most one line with a given slope μ through a given point p* [1].

For rational μ this follows from the fact that the parallel axiom holds for rational lines. For if $\mu = n/m$ and if the line L containing the points $p\Psi(\nu m, \nu n)$ has the equation $y = f(x)$, then any other line L' through $p = (x_0, y_0)$ has an equation $y = f'(x)$, with $f'(x) > f(x)$ for $x > x_0$ say. Because L is parallel to $L\Psi(0, 1)$, which has the equation $y = f(x) + 1$, the line L' must intersect $L\Psi(0, 1)$ for some $x' > x_0$.

Then for a suitable $\nu > 0$ the point $p\Psi(\nu m, \nu n)\,\Psi(0, 1)$ lies on $L\Psi(0, 1)$ and between L' and L. The line L'' through p and $p\Psi(\nu m, \nu n + 1)$ has slope $\dfrac{\nu n + 1}{\nu m} > \mu$ and the slope of L' cannot be smaller than the slope of L''.

Now, let μ be irrational, and assume for an indirect proof that there are two different lines L, K through p with slope μ. We may assume that p is the origin and that the two lines have equations of the form

$$L\colon\ y = f(x),\quad K\colon\ y = g(x) \quad \text{with} \quad g(x) > f(x) \quad \text{for} \quad x > 0.$$

Then for integral $n > 0$:

$$0 < g(n) - f(n) < 1 \tag{6}$$

because, otherwise the segment S_n connecting $(n, f(n))$ to $(n, g(n))$ would contain a point of the form (n, m) with integral m, and the rational line through p and (n, m) would lie between L and K. By the first part of this proof the slope of L would be smaller, and the slope of K would be greater, than n/m. Because the distance is invariant under the $\Psi(m, n)$ it follows from (22.23) that a $\delta > 0$ exists such that

$$g(n) - f(n) > \delta \qquad \text{for} \qquad n \geqslant 1.$$

For a given integral $\varkappa \geqslant 3$ determine the integer $m_\varkappa$ by

$$m_\varkappa \delta \geqslant \varkappa + 1 > (m_\varkappa - 1)\,\delta. \tag{7}$$

Then $\bigcup_{i=1}^{m_\varkappa} S_i$ contains $\varkappa + 1$ points p_i which represent the same point on the torus, or whose ordinates differ by integers. We distinguish two cases:

(a) There are for some $\varkappa$ four points p_i no three of which lie on the same geodesic. A familiar argument from elliptic functions shows that the convex closure of these 4 points in terms of P would then contain a "period parallelogram" Q whose sides are formed by segments of P as straight plane. Since the domain bounded by $y = f(x)$ and $y = g(x)$ for $x \geqslant 0$ is convex, Q would lie in this domain; on the other hand Q would contain a point equivalent to p, that is, of the form (m, n) which was already seen to be impossible.

(b) For every $\varkappa$ at least $\varkappa$ of the $\varkappa + 1$ points p_i lie on a geodesic $H^\varkappa$. Then $H^\varkappa$ is rational and has a rational slope $\rho_\varkappa$. Since no two of the $\varkappa$ points lie on the same S_i the abscissas $n_i^\varkappa$ of the $\varkappa$ points are different. Let $n_i^\varkappa < n_{i+1}^\varkappa$. Then $n_\varkappa^\varkappa - n_1^\varkappa \geqslant \varkappa - 1$, hence because of (7), $n_1^\varkappa/n_\varkappa^\varkappa \leqslant 1 - (\varkappa - 1)/m_\varkappa \leqslant 1 - \delta/4$. ($n_\varkappa \leqslant m_\varkappa$ follows from $p_{n_\varkappa} \in \bigcup_{i=1}^{m_\varkappa} S_i$). Since

$$\frac{f(x_1) - f(x_2)}{x_1 - x_2} = \frac{f(x_1)}{x_1} + \frac{(x_2/x_1)\ [f(x_1)/x_1 - f(x_2)/x_2]}{1 - x_2/x_1}$$

it follows that for $x_1 \to \infty$ and $0 < x_2/x_1 \leqslant \theta < 1$

$$\lim_{x_1 \to \infty} \frac{f(x_1)}{x_1} = \lim_{x_1 \to \infty} \frac{f(x_1) - f(x_2)}{x_1 - x_2},$$

where x_2 may, or may not, be bounded.

Hence it follows in the present case from $n_1^\varkappa/n_\varkappa^\varkappa \leqslant 1 - \delta/4$ that

$$\lim_{\varkappa \to \infty} \frac{f(n_\varkappa^\varkappa) - f(n_1^\varkappa)}{n_\varkappa^\varkappa - n_1^\varkappa} = \lim_{\varkappa \to \infty} \frac{g(n_\varkappa^\varkappa) - g(n_1^\varkappa)}{n_\varkappa^\varkappa - n_1^\varkappa} = \mu,$$

and therefore from (6) that also

$$\lim_{\varkappa \to \infty} \frac{f(n_\varkappa^\varkappa) - g(n_1^\varkappa)}{n_\varkappa^\varkappa - n_1^\varkappa} = \lim_{\varkappa \to \infty} \frac{g(n_\varkappa^\varkappa) - f(n_1^\varkappa)}{n_\varkappa^\varkappa - n_1^\varkappa} = \mu. \tag{8}$$

On the other hand,

$$(\varkappa - 1) \min_j (n_j^\varkappa - n_{j-1}^\varkappa) \leqslant n_\varkappa^\varkappa - n_1^\varkappa \leqslant m_\varkappa - 1$$

and by (7)

$$\min_j (n_j^\varkappa - n_{j-1}^\varkappa) \leqslant (m_\varkappa - 1)/(\varkappa - 1) < 2/\delta.$$

Therefore the denominator of the slope $\rho_\varkappa$ of $H^\varkappa$ (if reduced) cannot surpass $2/\delta$. Since μ is irrational there is an $\varepsilon > 0$ independent of $\varkappa$, such that $|\rho_\varkappa - \mu| \geqslant \varepsilon$. But if $y = h(x)$ represents $H^\varkappa$, since the $\varkappa$ points lie between L and K,

$$\frac{g(n_\varkappa^*) - f(n_1^*)}{n_\varkappa^* - n_1^*} \geqslant \frac{h(n_\varkappa^*) - h(n_1^*)}{n_\varkappa^* - n_1^*} = \rho_\varkappa \geqslant \frac{f(n_\varkappa^*) - g(n_1^*)}{n_\varkappa^* - n_1^*}$$

which in conjunction with (8) contradicts $|\rho_\varkappa - \mu| \geqslant \varepsilon$. This completes the proof of (33.1).

It implies for the torus:

(33.2) *In a metrization of the torus without conjugate points the closed geodesics are dense among all geodesics.* More precisely: If $p(\tau)$ represents an arbitrary geodesic and $N > 0$ and $\varepsilon > 0$ are given, then a representation $q(\tau)$ of a closed geodesic exists such that $p(\tau)q(\tau) < \varepsilon$ for $|\tau| \leqslant N$.

Sufficiency of the conditions

We now show that the properties (which we found) characterize the curve systems which can occur as geodesics in metrizations of the torus without conjugate points.

•(33.3) **Theorem.** *In the* (x, y)*-plane* P *let a system* S *of curves be given with the following properties:*

I. *Each curve in* S *is a topological image* $(x(t), y(t))$, $-\infty < t < \infty$ *of the real axis with* $x^2(t) + y^2(t) \to \infty$ *for* $|t| \to \infty$.

II. *Two given distinct points of* P *lie on exactly one curve in* S.

III. *The system* S *goes into itself under the translations* $\Psi(m, n) : x' = x + m$, $y' = y + n$, m, n *integers.*

IV. *If a curve in* S *contains* q *and* $q\Psi(m, n)$ *then it contains* $q\Psi(\nu m, \nu n)$ *for* $\nu = \pm 1, \pm 2, \ldots$.

V. *The system* S *satisfies the parallel axiom.*

Then P *can be metrized as a straight space for which the curves in* S *are the geodesics and the translations* $\Psi(m, n)$ *are motions.*

This theorem is of a nature similar to (11.2) and we are going to use the topological consequences of I and II obtained in the proof of (11.2), also the main idea in the construction of the metric is the same. Take any pair m, n with $m > 0$ and denote by L_p the curve in S containing the points $p\Psi(\nu m, \nu n)$. For any p, q the lines L_p and L_q are either parallel or identical. Let $y = f_p(x)$ represent L_p (since $n = 0$ is admitted, $f_p(x)$ may be constant). That $\Psi(m, n)$

carries L_p into itself implies $f_p(x+m) = f_p(x) + n$ for all x, therefore $f_p(x) - f_q(x)$ is periodic with period m and the area

$$d_{m,n}(p, q) = \int_{x_0}^{x_0+m} |f_p(x) - f_q(x)|\, dx$$

of the "parallelogram" Q bounded by L_p, L_q and $x = x_0$, $x = x_0 + m$ is independent of x_0. An arbitrary translation $\Psi' = \Psi(m', n')$ carries Q into a parallelogram which has the same relation to $p\Psi'$ and $q\Psi'$ as Q has to p and q. But Ψ' leaves area invariant, hence

(9) $$d_{m,n}(p, q) = d_{m,n}(p\Psi', q\Psi').$$

Clearly $d_{m,n}(p, q) = d_{m,n}(q, p)$ and[2]

(10) $$d_{m,n}(p, q) = 0 \quad \textit{if and only if } L_p = L_q$$

The arbitrariness of x_0 yields

(11) $d_{m,n}(p, q) + d_{m,n}(q, r) = d_{m,n}(p, r)$ *if and only if the line* L_q *lies in the closed strip bounded by* L_p *and* L_r;

(12) $d_{m,n}(p, q) + d_{m,n}(q, r) > d_{m,n}(p, r)$ *if* L_q *does not lie in this strip.*

Let δ be the difference of the ordinates of p and q and determine the integer k by $k - 1 \leqslant |\delta| < k$. Then $\Psi(0, \pm k)$ carries L_p into a line L_r for which L_q (if different from L_p) lies between L_p and L_r. Then

(13) $$d_{m,n}(p, q) < d_{m,n}(p, r) = k\, d_{m,n}(p, p\Psi(0, 1)) < (|\delta| + 1)\, \lambda_{m,n}$$

where $\lambda_{m,n}$ depends only on m and n. A distance which satisfies our requirements will be

(14) $$pq = \sum{}' d_{m,n}(p, q)\, \lambda_{m,n}^{-1}\, 2^{-m-n},$$

where the prime indicates that the summation is extended over all pairs m, n with $m > 0$ and all n, but such that $n/m \neq n'/m'$ for different pairs m, n and m', n'.

If p and q are given and have ordinate difference δ, then (13) implies $d_m(p, q)\, \lambda_{m,n}^{-1} < |\delta| + 1$, for all m, n, so that pq is always finite. (9) shows that pq is invariant under all $\Psi(m', n')$, and (11) and (12) imply that pq satisfies the triangle inequality. $pp = 0$ by (10), and $pq = qp > 0$ for $p \neq q$ follows from $d_{m,n}(p, q) = d_{m,n}(q, p)$ and from (10) because for all pairs m, n, with the exception of at most one, the lines L_p and L_q (in the previous notation) will be different.

Thus pq satisfies the axioms for a metric space. To see that the curves in S are the geodesics it must be shown: that for three different points p, q, r

(15) $pq + qr = pr$ *if* q *lies on the segment* σ *of the curve of* S *through* p *and* r.

(16) $pq + qr > pr$ *if* q *does not lie on* σ.

If q lies on σ, then for any m, n the line L_q will either contain p and r or L_q lies between L_p and L_r. Hence it follows from (10) and (11) that (15) holds.

If finally q does not lie on σ, let L be the curve in S through two arbitrary interior points q' and q'' of the segments (in the sense of S) from q to p and r respectively. Then L separates σ from q. If L contains for suitable m, n with $m > 0$ the points $q'\Psi(vm, vn)$, then (16) follows from (12). If L does not have this property (i. e., is either a line $x =$ const. or not rational) then the parallel axiom implies the existence of $m > 0$ and n such that the line L' containing the points $q'\Psi(vm, vn)$ is so close to L that it also separates σ, and therefore p and r, from q. Then (16) follows again from (12).

That the distance pq is equivalent to the euclidean distance is easily derived from either the analytic definition of pq or the geometric properties of S. The finite compactness of pq follows from its invariance under $\Psi(m', n')$.

Examples

The construction of the distance pq in the preceding section happens to yield a Minkowski metric if the curves in S are the euclidean lines $ax + by + c = 0$. This is, however, accidental because other functions $d_{m,n}(p, q)$ than the area could have been used. For instance, if $p_i = (x_i, y_i)$ then

$$(33.4)\quad p_1p_2 = [(x_1 - x_2)^2 + (y_1 - y_2)^2]^{1/2} + |7y_1 + \sin 2\pi y_1 - 7y_2 - \sin 2\pi y_2|$$

yields a metric for which the euclidean lines are the geodesics, because $7\,y + \sin 2\pi\,y$ increases monotonically. Moreover, this metric is invariant under the $\Psi(m, n)$. Instead of $7\,y + \sin 2\pi\,y$ many other functions could have been used, a similarly formed term in the x_i could have been added, the euclidean distance occurring in the definition of p_1p_2 could have been replaced by an arbitrary Minkowski distance. *Hence there is so much choice that the problem of determining all metrics which belong to a given system of curves becomes uninteresting.*

One might ask whether conditions I, II and III do not imply either IV or V. The examples under (23.5) show that this is not the case[3].

Finally we give an example which confirms the assertion of the introduction to this chapter that *the curves of a system S satisfying conditions I to V need not be Desarguesian.*

To construct such a system S, we first define certain functions $f_t(x)$, $t \geqslant 1$ in the interval $0 \leqslant x \leqslant 1$. Put

$$f_1(x) = x$$

$$f_n(x) = \begin{cases} a_n x & \text{for} \quad 0 \leqslant x \leqslant \frac{1}{2} \\ b_n(x - \frac{1}{2}) + \frac{1}{2} a_n & \text{for} \quad \frac{1}{2} \leqslant x \leqslant 1 \end{cases} \quad \text{if} \quad n > 1$$

where

$$a_n = 2n - 1 - 2c_n, \qquad b_n = 1 + 2c_n, \qquad c_n = \sum_{\nu=1}^{n} 10^{-\nu}.$$

Then $f_n(1) = n$ and $f'_{n+1}(x) - f'_n(x) > 0$ for $x \neq \frac{1}{2}$, so that $f_{n+1}(x) - f_n(x)$ increases. Moreover, put

$$f_t(x) = (n + 1 - t)\, f_n(x) + (t - n)\, f_{n+1}(x) \qquad \text{if} \qquad n < t < n + 1.$$

Then it is easily seen that $f_{t_2}(x) - f_{t_1}(x)$ increases for $1 \leqslant t_1 < t_2$, and that $f_t(1) = t$.

Now define $g_t^b(x)$ for all x, all $t \geqslant 1$ and all b by

$$g_t^b(x) = f_t(x - m) + mt + b \qquad \text{for} \qquad m \leqslant x < m + 1.$$

Since

$$(g_{t_2}^b(x) - g_{t_1}^b(x))' = f'_{t_2}(x - m) - f'_{t_1}(x - m)$$

the difference $g_{t_2}^b(x) - g_{t_1}^b(x)$ increases for $t_2 > t_1$. Hence the two curves $y = g_{t_2}^b(x)$ and $y = g_{t_1}^b(x)$ intersect at most once. Moreover $y = g_t^b(x)$ has slope t in the sense of (2).

The system S is defined as consisting of all curves $y = g_t^b(x)$, all lines $y = tx + b$ with $t < 1$, and the lines $x = \text{const}$. Because $(g_t^b(x))' \geqslant 1$ each line in S intersects each $y = g_t^b(x)$ exactly once.

Through every point of the plane there is exactly one line with a given slope (2). Hence the parallel axiom holds. It is easily verified that two distinct points of the plane lie on exactly (and not only at most) one curve in S.

To show that the system S has property IV it suffices to prove the following: If $g_t^b(x') = y'$ and $g_t^b(x' + m) = y' + n$, where m and n are integers, then $g_t^b(x' + \nu m) = y' + \nu n$. Determine the integer k by $k \leqslant x' < k + 1$. Then putting $f_t(x' - k) + b = W$,

$$y' = g_t^b(x') = W + kt, \qquad g_t^b(x' + m) = W + (k + m)\, t = W + kt + n,$$

hence $t = n/m$. Moreover

$$g_t^b(x' + \nu m) = W + (k + \nu m)\, t = y' + \nu\, m\, t = y' + \nu n.$$

That the Theorem of Desargues does not hold is seen in the usual way: Two triangles, which are in the Desarguesian relation in the ordinary sense, are placed so that all but one of the lines entering the theorem have slope less than 1 and hence are also curves of S. The last line L has slope greater than 1. The curve $y = g_t^b(x)$ through two of the three points of the Desarguesian configuration on L will in general not contain the third point.

The system S goes into itself not only under the $\Psi(m, n)$ but under all translations $x' = x + m$, $y' = y + b$ where m is an integer, but b is an arbitrary real number. It is easily verified that the metric constructed above is invariant under all these translations. Thus:

(33.5) **Theorem.** *There are metrizations of the torus without conjugate points which have a one-parameter group of motions and for which the geodesics of the universal covering space do not satisfy Desargues' Theorem. Even if this theorem holds the metric need not be Minkowskian* (*see* (33.4)).

The system S is so constructed that it also exhibits another phenomenon, which is surprising at first sight. The curves $y = g_n^0(x)$ of S approach for $n \to \infty$ the curve $x = 0$, but there is an $\varepsilon > 0$ independent of n and a circular disk with radius ε (whose center depends on n) such that $g_n^0(x)$ does not enter this disk.

34. Transitive geodesics on surfaces of higher genus

We now turn to surfaces of higher genus. In contrast to the torus the emphasis here is on the analogy to the behaviour of the geodesics on more general surfaces with that on surfaces with constant negative curvature. There is very much interesting and deep material which can be extended to G-spaces, so that this section represents nothing but a small sample which, we hope, will enable the reader interested in these questions to extend other parts of the theory.

Poincaré Models

Let R be a compact orientable G-space of genus $\gamma \geqslant 2$ without conjugate points, i. e., such that its universal covering space is a straight plane P. Following Poincaré's procedure for surfaces of constant negative curvature we realize P in a specific way which we call a *Poincaré Model* of P. Since we are going to operate far more in P than in R we denote points in P by $x, y, \ldots$ and points in R by $x', y', \ldots$. Using the method outlined in Section 30 we metrize R also as a locally hyperbolic surface R_h with distance $\delta(x', y')$. The universal covering space of R_h is the hyperbolic plane H with distance $\delta(x, y)$, which we consider as the interior of the unit circle C of a euclidean plane with distance $e(x, y)$ such that the hyperbolic straight lines appear as euclidean circular arcs orthogonal to C. The euclidean center p_1 of C is the hyperbolic center of the fundamental set $H(p_1)$ with closure $F = F(p_1)$ whose boundary $F(p_1) - H(p_1)$ is a regular 4γ-gon in the hyperbolic sense. We denote the elements of the fundamental group of R_h by $\Xi_1 = \Xi, \Xi_2, \Xi_3, \ldots$ and the locally isometric mapping of H on R_h by Ω.

By using Ω^{-1} locally we define in the interior of C a new distance xy derived from the distance $x'y'$ in R. With this metric the interior of C becomes the universal covering space P of R. The $\Xi_\varkappa$ as transformations of P on itself are locally isometric mappings, hence by (27.17) motions of P. Since $\Xi_\varkappa \Omega = E$ it follows that the $\Xi_\varkappa$ represent the fundamental group of R in terms of motions of P. We call this realization of P a Poincaré model of P. It has the following useful property:

(34.1) *Given two positive numbers $\alpha < \beta$, then positive numbers $\alpha' < \beta'$ and $\alpha'' < \beta''$ exist such that*

$$\alpha' \leqslant \delta(x, y) \leqslant \beta' \quad \textit{implies} \quad \alpha \leqslant xy \leqslant \beta \qquad \textit{and}$$
$$\alpha'' \leqslant \; xy \; \leqslant \beta'' \quad \textit{implies} \quad \alpha \leqslant \delta(x, y) \leqslant \beta.$$

The argument is nearly the same in all cases. By way of an example, we show the existence of β'. It suffices to choose

$$\beta' = \inf \delta(x, y), \qquad \text{where} \quad x \in F \qquad \text{and} \qquad xy \geqslant \beta.$$

In the first place β' is positive. Otherwise a sequence of pairs x_ν, y_ν with $x_\nu \in F$, $x_\nu y_\nu \geqslant \beta$ and $\delta(x_\nu, y_\nu) \to 0$ would exist. For a suitable subsequence $\{\mu\}$ of $\{\nu\}$ the sequences $\{x_\mu\}$ and $\{y_\mu\}$ converge to a point x in F and $y \neq x$ but $\delta(x, y) = 0$.

If now x, y is any pair of points in H with $\delta(x, y) \leqslant \beta'$, then a Ξ_k exists such that $x\,\Xi_k \in F$, see (29.4 e). Then $\delta(x, y) = \delta(x\,\Xi_k, y\,\Xi_k) \leqslant \beta'$ and $xy = x\,\Xi_k\, y\,\Xi_k \leqslant \beta$. For if $x\,\Xi_k\, y\,\Xi_k > \beta$ then a point z on the hyperbolic segment from $x\,\Xi_k$ to $y\,\Xi_k$ close to $y\,\Xi_k$ would satisfy $\delta(x\,\Xi_k, z) < \beta'$ and $x\,\Xi_k\, z > \beta$.

Any oriented line $\mathfrak{g}_h^+$ in H is in the euclidean sense a circle orthogonal to C, hence a point traversing $\mathfrak{g}_h^+$ in the positive (negative) sense tends to a definite point on C, which we call the "*positive (negative) endpoint*" of $\mathfrak{g}_h^+$. If a point x traverses an oriented line $\mathfrak{g}^+$ in P in the positive (negative) sense it is clear that any accumulation point of x for $p_1 x \to \infty$ in the euclidean sense must lie on C. It is not immediately obvious, but will be shown presently, that x actually converges. The limit point will be called the positive (negative) endpoint of $\mathfrak{g}^+$.

Every motion $\Xi_\varkappa \neq \Xi$ in H is axial. Its oriented axis $\mathfrak{g}_h^+$ lies over a closed geodesic $\mathfrak{g}_t'^+$ in R_h which cannot be contracted to a point (because $\Xi_\varkappa \neq \Xi$). Since R is compact the free homotopy class of $\mathfrak{g}_h'^+$ contains at least one closed geodesic $\mathfrak{g}'^+$. By (32.6) there is an axis $\mathfrak{g}^+$ of $\Xi_\varkappa$ over $\mathfrak{g}'^+$, and any axis $\mathfrak{h}^+$ of $\Xi_\varkappa$ in P lies over a closed geodesic in R freely homotopic to $\mathfrak{g}'^+$.

(34.2) *Any $\Xi_\varkappa \neq \Xi$ is axial both as a motion of P and of H. If the hyperbolic axis $\mathfrak{g}_h^+$ of $\Xi_\varkappa$ has $u_\varkappa^+$ and $u_\varkappa^-$ as positive and negative endpoints then any axis $\mathfrak{g}^+$ of $\Xi_\varkappa$ as motion of P also has u_k^+ and u_k^- as positive and negative endpoints.*

For any set Q in P or H we denote by $S(Q, \rho)$ and $S_h(Q, \rho)$ the sets of points x with $xQ < \rho$ and $\delta(x, Q) < \rho$ respectively. To prove the second statement in (34.2) let $q \in \mathfrak{g}^+$. Then

$$\mathfrak{g}^+ = \bigcup_{\nu=-\infty}^{\infty} T(q\,\Xi_k^\nu, q\,\Xi_k^{\nu+1}) \qquad \text{and} \qquad T(q, q\,\Xi_k) \subset S_h(\mathfrak{g}_h^+, \rho)$$

for a suitable ρ. Hence

$$\mathfrak{g}^+ \subset S_h(\mathfrak{g}_h^+, \rho).$$

(34.2) now follows from the fact that the points x for which $\delta(x, \mathfrak{g}_h^+) = \rho$ form two — in the hyperbolic sense — convex curves with u_k^+ and u_k^- as endpoints.

The endpoints of the axes either in the sense of P or of H depend therefore on the Ξ_k only. Following Nielsen [2, p. 210] we now establish an important fact regarding their distribution:

(34.3) *Given any two proper subarcs J^+ and J^- of C there is a $\Xi_k \neq \Xi$ such that the positive endpoint u_k^+ of an axis of Ξ_k lies on J^+ and the negative, u_k^-, on J^-.*

Let q_1 be a vertex of the hyperbolic polygon bounding the fundamental domain $H(p_1)$. Since the points $q_i = q_1\,\Xi_i$ have every point of C as accumulation point, it suffices to see that any hyperbolic line $\mathfrak{g}$ through two distinct points q_j, q_k in $Q = \bigcup q_i$ is the hyperbolic axis of some Ξ_ν (we do not claim that $q_j\,\Xi_\nu^{\pm 1} = q_k$).

For brevity we call — in this proof only — *polygon side* any hyperbolic line carrying a side on the boundary of some $H(p_i)$. There are 2γ polygon sides through q_j since the hyperbolic rotation Ψ about q_j through π carries a polygon side again into a polygon side. Ψ carries $\mathfrak{g}$ into itself and q_k into another point q_n in Q. A polygon side $\mathfrak{h}'$ through q_k goes under Ψ into a polygon side $\mathfrak{h}$ through q_n. The hyperbolic translation Φ along $\mathfrak{g}$ which carries q_n into q_k carries $\mathfrak{h}$ into $\mathfrak{h}'$ (but if $\mathfrak{h}'$ is oriented, then $\mathfrak{h}'\Psi\Phi$ has the opposite orientation from $\mathfrak{h}'$). Therefore Φ carries Q and $\mathfrak{g}$ into themselves and the same holds for the powers Φ^μ of Φ.

Since there are only 2γ polygon sides through q_n, the line $\mathfrak{h}\Phi^{\mu_0}$ must for some μ_0, $1 \leqslant \mu_0 < 2\gamma$, have the form $\mathfrak{h}\,\Xi_{\nu_0}$. Taking the orientation of $\mathfrak{h}\,\Xi_{\nu_0}$ into account we see that either $\Phi^{\mu_0} = \Xi_{\nu_0}$ or, if not, then $\Phi^{2\mu_0} = \Xi^2_{\nu_0}$. In any case, $\Xi_\nu = \Xi^2_{\nu_0}$ has $\mathfrak{g}$ as axis.

This permits us to prove

(34.4) **Theorem.** *Any ray $\mathfrak{r}$ in P possesses an endpoint on C. Any co-ray to $\mathfrak{r}$ has the same endpoint as $\mathfrak{r}$. Given two distinct points a^+, a^- on C there is a line in P with a^+ and a^- as endpoints. Given a point q in P and a point a^+ on C, there is a line through q with a^+ as an endpoint.*

The first statement means more precisely: If $x(\tau)$, $\tau \geqslant 0$, represents $\mathfrak{r}$, then the point $x(\tau)$ tends for $\tau \to \infty$ in the euclidean sense to a point of C. If two sequences $\tau_\nu \to \infty$ and $\tau_\nu' \to \infty$ existed such that $\lim x(\tau_\nu) = a \neq a' = \lim x(\tau_\nu')$, then (34.3) would guarantee the existence of a Ξ_k such that the corresponding points u_k^+ and u_k^- separate a from a'. An axis $\mathfrak{g}$ of Ξ_k as motion of P would then evidently intersect the segment $T[x(\tau_\nu), x(\tau_\nu')]$ on $\mathfrak{r}$ for every large ν, so that $\mathfrak{r}$ and $\mathfrak{g}$ would have infinitely many intersections.

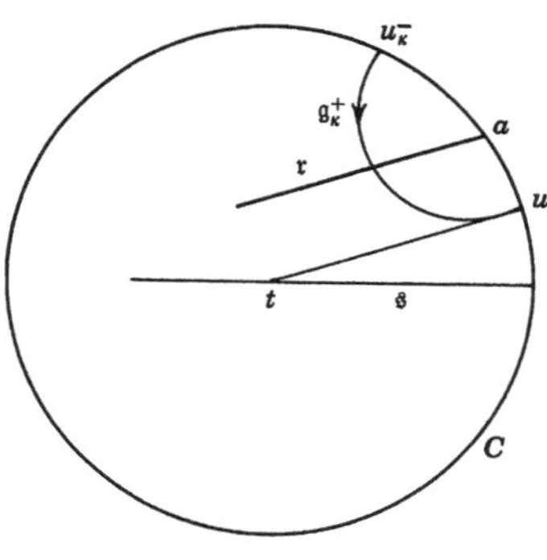

Fig. 38.

If $\mathfrak{r}$ is a subray of an axis, then the distance $x\mathfrak{r}$ of a variable point x on a co-ray $\mathfrak{s}$ to $\mathfrak{r}$ is bounded by (32.16). It follows from (34.1) that $\delta(x, \mathfrak{r})$ is also bounded hence x must tend to the endpoint of $\mathfrak{r}$ when it traverses $\mathfrak{s}$.

If $\mathfrak{r}$ is an arbitrary ray with endpoint a and $\mathfrak{s}$ any ray whose endpoint is different from a, then an axis $\mathfrak{g}_\varkappa^+$ of a $\Xi_\varkappa$ in P exists, whose endpoints $u_\varkappa^+$, $u_\varkappa^-$ separate the endpoint of $\mathfrak{s}$ from a on C. Choose the notation such that $u_\varkappa^+$

lies on the same side (in P) of the oriented line which carries $\mathfrak{r}$ as positive subray, as points of $\mathfrak{s}$ sufficiently close to C. The asymptote to $\mathfrak{g}_\varkappa^+$ from such a point t of $\mathfrak{s}$ passes through u_k^+, the ray from t through $x(\tau)$ can therefore not tend to $\mathfrak{s}$ for $\tau \to \infty$.

For a proof of the remaining statements in (34.4) let C_1 and C_2 be the two open subarcs of C with endpoints a^+ and a^-. Choose $\Xi_{\varkappa_\nu}$ and Ξ_{λ_ν} such that $u_{\varkappa_\nu}^+, u_{\varkappa_\nu}^- \in C_1$, $u_{\lambda_\nu}^+, u_{\lambda_\nu}^- \in C_2$ and monotonically $u_{\varkappa_\nu}^+ \to a^+$, $u_{\varkappa_\nu}^- \to a^-$, $u_{\lambda_\nu}^+ \to a^+$, $u_{\lambda_\nu}^- \to a^-$.

Finally let $\mathfrak{k}$ be an axis of some Ξ_σ whose endpoints separate a^+ from a^- and therefore also $u_{\varkappa_\nu}^+$ from $u_{\varkappa_\nu}^-$ and $u_{\lambda_\nu}^+$ from $u_{\lambda_\nu}^-$ of $\nu \geqslant \nu_0$ say. Then axes $\mathfrak{g}_{1\nu}^+$ and $\mathfrak{g}_{2\nu}^+$ of $\Xi_{\varkappa_\nu}$ and Ξ_{λ_ν} intersect $\mathfrak{k}$ for $\nu \geqslant \nu_0$ in points $p_{1\nu}$ and $p_{2\nu}$ which lie on $T(p_{i\nu_0}, p_{2\nu_0})$. Therefore a subsequence $\{\mathfrak{g}_{1\mu}^+\}$ of $\{\mathfrak{g}_{1\nu}^+\}$ exists which converges to a line $\mathfrak{h}^+$. Since $\{\mathfrak{g}_{1\mu}^+\}$ lies between $\mathfrak{g}_{1\nu}^+$ and $\mathfrak{g}_{2\nu}^+$ for $\mu > \nu$ it follows that a^+ and a^- must be the endpoints of $\mathfrak{h}^+$.

An asymptote through q to $\mathfrak{h}^+$ will by the preceding result have a^+ as an endpoint.

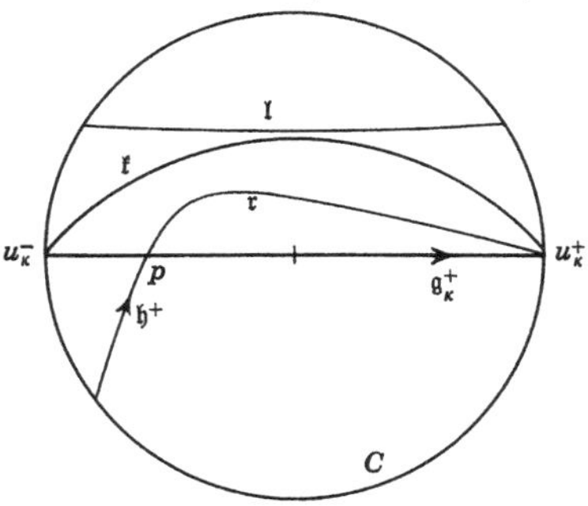

Fig. 39.

Next we show

(34.5) *A line $\mathfrak{h}^+$ through a point p of an axis $\mathfrak{g}_\varkappa^+$ of a $\Xi_\varkappa \neq \Xi$ and with the same positive endpoint $u_\varkappa^+$ as $\mathfrak{h}^+$ coincides with $\mathfrak{g}_k^+$.*

For if this were not correct, let $\mathfrak{r}$ be the positive subray of $\mathfrak{h}^+$ with origin p. Take any axis $\mathfrak{l}$ whose endpoints lie on the same side of $\mathfrak{g}_\varkappa^+$ as $\mathfrak{r}$. Then $\mathfrak{r}\,\Xi_\varkappa^\nu$ lies except for p for all integral ν between $\mathfrak{l}$ and $\mathfrak{g}_\varkappa$, because this ray has $p\,\Xi_\varkappa^\nu$ as origin and ends at $u_\varkappa^+$. Also, $\mathfrak{r}\,\Xi_\varkappa^\nu$ lies always between $\mathfrak{r}\,\Xi_\varkappa^{\nu+1}$ and $\mathfrak{l}$. Therefore $\mathfrak{r}\,\Xi_\varkappa^\nu$ converges for $\nu \to -\infty$ to a line $\mathfrak{k}$. This line goes into itself under $\Xi_\varkappa$ hence is an axis of $\Xi_\varkappa$. It would then be parallel to $\mathfrak{g}_\varkappa$, but the existence of $\mathfrak{h}$ shows that this is not the case.

This has as consequence

(34.6) ***Theorem.*** *The oriented lines which have the same positive endpoint u^+ as an oriented axis are all asymptotes to each other.*

For if u^+ is the positive endpoint of the axis $\mathfrak{g}^+$ of $\Xi_k \neq \Xi$ and $\mathfrak{r}$ is any ray with u^+ as endpoint, then for any point p on $\mathfrak{g}^+$ the positive subray $\mathfrak{s}$

of $\mathfrak{g}^+$ with origin p must by (34.4) be a co-ray to $\mathfrak{r}$ since it is by (34.5) the only ray with origin p and endpoint u^+. It follows then from (32.16) that $\mathfrak{r}$ is a co-ray to $\mathfrak{s}$.

This result in conjunction with (32.16) gives us complete information regarding the axes and their asymptotes, which corresponds to (33.1) in case of the torus. A characterization of all systems of curves which can occur as the straight lines of the universal covering spaces of compact orientable surfaces without conjugate points is probably possible, but intricate and delicate, since the treatment of the lines which are not axes proved already rather involved in the simpler case of the torus.

Conditions for the uniqueness of a geodesic through two points of $P \cup C$

However, for the reasons indicated above and in Section 26, we discuss only the special case where any two points in $P \cup C$ determine exactly one geodesic. We observe first

(34.7) *If the endpoints of a geodesic determine the geodesic uniquely then the two endpoints of a geodesic are distinct.*

The geodesic through a point of P and a point of C is unique. A geodesic depends continuously on its endpoints.

Note. The last assertion means more precisely one of the following two equivalent statements:

1) If $u_\nu^+ \to u^+$, and $u_\nu^- \to u^-$, where $u_\nu^+ \neq u_\nu^-$, $u^+ \neq u^-$, are points of C, then the geodesic $\mathfrak{g}_\nu$ with endpoints u_ν^- and u_ν^+ tends to the geodesic with endpoints u^+, u^- in the sense of Hausdorff's limit.

2) If $\varepsilon > 0$, $N > 0$ and a representation $x(\tau)$ of the oriented geodesic $\mathfrak{g}^+$ with endpoints u^+, u^- are given, then on C, intervals J^+ and J^- about u^+ and u^- respectively exist, such that any oriented geodesic $\mathfrak{h}^+$ with negative endpoint in J^- and positive endpoint in J^+ has a representation $y(\tau)$ such that

$$x(\tau)\, y(\tau) < \varepsilon \qquad \text{for} \qquad |\tau| \leqslant N.$$

If the two endpoints of the geodesic $\mathfrak{g}$ coincided at u then the endpoints of any asymptote $\mathfrak{a}$ to $\mathfrak{g}$ through a point in the interior of $\mathfrak{g} \cup u$ as curve in the euclidean plane would also coincide with u, so that $\mathfrak{a}$ and $\mathfrak{g}$ have the same endpoints.

The continuity in the sense of the first formulation follows from uniqueness and the proof of (34.4). That this is equivalent to the second formulation is then a consequence of (8.12) and (9.10).

Assume there are two distinct geodesics $\mathfrak{g}$, $\mathfrak{h}$ through the point p of P with a common endpoint u. Let v be the other endpoint of $\mathfrak{g}$. A line $\mathfrak{k}_x$ through a variable point x between p and v on $\mathfrak{g}$ and a fixed point y of $\mathfrak{h}$ between p and u, must have u as one endpoint. For $px \to \infty$ it would tend to a line through y, therefore different from $\mathfrak{g}$, with the same endpoints as $\mathfrak{g}$.

The assumption that the endpoints determine a geodesic in P uniquely is not satisfactory because its formulation depends on the Poincaré model. We therefore look for equivalent conditions which do not depend on the model and show first:

(34.8) *There is a positive η such that* 1) *a domain D_1 in P corresponding to the euclidean domain bounded by two distinct rays $\mathfrak{r}$, $\mathfrak{s}$ with the same origin and the same endpoint cannot contain a circular disk of radius η*; 2) *a domain D_2 in P bounded by two non-intersecting lines $\mathfrak{g}$, $\mathfrak{h}$ with the same endpoints cannot contain a circle $S(x, \eta)$.*

The proof of this lemma can be formulated concisely by introducing the term "diverging pentagon" for a convex pentagon in P for which non-adjacent sides do not intersect in P and have no common points on C.

Diverging pentagons exist: we merely have to choose points $u_i^{\pm}$, $i=1,\ldots,5$, in the cyclical order $u_1^-, u_5^+, u_2^-, u_1^+, u_3^-, u_2^+, u_4^-, u_3^+, u_5^-, u_4^+, u_1^-$ on C. Then five lines with endpoints u_i^- and u_i^+ will be sides of a diverging pentagon.

Neither D_1 nor D_2 can contain a diverging pentagon, because the arrangement of the endpoints of the sides on C implies that one of the rays $\mathfrak{r}$, $\mathfrak{s}$ (or one of the lines $\mathfrak{g}$, $\mathfrak{h}$), would intersect at least one of the sides of the pentagon twice. This argument explains why we chose pentagons: it would not work for triangles or quadrangles, and a diverging n-gon with $n > 5$ contains a diverging pentagon.

Let $S(y_1, \eta_1)$ be a circle containing a diverging pentagon and choose η_2 such that any $S(p, \eta_2)$ contains for any point p a point $y_k = y_1 \Xi_k$. The number $\eta = \eta_1 + \eta_2$ will satisfy (34.8).

For, if D_i contained a circle $S(x, \eta)$, then $S(x, \eta_2)$ would contain a point y_k, and $S(y_k, \eta_1)$ would contain the image of the diverging pentagon in $S(y_1, \eta_1)$, which is also a diverging pentagon. The latter would lie in D_i, because

$$S(x, \eta) \supset S(y_k, \eta - xy_k) \supset S(y_k, \eta_1).$$

This observation suggests introducing a property which will be of great importance also in the next chapter:

A straight space has the "*divergence property*" if, for any two distinct rays $\mathfrak{r}$, $\mathfrak{s}$ with the same origin, the distance $x\mathfrak{s}$ from $\mathfrak{s}$ of a point x traversing $\mathfrak{r}$ tends to infinity[1].

If two distinct rays $\mathfrak{r}$, $\mathfrak{s}$ with the same origin have different endpoints, and $x(\tau)$ represents $\mathfrak{r}$, then obviously $x(\tau)\mathfrak{s} \to \infty$, in fact, it is readily verified that $0 \leqslant \tau - x(\tau)\mathfrak{s} < M$. The interest lies in the converse:

(34.9) *If P has the divergence property then in a Poincaré model of P two distinct rays with the same origin have different endpoints.*

For, if two distinct rays $\mathfrak{r}$, $\mathfrak{s}$ with the same origin q and endpoint u existed and D_1 is the euclidean domain bounded by $\mathfrak{r} \cup \mathfrak{s} \cup u$, then a ray $\mathfrak{q}$ from q through an interior point of D_1 would also have u as endpoint. If $q(\tau)$ represents $\mathfrak{q}$, then by the divergence property $q(\tau)\mathfrak{r} \to \infty$ and $q(\tau)\mathfrak{s} \to \infty$, hence $S(q(\tau), \eta) \subset D_1$ for large τ, which contradicts (34.8).

We summarize the results of our analysis:

(34.10) **Theorem.** *If the universal covering space P of a compact orientable G-surface of genus greater than 1 is straight, then the following properties of P are equivalent:*

A_1. *A domain bounded by two non-intersecting lines contains circular disks of arbitrarily large radii (or only a circular disk of radius η, where η is determined in* (34.8)).

A_2. *P has the divergence property and contains no pairs of parallel lines.*

A_3. *In a Poincaré model of P the endpoints of a geodesic (and hence a point of P and an endpoint) determine a geodesic uniquely.*

For, the divergence property, (34.9) and (34.4) show that the ray $\mathfrak{r}$ is a co-ray to the ray $\mathfrak{s}$ if, and only if, $\mathfrak{r}$ and $\mathfrak{s}$ have the same endpoint (hence the asymptote relation is symmetric). Therefore two lines with the same endpoints are parallel and A_2 implies A_3. The converse is obvious.

If two non-intersecting lines have at most one common endpoint, then the domain bounded by them contains hyperbolic circles of arbitrarily large radii, consequently by (34.1) also circles of arbitrarily large radii in the sense of P. Thus A_3 implies A_1. The converse follows from (34.8).

In the next chapter we will see that the universal covering space of a G-surface with negative curvature, in the general sense defined there, is a straight plane which satisfies A_i.

THE RESULTS OF NIELSEN

We conclude from (34.3) and (34.7) that the axes of the Ξ_k are dense among all geodesics in P. This implies for R:

(34.11) *Under the conditions A_i, the closed geodesics on R are dense among all geodesics.*

In the following considerations we need the lemma

(34.12) Given two proper subarcs J_1 and J_2 of C, there is a Ξ_k for which $J_2 \Xi_k$ lies in the interior of J_1.

For a proof we select by (34.3) a Ξ_ν whose positive endpoint u_ν^+ is an interior point of J_1 and whose negative endpoint lies on $C - J_2$, and has therefore positive (euclidean) distance from J_2. Then $\lim_{j\to\infty} J_2 \Xi_\nu^j = u_\nu^+$. Hence if j_0 is large enough, $J_2 \Xi_k$ lies for $\Xi_k = \Xi_\nu^{j_0}$ in the interior of J_1.

Nielsen's principal step is the proof of the following fact:

(34.13) *If P satisfies A_i then an oriented geodesic $\mathfrak{h}^+$ with the following property exists: Let $y(\tau)$ represent $\mathfrak{h}^+$. Given $N > 0$, $\varepsilon > 0$, and an oriented segment T^+ represented by $x(\tau)$, $0 \leqslant \tau \leqslant \lambda$, then a motion Ξ_m in $\mathfrak{F}$ and a number $\alpha > N$ exist such that*

$$y(\tau)\, x(\tau - \alpha)\, \Xi_m < \varepsilon \qquad \textit{for} \qquad \alpha \leqslant \tau \leqslant \alpha + \lambda.$$

Proof. Denote by $\mathfrak{g}_i^+$ the axis of Ξ_i, $i > 1$ in the sense of P. On $\mathfrak{g}_i^+$ choose a segment T_i^+ oriented as $\mathfrak{g}_i^+$ with length $\gamma(\Xi_i) = \min xx\Xi_i$. Let $x_i(\tau)$ represent $\mathfrak{g}_i^+$, and T_i^+ for $|\tau| \leqslant \gamma(\Xi_i)/2$. By (34.7) there are disjoint arcs B_i^+ and B_i^- on C containing the endpoints u_i^+ and u_i^- of $\mathfrak{g}_i^+$ as interior points such that any geodesic $\mathfrak{k}^+$ connecting a point of B_i^- to a point of B_i^+ has a representation $z(\tau)$ with $z(\tau)x_i(\tau) < 1/i$ for $|\tau| \leqslant \gamma(\Xi_i)/2$. We briefly say $\mathfrak{k}^+$ approximates T_i^+ within $1/i$. Denote by B_i the arc of C distinct from B_i^- with the same endpoints as B_i^-.

Put $\Xi_2' = \Xi_2$ and choose $\Xi_3', \Xi_4', \ldots$ in $\mathfrak{F}$ successively as follows: By the preceding lemma there is a Ξ_3' in $\mathfrak{F}$ which carries B_3 and hence also B_3^+, into the interior of B_2^+. Then $B_3^- \Xi_3'$ contains B_2^- in its interior. Any geodesic that connects a point of B_2^- to a point of $B_3^+ \Xi_3'$ then approximates T_2^+ within $\frac{1}{2}$ and $T_3^+ \Xi_3'$ within $1/3$, because approximation within ε is invariant under motions.

We continue in the same manner: we choose Ξ_4' in $\mathfrak{F}$ such that $B_4 \Xi_4'$ lies in the interior of $B_3^+ \Xi_3'$. Then $B_4^+ \Xi_4'$ lies in $B_3^+ \Xi_3'$ and $B_4^- \Xi_4'$ contains $B_3^- \Xi_3'$. A line connecting a point of B_2^- to a point of $B_4^+ \Xi_4'$ then approximates T_2^+ within $1/2$, $T_3^+ \Xi_3'$ within $1/3$, and $T_4^+ \Xi_4'$ within $1/4$.

Since $\gamma(\Xi_\varkappa) > M > 0$, for $\varkappa > 1$, it follows that the arcs $B_i^+ \Xi_i'$ shrink to a point e^+.

Any line $\mathfrak{h}^+$ which connects a point e^- of B_1^- to e^+ will satisfy the assertion. For let T^+ represented by $x(\tau)$, $0 \leqslant \tau \leqslant \lambda$, $\varepsilon > 0$, and $N > 0$ be given. Because of (34.11) there is a $\mathfrak{g}_j^+$ with a representation $v(\tau)$ such that $x(\tau)\, v(\tau) < \varepsilon/2$ for $0 \leqslant \tau \leqslant \lambda$. Let S^+ be the segment $v(\tau)$, $0 \leqslant \tau \leqslant \lambda$.

The powers Ξ_j^k occur among the Ξ_ν, say $\Xi_{i_k} = \Xi_j^k$. Then $i_k \to \infty$ and $\gamma(\Xi_{i_k}) = k\gamma\,(\Xi_j) \to \infty$. If $k\,\gamma(\Xi_j) > 2\,\lambda + \gamma(\Xi_j)$ then for a suitable power Ξ_j^n of Ξ_j the segment $S^+ \Xi_j^n$ will be contained in $T_{i_k}^+$. Let $\Xi_j^n\, \Xi'_{i_k} = \Xi_m$. Then $S^+ \Xi_m$ is approximated within $1/i_k$ by $\mathfrak{h}^+$, or if $y(\tau)$ represents $\mathfrak{h}^+$ then a representation $z(\tau)$ of $S^+ \Xi_m$ exists such that $y(\tau) z(\tau) < 1/i_k$ for $\alpha_k \leqslant \tau \leqslant \alpha_k + \lambda$ with a suitable α_k. But $T_{i_k}^+ \Xi'_{i_k}$ tends by construction to e^+ when $k \to \infty$, hence $\alpha_k \to \infty$. We may therefore choose k so large that $1/i_k < \varepsilon/2$ and $\alpha = \alpha_k > N$.

Now $v(\tau)\, \Xi_m$ is a representation of $S^+ \Xi_m$, hence $z(\tau) = v(\tau - \alpha) \Xi_m$, $\alpha \leqslant \tau \leqslant \alpha + \lambda$. Since $x(\tau - \alpha)$, $\alpha \leqslant \tau < \alpha + \lambda$ represents T^+ we have

$$y(\tau)\, x(\tau)\, \Xi_m \leqslant y(\tau)\, v(\tau) + v(\tau)\, x(\tau) < \varepsilon \qquad \text{for} \qquad \alpha \leqslant \tau \leqslant \alpha + \lambda.$$

This result implies for R that the image $\mathfrak{h}^+\Omega$ of $\mathfrak{h}^+$ is *transitive*, i. e., approximates every geodesic curve within any given ε:

(34.14) **Theorem.** *A compact orientable G-surface R of genus greater than* 1 *whose universal covering space is straight and satisfies one of the conditions* A_1, A_2, A_3, *possesses a geodesic* $y'(\tau)$ *with the property*:

If a geodesic curve $x'(\tau)$, $0 \leqslant \tau \leqslant \lambda$, *and numbers* $\varepsilon > 0$, $N > 0$ are given *there is an* $\alpha > N$ *such that*

$$y'(\tau)\, x'(\tau - \alpha) < \varepsilon \qquad \textit{for} \qquad \alpha \leqslant \tau \leqslant \alpha + \lambda.$$

The number of distinct geodesics $y'(\tau)$ *on R with this property has the power of the continuum.*

The last remark follows from the observation that e^- is arbitrary on B_1^{-1} and the number of geodesics in P over the same geodesic in R is countable.

The theorem implies, of course, that for any N, the point set traversed by $y'(\tau)$ for $\tau > N$ is dense in R. Since $y'(\tau)$ can be approximated by closed geodesics, we see that for a given $\varepsilon > 0$ a closed geodesic exists which intersects every disk $S(p, \varepsilon)$ on R. Obviously the following much stronger statement holds:

(34.15) *Corollary. Under the hypothesis of* **(34.14)** *let a finite number of geodesic curves* $z_i'(\tau)$, $0 \leqslant \tau \leqslant \lambda_i$, $i = 1, 2, \ldots, m$ *and an* $\varepsilon > 0$ *be given. Then a closed geodesic* $y'(\tau)$ *and numbers* $\alpha_1, \ldots, \alpha_m$ *exist such that*

$$y'(\tau)\, z_i'(\tau - \alpha_i) < \varepsilon \qquad \text{for} \qquad \alpha_i \leqslant \tau \leqslant \alpha_i + \lambda_i.$$

CHAPTER V

THE INFLUENCE OF THE SIGN OF THE CURVATURE ON THE GEODESICS

35. Introduction

The present chapter penetrates, according to the usual notions, deeply into the realm of differential geometry proper because of its connection with curvature. For Riemann spaces it contains no new contribution beyond the elimination of differentiability hypotheses and a resulting deeper geometric understanding.

For Finsler spaces our results go, however, beyond anything ever attempted by the standard tensor approach. They refute forever the prejudice that non-Riemannian spaces are too general for beautiful concrete properties: every result in one of the show pieces of Riemannian geometry, namely the theory of spaces with negative curvature which was started by Hadamard [1] in 1898, extends to Finsler spaces, and so does almost all of Cohn-Vossen's work on integral curvature and the behaviour of geodesics on surfaces.

The explanation for these at first sight surprising facts is this: *In Riemann spaces curvature has many different functions.* It is not plausible that in Finsler spaces a single concept will suffice for all these functions; it is rather to be expected that different concepts, which happen to coincide in the Riemannian case, correspond to different functions.

The great majority of the investigations on intrinsic geometry *exploit, or can be modified so as to exploit, only one of the functions, and can therefore be extended to Finsler spaces.*

If a Riemann space has non-positive (negative) curvature then in a small geodesic triangle the line connecting the midpoints of two sides is at most (less than) half as long as the third side, and the capsules, i. e., the loci equidistant to segments, are locally (strictly) convex. In a Finsler space the first property is stronger than the second, see Section 18. We therefore say that a G-space has non-positive (negative) curvature if it has the first property, and that it has (strictly) convex capsules if it has the second. The *local implica-*

tions of these properties are discussed in Section 36. In Sections 37–41 we show that *all the principal results on Riemann spaces with non-positive curvature in the classical sense hold for G-spaces with non-positive curvature,* that many extend to spaces with convex capsules, and that most of them hold in the latter case when the universal covering space has the Divergence Property formulated in the last section. The latter follows from non-positive curvature but not from the convexity of capsules, see Figure 16[1].

More particularly, Section 37 treats the *theory of parallels* and shows, for instance, that non-positive curvature implies symmetry and transitivity of the asymptote relation. Then we establish the fundamental fact that of a G-space with convex capsules and the property of domain invariance has a straight universal covering space. In Section 39 we obtain statements on the *fundamental groups of spaces with convex capsules,* for instance: An abelian subgroup (of more than one element) of the fundamental group of a G-space with strictly convex capsules and domain invariance is infinite cyclic. For analytic Riemann spaces this fact was discovered by Preissmann [1]. Then we discuss some properties of geodesics in spaces with negative curvature, most of which go back to Hadamard [1]. Finally (Section 41) we show that *for Riemann spaces non-positive curvature in the classical sense, in the present sense, and convexity of the capsules are equivalent,* and also that on Finsler surfaces the invariant, which is there called curvature, is non-positive if the capsules are convex. In addition, some well known inequalities for volume and area of spheres are extended to Finsler spaces.

We then switch to an entirely different aspect of curvature: on Riemannian surfaces the Gauss-Bonnet Theorem relates the curvature to the excess of a geodesic triangle (i. e., the sum of the angles minus π), and hence to *angular measure.* Since there is no distinguished angular measure in Finsler spaces and none which shares many properties with the euclidean angular measure, the path to the results based on the Gauss-Bonnet Theorem seems barred. However, careful scrutiny reveals that many of these investigations use no other properties of angular measure than finite additivity for angles with the same vertex and that the measure π characterizes straight angles. Hardly any use more than the additional fact that nearly straight angles have, in a uniform way, measures close to π. It is quite easy to define natural angular measures with these properties on Finsler surfaces. In Section 42 we *introduce angular measure axiomatically,* derive its elementary properties and apply them to the theory of parallels, which requires, by the way, continuity of angular measure and a new property, called non-extendability.

By far the deepest and most beautiful applications of the Gauss-Bonnet Theorem are found in Cohn-Vossen [1, 2]. Sections **43** and **44** are devoted to these papers, although we also derive briefly, for comparison, the main properties of surfaces with negative curvature from the assumption that the excess is negative. For polygonal regions D on a G-surface there is a well known relation between the excess $\varepsilon(D)$, i. e., the sum of the excesses of the triangles in a simplicial decomposition of D, and the characteristic of D, which we derive in Section **43**. Cohn-Vossen's main idea is to utilize this relation also for non-compact surfaces by constructing domains D with particularly simple boundaries. High points among Cohn-Vossen's many results are: *the non-existence of straight lines in G-planes with positive excess, a description of the shape of the geodesics in such planes, and the theorem: if the total excess* (= integral curvature in the Riemannian case) *of a G-plane exists and exceeds* π, *then every point outside of a suitable domain is vertex of a simple geodesic monogon containing a given domain.*

36. Local properties

It will prove useful to denote a midpoint of a and b by m_{ab} or $m[a, b]$ so that

$$am_{ab} = m_{ab}b = ab/2.$$

Because of (8.6) and (8.7) the point m_{ab} is unique if and only if $T(a, b)$ is unique.

Non-positive curvature

The rigorous definition of spaces of non-positive, vanishing or negative curvature may then be phrased as follows:

The G-space R has "*non-positive curvature*" if every point p has a neighborhood $S(p, \gamma_p)$, where $0 < \gamma_p < \rho_1(p)$, such that for any three points a, b, c in $S(p, \gamma_p)$ the relation

(36.1) $$2m_{ab}\, m_{ac} \leqslant bc$$

holds.

If, instead of (36.1) always

(36.2) $$2m_{ab}\, m_{ac} = bc,$$

we say that R has "*curvature* 0", and if

(36.3) $2m_{ab}m_{ac} < bc$ when a, b, c are not on one segment, the space R has "*negative curvature*".

The, at first sight, mild convexity requirement (36.1) implies a much stronger convexity property:

(36.4) **Theorem.** *Let $x(\tau)$ and $y(\tau)$ represent geodesics in a space with non-positive curvature. If for suitable real $\alpha_1 < \alpha_2$, β, δ the segment $T[x(\tau), y(\beta\tau+\delta)]$, $\alpha_1 \leqslant \tau \leqslant \alpha_2$, is unique then $\varphi(\tau) = x(\tau)y(\beta\tau + \delta)$ is in $[\alpha_1, \alpha_2]$ a convex function of τ.*

Proof. We conclude from (7.5) that the least upper bound $\gamma(p)$ of the γ_p satisfying (36.1) also satisfies (36.1) and is continuous. (The present theorem implies $\gamma(p) = \rho_1(p)$, see (36.7).) Put $x'(\tau) = y(\beta\tau + \delta)$. The set $V = \bigcup_{\alpha_1 \leqslant \tau \leqslant \alpha_2} T[x(\tau), x'(\tau)]$ is bounded and closed because the uniqueness of $T[x(\tau), x'(\tau)]$ implies that it depends continuously on τ. Therefore $\gamma = \inf_{p \in V} \gamma(p) > 0$. Since the theorem is trivial for $x(\tau) \equiv x'(\tau)$ we may assume that

$$M = x(\tau^*)\, x'(\tau^*) = \max_{\alpha_1 \leqslant \tau \leqslant \alpha_2} x(\tau)\, x'(\tau) > 0.$$

Let $z(\tau^*, u)$, $0 \leqslant u \leqslant M$, represent the oriented segment $\mathfrak{s}[x(\tau^*), x'(\tau^*)]$, (that means $z(\tau^*, 0) = x(\tau^*)$, $z(\tau^*, M) = x'(\tau^*)$).

Denote generally as *linear* the mapping $x \to x'$ of $\mathfrak{s}(a, b)$ on $\mathfrak{s}(a', b')$ for which $ax : ab = a'x' : a'b'$ (or $x' = a'$ if $a' = b'$). If then $z(\tau, u)$ is the image of $z(\tau^*, u)$ under the linear mapping of $\mathfrak{s}[x(\tau^*), x'(\tau^*)]$ on $\mathfrak{s}[x(\tau), x'(\tau)]$, the definition of τ^* implies

(a) $$z(\tau, u_1)\, z(\tau, u_2) \leqslant z(\tau^*, u_1)\, z(\tau^*, u_2) = |u_1 - u_2|, \qquad 0 \leqslant u_i \leqslant M.$$

The point $z(\tau, u)$ depends continuously on both variables τ, u. Therefore $\varepsilon > 0$ exists such that

(b) $$z(\tau_1, u)\, z(\tau_2, u) < \gamma/2 \qquad \text{for} \qquad |\tau_1 - \tau_2| < \varepsilon.$$

Because $\varphi(\tau)$ is continuous, (36.4) is proved if we can show that

(c) $$\varphi\left(\frac{\tau_1 + \tau_2}{2}\right) \leqslant \frac{\varphi(\tau_1) + \varphi(\tau_2)}{2} \qquad \text{for} \qquad |\tau_1 - \tau_2| < \varepsilon$$

Let τ_1 and τ_2 be given such that $|\tau_1 - \tau_2| < \varepsilon$, choose $n > 2\,M/\gamma$ and put

$$a_i = z(\tau_1, i\, n^{-1}M), \qquad b_i = z(\tau_2, i\, n^{-1}M), \quad i = 0, \ldots, n, \text{ so that}$$

$$a_0 = x(\tau_1), \qquad a_n = x'(\tau_1), \qquad b_0 = x(\tau_2), \qquad b_n = x'(\tau_2).$$

The inequalities (a), (b) show that

$$a_i\, b_i < \gamma/2, \qquad a_i\, b_{i \pm 1} < \gamma/2 + n^{-1} M < \gamma$$

so that a_i, b_i, b_{i+1} lie in $S(a_i, \gamma)$ and b_{i+1}, a_i, a_{i+1} in $S(b_{i+1}, \gamma)$. Therefore (36.1) applies to these two triples:

$$\text{(d)} \qquad \begin{aligned} 2\, m[a_i, b_i]\, m[a_i, b_{i+1}] &\leqslant b_i b_{i+1} \\ 2\, m[a_i, b_{i+1}]\, m[a_{i+1}, b_{i+1}] &\leqslant a_i a_{i+1}. \end{aligned}$$

Adding these inequalities we find

$$2\, m[a_0, b_0]\, m[a_n, b_n] \leqslant \sum b_i b_{i+1} + \sum a_i a_{i+1} = a_0 a_n + b_0 b_n$$

which is (c), because $m[a_0, b_0] = x((\tau_1 + \tau_2)/2)$, $m[a_n, b_n] = x'(\tau_1 + \tau_2)/2$.

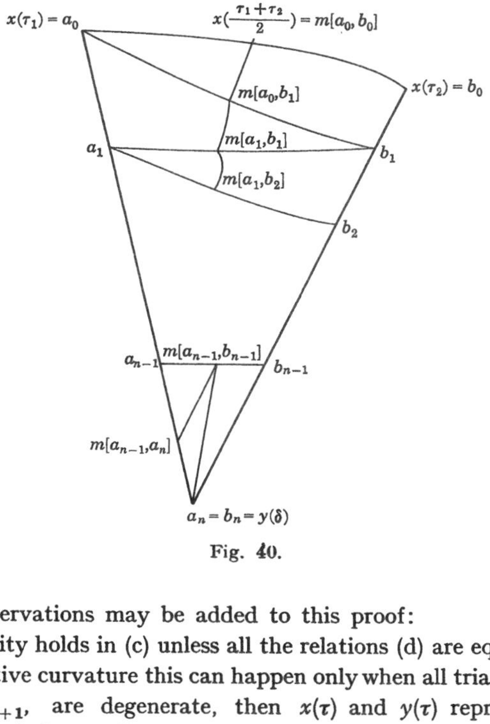

Fig. 40.

Several observations may be added to this proof:

The inequality holds in (c) unless all the relations (d) are equalities. If the space has negative curvature this can happen only when all triangles a_i, b_i, b_{i+1} and $b_{i+1}, a_i\, a_{i+1}$, are degenerate, then $x(\tau)$ and $y(\tau)$ represent the same geodesic provided $\beta \neq 0$. So we have:

(36.5) *If $x(\tau)$ and $y(\tau)$ represent different geodesics in a space with negative curvature and $T[x(\tau), y(\beta\tau + \delta)]$, $\beta \neq 0$ is unique for $\alpha_1 \leqslant \tau \leqslant \alpha_2$, then $x(\tau)\, y(\beta\tau + \delta)$ is in $[\alpha_1, \alpha_2]$ a strictly convex function of τ.*

If $\beta = 0$ in (36.4), then $a_n = b_n = m[a_n, b_n] = y(\delta)$ and we may strengthen the result by adding the inequalities (d) except for the last two, obtaining

$$2\,m[a_0, b_0]\,m[a_{n-1}, b_{n-1}] \leqslant a_0\,a_{n-1} + b_0 b_{n-1}.$$

If the triangle $a_n a_{n-1} b_{n-1}$ is non-degenerate we may then continue:

$$2\,m[a_{n-1}, b_{n-1}]\,a_n < 2\,m[a_{n-1}, b_{n-1}]\,m[a_{n-1}, b_n] + 2\,m\,[a_{n-1}, a_n]\,a_n$$
$$< b_{n-1}\,b_n + a_{n-1}\,a_n.$$

Adding this to the last inequality yields

$$2\,x\left(\frac{\tau_1 + \tau_2}{2}\right) y(\delta) < x(\tau_1)\,y(\delta) + x(\tau_2)\,y(\delta).$$

Hence:

(36.6) *In a space with non-positive curvature, if the segment* $T[p, x(\tau)]$ *is unique for* $\alpha_1 \leqslant \tau \leqslant \alpha_2$ *and any two of the segments have only the point* p *in common, then* $px(\tau)$ *is in* $[\alpha_1, \alpha_2]$ *a strictly convex function of* τ.

Finally we conclude from (36.4)

(36.7) *In a space with non-positive curvature* (36.1) *holds for any triple* a, b, c *in* $S(p, \rho_1(p))$. *The set* $S(p, \rho_1(p))$ *contains* $T(a, b)$ *if it contains* a *and* b, *and the spheres* $K(p, \sigma)$ *with* $\sigma < \rho_1(p)$ *are strictly convex.*

All the statements will follow immediately from (36.4) and (36.6) if $S(p, \eta)$ is convex for $\eta < \rho_1(\rho)$. Denote by δ the least upper bound of the δ' for which $a, b \in S(p, \delta')$ implies $T(a, b) \subset S(p, \rho_1(p))$. The result (36.6) implies that each $S(p, \delta')$ is convex and we know from (6.9) that $\delta \geqslant \rho_1(p)/2$. Assume for an indirect proof that $\delta < \rho_1(p)$. Then sequences $\{a_\nu\}$, $\{b_\nu\}$ with $a_\nu, b_\nu \in S(p, \delta + \nu^{-1})$ exist such that $T(a_\nu, b_\nu)$ contains a point c_ν with $pc_\nu \geqslant \rho_1(p)$. We may assume that $a_\nu \to a$, $b_\nu \to b$, $c_\nu \to c$. There are sequences $a_\nu' \to a$ and $b_\nu' \to b$ with $pa_\nu' < \delta$, $pb_\nu' < \delta$. The segments $T(a_\nu, b_\nu)$ and $T(a_\nu', b_\nu')$ tend, because of $\delta < \rho_1(p)$, to the unique segment $T(a, b)$. Since $a_\nu', b_\nu' \subset S(p, \delta')$ with $\delta' < \delta$ the segment $T(a_\nu', b_\nu')$ lies in $S(p, \delta')$, hence no accumulation point of $\cup\ T(a_\nu', b_\nu')$ can have greater distance from p than δ, but $pc \geqslant \rho_1(p) > \delta$.

Minkowskian trapezoids imbedded in the space

Since segments with endpoints in $S(p, \rho_1(p))$ lie in $S(p, \rho_1(p))$ we conclude from (36.4): If $a \neq b$, a', b' lie in $S(p, \rho_1(p))$ and $x \to x'$ maps $\mathfrak{s}(a, b)$ linearly on $\mathfrak{s}(a', b')$, then

(36.8) $$xx' \leqslant \frac{xb}{ab}\,aa' + \frac{ax}{ab}\,bb',$$

The equality for a single x between a and b has remarkably strong and important consequences:

(36.9) **Theorem.** *In a G-space with non-positive curvature let* a, b, a', b' *be four non-collinear points in* $S(p, \rho_1(p))$ *with* $a \neq b$. *If* $x \to x'$ *maps* $\mathfrak{s}(a, b)$ *linearly on* $\mathfrak{s}(a', b')$ *and an* x_0 *with* (ax_0b) *exists for which*

$$x_0x_0' = \frac{x_0b}{ab} aa' + \frac{ax_0}{ab} bb',$$

then $V = \cup_x T(x, x')$ *is isometric to a trapezoid in a Minkowski plane with* $T(a, a')$ *and* $T(b, b')$ *as parallel sides* (the trapezoid is a triangle when $a = a'$ or $b = b'$).

Proof. For $z \in T(x, x')$ put $ax = \xi_z$, $xz = \eta_z$. We know from (36.4) that $\eta_{x'}$ is a convex function of ξ_x. The hypothesis states that one point of the graph of this function (in a (ξ, η)-plane) other than the endpoints lies on the chord connecting the endpoints. Hence the graph coincides throughout with this chord, or $\eta_{x'}$ is a linear function of ξ_x. Therefore if $u, v \in V$ and $\xi_u \neq \xi_v$, $u \in T(\bar{u}, \bar{u}')$, $v \in T(\bar{v}, \bar{v}')$ (hence $\xi_u = \xi_{\bar{u}} = \xi_{\bar{u}'}$, $\xi_v = \xi_{\bar{v}} = \xi_{\bar{v}'}$) and $(\bar{u}\, x\, \bar{v})$

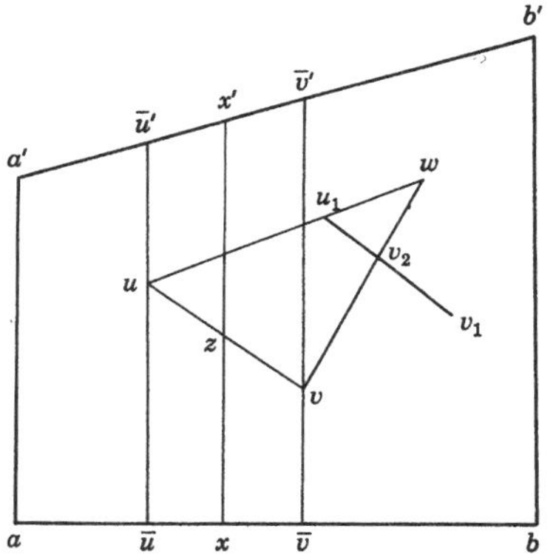

Fig. 41.

$$\eta_{x'} = \frac{\xi_v - \xi_x}{\xi_v - \xi_u} \eta_{\bar{u}'} + \frac{\xi_x - \xi_u}{\xi_v - \xi_u} \eta_{\bar{v}'}.$$

(36.10)

Denote by z the image of x under the linear mapping of $\mathfrak{s}(\bar{u}, \bar{v})$ on $\mathfrak{s}(u, v)$ and hence also the image of x' under the linear mapping of $\mathfrak{s}(\bar{u}', \bar{v}')$ on $\mathfrak{s}(u, v)$. Then by definition

(36.11) $uz : zv : uv = \bar{u}'x' : x'\bar{v}' : \bar{u}'\bar{v}' = \bar{u}x : x\bar{v} : \bar{u}\bar{v} = (\xi_x - \xi_u) : (\xi_v - \xi_x) : (\xi_v - \xi_u)$

and we conclude from (36.8) that

$$xz \leqslant \frac{\xi_v - \xi_x}{\xi_v - \xi_u} \eta_u + \frac{\xi_x - \xi_u}{\xi_v - \xi_u} \eta_v$$

$$x'z \leqslant \frac{\xi_v - \xi_x}{\xi_v - \xi_u} (\eta_{\bar{u}'} - \eta_u) + \frac{\xi_x - \xi_u}{\xi_v - \xi_u} (\eta_{\bar{v}'} - \eta_v)$$

Since $\eta_{x'} = xx' \leqslant xz + zx'$, adding these two inequalities and comparing with (36.10) yields $xx' = xz + zx'$ or $z \in T(x, x')$, hence $\xi_x = \xi_z$, $xz = \eta_z$ and

$$\eta_z = \frac{\xi_v - \xi_z}{\xi_v - \xi_u}\eta_u + \frac{\xi_z - \xi_u}{\xi_v - \xi_u}\eta_v.$$

Moreover, we see that $u, v \in V$ and $z \in T(u, v)$ implies $T(u, v) \subset V$, so that V is a convex set. Interpreting (ξ_z, η_z) as a point of a cartesian (ξ, η)-plane V becomes the trapezoid

$$0 \leqslant \xi \leqslant ab = \xi_b, \qquad 0 \leqslant \eta \leqslant \frac{\xi_b - \xi}{\xi_b}\eta_{a'} + \frac{\xi}{\xi_b}\eta_{b'}$$

If V is remetrized by the euclidean distance

$$e(u, v) = [(\xi_u - \xi_v)^2 + (\eta_u - \eta_v)^2]^{\frac{1}{2}}$$

then

$$e(u, v) = |\eta_u - \eta_v| \quad \text{if } \xi_u = \xi_v, \text{ and by (36.11):}$$

$$\frac{uz}{uv} = \frac{\xi_z - \xi_u}{\xi_v - \xi_u} = \frac{e(u, z)}{e(u, v)} \quad \text{if } (uzv) \text{ and } \xi_u \neq \xi_v.$$

This means that for every non-degenerate segment T in V a factor $\varphi > 0$ exists such that $uv = \varphi\, e(u, v)$ for $u, v \in T$. Since we may choose T as the entire intersection of a cartesian line with V the factor φ is the same for segments on the same line. To show that the metric uv is Minkowskian it remains to be shown that for two non-degenerate segments $T(u, v)$ and $T(u_1, v_1)$ on distinct parallel lines the corresponding factors φ and φ_1 are equal. For reasons of continuity we may assume that both segments lie in the interior of V (considered as space). Let the notation be such that $\mathfrak{s}(u, v)$ and $\mathfrak{s}(u_1, v_1)$ have equal (and not opposite) directions. To show $\varphi_1 \leqslant \varphi$ say, choose w with (uu_1w) so close to u_1 that $T(w, v)$ intersects $T(u_1, v_1)$ in a point v_2. Then by (36.8)

$$\varphi_1\, e(u_1, v_2) = u_1 v_2 \leqslant \frac{w\, u_1}{w\, u}\, u\, v = \varphi\, \frac{e(w, u_1)}{e(w, u)}\, e(u, v) = \varphi\, e(u_1, v_2).$$

A last corollary of (36.4) and (36.9) is:

(36.12) *If in a space with non-positive curvature a, b, c, d are points of $S(p, \rho_1(p))$, then the foot f_x of a point $x \in T(a, b)$ on $T(c, d)$ is unique and xf_x is a convex function of ax. If the space has negative curvature then xf_x takes a given positive value at most twice.*

If x, c, d lie on one segment, then xy reaches for $y \in T(c, d)$ its minimum either at x or at c or at d and nowhere else, hence f_x is unique. If x, c, d do not lie on a segment, then (36.6) proves the uniqueness of f_x.

Theorem (36.4) yields for $x, y \in T(a, b)$

(36.13)
$$2\, m_{xy} f_{m_{xy}} \leqslant 2\, m_{xy}\, m\, [f_x, f_y] \leqslant x f_x + y f_y,$$

which shows that $x f_x$ is a convex function of ax.

Assume

$$\alpha = u f_u = v f_v = w f_w > 0 \qquad \text{for} \qquad u, v, w \in T(a, b) \quad \text{and} \quad (uvw).$$

Then the argument at the beginning of the proof of (36.9) shows that $x f_x$ is linear for $x \in T(u, w)$, hence $x f_x \equiv \alpha$. Now (36.13) yields for $x, y \in T(u, w)$ that

$$m_{xy} f_{m_{xy}} = m_{xy}\, m[f_x, f_y].$$

Since the foot $f_{m_{xy}}$ of m_{xy} is unique, it must coincide with $m[f_x, f_y]$ so that $x \to f_x$ is the linear mapping of $T(u, w)$ on $T(f_u, f_w)$. It follows from $x f_x \equiv \alpha$ and (36.9) that $\bigcup_{x \in T(u,w)} T(x, f_x)$ is isometric to a trapezoid (in this case a parallelogram) of a Minkowski plane. The curvature of the space can therefore not be negative.

Non-positive curvature and convex capsules

This theorem and its proof imply:

(36.14) *If the segments $T(a, b)$ and $T = T(c, d)$ of a space with non-positive curvature lie in $S(p, \rho_1(p))$, then xT is for $x \in T(a, b)$ a peakless function of ax. This function is strictly peakless if $T(a, b) \cap T(c, d)$ is not a proper segment and the curvature is negative.*

We may give this statement a more intuitive form by introducing the notion of the "*capsule*" C_T^α with radius $\alpha > 0$ and the segment T as axis. C_T^α consists by definition of the points x with $xT \leqslant \alpha$. The following lemma is needed in the discussion:

(36.15) If $C_T^\alpha \subset S(p, \rho(p))$ and $C_T^{\alpha'}$ is convex for $0 < \alpha' \leqslant \alpha$ then the points x with $xT = \alpha$ are the boundary points of C_T^α.

For the set $xT < \alpha$ is open and lies in C_T^α, consequently only points with $xT = \alpha$ can be boundary points of C_T^α. If a point x with $xT = \alpha$ were an interior point of C_T^α, let f_x be a foot of x on T. Points z with (zxf_x) exist and lie for sufficiently small zx also in C_T^α. If f_z is a foot of z on T, then $zf_z \leqslant \alpha$, hence $f_z \neq f_x$. For (zuf_z) and sufficiently small xz and uz choose u' with (uxu') and $xu' = \frac{1}{2}\min(\rho(x), \alpha)$. For $u \to z$ the point u' tends to the point v on $T(x, f_z)$ with $xv = xu'$. Then for sufficiently small uz and zx

$$u'T = u'f_x \leqslant u'v + vf_x < xf_x - xz = \alpha' < \alpha$$

and $uT = zf_z - uz = \alpha'' < \alpha$. Then the capsule $C_T^{\max(\alpha', \alpha'')}$ would not be convex, because it contains u and u', without containing x.

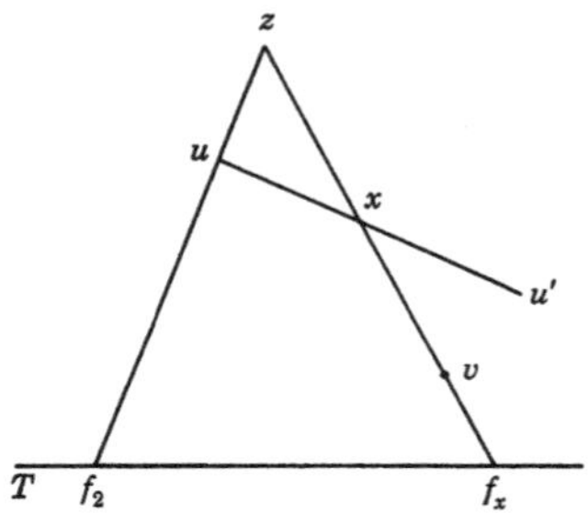

Fig. 42.

We now show:

(36.16) *The following two properties are equivalent for an arbitrary point p of a G-space:*

A. *There is a positive $\eta \leqslant \rho_1(p)$ such that for any two segments $T(a, b)$ and $T = T(c, d)$ in $S(p, \eta)$ the function $\varphi(ax) = xT$ is, for $x \in T(a, b)$, peakless (strictly peakless, unless $T \cap T(a, b)$ is a proper segment.* If this is always the case we speak of the strict property A.)

B. *There is a positive $\eta' \leqslant \rho_1(p)$ such that every capsule $C_T^\alpha \subset S(p, \eta')$ is (strictly) convex.*

Let A hold and put $\eta' = \eta/2$. If $C_T^\alpha \subset S(p, \eta')$ and $a, b \in C_T^\alpha$, then $T(a, b) \subset S(p, \eta)$. Since obviously $T \subset S(p, \eta)$ the function $\varphi(ax)$ is peakless so that $xT \leqslant \max(aT, bT) \leqslant \alpha$ or $T(a, b) \subset C_T^\alpha$. If a, b are boundary points of C, then the lemma shows $aT = bT = \alpha$. The set $T \cap T(a, b)$ is then not a proper segment, so that $xT < \alpha$, when $\varphi(ax)$ is strictly peakless and $(a\,x\,b)$. Thus C_T^α is strictly convex.

Let B hold and put $\eta = \frac{1}{2}\min(\eta', \rho(p))$. If $T(a, b)$ and $T = T(c, d)$ lie in $S(p, \eta)$ and $\varphi(ax)$ were not peakless, then $T(a, b)$ would contain points u, v, w with (uvw) and either $\alpha = vT > \max(uT, wT) = \alpha_0$ or $\alpha = vT = uT > wT$. Since $\alpha \leqslant vc \leqslant 2\eta'$, the set $C_T^{\alpha'}$ lies for $\alpha' \leqslant \alpha$ in $S(p, \min(\rho(p), \eta))$ and is convex. The first case is impossible because $C_T^{\alpha_0}$ would contain u and w but leave out v. In the second case $u \in C_T^\alpha$, and w is an interior point of C_T^α. By (20.1) v would be an interior point of C_T^α, but (36.15) shows that v is a boundary point. The strengthening for strictly convex C_T^α is obvious.

IMPLICATIONS OF PROPERTY A

Property B has an intuitively appealing content, but A is easier to handle. We now establish a fact with an import similar to that of (36.4).

(36.17) **Theorem.** *Let Property A hold for every point of a G-space. If $Q = T(a, b)$ and $T = T(c, d)$ are segments such that $T(x, y)$ is unique for $x \in Q$ and $y \in T$, then $\varphi(ax) = x\,T$ is for $x \in T(a, b)$ a peakless function. $\varphi(ax)$ is strictly peakless when the Strict Property A holds and $Q \cap T$ is not a proper segment.*

We prove at the same time:

(36.18) *If Property A holds for every point of a G-space and the segments $T(p, x)$ are unique for $x \in T(a, b)$, then px is for $x \in T(a, b)$ a strictly peakless function of ax. Hence the foot of p on $T(a, b)$ is unique.*

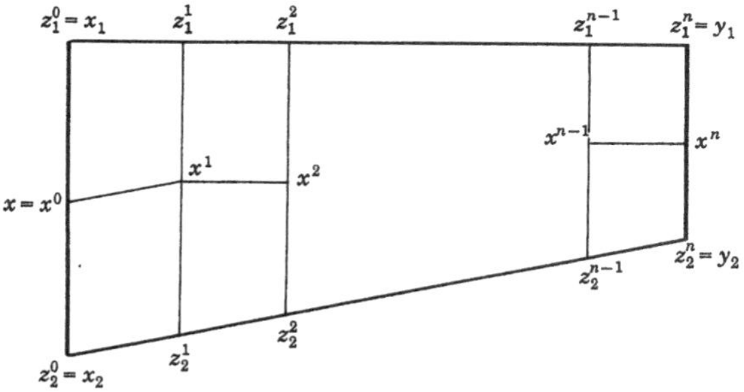

Fig. 43.

If for every point p of the space a positive $\eta_p \leqslant \rho_1(p)$ exists such that A holds in $S(p, \eta_p)$ then (7.5) shows that the least upper bound $\eta(p)$ of all $\eta_p \leqslant \rho_1(p)$ satisfying A still satisfies A and that either $\eta(p) \equiv \infty$ or $|\eta(p) - \eta(q)| \leqslant pq$.

The set $V = \bigcup_{\substack{x \in Q \\ y \in T}} T(x, y)$ is compact, hence $3\,\eta = \inf_{p \in V} \eta(p) > 0$.

There is an ε, $0 < \varepsilon < \eta$, such that for $x_1, x_2 \in Q$, $x_1x_2 < \varepsilon$ and $y_1, y_2 \in T$, $y_1y_2 < \varepsilon$ the distance of any point z_1 on $T(x_1, y_1)$ from its image z_2 under the linear mapping of $\mathfrak{s}(x_1, y_1)$ on $\mathfrak{s}(x_2, y_2)$ is smaller than η.

Let $x, x_1, x_2 \in Q$ and (x_1xx_2) with $x_1x_2 < \varepsilon$, and denote by T_0 any subsegment of T with length less than ε. Let y_1, y, y_2 be feet on T_0 of x_1, x, x_2 respectively. Let $n > 3\,\eta^{-1} \max(x_1y_1, x_2y_2)$. If $z_i(\tau)$ represents $\mathfrak{s}(x_i, y_i)$, $i = 1, 2$ put

$$z_i^k = z_i(kx_iy_i/n), \qquad i = 1, 2, \qquad k = 0, \ldots, n.$$

Then $T(z_1^{i+1}, z_2^{i+1}) \subset S(z_1^i, 3\eta)$. Finally put $x^0 = x \in T(x_1, x_2) = T(z_1^0, z_2^0)$ and for $i \geqslant 0$ let x^{i+1} be a foot of x^i on $T(z_1^{i+1}, z_2^{i+1})$.

Property A applied to $S(x^i, 3\eta)$ yields

(a) $$x^i x^{i+1} = x^i T(z_1^{i+1}, z_2^{i+1}) \leqslant \max(z_1^i z_1^{i+1}, z_2^i z_2^{i+1}) = \max(x_1y_1, x_2y_2)/n$$

and equality implies

$$x^i x^{i+1} = x_1y_1/n = x_2y_2/n.$$

By addition we obtain

(b) $$x\, T_0 \leqslant xx^n \leqslant \sum_{i=0}^{n-1} x^i x^{i+1} \leqslant \max(x_1y_1, x_2y_2) = \max(x_1T_0, x_2T_0);$$

equality implies equality in each relation (a) and $xT_0 = x_1T_0 = x_2T_0$.

It now follows that xT_0 is a peakless function of ax for $x \in Q$. For otherwise points u, v, w with (uvw) on Q would exist such that either $vT_0 > \max(uT_0, wT_0)$ or $uT_0 = vT_0 > wT_0$. In the first case let x be the last point on $\mathfrak{s}(u, w)$ at which $x'T_0$ reaches its maximum for $x' \in \mathfrak{s}(u, w)$. Then $xT_0 \geqslant vT_0$, hence $x \neq u, w$. Choose x_1, x_2 on $\mathfrak{s}(u, w)$ such that (x_1xx_2) and $x_1x_2 < \varepsilon$. Then $x_1T_0 < zT_0$, $x_2T_0 < xT_0$ which contradicts (b).

This implies for the second case that $uT_0 = vT_0$ is the maximum of $x'T_0$ for $x' \in \mathfrak{s}(u, v)$. If x is chosen as before we obtain the same contradiction.

Applying this to the case where T_0 degenerates into a point p we find that px is a peakless function of ax for $x \in Q$. But (36.18) states that px is strictly peakless. To see this observe that the assumptions of (36.17) are satisfied in $S(p, \eta(p)/2)$. It follows from (20.5) that $K(p, \sigma)$ is strictly convex for sufficiently small positive σ.

If px were not strictly peakless it would reach its minimum on a nondegenerate subinterval of Q. If $T(x_1, x_2)$ lies on this interval and $0 < x_1x_2 < \varepsilon$, then $px_1 = px_2$. Increasing the number n of the last proof, if necessary, we reach that $pz_i^{n-1} = \sigma$ with strictly convex $K(p, \sigma)$. If x^{n-1} is the foot p on $T(z_1^{n-1}, z_2^{n-1})$, then $px^{n-1} < pz_1^{n-1} = pz_2^{n-1}$, and if x^{i-1} is for $i \leqslant n - 1$ a foot of x^i on $T(z_1^{i-1}, z_2^{i-1})$ we obtain by addition

$$px_0 < px_1 = px_2$$

so that px is not constant on $T(x_1, x_2)$.

It is now easy to complete the proof of (36.17). Since the hypotheses regarding Q and T are symmetric it follows from (36.18) that every point u

of Q has a unique foot f_u on T. Then f_u depends continuously on u and we can find a positive ε' such that

$$f_u f_v < \varepsilon/3 \qquad \text{for} \qquad uv < \varepsilon'.$$

If $\varphi(ax) = xT$ were not peakless then, as above, three points x, x_1, x_2 on T with $(x_1 x x_2)$ and $x_1 x_2 < \varepsilon'$ would exist such that $x_1 T \leqslant xT$, $xT > x_2 T$. The choice of ε' guarantees that the feet f_u lie for $u \in T(x_1, x_2)$ in $S(x_1, \varepsilon/3)$ hence on a subsegment T_0 of T with length less than ε. Then $uT = uf_u = uT_0$ would not be peakless, contrary to what has already been proved.

The remark in (36.17) on strict peaklessness is obvious because inequality in a single relation (a) produces inequality in (b).

Exactly the same argument as for (36.7) shows:

(36.19) *If Property A holds for every point, then $\eta(p) = \rho_1(p)$. The set $S(p, \rho_1(p))$ contains with a and b the segment $T(a, b)$ and $K(p, \sigma)$ is strictly convex for $\sigma < \rho_1(p)$.*

We now reformulate these results in terms of Property B. If this property holds for every point then (36.15) shows that A holds everywhere. (36.19) guarantees that the assumptions of (36.17) are satisfied for segments $T(a, b)$ and $T(c, d)$ with endpoints in $S(p, \rho_1(p))$, hence every capsule C_T^α in $S(p, \rho_1(p))$ is convex (and using (36.15), strictly convex if they are locally strictly convex).

If every point of a G-space has a neighborhood in which the capsules are (strictly) convex we simply say that *the space has convex capsules.* Thus we have proved:

(36.20) **Theorem.** *If every point p has a neighborhood $S(p, \eta_p')$, $0 < \eta_p' \leqslant \rho_1(p)$ such that the capsules C_T^α in $S(p, \eta_p')$ are (strictly) convex, then any capsule contained in a sphere $S(p, \rho_1(p))$ is (strictly) convex. Each sphere $K(p, \sigma)$ with $0 < \sigma < \rho_1(p)$ is strictly convex.*

A space with non-positive (negative) curvature has (strictly) convex capsules.

The following remark further emphasizes the implications of Properties A and B.

(36.21) *If in a G-space the capsules C_T^α in $S(p, \eta)$, $0 < \eta \leqslant \rho_1(p)$, are convex and $S(Q, \rho) \subset S(p, \eta)$, where Q is a convex set, then $S(Q, \rho)$ is convex. If the closure of $S(Q, \rho)$ lies in $S(p, \eta)$ and the C_T^α are strictly convex, then $S(Q, \rho)$ is strictly convex.*

For if $a, b \in S(Q, \rho)$, then points a', b' in Q with $aa' < \rho$ and $bb' < \rho$ exist. We may assume that $\alpha = \max(aa', bb') > 0$. Because Q is convex $T = T(a', b') \subset Q$, and because $\alpha < \rho$ we have $C_T^\alpha \subset S(Q, \rho) \subset S(p, \eta)$. Therefore C_T^α is convex and $T(a, b) \subset C_T^\alpha \subset S(Q, \rho)$, so that $S(Q, \rho)$ is convex.

The addition on strict convexity is readily verified.

37. Non positive curvature in the theory of parallels

For a straight space R all the preceding statements hold in the large since $\rho(p) \equiv \rho_1(p) \equiv \infty$. They enable us to establish a satisfactory theory of parallels for straight spaces with non-positive curvature, and for spaces with convex capsules which satisfy an additional condition.

The general case

We begin by showing

(37.1) ***Theorem.*** *Let R be a straight space with convex capsules. If $y(\tau)$ represents the oriented asymptote $\mathfrak{a}^+$ to $\mathfrak{g}^+$ then $y(\tau)\,\mathfrak{g}$ is decreasing or constant. The latter case enters when $\mathfrak{a}$ is parallel to $\mathfrak{g}$. If $\mathfrak{r}$ and $\mathfrak{s}$ are positive subrays of $\mathfrak{g}^+$ and $\mathfrak{a}^+$, then $x\mathfrak{s}$ and $y\mathfrak{r}$ are bounded for $x \in \mathfrak{r}$ and $y \in \mathfrak{s}$.*

If the capsules are strictly convex then no parallel to $\mathfrak{g}$ (other than $\mathfrak{g}$) exists.

Proof. Let $y_\nu(\tau)$ represent the line $\mathfrak{g}^+(y(0), x(\nu))$ such that $y_\nu(0) = y(0)$ and put $y(0)x(\nu) = \tau_\nu$. Since $\mathfrak{g}^+(y(0), x(\nu))$ tends to $\mathfrak{a}^+$, we conclude from (8.12) that $y_\nu(\tau) \to y(\tau)$ for all τ. The functions $y(\tau)\mathfrak{g}$ and $y_\nu(\tau)\mathfrak{g}$ are non-negative and peakless (compare the proof of (18.12)), moreover $y_\nu(\tau_\nu)\mathfrak{g} = 0$. We may assume that $y(0)$ does not lie on $\mathfrak{g}$, otherwise $\mathfrak{a}^+ = \mathfrak{g}^+$. Then $y(0)\mathfrak{g} > 0$, hence $y(\tau)$ decreases for $-\infty < \tau \leqslant \tau_\nu$. It follows from $\tau_\nu \to \infty$, $y_\nu(\tau) \to y(\tau)$ and the peaklessness of $y(\tau)\mathfrak{g}$ that $y(\tau)\mathfrak{g}$ decreases or is constant.

When $\mathfrak{a}$ is parallel to $\mathfrak{g}$ then this result shows that $y(\tau)\mathfrak{g}$ is both non-decreasing and non-increasing, hence constant. If the capsules are strictly convex; then $y(\tau)\mathfrak{g}$ is strictly peakless, hence cannot be constant.

Since the sets $S(\mathfrak{r}, \rho)$ are also convex it follows that $y(\tau)\mathfrak{r}$ is peakless and then as above, that it is non-increasing, hence bounded for $\tau \geqslant 0$. If $x(\pi_\tau)$ is the foot of $y(\tau)$ on $\mathfrak{r}$ then for $\tau \to \infty$

$$\pi_\tau = x(0)\,x(\tau_\tau) \geqslant y(0)\,y(\pi) - x(0)\,y(0) - y(\tau)\,x(\pi_\tau) \to \infty$$

because $y(\tau)x(\pi_\tau)$ is bounded. The uniqueness of the foot $x(\pi_\tau)$ of $y(\tau)$ implies that π_τ depends continuously on τ, hence takes all values τ from a certain τ_1 on, or, if $\tau \geqslant \tau_1$ is given then τ' exists such that $\pi_{\tau'} = \tau$. Then

$$x(\tau)\,\mathfrak{s} \leqslant x(\tau)\,y(\tau') = y(\tau')\,\mathfrak{r}$$

shows that $x(\tau)\mathfrak{a}$ is bounded for $\tau \geqslant 0$.

The boundedness of $x\mathfrak{s}$ for $x \in \mathfrak{r}$ and of $y\mathfrak{r}$ for $y \in \mathfrak{s}$ does however not imply that one of them is a co-ray to the other, as the examples (25.3 b c) show. Also, in a Hilbert geometry with one segment T on the boundary of the domain any two rays $\mathfrak{r}$, $\mathfrak{s}$ with the same origin and interior points of T as endpoints have the property that for $x \in \mathfrak{r}$ and $y \in \mathfrak{s}$ the distances $x\mathfrak{s}$ and $y\mathfrak{r}$ are bounded, compare Section 18.

Implications of the Divergence Property

It suffices to exclude the latter case to develop the theory of parallels. We introduce the Divergence Property formulated in Section 34. In the present case it is equivalent to the weaker condition: If $\mathfrak{r}$ and $\mathfrak{s}$ are distinct rays with the same origin, then $x\mathfrak{s}$ is not bounded for $x \in \mathfrak{r}$.

This property holds in a Hilbert geometry defined in a strictly convex domain of the affine space. Also

(37.2) *A straight space with non-positive curvature has the divergence property.*

For if $\mathfrak{g}$ is the line carrying $\mathfrak{s}$ and $x(\tau)$, $\tau \geqslant 0$, represents $\mathfrak{r}$ then $x(\tau)\mathfrak{g}$ is by (36.12) a convex function which vanishes for $\tau = 0$ and is, unless $\mathfrak{s}$ is the opposite ray to $\mathfrak{r}$ in which case $x(\tau)\mathfrak{s} \to \infty$ is obvious, positive for $\tau \neq 0$. It follows that $x(\tau)\mathfrak{s} \geqslant x(\tau)\mathfrak{g} \to \infty$ for $\tau \to \infty$.

We now prove:

(37.3) *In a straight space with convex capsules and the divergence property the ray $\mathfrak{s}$ is a co-ray to the ray $\mathfrak{r}$ if and only if $y\mathfrak{r}$ is bounded for $y \in \mathfrak{s}$.*

We proved already that $y\mathfrak{r}$ is bounded when $\mathfrak{s}$ is a co-ray to $\mathfrak{r}$. To see the converse let $\mathfrak{s}'$ be the co-ray to $\mathfrak{r}$ from the origin p of $\mathfrak{s}$. We have to show that $\mathfrak{s}' = \mathfrak{s}$. If x is the foot of the arbitrary point $y' \in \mathfrak{s}'$ on $\mathfrak{r}$, and y is the foot of x on $\mathfrak{s}$, then we know from (37.1) that both $y'x$ and xy are bounded, hence $y'y \leqslant y'x + xy$ is bounded and $\mathfrak{s}' = \mathfrak{s}$ by the divergence property.

It is now easy to prove:

(37.4) **Theorem.** *In a straight space with convex capsules and the divergence property — and therefore in a space with non-positive curvature — the asymptote relation is symmetric and transitive.*

Let $x(\tau)$ and $y(\tau)$ represent the oriented lines $\mathfrak{g}^+$ and $\mathfrak{h}^+$, then $\mathfrak{g}^+$ and $\mathfrak{h}^+$ are asymptotes to each other if and only if one of the following two conditions holds:

(a) $x(\tau)y(\tau)$ *is bounded for* $\tau \geqslant 0$.

(b) $x(\tau)\mathfrak{h}$ *and* $y(\tau)\mathfrak{g}$ *are bounded for* $\tau \geqslant 0$.

We begin with (a): if (a) holds and $\mathfrak{r}$, $\mathfrak{s}$ are the subrays $\tau \geqslant 0$ of $\mathfrak{g}^+$ and $\mathfrak{h}^+$, then $x(\tau)\mathfrak{s}$ and $y(\tau)\mathfrak{r}$ are bounded for $\tau \geqslant 0$, hence $\mathfrak{r}$ and $\mathfrak{s}$ are by (37.3) co-rays to each other.

Let $\mathfrak{g}^+$ be an asymptote to $\mathfrak{h}^+$. If $y(\pi_\tau)$ is the foot of $x(\tau)$ on $\mathfrak{s}$, then $x(\tau)y(\pi_\tau)$ is by (37.3) bounded for $\tau \geqslant 0$, say $x(\tau)y(\pi_\tau) \leqslant M$. Then

$$(37.5) \qquad y(\tau)y(\pi_\tau) = |\tau - \pi_\tau| = x(0)x(\tau) - y(0)y(\pi_\tau) \leqslant$$
$$\leqslant x(0)y(0) + x(\tau)y(\pi_\tau) \leqslant x(0)y(0) + M$$

and

$$x(\tau)y(\tau) \leqslant x(\tau)y(\pi_\tau) + y(\pi_\tau)y(\tau) \leqslant 2M + x(0)y(0).$$

The symmetry of the asymptote relation is now obvious and transitivity too, because, if $x(\tau)$ represents an asymptote $\mathfrak{g}^+$ to $\mathfrak{h}^+$, and $\mathfrak{h}^+$, represented by $y(\tau)$, is an asymptote to $\mathfrak{k}^+$ given by $z(\tau)$, then

$$x(\tau)\, z(\tau) \leqslant x(\tau)\, y(\tau) + y(\tau)\, z(\tau)$$

is bounded, hence $\mathfrak{g}^+$ is an asymptote to $\mathfrak{k}^+$.

The necessity of (b) is also clear. A little caution must be exerted in the proof of the sufficiency because of the possibility that $\mathfrak{g}$ and $\mathfrak{h}$ are parallel. Let $y(\pi_\tau)$ be the foot of $x(\tau)$ on $\mathfrak{h}$ and $x(\tau)y(\pi_\tau) \leqslant M$. Then it is seen as in (37.5) that

$$\bigl|\tau - |\pi_\tau|\bigr| \leqslant x(0)\, y(0) + M.$$

Since π_τ depends continuously on τ, either $\pi_\tau \to \infty$ or $\pi_\tau \to -\infty$ for $\tau \to \infty$. In the first case $x(\tau)\,\mathfrak{s}$ is bounded for $\tau \geqslant 0$, hence $\mathfrak{r}$ is by (37.3) a co-ray to $\mathfrak{s}$. Because of the symmetry, $\mathfrak{g}^+$ and $\mathfrak{h}^+$ are asymptotes to each other.

In the second case denote by $\mathfrak{s}'$ the ray $y'(\tau) = y(-\tau)$, $\tau \geqslant 0$ opposite to $\mathfrak{s}$. Then $\mathfrak{s}'$ and $\mathfrak{r}$ are co-rays to each other. Since $y(\tau)\mathfrak{g}$ is by assumption also bounded for $\tau \geqslant 0$, it follows that $\mathfrak{s}$ is a co-ray to either $\mathfrak{r}$ or to the opposite ray $\mathfrak{r}'$. If $\mathfrak{s}$ were a co-ray to $\mathfrak{r}$, then it would by transitivity also be a co-ray to $\mathfrak{s}'$, which is impossible, hence $\mathfrak{s}$ is a co-ray to $\mathfrak{r}'$ and $\mathfrak{g}$ and $\mathfrak{h}$ are parallels.

Notice the corollary:

(37.6) *In a straight space with convex capsules and the divergence property the two lines* $\mathfrak{g}$ *and* $\mathfrak{h}$ *are parallels, if and only if they have one (and then all) of the following properties*

(a′) $x(\tau)y(\tau)$ *is bounded for suitable representations* $x(\tau)$ *of* $\mathfrak{g}$ *and* $y(\tau)$ *of* $\mathfrak{h}$.

(b′) $x\mathfrak{h}$ *and/or* $y\mathfrak{g}$ *is bounded for* $x \in \mathfrak{g}$ *and* $y \in \mathfrak{h}$.

(c′) $x\mathfrak{h}$ *and/or* $y\mathfrak{g}$ *is constant for* $x \in \mathfrak{g}$ *and* $y \in \mathfrak{h}$.

It suffices to know that $x\mathfrak{h}$ is bounded because the argument in the last proof shows that for a suitable representation $y(\tau)$ of $\mathfrak{h}$ both $\pi_\tau \to \infty$ for $\tau \to \infty$ and $\pi_\tau \to -\infty$ for $\tau \to \infty$. (c′) is a consequence of (37.1).

In a space with non-positive curvature (37.6) may be strengthened.

(37.7) *If* $x(\tau)$ *and* $y(\tau)$ *represent parallels in a straight space with non positive curvature, then for a suitable* $\eta = \pm 1$ *the function* $x(\tau)y(\eta\,\tau)$ *is constant and* $\bigcup_\tau T[x(\tau), y(\eta\,\tau)]$ *is isometric to a strip of a Minkowski plane bounded by parallel lines.*

For $x(\tau)y(\eta\,\tau)$ is convex, and since it is bounded it must be constant. The second statement is then a consequence of (36.9).

We will later need the following lemma which slightly strengthens the preceding results:

(37.8) In a straight space with convex capsules and the divergence property, if $px_\nu \to \infty$ and $\mathfrak{g}^+(p, x_\nu) \to \mathfrak{g}^+$ then $\mathfrak{g}^+(q, x_\nu)$ tends for any point q to an asymptote to $\mathfrak{g}^+$.

Since the asymptote to $\mathfrak{g}^+$ through q is unique, it suffices to prove (37.8) for a given subsequence $\{\lambda\}$ of $\{\nu\}$ for which $\mathfrak{g}^+(q, x_\lambda)$ converges to a line $\mathfrak{h}^+$. Let $y_\lambda(\tau)$ represent $\mathfrak{g}^+(q, x_\lambda)$ with $y_\lambda(0) = q$. If $\mathfrak{r}_\lambda$ is the positive subray of $\mathfrak{g}^+(p, x_\lambda)$ with origin p, then $\mathfrak{r}_\lambda$ tends to the positive subray $\mathfrak{r}$ of $\mathfrak{g}^+$ with origin p. The function $y_\lambda(\tau)\mathfrak{r}_\lambda$ is non-increasing for $0 \leqslant \tau \leqslant qx_\lambda$ because it is peakless and reaches for $\tau = qx_\lambda$ its minimum 0. But $\lim y_\lambda(\tau) = y(\tau)$ exists and represents $\mathfrak{h}^+$, and $\lim y_\lambda(\tau)\mathfrak{r}_\lambda = y(\tau)\mathfrak{r}$. Since $y(\tau)\mathfrak{r}$ is non-increasing for $\tau \geqslant 0$, it is bounded for $\tau \geqslant 0$, so that $y(\tau)$, $\tau \geqslant 0$ is a co-ray to $\mathfrak{r}$ and $\mathfrak{h}^+$ is an asymptote to $\mathfrak{g}^+$.

Axial motions

We now combine the present results with those of Section 32. We conclude from (32.4) and (37.1):

(37.9) *An axial motion of a straight space with strictly convex capsules has exactly one axis.*

Stronger statements can be made when the space has non-positive or negative curvature.

(37.10) *If $\Phi \neq E$ is a motion of a space R with non-positive curvature for which $xx\Phi$ is bounded, then $xx\Phi$ is constant. Every point p lies on an axis $\mathfrak{g}_p$. A point q not on $\mathfrak{g}_p$ determines with $\mathfrak{g}_p$ a Minkowski plane P in R and Φ^k is a translation of P along $\mathfrak{g}_p$.*

If $x(\tau)$ represents any line $\mathfrak{g}$, then $x(\tau)x(\tau)\Phi$ is bounded, and since it is convex it is constant, also, $\mathfrak{g}$ is parallel to $\mathfrak{g}\Phi$. Since any two points can be connected by a line it follows that $xx\Phi$ is constant, but not 0 because $\Phi \neq E$. Therefore trivially $pp\Phi = \min xx\Phi$, so that p lies on an axis $\mathfrak{g}_p$. If q does not lie on $\mathfrak{g}_p$, then $\mathfrak{g}(p, q)\,\Phi = \mathfrak{g}(p\Phi, q\Phi)$ is parallel to $\mathfrak{g}(p, q)$. The two lines bound by (37.7) a strip S of a Minkowski plane, and $\cup S\Phi^k$ is a Minkowski plane.

Corollary. A two-dimensional straight space with non-positive curvature is Minkowskian if a motion $\Phi \neq E$ exists for which $xx\Phi$ is bounded.

That the corresponding statement for higher dimensions is not correct may be seen by choosing the space R preceding (25.4) as a hyperbolic space of any dimension $\geqslant 2$. Then the motions Φ of R^* of the form $x_\tau \to x_{\tau+\alpha}$, $\alpha \neq 0$, satisfy $x_\tau x_\tau \Phi = x_\tau x_{\tau+\alpha} = |\alpha|$, but R^* is not Minkowskian.

The following lemma will be needed in the further discussion:

(37.11) *For a motion Φ of a straight space R with convex capsules and the divergence property let a point z and a sequence $\{x_\nu\}$ exist such that $x_\nu x_\nu \Phi$ is bounded, while $zx_\nu \to \infty$. If $\mathfrak{g}^+(z, x_\nu) \to \mathfrak{g}^+$ then Φ^k transforms (for any integer k) any asymptote to $\mathfrak{g}^+$, in particular $\mathfrak{g}^+$ itself, into an asymptote to $\mathfrak{g}^+$.*

Note: Φ satisfies the hypothesis if no point y with $yy\Phi = \inf_{x \in R} xx\Phi$ exists.

It may be assumed that Φ is not the identity. Let $x_\nu(\tau)$ represent $\mathfrak{g}^+(z, x_\nu)$ with $x_\nu(0) = z$. Then $x_\nu(\tau)$ tends to a representation $x(\tau)$ of $\mathfrak{g}^+$. If $\mathfrak{r}_\nu$ and $\mathfrak{r}$ are the subrays $\tau \geqslant 0$ of $\mathfrak{g}_\nu^+$ and $\mathfrak{g}^+$, then $x_\nu(\tau)\mathfrak{r}_\nu\Phi$ equals for $\tau = 0$ and $\tau = zx_\nu$ at most $zz\Phi$ and $x_\nu x_\nu \Phi$ respectively. Because $x_\nu(\tau)\,\mathfrak{r}_\nu\Phi$ is peakless

$$x_\nu(\tau)\mathfrak{r}_\nu\Phi \leqslant \beta = \sup\,(zz\Phi, x_\nu x_\nu \Phi) \quad \text{for} \quad 0 \leqslant \tau \leqslant zx_\nu$$

hence $x(\tau)\mathfrak{r}\Phi \leqslant \beta$ for $0 \leqslant \tau < \infty$, so that $\mathfrak{r}\Phi$ is a co-ray to $\mathfrak{r}$. Hence $\mathfrak{r}\Phi^2$ is a co-ray to $\mathfrak{r}\Phi$, and by transitivity also to $\mathfrak{r}$. Similarly $\mathfrak{r}$ is a co-ray to $\mathfrak{r}\Phi^{-1}$ hence by symmetry $\mathfrak{r}\Phi^{-1}$ is a co-ray to $\mathfrak{r}$. Thus for any integral k, the ray $\mathfrak{r}\Phi^k$ is a co-ray to $\mathfrak{r}$.

If $\mathfrak{h}^+$ is any asymptote to $\mathfrak{g}^+$, then $\mathfrak{h}^+\Phi^k$ is an asymptote to $\mathfrak{g}^+\Phi^k$ because Φ^k is a motion; by transitivity $\mathfrak{h}^+\Phi^k$ is an asymptote to $\mathfrak{g}^+$.

(37.12) *If in addition to the assumptions of* (37.11) *the space has non-positive curvature and z and $\mathfrak{g}^+$ have the property that* $zz\Phi = \inf_{x \in R} xx\Phi$ *and* $\mathfrak{g} \neq \mathfrak{g}(z, z\Phi)$, *then* $\mathfrak{g}(z, z\Phi)$ *bounds a Minkowskian half plane imbedded in* R.

For if $x(\tau)$ represents $\mathfrak{g}^+$ with $x(0) = z$ and $y(\tau) = x(\tau)\,\Phi$, then $x(\tau)\,y(\tau)$ is non-increasing because $x(\tau)$ and $y(\tau)$ are asymptotes. On the other hand $x(\tau)y(\tau)$ reaches a minimum at $\tau = 0$, hence $x(\tau)y(\tau)$ is constant for $\tau \geqslant 0$. The rays $x(\tau)$, $y(\tau)$, $\tau \geqslant 0$, bound therefore by (36.9) together with $T(z, z\Phi)$ a halfstrip V of a Minkowski plane, and $\cup\, V\Phi^i$ is a Minkowski half-plane.

That the situation of this theorem may actually occur without $\cup\, V\Phi^i$ being part of a Minkowski plane in R may be seen by considering the surface of revolution in ordinary (ξ, η, ζ)-space with the ζ-axis as axis, and with $\xi = 1$ for $\zeta \leqslant 0$ and $\xi = \zeta^2 + 1$ for $\zeta \geqslant 0$, in the plane $\eta = 0$, as meridian. If the ordinary geodesic distance is taken as distance, then the universal covering space of this surface with non-positive curvature which is topologically a cylinder, is topologically speaking a plane. The part corresponding to $\zeta \leqslant 0$ is isometric to an ordinary half plane, but the other part is not.

We show that this situation is, in a sense, general.

(37.13) If for an axial motion Φ of a straight space R with non-positive curvature an axis $\mathfrak{g}$ and a sequence $\{x_\nu'\}$ exist for which $x_\nu'\mathfrak{g} \to \infty$ and $x_\nu' x_\nu'\Phi$ is bounded, then $\mathfrak{g}$ bounds a Minkowski half plane imbedded in R.

Let f_ν' be the foot of x_ν' on $\mathfrak{g}$ and b be any point of $\mathfrak{g}$. Choose k_ν such that $f_\nu = f_\nu'\,\Phi^{k_\nu} \in T(b, b\Phi)$. Then $x_\nu = x_\nu'\Phi^{k_\nu}$ has f_ν as foot on $\mathfrak{g}$ and $x_\nu' f_\nu' = x_\nu f_\nu = x_\nu\mathfrak{g}$ moreover $x_\nu x_\nu\Phi = x_\nu'\Phi^{k_\nu}\, x_\nu'\Phi^{k_\nu+1} = x_\nu' x_\nu'\Phi$ is bounded.

For a suitable subsequence $\{\lambda\}$ of $\{\nu\}$ the sequence $\{f_\lambda\}$ converges to a point z of $T(b, b\Phi)$, and since $\mathfrak{g}(f_\lambda, x_\lambda)$ is a perpendicular to $\mathfrak{g}$, the line $\mathfrak{g}(z, x_\lambda)$ tends to a perpendicular $\mathfrak{h}$ to $\mathfrak{g}$. The assumptions of (37.12) hold for $\mathfrak{h}$, $\mathfrak{g} = \mathfrak{g}(z, z\Phi)$ and $\{x_\lambda\}$, and (37.13) is proved.

An important corollary is:

(37.14) **Theorem.** *An axial motion in a straight space with negative curvature has exactly one axis* $\mathfrak{g}$, *and* $x_\nu x_\nu\Phi \to \infty$ *if* $x_\nu\mathfrak{g} \to \infty$.

38. Straightness of the universal covering space

The importance of straight spaces with non-positive curvature is that the universal covering space of any space with non-positive curvature and with the property of domain invariance is straight, so that any space of the latter type is locally isometric to a straight space.

Domain invariance

This property may be formulated for an arbitrary Hausdorff space R as follows

(38.1) *If one of two homeomorphic subsets of R is open, then the other is, too.*

By Brouwer's famous theorem every topological manifold has this property, hence domains are invariant in any two-dimensional G-space, see (10.4), and in every Finsler space in the usual sense, see Section 15. According to a communication from Prof. D. Montgomery every finite dimensional G-space satisfies (38.1). The existence of the function $\rho(p)$ and finite compactness make it appear likely that every G-space has finite dimension but, at present, the problem seems hard to approach. There is even a strong possibility that every G-space is a topological manifold, but if this is correct, a proof will be quite inaccessible as long as no topological characterization of E^n exists.

The main theorem

If a G-space has convex capsules, strictly convex capsules non-positive curvature, vanishing curvature, or negative curvature, then any covering space has the same property, because each is a local property whose absence is preserved under locally isometric mappings. This must be kept in mind for the following principal result of the present section, which in the Riemannian case was stated by E. Cartan [2, p. 260], but whose proof can hardly be called adequate.

(38.2) **Theorem.** *The universal covering space of a G-space with convex capsules and domain invariance is straight.*

Proof. Since a straight space is simply connected and therefore its own universal covering space, we may assume that $\rho(x) < \infty$ for the given space R.

Select a point p and put $V = K(p, \rho(p)/2)$. There is a representation $x(u, \tau)$, $0 \leqslant \tau \leqslant \infty$, $u \in V$, of a half geodesic such that $x(u, \tau)$ represents $\mathfrak{s}(p, u)$ for $0 \leqslant \tau \leqslant \rho(p)/2$. For every point q of the space there is at least one pair (u, τ), $u \in V$, $\tau \geqslant 0$ such that $q = x(u, \tau)$. The main point of the proof, which would clearly break down on a sphere, is to show that the mapping $(u, \tau) \to q$ is locally topological, that is, for a given pair (u_0, τ_0), $\tau_0 > 0$, there is an $\varepsilon > 0$ such that $(u, \tau) \to q$ is topological for $uu_0 < \varepsilon$ and $|\tau - \tau_0| < \varepsilon$.

For this purpose let $\alpha > \rho(p)/2$ be given. Put

(a) $$\beta_\alpha = \tfrac{1}{4} \inf \rho(x), \text{ where } x \in S(p, 2\alpha). \text{ Then } \beta_\alpha > 0.$$

Because of (8.12) $x(u, \tau)$ is continuous in both variables u, τ for $u \in V$ and $0 \leqslant \tau \leqslant 2\alpha$. Therefore an $\varepsilon_\alpha > 0$ exists such that

(b) $$x(u_1, \tau_1)\, x(u_2, \tau_2) < \beta_\alpha \text{ if } u_i \in V,\ u_1u_2 < 2\varepsilon_\alpha,\ 0 \leqslant \tau_i \leqslant 2\alpha \text{ and } |\tau_1 - \tau_2| < \varepsilon_\alpha.$$

Select two points u_1, u_2 on V such that $u_1u_2 < \varepsilon_\alpha$. Put $x_i(\tau) = x(u_i, \tau)$. Denote by U_{τ_0} and U'_{τ_0} the segments represented by $x_2(\tau)$ for $|\tau - \tau_0| \leqslant 2\beta_\alpha$ and $|\tau - \tau_0| \leqslant \beta_\alpha$ respectively, where $2\beta_\alpha \leqslant \tau_0 \leqslant \alpha$ and define

$$\varphi(\tau) = x_1(\tau)\, U_\tau, \qquad 2\beta_\alpha \leqslant \tau \leqslant \alpha.$$

We are going to show that $\varphi(\tau)$ increases for $u_1 \neq u_2$. (If $u_1 = u_2$ then $\varphi(\tau) \equiv 0$.) Obviously

$$\varphi(\tau) \leqslant x_1(\tau)\, x_2(\tau) < \beta_\alpha$$

and, for $|\tau - \tau_0| < \varepsilon_\alpha$,

$$x_1(\tau_0)\, x_2(\tau) \geqslant x_1(\tau_0)\, x_1(\tau) - x_1(\tau)\, x_2(\tau) > |\tau - \tau_0| - \beta_\alpha.$$

Hence the foot of $x_1(\tau_0)$ on U_{τ_0} lies in the interior of U'_{τ_0}.

It is clear that $\varphi(\tau)$ increases for $2\beta_\alpha \leqslant \tau \leqslant 3\beta_\alpha (< \rho(p))$, because the foot of $x_1(\tau)$ on U_τ then coincides with the foot of $x_1(\tau)$ on $T(p, u_2)$, so that $x_1(\tau) U_\tau = x_1(\tau) T(p, u_1)$. The latter vanishes for $\tau = 0$, is peakless and positive for $\tau > 0$ (because $u_1 \neq u_2$). Therefore $\varphi(\tau)$ increases in $[2\beta_\alpha, 3\beta_\alpha]$.

If $\varphi(\tau)$ did not increase in the whole interval $[2\beta_\alpha, \alpha]$ then numbers τ_1, τ_2, τ_3 with $2\beta_\alpha < \tau_1 < \tau_2 < \tau_3 < \alpha$ and $\tau_3 - \tau_1 < \beta_\alpha$ would exist such that $\varphi(\tau_1) < \varphi(\tau_2)$, $\varphi(\tau_2) \geqslant \varphi(\tau_3)$. Because the foot of $x_1(\tau_i)$ on U_{τ_i} falls in the interior of U'_{τ_i} the feet of $x(\tau_i)$ on U_{τ_i} fall in the interior of U_{τ_2} and are feet of $x_1(\tau_i)$ on U_{τ_2}, because peaklessness precludes a relative minimum which is not the absolute minimum. But then

$$\varphi(\tau_i) = x_1(\tau_i)\, U_{\tau_2}$$

and $x_1(\tau)\, U_{\tau_2}$ would not be peakless.

We can now show that $(u, \tau) \to x(u, \tau)$ is a locally topological mapping. For $\alpha \geqslant 3\rho(p)/4$ and $v \in V$ define $W(v, \alpha)$ as the set of points $x(u, \tau)$ for which $u \in V \cap S(v, \varepsilon_\alpha) = V_{\alpha v}$ and $|\tau - \rho(p)/2| < \varepsilon_\alpha/M$ where $M = 2\alpha/\rho(p)$. The set $W(v, \alpha)$ is open and

$$x(u, \tau) \to x(u, M\tau), \qquad u \in V_{\alpha v}, \quad |\tau - \rho(p)/2| < \varepsilon_\alpha/2M,$$

maps $W(v, \alpha)$ continuously on the set $W'(v, \alpha)$ consisting of the points $x(u, \tau)$ for which $u \in V_{\alpha v}$ and $|\tau - \alpha| < \varepsilon_\alpha$. To show that the mapping of $W(v, \alpha)$ on $W'(v, \alpha)$ is topological we observe that for $u_1, u_2 \in V_{\alpha v}$ and $\varphi(\tau)$ defined as above, the distance

$$x(u_1, M\tau_1)\, x(u_2, M\tau_2) \geqslant \varphi(\tau_1),$$

hence cannot vanish unless $u_1 = u_2$ and $M\tau_1 = M\tau_2$ or $\tau_1 = \tau_2$. The mapping is therefore one-to-one. If $x(u_1, M\tau_1)\, x(u_2, M\tau_2) \to 0$ for fixed u_1 and τ_1, then $\varphi(\tau_2) \to 0$, hence $u_2 \to u_1$ and hence $M\tau_2 \to M\tau_1$ or $\tau_2 \to \tau_1$. *We now use the domain invariance to assert that $W'(v, \alpha)$ is an open set.* Because $(u, \tau) \to x(u, \tau)$ maps the (u, τ) set: $u \in V_{\alpha v}$, $|\tau - \rho(p)/2| < \alpha/2M$ with metric $u_1u_2 + |\tau_1 - \tau_2|$ say, topologically on $W(v, \alpha)$, the mapping $(u, \tau) \to x(u, \tau)$ maps the set $W^*(v, \alpha)$ consisting of the u, τ with $u \in V_{\alpha v}$, $|\tau - \alpha| < \varepsilon_\alpha$ topologically on $W'(v, \alpha)$. Let $S(x(v, \alpha), \eta_{\alpha v})$ be the sphere about $x(v, \alpha)$ of maximal radius in $W'(v, \alpha)$. Then $\eta_{\alpha v} > 0$ because $W'(v, \alpha)$ is open and the sphere is convex, because $\eta_{\alpha v} < \rho_1(x(v, \alpha))$. Denote by $U^*(v, \alpha)$ the corresponding subset of $W^*(v, \alpha)$.

The universal covering space is the set $\bar{R}$ of all pairs (u, τ), $v \in V$, $\tau \geqslant 0$, (with the pairs $(u, 0)$ identified) locally metrized as follows:

$$(u_1, \tau_1)\,(u_2, \tau_2) = x(u_1, \tau_1)\, x(u_2, \tau_2) \begin{cases} \text{for } 0 \leqslant \tau_i < \rho(p) \\ (u_i, \tau_i) \in U^*(v, \alpha). \end{cases}$$

This definition is consistent in that whenever the right side defines $(u_1, \tau_1)\,(u_2, \tau_2)$ twice, the result is the same. Since distances are now locally defined, length is defined for any curve in $\bar{R}$. We metrize $\bar{R}$ in the large by defining the distance of two points as the greatest lower bound of the lengths of all curves connecting them. The convexity of $U(v, \alpha)$ and $S(p, \rho(p))$ guarantees that distances already defined do not alter.

If we can show that $\bar{R}$ is finitely compact, then it follows from the definition of distance in $\bar{R}$ and (5.18) *that $\bar{R}$ is M-convex,* because the curve (u_1, τ), $0 \leqslant \tau \leqslant \tau_1$, has length τ_1, hence (u_1, τ_1) and (u_2, τ_2) can be connected by a curve of length $\tau_1 + \tau_2$. Since $U(v, \alpha)$ and $S(p, \rho(p))$ are isometric to $U^*(v, \alpha)$ and $0 \leqslant \tau < \rho(p)$

respectively the mapping $(u, \tau) \to x(u, \tau)$ is locally isometric and $\overline{R}$ satisfies the axioms of local prolongability and the hypothesis of (8.2), hence is a G-space.

The finite compactness of $\overline{R}$ will follow if we can show that (u, τ), $\tau \geqslant 0$, is for fixed u a ray or, since τ is arclength on (u, τ), that the arc (u, τ), $0 \leqslant \tau \leqslant \sigma$, is for every σ a shortest connection of $(u, 0) = \overline{p}$ and (u, σ).

We call s admissible if for every $u \in V$ the arc (u, τ), $0 \leqslant \tau \leqslant s$, is a shortest connection of $\overline{p}$ and (u, s). Let ρ denote the least upper bound of all admissible s. Obviously $\rho \geqslant \rho(p)/2$, but we must show $\rho = \infty$.

For a given finite s the number $8\,\eta_s = \inf\limits_{u \in V, \alpha \leqslant s} \eta_{u\alpha}$ is positive. If all $s < s_0$ are admissible then s_0 is admissible. Therefore it sufficies to show that with any s the number $s' = s + \eta_s$ is also admissible. For a given $v \in V$ let $\overline{q} = (v, s')$ and $p_\nu(\sigma)$, $0 \leqslant \sigma \leqslant \sigma_\nu$, a sequence of curves from $\overline{p}$ to $\overline{q}$ referred to the arc length σ as parameter whose length σ_ν tends to the distance $\overline{p}\,\overline{q}$ of $\overline{p}$ and $\overline{q}$ in $\overline{R}$. Since the arc (v, τ), $0 \leqslant \tau \leqslant s'$, has length s' it may be assumed that $\sigma_\nu \leqslant s'$.

The curve $p_\nu(\sigma)$ contains a point $p_\nu(\sigma_\nu^0)$ of the form (u_ν, s). If the arc $0 \leqslant \sigma \leqslant \sigma_\nu^0$ of $p_\nu(\sigma)$ is replaced by the arc (u_ν, τ), $0 \leqslant \tau \leqslant s$, the new curve is, because of the admissibility of s, not longer than $p_\nu(\sigma)$. Therefore we may further assume that $p_\nu(\sigma)$ represents for $0 \leqslant \sigma \leqslant s$ the arc (u_ν, τ), $0 \leqslant \tau \leqslant s$.

No point of $p_\nu(\sigma)$ with $\sigma \geqslant s$ can lie outside the sphere $S((v, s), 2\,\eta_s)$ since this sphere is isometric to $S(x(v, s), 2\,\eta_s)$ and the length of the arc $s \leqslant \sigma \leqslant \sigma_\nu$ of $p_\nu(\sigma)$ would be at least $2\,\eta_s$. Consequently, the arc $s \leqslant \sigma \leqslant \sigma_\nu$ of $p_\nu(\sigma)$ may be replaced by the segment from $p_\nu(s)$ to $\overline{q}$ without increasing the length of $p_\nu(\sigma)$. Then a subsequence of the new curves $p_\nu(\sigma)$ will tend to a curve $\overline{p}(\sigma)$ from $\overline{p}$ to $\overline{q}$ of length $\overline{p}\,\overline{q} \leqslant s'$ which consists of an arc (u_0, τ), $0 \leqslant \tau \leqslant s$, and a segment from (u_0, s) to $\overline{q}$.

The minimizing property of $\overline{p}(\sigma)$ (it has by (5.16) length $\overline{p}\,\overline{q}$) shows that the segment from (u_0, s) to $\overline{q}$ must be a continuation of the arc (u_0, τ). By construction (u_0, τ) and (v, τ) have common points different from $\overline{p}$ only for $u_0 = v$. This shows that (v, τ), $0 \leqslant \tau \leqslant s'$, is a shortest connection of $\overline{p}$ to (v, s').

Since every half-geodesic in $\overline{R}$ issuing from $\overline{p}$ is a ray, every bounded set can be contracted to $\overline{p}$, therefore $\overline{R}$ is simply connected, hence the universal covering space of R. By the motions of the fundamental group, $\overline{p}$ can be moved into any other point over p, and p was arbitrary in R. Therefore the half-geodesics issuing from any point in $\overline{R}$ are rays. This implies that $\overline{R}$ is straight and proves the theorem.

It would be desirable to prove directly that the universal covering space of R is straight instead of constructing $\overline{R}$. The difficulty common to all such cases is the lack of a priori information about $\overline{R}$.

39. The fundamental groups of spaces with convex capsules

Theorem (38.2) has very many applications some of which we discuss now, others will be found in Section 40.

First consequences

(39.1) *A simply connected G-space with convex capsules and domain invariance is straight. If the space is an n-dimensional topological manifold then it is homeomorphic to* E^n.

The latter assertion follows because a straight space can be mapped topologically on a set interior to any sphere $S(p, \rho)$. Since a compact space is never straight we conclude from (39.1) and (29.6) that

(39.2) *A compact G-space with convex capsules and domain invariance is not simply connected but has finite connectivity.*

This theorem implies for instance, that spheres of dimension $n \geqslant 2$ and topological products of such spheres, cannot be metrized such that they become G-spaces with convex capsules, compare Seifert and Threlfall [1, § 43].

Next we prove a theorem which is due to E. Cartan (for the Riemannian case, but his proof carries over to the present case, compare [2, p. 266]).

(39.3) ***Theorem.*** *The fundamental group of a G-space with domain invariance and non-positive curvature has no finite sub group* (except the group formed by the identity).

Since an element of the fundamental group of any G-space other than the identity considered as motion of the universal covering space has no fixed point, Cartan's theorem follows from (38.2) and the following fact:

(39.4) If $\mathfrak{G}$ is a finite group of motions in a straight space with non-positive curvature, then a point p exists which remains fixed under all motions in $\mathfrak{G}$.

Let p_1 be any point of the space and $p_2, \ldots, p_n$ its images under the motions of $\mathfrak{G}$. The set $p_1, \ldots, p_n$ goes into itself under all motions of $\mathfrak{G}$.

The function $\sum_{i=1}^{n} xp_i^2$ reaches, as x varies over the space, its minimum at exactly one point q. It is clear that a minimum is reached. If it were attained at two distinct points q_1, q_2, let $x(\tau)$ represent the line $\mathfrak{g}(q_1, q_2)$ with $x(\tau_k) = q_k$, $k = 1, 2$. Then $\varphi(\tau) = \sum_{i=1}^{n} x(\tau)p_i^2$ reaches a minimum for both τ_1 and τ_2.

But $x(\tau)p_i$ is a convex function of τ and is nowhere constant. (It is not strictly convex when $p_i \in \mathfrak{g}(q_1, q_2)$). Therefore $x(\tau)p_i^2$ is strictly convex and so is $\varphi(\tau)$. But then it attains its minimum only once.

Since any motion in $\mathfrak{G}$ transforms the set $p_1, \ldots, p_n$ into itself it leaves q fixed, which proves (39.4).

Abelian subgroups of the fundamental group

An immediate corollary of (38.2), (37.9) and (32.11) and its proof is the following theorem which was proved by Preissmann [1] for analytic Riemann spaces with negative curvature.

(39.5) **Theorem.** *A compact G-space with domain invariance and strictly convex capsules cannot have an abelian fundamental group.*

Because group spaces have abelian fundamental groups (see Pontrjagin [1, p. 228]) we have the following two corollaries:

(39.6) *A compact group space of dimension greater than* 1 *cannot be metrized as a G-space with strictly convex capsules.*

Notice that we do not require that the metric be invariant under the group.

(39.7) *There is no compact G-space with domain invariance, strictly convex capsules and a simply transitive group of motions.*

(39.5) was not actually stated by Preissmann, but it is closely related to the following fact (proved by him under the assumption that the space be analytic and Riemannian):

(39.8) **Theorem.** *A (non-trivial) abelian subgroup of the fundamental group of a compact G-space (of dimension greater than* 1*) with domain invariance and strictly convex capsules is infinite cyclic.*

To prove this theorem we notice first the trivial but useful lemma:

(39.9) *If Φ is an axial motion of the straight space R and Ψ is any motion of R, then $\Phi' = \Psi^{-1}\Phi\Psi$ is axial and $\gamma(\Phi) = \gamma(\Phi')$. If $\mathfrak{g}$ is an axis of Φ then $\mathfrak{g}\Psi$ is an axis of Φ'.*

For the definition of $\gamma(\Phi)$ compare Section 32.

For, Φ' has no fixed point p since this would imply $p\,\Psi^{-1}\Phi = p\,\Psi^{-1}$ or that $p\,\Psi^{-1}$ is a fixed point of Φ. Moreover, $\mathfrak{g}\,\Psi\,\Phi' = \mathfrak{g}\,\Phi\,\Psi = \mathfrak{g}\,\Psi$, hence $\mathfrak{g}\,\Psi$ is an axis of Φ'. If $x \in \mathfrak{g}\,\Psi$ then $x = z\Psi$ for a suitable $z \in \mathfrak{g}$ and

$$\gamma(\Phi') = xx\Phi' = z\Psi z\Psi\Phi' = z\Psi z\Phi\Psi = zz\Phi = \gamma(\Phi).$$

Next we establish:

(39.10) Let $\mathfrak{A}$ be an abelian group of motions of a straight space with strictly convex capsules such that no motion in $\mathfrak{A}$, except E, has fixed points. If one motion Φ in $\mathfrak{A}$ has an axis $\mathfrak{g}$, then all motions in $\mathfrak{A} - E$ are axial with (the unique) axis $\mathfrak{g}$. If $\mathfrak{A}$ is discrete then it is infinite cyclic.

For if $\Psi \in \mathfrak{A} - E$ then $\Phi = \Psi^{-1}\Phi\Psi$ is by (39.9) axial and has axis $\mathfrak{g}\,\Psi$. Since Φ has only one axis, see (37.9), it follows that $\mathfrak{g}\,\Psi = \mathfrak{g}$ so that Ψ is axial and has axis $\mathfrak{g}$.

If $\mathfrak{A}$ is discrete then $\mathfrak{A} - E$ contains an element Φ_0 for which $\gamma(\Phi_0)$ is minimal. Otherwise Φ with arbitrary small $\gamma(\Phi)$ would exist, hence also a sequence $\{\Phi_\nu\}$ in $\mathfrak{A} - E$ with $\gamma(\Phi_\nu) \to 0$ and for which Φ_ν converges. The limit would lie in $\mathfrak{A}$, because as a discrete group $\mathfrak{A}$ is closed, but then $\mathfrak{A}$ cannot be discrete. A standard argument shows that $\gamma(\Phi)$ is for any Φ in $\mathfrak{A} - E$ an integral multiple of $\gamma(\Phi_0)$. Therefore a k exists such that $x\Phi_0^k = x\Phi$ for $x \in \mathfrak{g}$.

Since no motion in $\mathfrak{A} - E$ has fixed points it follows that $\mathfrak{A}$ is cyclic. It is infinite because $x\Phi_0^k \neq x\Phi_0^j$ for $k \neq j$ and $x \in \mathfrak{g}$.

A corollary of (39.10) is

(39.11) *If an abelian subgroup $\mathfrak{A}$ of the fundamental group of a straight space with domain invariance and strictly convex capsules contains an axial motion, then $\mathfrak{A}$ is infinite cyclic.*

Theorem (39.8) now follows: Because R is not simply connected (see (39.2)), there are non-trivial free homotopy classes. Since R is compact, each motion in the fundamental group of R is axial by (32.7).

Theorem (39.8) shows that *many manifolds cannot be metrized as G-spaces with strictly convex capsules.* For instance, the product of two compact manifolds cannot be so metrized. A consequence of (39.10) is the following observation of Preissmann: If the fundamental group of a G-space with strictly convex capsules is cyclic then all closed geodesics are multiples of one great circle.

SPACES WITH CURVATURE 0

For spaces of curvature 0 domain invariance need not be assumed, it can easily be proved: If (36.2) holds in $S(p, \gamma_p)$ it follows from (36.9) that any three points a, b, c in $S(p, \gamma_p)$ which do not lie on one segment, span a set isometric to a triangle in a Minkowski plane. Let T_ν or T_ν', $\nu = 1, 2, \ldots$; mean a segment of length γ_ρ and with center p.

Choose any T_1. Unless the space has dimension 1, there is a $T_2 \neq T_1$. Then T_1 and T_2 determine a set C_2 of points x with $px \leqslant \gamma_p/2$ isometric to a circular disk in a Minkowski plane. If the dimension of the space is greater than 2, then a T_3 not in C_2 exists. It determines as before with every T_3' in C_2 a set isometric to a circular disk in a Minkowski plane, hence T_3 and C_2 determine together a set C_3 of points x with $px \leqslant \gamma_p/2$ isometric to a solid sphere of a three-dimensional Minkowski space. It is clear that this procedure may be continued, yielding sets $C_1 = T_1, C_2, C_3, \ldots$,such that C_ν is isometric to a solid sphere in a ν-dimensional Minkowski space, as long as $T_{\nu+1}$ not in C_ν exists. The process breaks off after a finite number of steps. For, as a set isometric to a sphere in a Minkowski space, C_ν contains for $\nu > 1$ a point q_ν with $q_\nu p = q_\nu C_{\nu-1} = \gamma_p/2$, so that the finite compactness guarantees the existence of a last ν, say $\nu = n$. Then $C_n = \overline{S}(p, \gamma_p/2)$ shows that the space R is locally Minkowskian. R is therefore a topological manifold and has the property of domain invariance. We proved, however, already in (30.1), and it follows again from (38.2), that the universal covering space of R is a Minkowski space. Thus we have found:

(39.12) *The spaces with curvature* 0 *are locally Minkowskian spaces.*

In certain cases the assumption that the curvature is non-positive implies that the curvature vanishes.

(39.13) *If a torus R of arbitrary dimension n is metrized as a G-space with non-positive curvature then it has curvature* 0.

R is by definition, topologically, the product of n circles $K_2, \ldots, K_{n+1}$. It is compact and its fundamental group $\mathfrak{F}$ is abelian. By (32.11) and its proof, there is a closed geodesic $\mathfrak{g}_\nu$ in R freely homotopic to K_ν, and $\mathfrak{g}_\nu$ determines a class of conjugate elements in $\mathfrak{F}$ (see Section 32), which reduces to one element $\Xi_\nu \neq \Xi$ because $\mathfrak{F}$ is abelian; $\overline{y}\overline{y}\Xi_\nu$ is constant for $\overline{y} \in \overline{R}$, the universal covering space of R, and every point of the space lies on an axis of Ξ_ν.

If $\overline{p}$ is now an arbitrary point of $\overline{R}$ and $\overline{\mathfrak{g}}_\nu$ the axis of Ξ_ν through $\overline{p}$, then (37.10) applied repeatedly yields that $\overline{\mathfrak{g}}_2$ and $\overline{\mathfrak{g}}_3$ determine a Minkowski plane P_2 in $\overline{R}$, then $\overline{\mathfrak{g}}_3$ and P_2 a Minkowski space, etc. Thus R is a Minkowski space.

It follows from (39.13) that every space with non-positive curvature which has a torus as covering space, has curvature 0. Thus, for instance, a Klein Bottle with non-positive curvature has curvature 0.

Very little is known about a torus metrized as a G-space with convex capsules. It was proved in (33.1) that if a two-dimensional torus R is metrized such that its universal covering space $\overline{R}$ (the ordinary plane) is straight, the parallel axiom holds in $\overline{R}$. If R has convex capsules, then $\overline{R}$ has convex circles, and we meet a problem similar to the one left unsolved in Section 25. The parallels to a line $\mathfrak{g}$ are equidistant to $\mathfrak{g}$, any line $\mathfrak{h}$ intersecting $\mathfrak{g}$ intersects all parallels to $\mathfrak{g}$. Therefore $\overline{R}$ has the divergence property.

Metrization of surfaces

Making use of the fact that every two-dimensional manifold of finite connectivity, except the sphere, the projective plane, the ordinary and one-sided torus, can be provided with a locally hyperbolic metric, we now have complete information on the question, which surfaces can carry metrics with convex capsules, negative curvature, etc. The sphere cannot carry a metric with convex capsules because of (39.2). The same holds for the projective plane because it has the sphere as covering space. The results of this section and (30.7) then permit us to formulate the following theorem, keeping in mind that non-positive and negative curvature imply convex and strictly convex capsules respectively:

(39.14) ***Theorem.*** *All two-dimensional manifolds of finite connectivity can be metrized as G-spaces with convex capsules except the sphere and the projective plane.*

The plane, cylinder and Moebius strip are the only manifolds that can be metrized both as G-spaces with curvature 0 *and with negative curvature.*

A torus or a one-sided torus cannot carry a metric with strictly convex capsules. If they have non-positive curvature, the curvature vanishes.

All other than these five types can be metrized with negative curvature.

40. Geodesics in spaces with negative curvature.

The idea of drawing conclusions on the shape of the geodesics in a manifold from its topological structure is one of Poincaré's great contributions to modern mathematical thought. That for two-dimensional Riemannian manifolds with negative curvature this connection is particularly transparent and leads to far-reaching results was discovered by Hadamard in his well-known paper [1]:

Les surfaces à courbures opposées et leurs lignes géodésiques. There is, perhaps, no stronger proof of the truly geometric content of Hadamard's paper than that much of it can be extended to higher dimensions and nearly all, that is not of a predominantly analytical character, to two-dimensional manifolds with negative curvature in the present sense. There is so much work on spaces with negative curvature and related work like symbolic dynamics and ergodic theory, that we can only give a few samples here.

Geodesic connections and closed geodesics

The theorems (32.2), (32.8), (32.9) and (38.2) contain:

(40.1) **Theorem.** *In a G-space with convex capsules and domain invariance a given class of homotopic curves from p to q contains exactly one geodesic curve, and there are no closed geodesics or monogons freely contractable to a point.*

A consequence of (32.1, 2, 6) and (38.2) is

(40.2) **Theorem.** *If the universal covering space of a G-space R with strictly convex capsules is straight then a given class K of freely homotopic curves contains at most one closed geodesic $\mathfrak{g}$ (exactly one if R is compact). If the curvature is negative then the length of a geodesic monogon in K tends to ∞ when the distance of its vertex from $\mathfrak{g}$ tends to ∞.*

Asymptotic geodesics

Let R be a G-space with a straight covering space $\overline{R}$. Let $x(\tau)$ represent the oriented geodesic $\mathfrak{g}^+$ and connect a given point p to $x(0)$ by a curve $\mathfrak{c}$. Using the same notations as in Section 32, let $\mathfrak{h}_s$ be the unique geodesic curve from p to $x(s)$ which is homotopic to $\mathfrak{c}x(0, s)$. If $y_s(\tau)$ represents the geodesic carrying $\mathfrak{h}_s$ such that $y_s(\tau)$ represents $\mathfrak{h}_s$ for $0 \leqslant \tau \leqslant$ *length* $\mathfrak{h}_s$, then $y_s(\tau)$ converges for $s \to \infty$ to a geodesic $y(\tau)$. We call the oriented geodesic $\mathfrak{h}^+$ represented by $y(\tau)$ the "*asymptote of type* $\mathfrak{c}$ *to* $\mathfrak{g}^+$ *through* p".

That $y_s(\tau)$ converges follows by translating this procedure into operations in $\overline{R}$: If $\overline{p}$ is any point over p and $\overline{\mathfrak{c}}$ the curve over $\mathfrak{c}$ beginning at $\overline{p}$, then $\overline{\mathfrak{c}}$ ends at a point $\overline{x}$ over $x(0)$ and there is exactly one geodesic $\overline{x}(\tau)$ over $x(\tau)$ (i. e. $\overline{x}(\tau)\,\Omega = x(\tau)$) with $\overline{x}(0) = \overline{x}$. The curve over $\mathfrak{h}_s$ beginning at $\overline{p}$ is then simply the segment $T(\overline{p}, \overline{x}(\tau))$, and if $\overline{y}_s(\tau)$ lies over $y_s(\tau)$ with $\overline{y}_s(0) = \overline{p}$, then $\overline{y}_s(\tau)$ tends for $s \to \infty$ to the asymptote $\overline{\mathfrak{h}}^+$ to $\overline{\mathfrak{g}}^+$ through $\overline{p}$, hence $y_s(\tau) = \overline{y}_s(\tau)\,\Omega$ converges too and represents $\overline{h}^+\Omega$.

Assume in addition to the straightness of $\overline{R}$ that R has convex capsules so that the spheres in $\overline{R}$ are convex. Let $x(\tau)$ and $y(\tau)$ represent geodesics in R and let $\mathfrak{c}$ connect $y(0)$ to $x(0)$. If $\overline{\mathfrak{c}}\Omega = \mathfrak{c}$ and $\overline{\mathfrak{c}}$ begins at $\overline{y}$ and ends at $\overline{x}$, let $\overline{x}(\tau)\,\Omega = x(\tau)$ with $\overline{x}(0) = \overline{x}$ and $\overline{y}(\tau)\,\Omega = y(\tau)$ with $\overline{y}(0) = \overline{y}$. If finally $\overline{x}\,(\pi(\tau))$ is the foot of $\overline{y}(\tau)$ on the geodesic $\overline{\mathfrak{g}}$ represented by $\overline{x}(\tau)$, then $T_\tau = T[\overline{y}(\tau), \overline{x}\,(\pi(\tau))]\,\Omega$ is a geodesic curve in R which is (locally) perpendicular to the geodesic $\mathfrak{g}$ represented by $x(\tau)$ and has the property that $y(0, \tau)\;T_\tau \sim \mathfrak{c}x\,(0, \pi(\tau))$: The curve T_τ is uniquely determined by these properties. The length $\delta\,(\mathfrak{c}, y(\tau), \mathfrak{g})$ of T_τ (or the number $\overline{y}(\tau)\,\overline{x}\,(\pi(\tau))$) is the "*distance of type* $\mathfrak{c}$ *from* $y(\tau)$ *to* $\mathfrak{g}$". This concept and that of asymptotic geodesics are due to Hadamard. We have:

(40.3) *In a space with convex capsules and domain invariance, if* $y(\tau)$ *is an asymptote of type* $\mathfrak{c}$ *to* $\mathfrak{g}^+$, *then* $\delta(\mathfrak{c}, y(\tau), \mathfrak{g})$ *is bounded for* $\tau \geqslant 0$. *If the capsules are strictly convex and* $y(0)y(\tau)$ *is bounded for* $\tau \geqslant 0$ *then* $\delta(\mathfrak{c}, y(\tau), \mathfrak{g}) \to 0$ *for* $\tau \to \infty$.

The first part follows from (37.1). For the second part let $y(0)y(\tau) < M$ and $2\,\beta = \inf \rho(x)$, where $x \in S(y(0), 2\,M)$. The function $\delta(\mathfrak{c}, y(\tau), \mathfrak{g})$ decreases by (37.1) and tends for $\tau \to \infty$ to a limit δ. The lengths of the curves $T_{\tau-\beta}$, T_τ, $T_{\tau+\beta}$ defined above tend to δ when $\tau \to \infty$. Because $y(0)y(\tau)$ is bounded there is by (8.12) a sequence $\tau_\nu \to \infty$ such that the segments $y(\tau_\nu - \beta, \tau_\nu + \beta)$ converge to a segment $z(\tau_0 - \beta, \tau_0 + \beta)$ on a geodesic $z(\tau)$ and the curve $x[\pi(\tau_\nu - \beta), \pi(\tau_\nu + \beta)]$ tends to a geodesic curve on a geodesic $\mathfrak{k}$. The limits of $T_{\tau_\nu-\beta}$, T_{τ_ν}, $T_{\tau_\nu+\beta}$ are geodesic curves of length δ which are perpendicular to $\mathfrak{k}$ from the points $z(\tau_0 - \beta)$, $z(\tau_0)$, $z(\tau_0 + \beta)$ of $z(\tau)$, of the same type. Translating this back into $\overline{R}$ we obtain three collinear points with the same distance δ from a geodesic over $\mathfrak{k}$. If the capsules are strictly convex this implies $\delta = 0$.

The following theorem is — as the preceding ones — due to Hadamard, its proof is, however quite different from Hadamard's (l. c. pp. 42, 65, 66) whose method is in this case strictly Riemannian.

(40.4) *In a G-space R with strictly convex capsules and domain invariance let* $y(\tau)$ *represent an oriented geodesic asymptote to the (also oriented) geodesic* $x(\tau)$ *both of type* $\mathfrak{c}_1$ *and* $\mathfrak{c}_2$, *where* $\mathfrak{c}_1$ *and* $\mathfrak{c}_2$ *connect* $y(0)$ *to* $x(0)$ *and are not homotopic. If, moreover,* $y(0)y(\tau)$ *is bounded for* $\tau \geqslant 0$ *and the free homotopy class of* $\mathfrak{c}_1\mathfrak{c}_2^{-1}$ *contains a closed geodesic* $\mathfrak{g}$, *then* $x(\tau)$ *and* $y(\tau)$ *are asymptotes to a suitable orientation of* $\mathfrak{g}$.

For a proof, define the geodesic curve $T_{i\tau}$, $i = 1, 2$ with respect to $y(\tau)$, $x(\tau)$ and $\mathfrak{c}^i$ as above T_τ with respect to $y(\tau)$, $x(\tau)$ and $\mathfrak{c}$. Let $T_{i\tau}$ begin at $y(\tau)$ and end at $x(\pi_i(\tau))$. Then

$$T_{1\tau}\, x[\pi_1(\tau), \pi_2(\tau)]\, T_{2\tau}^{-1} \sim \mathfrak{c}_1 \mathfrak{c}_2^{-1} \sim 0. \tag{40.5}$$

Because of (32.7) the curve on the left side has at least the length 2ρ of $\mathfrak{g}$. By (40.3) the length of $T_{i\tau}$ tends to 0 for $\tau \to \infty$, hence

$$x[\pi_1(\tau)]\, x[\pi_2(\tau)] \to 0, \qquad \text{but} \qquad |\pi_1(\tau) - \pi_2(\tau) > \rho| \text{ for large } \tau.$$

The function $y(\tau)x(\tau)$ is bounded for $\tau \geqslant 0$ because suitable lines $\bar{x}(\tau)$, $\bar{y}(\tau)$ over $x(\tau)$, $y(\tau)$ are asymptotes to each other and $\bar{y}(\tau)\bar{x}(\tau)$ ($\geqslant y(\tau)x(\tau)$) is bounded. Since $y(0)y(\tau)$ is bounded for $\tau \geqslant 0$, so is $y(0)x(\tau)$. Moreover $\pi_i(\tau) \to \infty$ for $\tau \to \infty$. Therefore a sequence $\tau_\nu \to \infty$ exists such that $x(\pi_1(\tau_\nu) + \sigma)$ tends for all σ to a representation $z(\sigma)$ of a geodesic. Since the left side of (40.5) is homotopic to $\mathfrak{g}$, the number $|\pi_1(\tau_\nu) - \pi_2(\tau_\nu)|$ is bounded. Therefore, possibly after passing to subsequence of $\{\nu\}$, $\pi_2(\tau_\nu)$ will tend to a finite value $|\sigma_2| \geqslant \rho$, say $\sigma_2 \geqslant \rho$. Then $z(0, \sigma_2)$ is a geodesic monogon freely homotopic to a suitable orientation $\mathfrak{g}^+$ of $\mathfrak{g}$.

If $\pi_1(\tau_\nu') = (1 - \theta)\,\pi_1(\tau_\nu) + \theta\,\pi_2(\tau_\nu)$, $0 < \theta < 1$, then $\pi_1(\tau_\nu) \to \theta\,\sigma_2$ and $\pi_2(\tau_\nu')$ tends to a value σ_3 such that $z(\theta\,\sigma_2, \sigma_3)$ is another geodesic monogon. Therefore $z(\sigma)$ either has a multiple point at $\theta\,\sigma_2$, or the two line elements corresponding to $\theta\,\sigma_2$ and σ_2 coincide. Since a geodesic has not more than a countable number of multiple points, these line elements must coincide for some θ. This means that $z(0, \sigma_2)$ is a closed geodesic, homotopic to $\mathfrak{g}^+$, hence must coincide with $\mathfrak{g}^+$ by (40.2). Since $x(\tau)$ is an asymptote to $\mathfrak{g}^+$, it follows from transitivity that $y(\tau)$ is too.

For not simply connected spaces the following unsolved question arises: If $x(\tau)$ and $y(\tau)$ represent asymptotic geodesics of some type which are rays for $\tau \geqslant 0$. Are these rays necessarily co-rays to each other?

The converse is certainly not true, that is, a space with non-positive curvature may contain a pair of rays $x(\tau)$, $y(\tau)$, $\tau \geqslant 0$ which are co-rays to each other, but the oriented geodesics which coincide with these rays for $\tau \geqslant 0$ are not asymptotic of any type.

To obtain an *example*, consider in the plane $\zeta = 0$ of a Cartesian (ξ, η, ζ)-space circular disks of the form

$$C_k : (\xi - 2k)^2 + \eta^2 = \rho_k^2, \quad k = 1, 2, \ldots,$$

where ρ_k is so small that the ordinary straight lines through $p = (0, 1)$ and $(2k \pm 1, 0)$ do not intersect C_k. The line through p and $(2k - 1, 0)$ intersects $\eta = -1$ in $(\tau_k, -1)$ with $\tau_k = 4k - 2$.

Our space will be a surface S in the (ξ, η, ζ)-space with the ordinary geodesic distance, constructed as follows: The plane $\zeta = 0$ except for the discs C_k belongs to S. Each C_k is replaced by a surface of revolution with $\xi = 2k$, $\eta = 0$ as axis, whose intersection with the half plane $\eta = 0$, $\xi > 2k$ is a curve of class C^m, $m \geqslant 2$, of the form

$$\zeta = \varphi_k(\xi) \geqslant 0, \quad 2k \leqslant \delta_k < \xi \leqslant 2k + \rho_k, \quad \varphi_k(\xi) \text{ convex},$$
$$\varphi_k(\xi) \to \infty \text{ for } \xi \to \delta_k, \varphi(2k + \rho_k) = \varphi'(2k + \rho_k) = \dots = \varphi^{(m)}(2k + \rho_k) = 0.$$

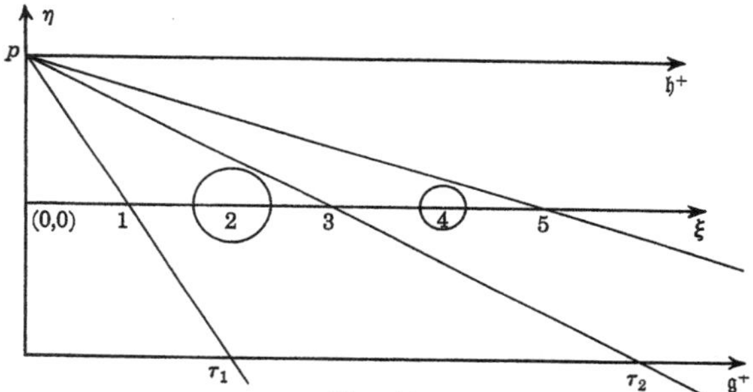

Fig. 44.

On this surface $(\tau, -1)$ represents an oriented geodesic $\mathfrak{g}^+$. The ordinary straight line through p and $(\tau_k, -1)$, $k = 1, 2 \dots$, is an oriented straight line $\mathfrak{h}_k^+$ on S as G-space and

$$\left(\frac{2k - 1}{\sqrt{4k^2 - 4k + 2}}\tau, 1 - \frac{\tau}{\sqrt{4k^2 - 4k + 2}}\right)$$

is a representation of $\mathfrak{h}_k^+$. For $k \to \infty$ this representation tends to the representation $(\tau, 1)$ of the oriented geodesic $\mathfrak{h}^+$. It follows that the ray $\tau \geqslant 0$ of $\mathfrak{h}^+$ is a co-ray to the ray $\tau \geqslant 0$ of $\mathfrak{g}^+$ and also conversely because the surface S has $\eta = 0$ as plane of symmetry.

Nevertheless, neither of the two geodesics $\mathfrak{h}^+$ and $\mathfrak{g}^+$ is asymptotic to the other of any type.

Compact surfaces with negative curvature

For geodesics on compact orientable surfaces with negative curvature we have the following theorem implied by (34.11), (34.14), (34.15), (37.1), (37.2) and (39.14).

(40.6) ***Theorem.*** *On a compact orientable surface with negative curvature the closed geodesics are dense among all geodesics in the following sense:*

Given a finite number of geodesic curves $z_i(\tau)$, $0 \leqslant \tau \leqslant \lambda_i$, $i = i, \ldots, m$, *then a closed geodesic* $y(\tau)$ *and numbers* α_i *exist such that* $y(\tau)\, z_i(\tau - \alpha_i) < \varepsilon$ *for* $\alpha_i \leqslant \tau \leqslant \alpha_i + \lambda_i$ *and all* $i = 1, 2, \ldots, m$.

There are alef open geodesics $y(\tau)$ *with the property: Given any geodesic curve* $z(\tau)$, $0 \leqslant \tau \leqslant \lambda$, *any* $\varepsilon > 0$ *and any* $N > 0$, *there exists an* $\alpha = \alpha(z, \varepsilon, N) > N$ *such that* $y(\tau)\, z(\tau - \alpha) < \varepsilon$ *for* $\alpha \leqslant \tau \leqslant \alpha + \lambda$.

This theorem can easily be extended to non-orientable compact surfaces by using the fact that such a surface has a compact orientable surface as two-sheeted covering space.

41. Relation to non-positive curvature in standard sense.

In the last five sections we introduced non-positive curvature and convexity of capsules and saw that all the theorems proved for Riemann spaces with non-positive or negative curvature hold with occasional, obviously necessary, changes. Therefore there must be a close connection between our and the classical concepts, which we are now going to discuss.

It is unavoidable that in this section first some facts on Riemannian geometry, and later on Finsler spaces and measure theory, will be used. However, omitting part, or all, of this section will not impair the understanding of the rest of the book.

The cosine inequality

We begin with a simple lemma:

(41.1) *If in a G-space with non-positive curvature* $x(\tau)$ *and* $y(\tau)$ *represent geodesics with* $x(0) = y(0)$ *then*

$$\lim x(\alpha\tau)y(\beta\tau)/\tau = \bar{\mu}(\alpha, \beta) \qquad \alpha, \beta \neq 0$$

exists, $\bar{\mu}(\alpha, \beta) \leqslant |\alpha| + |\beta|$, *and*

(41.2) $\quad x(\alpha\tau)\, y(\beta\tau) \geqslant \tau\bar{\mu}(\alpha, \beta)$ *when* $|\alpha|\tau < \rho_1(p)$, $|\beta|\tau < \rho_1(p)$ *and* $\tau \geqslant 0$.

For $x(\alpha\tau)\, y(\beta\tau)$ is by (36.5) a convex function of τ if $0 \leqslant \tau \leqslant \rho_1(x(0))/\max(|\alpha|, |\beta|)$. This function has at $\tau = 0$ a right hand derivative $\bar{\mu}(\alpha, \beta) \leqslant |\alpha| + |\beta|$. The relation (41.2) merely states that the function $x(\alpha\tau)y(\beta\tau)$ lies above its right-hand tangent.

We use this lemma to show that in Riemann spaces non-positive curvature in our sense is equivalent to another interesting elementary inequality:

(41.3) ***Theorem.*** *A Riemann space has non-positive curvature in the present sense, if and only if every point p has a neighborhood $S(p, \delta)$ such that any triangle with vertices a, b, c in $S(p, \delta)$ satisfies the "cosine inequality".*

$$\gamma^2 \geqslant \alpha^2 + \beta^2 - 2\,\alpha\,\beta \cos C, \tag{41.4}$$

where $\alpha = b\,c$, $\beta = c\,a$, $\gamma = a\,b$ and C is the angle at c.

Proof. Let R be a Riemann space with non-positive curvature in the present sense, and $a, b, c \in S(p, \rho_1(p))$. If $x(\tau)$, $y(\tau)$ are geodesics which represent the oriented segments $T^*(c, a)$ and $T^*(c, b)$ for $0 \leqslant \tau \leqslant \alpha$ and $0 \leqslant \tau \leqslant \beta$ respectively, then $x(0) = y(0) = c$ and, because R is Riemannian,

$$\overline{\mu}^2(\alpha, \beta) = \lim\,[x(\alpha\,\tau)\;y(\beta\,\tau)]^2\,\tau^{-2} = \lim\,(\alpha^2\tau^2 + \beta^2\tau^2 - 2\,\alpha\,\tau\,\beta\,\tau\cos C)\tau^{-2}$$
$$= \alpha^2 + \beta^2 - 2\,\alpha\,\beta\cos C$$

so that (41.4) follows from (41.2).

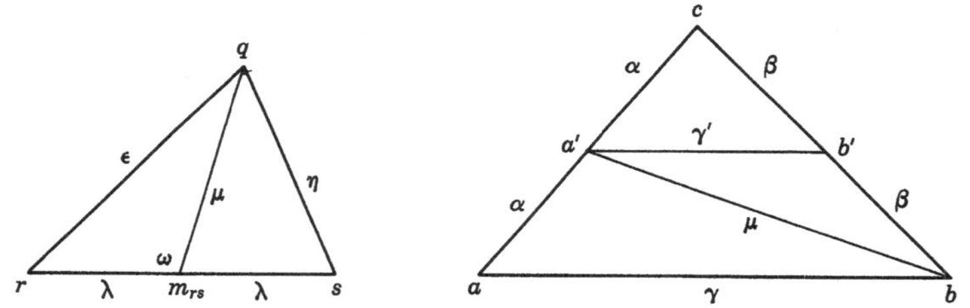

Fig. 45.

For the proof of the converse observe first (see Figure 45) that in any triangle qrs in $S(p, \delta)$ with $qr = \varepsilon$, $qs = \eta$, $rs = 2\,\lambda$, $qm[r, s] = \mu$ the relation

$$\varepsilon^2 + \eta^2 \geqslant 2(\lambda^2 + \mu^2) \tag{41.5}$$

holds. For if ω denotes the angle $\angle\, qm[r, s]\,r$ then by (41.4)

$$\varepsilon^2 \geqslant \mu^2 + \lambda^2 - 2\,\mu\,\lambda\cos\omega, \qquad \eta^2 \geqslant \mu^2 + \lambda^2 + 2\,\mu\,\lambda\cos\omega.$$

Consider now a triangle abc in $S(p, \delta/2)$ (compare Figure 45). Put

$a' = m[c, a]$, $b' = m[c, b]$, $ac = 2\,\alpha$, $bc = 2\,\beta$, $a'b' = \gamma'$, $a'b = \mu$, $ab = \gamma$.

Then $a', b' \in S(p, \delta)$. Applying (41.5) to the triangles $a'bc$ and bca yields

$$4(\gamma'^2 + \beta^2) \leqslant 2(\alpha^2 + \mu^2) \leqslant \gamma^2 + (2\beta)^2$$

hence $2\gamma' \leqslant \gamma$ or $2a'b' \leqslant ab$, which means non-positive curvature.

Equivalence of non-positive curvature and convex capsules for Riemann spaces

For Riemann spaces we can now clear up completely the relation between the concepts discussed in the preceding sections and the curvature tensor where we assume, of course, tacitly that the space is sufficiently smooth to form this tensor:

(41.6) ***Theorem.*** *For a Riemann space non-positive curvature in the usual sense, non-positive curvature in the present sense, and convexity of the capsules are equivalent conditions.*

Proof. If a Riemann space has non-positive curvature in the usual sense then the cosine inequality (41.4) holds locally. For E. Cartan [2, p. 261] proves (41.4) in the large for simply connected spaces, which implies that it holds locally for spaces of general topological structure. According to the preceding theorem the space has non-positive curvature in the present sense, and by (36.14) and (36.16) it also has convex capsules.

To complete the proof of (41.6) we must show that convexity of the capsules implies non-positive curvature in the usual sense. In a Riemann space perpendicularity and transversality are equivalent. To emphasize this fact we will use normality for both. That the space has non-positive curvature means, see Cartan [2, pp. 191–199]: if dP is an arbitrary two-dimensional surface element with origin p, then the surface formed by the geodesics through p and tangent to dP has at p a non-positive Gauss curvature K.

Take any two segments $T(p, q)$ and $T = T(p, r)$ with $q, r \in S(p, \rho(p))$, normal to each other, and tangent to dP. Under parallel displacement along $T(p, q)$ the segment T goes into a segment $T(q, s)$ normal to $T(p, q)$ at q, and hence tangent to the capsule C_T^{pq} at q. Since this capsule is convex, the tangent is supporting line, hence s lies outside or on C_T^{pq} so that $rs \geqslant pq$. Now a result of Levi-Civita, compare Cartan [2, p. 232], yields

$$K = \lim_{\substack{pq \to 0 \\ pr \to 0}} \frac{pq^2 - rs^2}{pr \cdot pq} \leqslant 0.$$

From (39.12) we see that Riemann spaces with curvature 0 in the present sense coincide with the locally euclidean spaces and hence with the spaces of vanishing curvature in the usual sense. For spaces with negative curvature the situation is slightly different:

(41.7) *A Riemann space with negative curvature in the usual sense has negative curvature in the present sense (and therefore strictly convex capsules). The converse need not hold.*

Theorems (36.9) and (41.6) imply the positive part of the assertion. However, the inequality $2m_{ab}m_{ac} < bc$ still holds for all non-degenerate small triangles abc if the curvature K defined above vanishes for certain or all dP at isolated points and is negative otherwise.

At any rate, our theorems on spaces with negative curvature or with strictly convex capsules, imply theorems on Riemann spaces with negative curvature in the classical sense.

Curvature in two-dimensional Finsler spaces

For non-Riemannian spaces a satisfactory theory of curvature comparable to that of Riemann spaces does not yet exist. The main difficulty in handling the existing concepts from our point of view is that they all use osculating Riemannian metrics, which do not lend themselves to our direct approach in terms of the given metric.

We therefore limit the discussion to a result of Pedersen [1, p. 88] on two-dimensional Finsler spaces. There the curvature does not only depend on the point, but is a function $K(x, y, dx, dy) = K(p, \varphi)$ of the line element. We show

(41.8) *If a two-dimensional Finsler space has convex capsules then its curvature is non-positive for every line element.*

The argument is similar to that in the proof of (41.6). Let a line element be given and choose the segment $T(p, q)$ such that it is tangent to the line element at p, where $q \in S(p, \rho(p))$. Let $T = T(p, r)$ and $T(q, s)$ be transversal to $T(p, q)$ at p and q, and lie on the same side of $T(p, q)$, where $r, s \in S(p, \rho(p))$. Since $T(p, q)$ is perpendicular to $T(q, s)$ the latter is tangent to C_T^{pq} at q, hence the supporting geodesic to C_T^{pq} at q is unique and contains $T(q, s)$. Therefore (pxr) and (qys) imply $xy \geqslant pq$ and any curve connecting x to y is at least as long as $T(p, q)$.

In particular the parallel curve to the geodesic $\mathfrak{g}$ carrying $T(p, q)$ in the sense of Finsler, see Finsler [1, p. 105] and Berwald [1], through x intersects $T(q, s)$ for small px in a point y_x. Denote the length of the arc of the parallel curve between x and y_x by λ_x.

By introducing on $T(p, x)$ a proper parameter $\varepsilon = \varepsilon(px)$ with $d\varepsilon/dpx > 0$, whose definition need not concern us here, the function λ_x becomes a function $\lambda(\varepsilon)$, such that

$$\left.\frac{d^2\lambda(\varepsilon)}{d\varepsilon^2}\right|_{\varepsilon=0} = -\int_0^{pq} K(x(\tau), \varphi_\tau)\, d\tau,$$

where $x(\tau)$ represents $T^+(p, q)$ and $(x(\tau), \varphi_\tau)$ is the line element of $T^+(p, q)$ at $x(\tau)$, see Berwald [1, § 3].

Since $\lambda(\varepsilon)$ reaches for $\varepsilon = 0$ a minimum, we have $\left.\frac{d^2\lambda(\varepsilon)}{d\varepsilon^2}\right|_{\varepsilon=0} \geqslant 0$, hence

$$\int_0^{pq} K(x(\tau), \varphi_\tau)\, d\tau \leqslant 0.$$

The arbitrariness of q and the — assumed — continuity of K imply $K(x(0), \varphi_0) \leqslant 0$, which proves the theorem.

Volume and area of spheres

We conclude the discussion of spaces with non-positive curvature by showing that certain inequalities for the volume and area of spheres which are well known for Riemann spaces hold, surprisingly enough, for G-spaces, under very weak differentiability hypotheses which we formulate as follows:

(*) *In $S(p, \rho(p))$ a Minkowski metric $\mu(a, b)$ topologically equivalent to the given metric ab can be introduced such that*

(41.9) $\qquad ab = \mu(a, b)$ *for points a, b collinear with p.*

(41.10) $\qquad$ *If $a_\nu \to p$, $b_\nu \to p$, $a_\nu \neq b_\nu$ then $a_\nu b_\nu / \mu(a_\nu, b_\nu) \to 1$.*

The first condition means, in the language of differential geometry, that cartesian coordinates of a euclidean metric associated to $\mu(a, b)$ (Section 17), are normal coordinates at p; the second condition is familar to us from Section 15.

The inequalities will follow from:

If the curvature is non-positive and (*) holds in $S(p, \rho(p))$, then

(41.11) $\qquad ab \geqslant \mu(a, b) \qquad$ for $\qquad a, b \in S(p, \rho(p))$.

The assertion follows from (41.9) if $a = p$ or $b = p$. Assume therefore $a \neq p$, $b \neq p$, $a \neq b$ and let $x(\tau)$, $y(\tau)$ represent the segments $T(p, a)$, $T(p, b)$ respectively with $x(0) = y(0) = p$. Put $\alpha = pa$, $\beta = pb$. Thus, because μ is Minkowskian,

$$\mu[x(\alpha\tau), y(\beta\tau)] = \tau\mu[x(\alpha), y(\beta)] = \tau\mu(a, b) \qquad \text{for} \qquad 0 \leqslant \tau \leqslant 1,$$

and by (41.10)

$$\lim x(\alpha\tau)\, y(\beta\tau)/\tau = \lim \mu[x(\alpha\tau), y(\beta\tau)]/\tau = \mu(a, b).$$

We conclude from the lemma (41.1) for $\tau = 1$ that

$$ab = x(\alpha)\, y(\beta) \geqslant \mu(a, b).$$

If the Minkowskian geometry, and hence the G-space, is n-dimensional we mean by volume $|M|_n$ of a set M the n-dimensional Hausdorff measure of M defined with arbitrary sets, and by area of a hypersurface S its $(n-1)$-dimensional Hausdorff measure $|S|_{n-1}$. We have

(41.12) *If an n-dimensional G-space has non-positive curvature and satisfies* (*), *then*

$$|S(p, \sigma)|_n \geqslant \varkappa^{(n)}\sigma^n, \qquad \text{for} \qquad 0 < \sigma < \rho(p),$$

where $\varkappa^{(n)} = \pi^{n/2}/\Gamma(n/2 + 1)$ *is the volume of the n-dimensional euclidean unit-sphere, and*

$$|K(p, \sigma)|_{n-1} \geqslant \varkappa^{(n-1)}\, \sigma^{n-1} \int_{\Omega} A^{-1}(u)\, d\omega_u\, .$$

Here Ω is the unit-sphere of a euclidean space associated to $\mu(a, b)$ and $d\omega_u$ its area element at the point with exterior normal u, $A(u)$ is the euclidean area of the section of the Minkowskian unit sphere $\mu(p, x) \leqslant 1$ with the plane through p with normal u.

The point sets $S(p, \sigma)$ and $\mu(p, x) < \sigma$ coincide by (41.9) and so do $K(p, \sigma)$ and $\mu(p, x) = \sigma$. The Minkowskian volume of $\mu(p, x) < \sigma$ equals $\varkappa^{(n)}\sigma^n$, and the Minkowskian area of $\mu(p, x) = \sigma$ is $\varkappa^{(n-1)}\,\sigma^{n-1} \int_{\Omega} A^{-1}(u)\, d\omega_u$, compare Busemann [4, p. 261]. On the other hand, Hausdorff measure satisfies Kolmogoroff's Principle: if a set N carries two different metrics ab and $\mu(a, b)$ and $ab \geqslant \mu(a, b)$ then its r-dimensional measure is for any r with respect to the first metric at least as large as with respect to the second, see l. c. p. 238. Thus (41.12) follows from (41.11). Notice that in the Riemannian case the Minkowskian geometry is euclidean, hence $A(u) = \varkappa^{(n-1)}$, so that the right side of the last inequality becomes the area of the euclidean sphere with radius σ.

42. Angular measure

The only metric invariant which we have hitherto discussed in G-spaces is distance, except in our very last remarks which indicate that volume and area may be significant in Finsler spaces. This is actually the case, for the following reasons:

The local geometry of a Finsler space is Minkowskian, see Section 15. A Minkowskian metric is invariant under the translations of an associated euclidean space (Section 17). The theory of Haar measure shows immediately that there is, up to a factor, only one measure invariant under the translations. This factor was determined by putting the volume of the unit sphere equal to that of the euclidean unit sphere with the same dimension. There are other reasons for this normalization, see Busemann [4]. Volume in Minkowskian geometry determines volume in Finsler spaces, since it yields the volume element at each point. It also determines area, because the metric of the space induces a Finsler metric on any surface immersed in the space. Since we wish area to be intrinsic, area must be defined as the volume of the surface considered as a Finsler space (of lower dimension). Sufficiently many results on these concepts have been obtained for Minkowskian geometry, to make us certain that they are important invariants for the geometry of Finsler spaces. However, the direction in which the theory has been developed so far is not immediately connected with the problems that concern us here.

Limitations of angular measures

The situation is quite different for angles. The measure of an angle in the euclidean plane is uniquely determined by requiring the basic properties α_1, α_2 below and invariance under rotations, see Theorem (42.7). The euclidean angular measure determines the measure in Riemann spaces.

The same requirements do in general not determine an angular measure in a Minkowski plane, because it possesses no rotations, except through π. Consequently, there is a great deal of arbitrariness in any choice of the angular measure in Minkowskian geometry, and any individual measure is bound to be satisfactory in some respects and deficient in others.

To obtain significant results we restrict ourselves to ***investigations which depend only on a few basic indispensable properties of angular measure, and thus apply to many individual measures.***

It may well appear doubtful whether a significant theory can be built on such a general concept of angle. However, it turns out that many beautiful results on Riemannian geometry, in particular those of Cohn-Vossen, do not use the special properties of euclidean angle at all. The remainder of the present chapter deals with questions of this type. But first angular measure must be rigorously defined.

DEFINITION OF ANGULAR MEASURE

Let p be a point of G-surface R. A *direction from, or with origin p,* is an oriented segment $T^+(p, q)$ of length $\sigma(p) = \min\left(\rho(p)/4,1\right)$. The circle $K_p = K(p, \sigma(p))$ is homeomorphic to an ordinary circle and each of its points lies on exactly one direction from p. We conceive of an angle A with vertex p as a set of directions with origin p passing through the points of a subarc of K. If s_1, s_2 are the endpoints of the arc, then the directions $\mathfrak{d}_i$ from p through s_i are the legs of the angle.

We distinguish two cases:

1) If $\mathfrak{d}_1$ and $\mathfrak{d}_2$ are not opposite, then exactly one of the angles with legs $\mathfrak{d}_1, \mathfrak{d}_2$ contains with some direction from p also the opposite direction. We call this angle the "*concave angle*" with legs $\mathfrak{d}_1, \mathfrak{d}_2$ and denote it by $(\mathfrak{d}_1, \mathfrak{d}_2)^{cav}$. The other angle is the "*convex angle*" with legs $\mathfrak{d}_1, \mathfrak{d}_2$ and denoted by $(\mathfrak{d}_1, \mathfrak{d}_2)^{vex}$. For convenience we include the case where $\mathfrak{d}_1 = \mathfrak{d}_2$. In that case $(\mathfrak{d}_1, \mathfrak{d}_2)^{vex}$ consists of $\mathfrak{d}_1$ alone and $(\mathfrak{d}_1, \mathfrak{d}_2)^{cav}$ contains all directions with origin p.

2) If $\mathfrak{d}_1$ and $\mathfrak{d}_2$ are opposite, we call each of the angles with legs $\mathfrak{d}_1, \mathfrak{d}_2$ "*straight*".

A measure for the angles at p is a non-negative function $|A|$ which is defined for all angles A with vertex p and has the following two properties:

α_1: *$|A| = \pi$ if and only if A is straight.*

α_2: *If $A_1 \cap A_2$ consists of exactly one direction, then*

$$|A_1 \cup A_2| = |A_1| + |A_2|.$$

The hypothesis in α_2 means that the arcs on K_p corresponding to A_1 and A_2 have an endpoint, but no other point, in common.

We list some consequences of this definition.

(42.1) *$|A| = 0$ if and only if A is a convex angle with coinciding legs.*

For let $A = (\mathfrak{d}_1, \mathfrak{d}_1)^{vex}$ and let A' be one of the two straight angles with $\mathfrak{d}_1$ as one leg. Then $A \cup A' = A'$ hence

$$\pi = |A'| = |A| + |A'| = |A| + \pi$$

or $|A| = 0$.

Conversely, let $|A| = 0$. The legs $\mathfrak{d}_1$, $\mathfrak{d}_2$ of A cannot be opposite because of α_1. Denote the opposite direction to $\mathfrak{d}_1$ by $\mathfrak{d}_3$. If $\mathfrak{d}_1$ and $\mathfrak{d}_2$ did not coincide, and $A = (\mathfrak{d}_1, \mathfrak{d}_2)^{cav}$ then by α_2

$$|A| = \pi + |(\mathfrak{d}_3, \mathfrak{d}_2)^{vex}| \geqslant \pi.$$

If $A = (\mathfrak{d}_1, \mathfrak{d}_2)^{vex}$, then $\mathfrak{d}_2$ and $\mathfrak{d}_3$ must be opposite, because otherwise

$$\pi = |A| + |(\mathfrak{d}_2, \mathfrak{d}_3)^{vex}| = |(\mathfrak{d}_2, \mathfrak{d}_3)^{vex}|$$

hence $\mathfrak{d}_1 = \mathfrak{d}_2$.

(42.2) *If A contains A' properly then $|A_1| > |A'|$.*

(42.3) *Convex angles have measure less than π and conversely.*
Concave angles have measure greater than π and conversely.

(42.4) *Vertical angles have the same measure.*

It is easy to construct all measures for angles with the same vertex p: Let the two opposite directions $\mathfrak{d}_1$, $\mathfrak{d}_2$ intersect K_p in s_1, s_2. Denote one of the semicircles of K_p with endpoints s_i by K' the other by K''. Let $f(s)$ be any strictly monotone function on K' with $f(s_1) = 0$ and $f(s_2) = \pi$. If $s \in K''$ then the diametrically opposite point s' to s lies on K' and we put $f(s) = \pi + f(s')$. For any angle A with vertex p and legs $\mathfrak{d}'$, $\mathfrak{d}''$ intersecting K_p in s', s'' we define

(42.5)
$$|A| = |f(s') - f(s'')| \quad \text{if } A \text{ contains } \mathfrak{d}_1,$$
$$|A| = 2\pi - |f(s') - f(s'')|, \quad \text{if } A \text{ does not contain } \mathfrak{d}_1.$$

Then α_1, α_2 are satisfied and $|A| \geqslant 0$.

Conversely, if $f(s)$ is defined as $|(\mathfrak{d}_1, \mathfrak{d})^{vex}|$ when $\mathfrak{d}$ intersects K', and as $|(\mathfrak{d}_1, \mathfrak{d})^{cav}|$ when it intersects K'', and $f(s_2) = \pi$, then $|A|$ satisfies (42.5).

Uniqueness of angular measure in elementary geometry

The great arbitrariness of angular measure in the general case makes it worthwhile to briefly discuss the reason for the universal agreement on the value of angular measure in the euclidean and non-euclidean geometries.

A "*rotation*" about p of the disc $S(p, \eta)$, $0 < \eta \leqslant \rho(p)/2$, in a two-dimensional G-space is an orientation preserving isometry of $S(p, \eta)$ on itself which leaves p fixed. ($S(p, \eta)$ is orientable because it is homeomorphic to the plane, see Section 10.)

(42.6) *Given two points q_1, q_2 with $0 < pq_1 = pq_2 < \eta$, there is at most one rotation of $S(p, \eta)$ about p which carries q_1 into q_2.*

For, if Φ_1 and Φ_2 are two rotations carrying q_1 into q_2 then $\Phi = \Phi_1\Phi_2^{-1}$ is a rotation which leaves q_1 fixed. Φ will leave every point of $S(p, \eta)$ fixed if it leaves every point of $S(p, \eta')$, $\eta' = \min(\eta, \rho(p)/16)$, fixed, compare (28.8).

Obviously Φ leaves the diameter B of $S(p, \eta')$ through p and q_1 pointwise fixed. Let r be any point of $S(p, \eta')$ not on B and (pqq_1), $pq < \eta'$. If p' is the midpoint of p and q, then

$$\rho(p') \geqslant \rho(p) - pp' > \rho(p) - \rho(p)/32 = 31\,\rho(p)/32.$$

Hence

$$pq < \rho(p')/4, \qquad rp < \rho(p')/8, \qquad rq < \rho(p')/8.$$

The point $r\Phi$ lies on the same side of B as r because Φ preserves the orientation, moreover $r\Phi\, p = rp$, $r\Phi\, q = rq$ because Φ is an isometry. It now follows from (10.11) that $r\Phi = r$.

Hence, if for some θ, $0 < \theta < \eta$, the rotations of $S(p, \eta)$ about p are transitive on $K(p, \theta)$, then they are simply transitive on $K(p, \theta)$, and also simply transitive on $K(p, \theta')$, if $0 < \theta' < \eta$. In this case we say that $S(p, \eta)$ *can be freely rotated about* p.

A rotation of $S(p, \eta)$ about p induces a one-to-one mapping of the directions with origin p on themselves and transforms an angle with vertex p into an angle with vertex p. We show:

(42.7) *If $S(p, \eta)$, $0 < \eta \leqslant \rho(p)/2$, can be freely rotated about p, then there is exactly one measure $|A|$ for the angles A with vertex p, which is invariant under rotation about p. For sufficiently small positive θ the disk $\overline{S}(p, \theta)$ is convex and the measure of A is proportional to the length of the arc of $K(p, \theta)$ which the directions in A intercept on $K(p, \theta)$.*

Proof: Let $0 < \theta'' < \eta/2$. In $S(p, \theta'')$ choose a convex domain D containing p as interior point and let $S(p, 2\,\theta') \subset D$. There is a segment T with endpoints on $K(p, 2\,\theta')$, hence in $S(p, \eta)$, which does not pass through p but contains points of $S(p, \theta')$.

Let q be a foot of p on T. Then $0 < \theta = pq < \theta'$ and T does not enter $S(p, \theta)$, hence separates no two points of $K(p, \theta)$. A rotation of $S(p, \eta)$ about p carries the point q into a point q_1 of $K(p, \theta')$ and T into a segment T_1 with endpoints on $K(p, \theta')$ through q_1 which does not separate any two points of $K(p, \theta)$. In the terminology of Section 25 the sets $T \cap D$ and $T_1 \cap D$ are supporting lines of $K(p, \theta)$. Since q_1 can be chosen anywhere on $K(p, \theta)$ this circle is convex, see (25.2), and has by (25.3) finite length 2λ.

Let K' be a semicircle of $K(p, \theta)$ with endpoints s', s'' and choose $s_0 = s'$, $s_1, \ldots, s_n = s''$ in this order on K' such that the arc of K' from s_i to s_{i+1} has length λ/n. If A_i^k is the convex angle with legs from p through s_i and s_k, it follows from the invariance of the measure under rotation about p that

$$|A_0^1| = |A_1^2| = \ldots = |A_{n-1}^n| \text{ and from } \alpha_1, \alpha_2 \text{ that}$$

$$\pi = \Big|\bigcup_{i=0}^{n-1} A_i^{i+1}\Big| = \sum_{i=0}^{n-1} |A_i^{i+1}|, \qquad \text{hence} \qquad |A_i^{i+1}| = \pi/n$$

$$\text{and } |A_i^k| = (k-i)\,\pi/n, \text{ again by } \alpha_2.$$

Thus, if σ is the length of the arc of K' from s' to s and A_σ is the convex angle with vertex p and legs s' and s, then

$$|A_\sigma| = \pi\sigma/\lambda \qquad \text{if} \qquad \sigma/\lambda \qquad \text{is rational.}$$

This equation extends to all real σ with $0 \leqslant \sigma \leqslant \lambda$ because $|A_\sigma|$ is by (42.2) a monotone function of σ. With $f(s) = |A_\sigma|$ the angular measure at p is determined as in (42.5).

Angles in Riemann spaces are uniquely determined by the requirement that the measure of the angles between two curves at a point equals that of the angle formed by tangents in the local euclidean geometry.

Excess

So far we have only considered angular measure at one point. We speak of an angular measure on a G-surface R if a definite angular measure has been defined at each of its points.

We say that *the angle* A_ν *with vertex* p_ν *tends to the angle* A with vertex p if $p_\nu \to p$ and the point-set carrying the directions in A_ν tends to the point-set carrying the directions in A. This is equivalent to requiring that the legs of A_ν tend with proper notations to the legs of A, and a converging sequence of direction $\mathfrak{d}_\nu \in A_\nu$ tends to a direction in A.

An angular measure is *continuous* if $A_\nu \to A$ implies $|A_\nu| \to |A|$. Owing to the combinatorial character of the methods, continuity of the angular measure is, surprisingly enough, not necessary for the generalization of many results on Riemannian geometry which are our primary subject here. Continuity is needed, however, for establishing the usual relation between angular measure and the theory of parallels.

In all these investigations the concept of excess of a triangle is fundamental. To define it, let a, b, c be such that $ab > 0$, $bc > 0$ and the segments $T(a, b)$ and $T(b, c)$ are unique. Denote by $\mathfrak{d}_a$ and $\mathfrak{d}_c$ the directions with origin b which have proper common subsegments with $T^+(b, a)$ and $T^+(b, c)$ respectively. We define the angle abc by $\angle\, abc = (\mathfrak{d}_a, \mathfrak{d}_c)^{vex}$ if $\mathfrak{d}_a$ and $\mathfrak{d}_c$ are not opposite, and put $|\angle\, abc| = |abc|$. If $\mathfrak{d}_a$ and $\mathfrak{d}_c$ are opposite we put $|abc| = \pi$.

Three distinct points a, b, c in a neighborhood $S(p, \rho(p)/8)$ determine a triangle or simplex abc with vertices a, b, c, sides $T(a, b)$, $T(b, c)$, $T(c, a)$, whose interior is the simply connected set bounded in $S(p, \rho(p)/2)$ (which is homeomorphic to the plane) by $T(a, b) \cup T(b, c) \cup T(c, a)$, provided the vertices do not lie on one segment. If the latter is the case we call the triangle *degenerate*. If abc is not degenerate a direction $T^+(b, x)$ in $\angle\, abc$ from b intersects $T(a, c)$, similarly for the other vertices a, c. We define the "*excess*" of any triangle abc by

$$\varepsilon(abc) = |abc| + |bca| + |cab| - \pi.$$

Degenerate triangles have excess 0.

(42.8) *If the angular measure is continuous and a_ν, b_ν, c_ν are distinct and tend to the same point p, then $\varepsilon(a_\nu b_\nu c_\nu) \to 0$.*

For a proof we may assume that the directions from a_ν through b_ν, from b_ν through c_ν, and from c_ν through a_ν converge. If two of these directions tend to opposite directions, (42.8) follows from the remark on degenerate triangles.

Otherwise the angle $\angle\, c_\nu a_\nu b_\nu$ and the vertical angles to $\angle\, a_\nu b_\nu c_\nu$ and $\angle\, b_\nu c_\nu a_\nu$ tend to three angles whose union is straight.

Most important, although trivial, is the additivity of excess:

(42.9) *If the triangle abc is simplicially decomposed into a (finite) number of triangles $a_\nu b_\nu c_\nu$, $\nu = 1, \ldots, n$, then*

$$\varepsilon(abc) = \sum_{\nu=1}^{n} \varepsilon(a_\nu b_\nu c_\nu).$$

We say that an *angular measure on a G-surface R has positive excess* if every non-degenerate triangle in R has positive excess. In the same way we define *angular measures with negative, non-positive, non-negative and zero excess.*

Sign of the excess and parallels

In non-euclidean geometry the sign of the excess is closely related to the theory of parallels. We are going to investigate briefly whether there are such relations in the case of general angular metrics.

Let P be a two-dimensional straight G-space, hence homeomorphic to a plane. We begin with a trivial remark:

(42.10) *If P satisfies the parallel axiom then it possesses continuous angular measures with zero excess.*

For we only have to choose an angular measure at one point p according to (42.5) such that $f(s)$ is a continuous function of s. At any other point q we determine the metric by putting the measure of an angle A equal to that of the angle at p whose directions are parallel to those in A.

The converse is not true without further assumptions on the angular metric: If in the Klein Model of hyperbolic geometry, i. e., a Hilbert geometry (Section 18) inside an ellipse, the euclidean angular measure is used instead of the hyperbolic measure, the excess will be zero although the hyperbolic parallel axiom holds.

Nor is it true, that the parallel axiom and constant sign of the excess imply that the excess vanishes: A hemisphere H' with diametrically opposite points on the bounding great circle identified is a well known model of the projective plane with the arcs of the great circles on H' as straight lines. If we leave the bounding great circle out, we obtain the projective plane without the line at infinity, hence the euclidean plane H. If we use the spherical angles in H instead of the euclidean angles, we obtain an angular metric with positive excess.

However, negative excess and the parallel axiom are incompatible:

(42.11) *If the excess is non-positive and the angular metric is continuous, then the parallel axiom implies zero excess.*

Let abc be a non-degenerate triangle. If (abx) and x traverses $\mathfrak{g}^+ = \mathfrak{g}^+(a, b)$ then $|bac| + |acx| + |cxa| \leqslant \pi$; $\mathfrak{g}^+(c, x)$ tends to the asymptote $\mathfrak{h}^+$ to $\mathfrak{g}^+$ through c. If y follows c on $\mathfrak{h}^+$, then $|acx| \to |acy|$ hence

$$|bac| + |acy| < \pi.$$

By the parallel axiom $\mathfrak{h}^-$ is the asymptote to $\mathfrak{g}^-$ through c, hence by the same argument

$$(\pi - |bac|) + (\pi - |acy|) < \pi$$

which shows together with the above inequality that

$$|cab| + |bca| + |bcy| = |bac| + |acy| = \pi$$

For the same reason

$$(\pi - |abc|) + |bcy| = |xbc| + |bcy| = \pi.$$

Subtracting this equation from the preceding one yields $|abc| + |cab| + |bca| = \pi$.

This proof makes it clear which property of angular measure would allow us to reverse the argument: we neglected the angle $|cxa|$, only if $|cxa| \to 0$ for $ax \to \infty$ can we retrace our steps.

The above example of a euclidean metric with an angular metric of positive excess violates the condition $|cxa| \to 0$. The part of H inside a (spherical or euclidean) circle may serve as model for hyperbolic geometry. With the spherical angles we obtain an example for the hyperbolic parallel axiom with positive excess, and the condition $|cxa| \to 0$ is, of course, still less satisfied.

The reason why the condition does not hold is that the spherical angular measure was only used in part of its natural range. This justifies the following definition:

An angular measure in a straight plane is called non-extendable if for any ray $x(\tau)$ and any point $y \neq x(0)$

$$\lim_{\tau \to \infty} |x(0)\, x(\tau)\, y| = 0.$$

With this new condition various facts suggested by non-euclidean geometry can be established.

(42.12) *If the excess of a non-extendable angular measure in a straight plane is non-negative then it vanishes.*

Take any non-degenerate triangle abc. Let (abx) and (bax'). Since excess is additive and non-negative

$$|cxa| + |cx'b| > |cxa| + |cx'b| + |x'cx| - \pi = \varepsilon(x'cx) \geqslant \varepsilon(abc) \geqslant 0.$$

Because the measure is non-extendable $|cxa| \to 0$ for $ax \to \infty$ and $|cx'b| \to 0$ for $bx' \to \infty$, so that $\varepsilon(abc) = 0$.

(42.13) *If a straight plane possesses a continuous non-extendable angular measure with zero excess then it satisfies the parallel axiom.*

Let c not lie on $\mathfrak{g}(a, b)$ and (abx), (bax'). For $ax \to \infty$ and $bx' \to \infty$ the lines $\mathfrak{g}^+(c, x)$ and $\mathfrak{g}^+(c, x')$ tend to the asymptotes $\mathfrak{h}^+$ to $\mathfrak{g}^+$ and $\mathfrak{k}^+$ to $\mathfrak{g}^- = \mathfrak{g}^+(b, a)$ through c respectively. Let y and z follow c on $\mathfrak{h}^+$ and $\mathfrak{k}^+$ respectively.

Because the excess vanishes $|cax| + |acx| + |axc| = \pi$, by non-extendability $|axc| \to 0$, and by continuity $|acx| \to |acy|$. Therefore $|cax| + |acy| = \pi$ and, similarly, $|cax'| + |acz| = \pi$. But $|cax| + |cax'| = \pi$, hence $|acy| + |acz| = \pi$, which means, because of α_1, that $\mathfrak{k}^+$ and $\mathfrak{h}^+$ are opposite orientations of the same line.

(42.14) *If a straight plane possesses a non-extendable angular measure with non-positive excess, then the asymptote relation is symmetric (and transitive).*

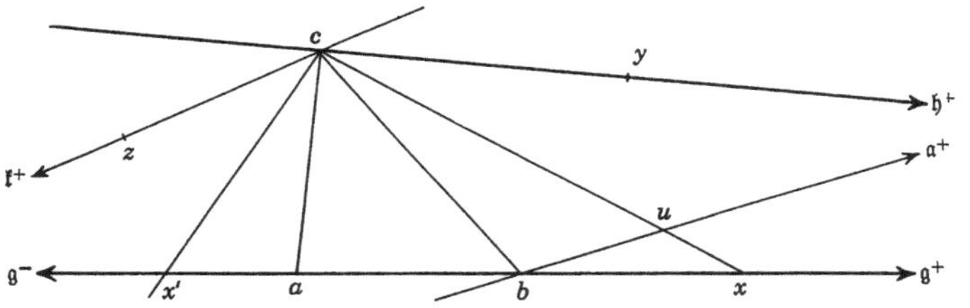

Fig. 46.

For, with the same notations as above let $\mathfrak{h}^+$ be the asymptote to $\mathfrak{g}^+(a, b) = \mathfrak{g}^+$ through c. Denote the asymptote through b to $\mathfrak{h}^+$ by $\mathfrak{a}^+$. Then $\mathfrak{a}^+$ intersects $T(c, x)$ in a point u, and $bu \to \infty$ for $ax \to \infty$, because $\mathfrak{g}^+(c, x) = \mathfrak{g}^+(c, u) \to \mathfrak{h}^+$ which does not intersect $\mathfrak{a}^+$. Non-extendability yields $|cub| \to 0$. Because the excess is non-positive $(\pi - |cub|) + |ubx| + |bxu| - \pi \leqslant 0$, hence

$$|cub| > |ubx|.$$

This shows $|ubx| = 0$ or $\mathfrak{a}^+ = \mathfrak{g}^+$ by (42.1).

For the sake of completeness, it should be mentioned here that a two-dimensional G-space of the elliptic type does not possess angular measures with non-positive excess. This is contained in Theorem (43.5).

RIEMANNIAN METRICS

The link between the present investigations on angular measure and the classical differential geometry of surfaces is the *Gauss-Bonnet Theorem* which states in its simplest form that the excess of a (geodesic) triangle equals for the Riemannian angular measure the integral of the Gauss curvature (assumed continuous) over the triangle, see Blaschke [1, p. 110]. Hence, if the Gauss curvature is positive, then the excess is positive, etc. Consequently, (42.11) is for simply connected two-dimensional Riemann spaces another proof of the fact contained in (37.7) that non-positive curvature and the parallel axiom imply that the metric is euclidean. We also notice:

(42.15) *The Riemannian angular measure of a straight Riemannian plane with non-positive curvature is non-extendable.*

For if $x(\tau)$ represents a ray and y is any point then the inequality (41.4), which holds in the large, see (41.2), yields with $yx(\tau) = \sigma$, $x(0)y = \beta$ that

$$\beta^2 \geqslant \tau^2 + \sigma^2 - 2\tau\sigma \cos |x(0)\ x(\tau)\ y| \quad \text{or}$$

$$\cos |x(0)\ x(\tau)\ y| \geqslant \frac{\tau^2 + \sigma^2 - \beta^2}{2\tau\sigma} = 1 + \frac{(\tau - \sigma)^2 - \beta^2}{2\tau\sigma}.$$

Since $|\tau - \sigma| \leqslant \beta$ the right side tends to 1 for $\tau \to \infty$ so that $|x(0)\ x(\tau)\ y| \to 0$.

Therefore (42.14) contains another proof for Riemannian planes with non-positive curvature, that the asymptote relation is symmetric and transitive.

43. Excess and characteristic

The main application of the Gauss-Bonnet Theorem is a basic relation between the integral curvature of a domain D and its topological invariants. This relation is of a purely combinatorial nature, if integral curvature is interpreted as the sum of the excesses of the triangles in a simplicial decomposition of D. It is therefore not surprising that this relation holds for general angular measures on G-surfaces. In the proof it is assumed that the reader is familiar with the elementary topology of two-dimensional manifolds, for instance, Kerékjártó [1, pp. 130–156].

THE BASIC RELATION

Let $T(a, b) \cup T(b, c)$, where a, b, c are distinct, be part of the boundary B of a closed domain D (i. e., the closure of a connected open set) on the G-surface R. Assume also that for a suitable $\delta < \min (ab, bc, \rho(b)/4, 1)$ the intersection $(D - B) \cap S(b, \delta)$ coincides with one of the sets into which $T(a, b) \cup T(b, c)$ decomposes $S(b, \delta)$. Then $D \cap K(b, \delta/2)$ is an arc. The directions from b through points of this arc form, by definition, *the angle of B in D at b*. A simple closed geodesic polygon is a Jordan curve of the form

$$\bigcup_{i=1}^{n} T(a_i, a_{i+1}), \quad a_i \neq a_{i+1}, \quad a_{n+1} = a_1$$

Let D be a compact domain on the G-surface R whose boundary B, if any, consists of k simple closed, mutually disjoint, geodesic polygons $P_1, \ldots, P_k$. Denote by $\beta_1, \ldots, \beta_r$ the measures of the non-straight angles of B in D at the vertices of all the P_j.

Since D is compact it can be simplicially decomposed into non-degenerate triangles $a_1b_1c_1, \ldots, a_tb_tc_t$. This may require subdivision of the sides of the P_j. Denote by v and s the total numbers of vertices and sides of the $a_\nu b_\nu c_\nu$. Then

$$v - s + t = \chi(D) \tag{43.1}$$

is independent of the simplicial decomposition of D. According to the now prevailing terminology *$-\chi(D)$ is the Euler characteristic of D*. (However, Cohn-Vossen calls $\chi(D)$ the Euler characteristic.)

$\chi(D)$ is a topological invariant of D and has the following interpretation

$$\chi(D) = \begin{cases} 2 - (2\gamma + k) & \textit{if } D \textit{ is orientable} \\ 2 - (\gamma + k) & \textit{if } D \textit{ is not orientable,} \end{cases} \tag{43.2}$$

where γ is the genus of D.

The *fundamental relation* between excess and characteristic is:

$$\sum_{\nu=1}^{t} \varepsilon(a_\nu b_\nu c_\nu) = 2\pi\, \chi(D) - \sum_{j=1}^{r} (\pi - \beta_j). \tag{43.3}$$

For a proof denote by v_b, s_b the number of vertices and sides on B. Then $v_b = s_b$, because the P_j are simple closed polygons. Put $v_i = v - s_b$, $s_i = s - s_b$. We observe first:

$$3\,t = s_b + 2\,s_i = v_b + 2\,s_i. \tag{43.4}$$

To see this, assume first that B is not empty and put

$$3\,t = s_b + 2\,s_i + \delta.$$

If $t > 1$, we remove from D a triangle $a_\mu b_\mu c_\mu$ with two sides on B if there are such, otherwise an $a_\mu b_\mu c_\mu$ with one side on B. We denote the new domain by D_1 and the corresponding numbers by t^1, s_b^1, s_i^1. In the first case the third side of $a_\mu b_\mu c_\mu$ becomes a boundary side, hence $s_b^1 = s_b - 1$, $s_i^1 = s_i - 1$. In the second case the other two sides of $a_\mu b_\mu c_\mu$ become boundary sides, hence $s_b^1 = s_b + 1$ and $s_i^1 = s_i - 2$. In either case

$$3(t-1) = 3\,t^1 = s_b^1 + 2\,s_i^1 + \delta.$$

If $t > 2$ we continue with D_1 as with D, obtaining D_2 and numbers

$$t^2 = t^1 - 1 = t - 2,\ s_b^2,\ s_i^2 \text{ which satisfy}$$

$$3(t-2) = 3\,t^2 = s_b^2 + 2\,s_i^2 + \delta.$$

After $t-1$ steps we arrive at a single triangle, where

$$3 = 3\,t^{t-1} = s_b^{t-1} + 2\,s_i^{t-1} + \delta = 3 + 2\cdot 0 + \delta,$$

hence $\delta = 0$.

If B is empty we remove $a_t b_t c_t$. The numbers t', s_b', s_i' for the remaining set D' are $t' = t - 1$, $s_b' = 3$, $s_i' = s_i - 3$ and satisfy

$$3\,t - 3 = 3\,t' = s_b' + 2\,s_i' = 2\,s_i - 3,$$

which proves (43.4).

Now $\Sigma\varepsilon(a_\nu b_\nu c_\nu)$ is the sum of the measures of all the angles in the various triangles $a_\nu b_\nu c_\nu$ minus $t\pi$. The angles with a given vertex in the interior add up to 2π, those with a vertex on B add up to π or β_i. If v_b' is the number of vertices on B where the angle sum is π, then

$$\sum_{v=1}^{t} \varepsilon(a_\nu b_\nu c_\nu) = 2\pi\, v_i + \pi\, v_b' + \sum_{j=1}^{r} \beta_j - \pi\, t = 2\pi\, v_i + \pi\, v_b - \sum_{j=1}^{r} (\pi - \beta_j) - \pi\, t.$$

Because $v_i = v - v_b = v - s_b$ we obtain from (43.4)

$$\sum_{\nu=1}^{t} \varepsilon(a_\nu b_\nu c_\nu) = 2\pi(v - s_b) + \pi(3\,t - 2\,s_i) - \pi\, t - \sum_{j=1}^{r} (\pi - \beta_j)$$

$$= 2\pi(v - s + t) - \sum_{j=1}^{r} (\pi - \beta_j).$$

which proves (43.3). This relation implies that

$\sum_{v=1}^{r} \varepsilon(a_v b_v c_v)$ *is independent of the decomposition of D into triangles,*

which also follows from the additivity, (42.9), of excess. We put

$$\varepsilon(D) = \sum_{v=1}^{t} \varepsilon(a_v b_v c_v),$$

and call it the "*total excess of D*".

Sign of the excess and structure of compact surfaces

Among the numerous applications of (43.3) we notice first that for compact R we may identify D with R. Then the boundary B is empty and we deduce from (43.2):

$$\varepsilon(R) = 2\pi\,\chi(R) = \begin{cases} 4\pi(1-\gamma) & \text{if } R \text{ is orientable} \\ 2\pi(2-\gamma) & \text{if } R \text{ is not orientable.} \end{cases}$$

In the orientable case, $\gamma = 0$ characterizes the sphere and $\gamma = 1$ the torus. In the non-orientable case, $\gamma = 1$ characterizes the projective plane and $\gamma = 2$ the one-sided torus. Thus $\varepsilon(R) > 0$ if R is — topologically — the sphere or the projective plane, $\varepsilon(R) = 0$ if R is the torus or one-sided torus and $\varepsilon(R) < 0$ in all other cases. This yields the necessity part in each of the following statements

(43.5) *A compact two-dimensional manifold M can be metrized as a G-space with angular measure which has*

a) *non-negative excess if, and only, if M is the sphere, projective plane, torus or one-sided torus,*

b) *non-positive excess if, and only if, M is not the sphere or projective plane,*

c) *positive excess if, and only if, M is the sphere or projective plane,*

d) *negative excess if, and only if, M is not the sphere, projective plane, torus or one-sided torus.*

A torus or one-sided torus with non-positive or non-negative excess has zero excess.

The sufficiency follows from the fact that a compact two-dimensional manifold can be provided with a Riemann metric of constant curvature.

Surfaces with negative excess

The last section and (43.5) contain analogues for spaces with non-positive excess to the results on spaces with non-positive curvature, compare (39.14). There are analogues to most theorems. We restrict ourselves to the two most important ones:

(43.6) *The universal covering space $\overline{R}$ of a G-surface R with non-positive excess is straight.*

The angular measure on R induces on $\overline{R}$ locally an angular measure with non-positive excess, which has non-positive excess in the large because of the additivity property (42.9). Since the sphere and the plane are the only simply connected surfaces, $\overline{R}$ must by (43.5 b) be homeomorphic to the plane.

If the geodesic through two points of $\overline{R}$ were not unique, then R would contain two geodesic curves $\mathfrak{c}_1$, $\mathfrak{c}_2$ lying on different geodesics with common distinct endpoints a, b. Since each $\mathfrak{c}_i$ consists of a finite number of segments, $\mathfrak{c}_1$ and $\mathfrak{c}_2$ intersect only in a finite number of points. Each $\mathfrak{c}_i$ is a Jordan arc. Otherwise a simple closed geodesic monogon would exist and bound a domain D' with $\chi(D') = 1$. But this contradicts (43.3) and $\varepsilon(D') \leqslant 0$.

Therefore a domain D exists bounded by subarcs of $\mathfrak{c}_1$ and $\mathfrak{c}_2$ which have their endpoints and no other point in common. Again $\chi(D) = 1$ and $\varepsilon(D) \leqslant 0$, but $\Sigma(\pi - \beta_i) < 2\pi$ because D has only two proper vertices, which contradicts (43.3).

Our last theorem on surfaces with negative excess is:

(43.7) *On a G-surface with negative excess a free homotopy class contains at most one closed geodesic.*

For two freely homotopic closed geodesics on R would give rise to two parallel axes $\mathfrak{g}_1$, $\mathfrak{g}_2$ on $\overline{R}$ of a motion Ξ' in the fundamental group of R. Let $a_i \in \mathfrak{g}_i$ and $a_i' = a_i \Xi'$. Then $T(a_1, a_2)\, \Xi' = T(a_1', a_2')$. Since the angular measure on $\overline{R}$ induced by that on R is invariant under Ξ', the angle sum in the quadrangle Q bounded by $T(a_1, a_1') \cup T(a_1', a_2') \cup T(a_2', a_2) \cup T(a_2, a_1)$ is 2π, hence $\varepsilon(Q) = \varepsilon(a_1 a_2 a_2') + \varepsilon(a_1 a_1' a_2') = 0$ so that at least one of these two triangles has non-negative excess.

Non-compact surfaces

If R is not compact, then any compact domain D on R has a boundary. In that case the term $-\Sigma(\pi - \beta_i)$ in (43.3) will in general be very disturbing because little can be said about it. Cohn-Vossen's principal idea is to establish the existence of special compact domains for which the term $\Sigma(\pi - \beta_i)$ takes a manageable form. Up to this point our considerations on angular measure have used only the topological properties of the geodesics, henceforth their metric properties will be essential.

A G-surface R of finite connectivity is topologically a compact manifold M of finite genus γ punctured at a finite number of points $z_1, \ldots, z_k$. For each z_i we construct a simple closed geodesic polygon P_j, that bounds on M a simply connected (closed) domain M_j which contains z_j in its interior and contains no other z_i either in the interior or on the boundary. The set $T_j = M_j - z_j$ on R is homeomorphic to a half cylinder. R being finitely compact, a point approaching z_j on M tends to infinity on R. We call T_j a "*tube*". These tubes are the new phenomena on non-compact as compared to compact manifolds.

We denote by $C_j(u)$, $u > 0$, the class of all curves on T_j which are freely homotopic to P_j and have distance at most u from P_j. For every curve $C \in C_j(u)$ there are points $a \in C$ and $b \in P_j$ such that

$$ab = CP_j \leqslant u.$$

Any segment $T(a, b)$ lies, except for a and b, in the interior of the domain on T_j bounded by P_j and C. Otherwise a proper subsegment of $T(a, b)$ would join C to P; and $ab > P_j C$.

The class $C_j(u)$ contains curves of finite length, e. g., P_j. Since T_j is a finitely compact set, we conclude from Section 5 that $C_j(u)$ contains an, in general not unique, curve $R_j(u)$ of minimal length $\lambda_j(u)$. We notice first that

$$(43.8)\qquad 0 \leqslant \lambda_j(u_1) - \lambda_j(u_2) \leqslant 2(u_2 - u_1) \qquad \textit{if} \qquad 0 < u_1 < u_2.$$

In the proof the subscript j will be omitted. It is obvious that $\lambda(u_1) \geqslant \lambda(u_2)$ because $C(u_1) \subset C(u_2)$.

There are points $a_2 \in R(u_2)$ and $b \in P$ such that $a_2 b = R(u_2)P \leqslant u_2$. If $a_2 b \leqslant u_1$ then $R(u_2) \subset C(u_1)$, hence $\lambda(u_1) \leqslant \lambda(u_2)$ and by the preceding remark $\lambda(u_1) = \lambda(u_2)$. If $a_2 b > u_1$, choose a_1 with $(b a_1 a_2)$ and $b a_1 = u_1$. We know that a_1 lies in T and

$$a_1 a_2 = a_2 b - a_1 b \leqslant u_2 - u_1.$$

Now the curve $T^+(a_1, a_2) \cup R(u_2) \cup T^+(a_2, a_1)$ is freely homotopic to P on T and has distance at most u_1 from P and length $\lambda(u_2) + 2\, a_1 a_2$. Therefore $\lambda(u_1) \leqslant \lambda(u_2) + 2\, a_1 a_2 \leqslant \lambda(u_2) + 2(u_2 - u_1)$.

Proof of Cohn-Vossen's Theorem

We call a vertex a_i of a simple polygon $\ldots \cup T(a_{i-1}, a_i) \cup T(a_i, a_{i+1}) \cup \ldots$ *proper* if for points a'_{i-1} with $(a_{i-1}\, a'_{i-1} a_i)$ on $T(a_{i-1}, a_i)$ and a'_{i+1} with $(a_i a'_{i+1} a_{i+1})$ on $T(a_i a_{i+1})$ the angle $\angle\, a'_{i-1} a_i a'_{i+1}$ is not straight, and prove

(43.9) ***Cohn-Vossen's Theorem.*** *$R_j(u)$, $u > 0$, is a simple closed geodesic polygon and hence bounds a subtube $T_j(u)$ of T_j. Moreover, $R_j(u)$ is either a closed geodesic; or all its proper vertices are also proper vertices of P_j and the angles of $R_j(u)$ in $T_j(u)$ at these vertices are concave; or $R_j(u)$ is a closed geodesic monogon with one proper vertex q in which case the angle of $R_j(u)$ at q in $T_j(u)$ is convex and every point of $R_j(u)$ other than q has distance greater than u from P_j.*

In the proof we again omit the subscript j. Represent $R(u)$ with the arc length σ as parameter: $x(\sigma)$, $0 \leqslant \sigma \leqslant \lambda(u)$. Take a point $x(\sigma_0)$ which is not a vertex of P, and for which either $x(\sigma_0)P > u$ or $x(\sigma_0)P \leqslant u$ and a $\sigma_1 \neq \sigma_0$ exists with $x(\sigma_1)P \leqslant u$. Then $x(\sigma)$ represents with a suitable positive δ a segment for $|\sigma - \sigma_0| \leqslant \delta$. (If $\sigma_0 = 0$ the last inequality is to be replaced by $0 \leqslant \sigma \leqslant \delta$ and $0 \leqslant \lambda(u) - \sigma \leqslant \delta$.)

For, replace the curve $x(\sigma)$, $|\sigma - \sigma_0| \leqslant \delta$, by a segment $T[x(\sigma_0 - \delta), x(\sigma_0 + \delta)]$ thus obtaining a new curve R'. If δ is sufficiently small then R' still lies in T, even if $x(\sigma_0)$ is a point of P other than a proper vertex. Moreover R' is still homotopic to P and has distance $\leqslant u$ from P.

This implies, first, that $R(u)$ is a geodesic polygon. Next: if $R(u)$ contains points with distance less than u from P, then it contains infinitely many; hence none of its points can be a proper vertex unless it is also a proper vertex of P. Thus $R(u)$ is either a closed geodesic, or all its proper vertices are vertices of P. Thirdly, if $R(u)$ is not a closed geodesic and does not contain points of P, then it has a proper vertex q, and q is the only point of $R(u)$ whose distance from P does not exceed u. Hence $qP = u$. If $R(u)$ were not a Jordan curve it would contain a simple closed polygon R' homotopic to P. Then $R'P > u$ otherwise $R' \in C(u)$, but its length $2\eta'$ would be smaller than $\lambda(u) = 2\eta$.

Choose the origin of arc length on $R(u)$ such that R' is traversed by $x(\sigma)$ for $\eta - \eta' \leqslant \sigma \leqslant \eta + \eta'$. By the preceding argument R' is a geodesic monogon with a proper vertex at $x(\eta - \eta') = x(\eta + \eta')$. For $x(\sigma)$ continues to represent the same geodesic unless a point with distance at most u from P is passed, so that $x(\sigma)$ would always stay on R' if R' were a closed geodesic.

At least one of the two curves traversed by $x(\sigma)$ for $0 \leqslant \sigma \leqslant \eta - \eta'$ and $\eta + \eta' \leqslant \sigma < 2\eta = \lambda(u)$ must contain a point with distance u from P, assume the first curve does. If we denote it by A, then $AR'A^{-1}$ or, more precisely, the curve $y(\sigma)$ defined by

$$y(\sigma) = \begin{cases} x(\sigma) & \text{for} \quad 0 \leqslant \sigma < \eta + \eta' \\ x(2\eta - \sigma) & \text{for} \quad \eta + \eta' \leqslant \sigma \leqslant \lambda(u) \end{cases}$$

is homotopic to P, lies in $C(u)$, has length $\lambda(u)$ and is therefore an $R(u)$. But it has a proper vertex for $\sigma = \eta + \eta'$ although $y(\eta' + \eta)\, P > u$.

Since $R(u)$ is a simple closed polygon it bounds a tube $T(u) \subset T$. The angle A of $R(u)$ in $T(u)$ at a proper vertex b of $R(u)$ on P must be concave. For if it were convex, then for points a, c on the two legs of A and sufficiently close to b, the segment $T(a, c)$ would lie in T. The curve originating from $R(u)$ by replacing $T(a, b) \cup T(b, c)$ with $T(a, c)$ would have smaller length than $R(u)$ and lie in $C(u)$ for $ab < u$.

Finally, take the case where $R(u)$ is a geodesic monogon with a single proper vertex q with $qP = u$. As previously observed, a segment $T(q, p)$ of length $qp = qP = R(u)P$ and connecting q to a point p of P lies, except for q and p, in the interior of the domain D bounded by $R(u)$ and P, hence does not enter the interior of $T(u)$. If the angle of $R(u)$ at q in $T(u)$ were concave, the angle A' of $R(u)$ at q in D would be convex. If the points a and c lie on the two legs of A' and sufficiently close to q, then $T(q, p)$ which lies in D, must cross $T(a, c)$ at a point r. Then $pr < u$, hence the polygon obtained from $R(u)$ through replacing $T(a, q) \cup T(q, c)$ by $T(a, c)$ would lie in $C(u)$, but have smaller length than $R(u)$. This completes the proof of (43.9).

Note. For a later application we observe that the hypothesis in (43.9) may be widened as follows: Instead of a simple closed polygon P_j, we consider a curve P_j' originating from P_j by adding a segment $T(p, q)$, where $p \in P_j$ and $T(p, q)$ lies, except for p, in the interior of T_j. The Tube T_j is interpreted as "cut" along $T(p, q)$, that is, the class $C'_j(u)$ consists of curves homotopic to P'_j or $P_j \cup T_j$ which do not cross $T(p, q)$ and have distance at most u from P'_j. A curve $R'_j(u)$ in $C'_j(u)$ of minimal length $\lambda'_j(u)$ has then the same properties as $R_j(u)$. The only not entirely obvious property is the simplicity of $R'_j(u)$; it is conceivable that $R'_j(u)$ contains part of the segment $T(p, q)$ traversed twice. However, then $R'_j(u)$ must, by the previous arguments, contain all of $T(p, q)$ traversed twice. Since at least one of the angles which $T(p, q)$ forms with P_j at p in T_j is convex, $R'_j(u)$ could be shortened as above.

Applications of Cohn-Vossen's Theorem

Let $u_1, \ldots, u_k$ be arbitrary positive numbers and construct $R_j(u_j)$ and $T_j(u_j)$ as before. Denote by $D(u_1, \ldots, u_k) = D_u$ the compact domain on R bounded by $R_1(u_1), \ldots, R_k(u_k)$. The angle of $R_j(u_j)$ at a proper vertex in D_u is concave (convex) if the angle in $T_j(u_j)$ is convex (concave), so that $R_j(u_j)$ yields a positive contribution to the term $-\Sigma(\pi - \beta_i)$ in (43.3) applied to D_u if it has a single convex angle in $T_j(u_j)$. Hence (43.9) and (43.3) yield the following estimate:

$$(43.10) \qquad \varepsilon(D_u) \leqslant 2\pi\, \chi(D_u) + \pi k,$$

with the equality sign only when all $R_j(u_j)$ are closed geodesics.

A first consequence of this inequality is:

(43.11) ***Theorem.*** *The only G-surfaces of finite connectivity which possess angular measures with non-negative, but not identically vanishing, excess are the sphere, the projective plane and the plane.*

Because of (43.5) we merely have to treat the case of a non-compact G-surface R of finite connectivity. If such a surface were not homeomorphic to the plane, its universal covering space $\overline{R}$ would be the plane which corresponds to $\gamma = 0$ and $k = 1$, and there are infinitely many points of $\overline{R}$ over a given point of R. Choose in R a triangle abc with $\varepsilon(a\,b\,c) > 0$. The triangles $\overline{a}_\nu \overline{b}_\nu \overline{c}_\nu$ over $a\,b\,c$ are all disjoint because by definition $a\,b\,c \subset S(p, \rho(p)/8)$ for a suitable p, and their excess equals $\varepsilon(a\,b\,c)$. Choose n so large that $\sum_{\nu=1}^{n} (\overline{a}_\nu \overline{b}_\nu \overline{c}_\nu) > 4\pi$. Let P_1 be a simple closed geodesic polygon in $\overline{R}$ containing $\bigcup_{\nu=1}^{n} \overline{a}_\nu \overline{b}_\nu \overline{c}_\nu$ in its interior. (That such P_1 exist is shown in (44.10).)

The unbounded domain in $\overline{R}$ bounded by P_1 is a tube T_1. For any $u_1 > 0$ we choose $R_1(u_1)$ in T_1. Then D_u is the finite part of R bounded by $R_1(u_1)$ and contains $\bigcup_{\nu=1}^{n} \overline{a}_\nu \overline{b}_\nu \overline{c}_\nu$. Because the excess is non-negative, $\varepsilon(D_u) \geqslant \Sigma\varepsilon(\overline{a}_\nu \overline{b}_\nu \overline{c}_\nu) \geqslant 4\pi$, but $2\pi\ \chi(D_u) + k\pi = 3\pi$, which contradicts (43.10).

For the further discussion of non-compact surfaces (i. e., $k > 0$) we improve the estimate (43.10). For that purpose we divide the tubes — following Cohn-Vossen — into three categories:

The tube T_j is "*contracting*" if it contains no curve freely homotopic to P_j of minimal length.

A tube is "*bulging*" if it is not itself contracting, but contains a contracting sub-tube.

A tube without a contracting sub-tube is "*expanding*".[1]

We notice:

(43.12) *A sub-tube of a contracting (expanding) tube is contracting (expanding).*

The statement on expanding tubes is trivial. If a contracting tube T_j with a non-contracting sub-tube T' existed, then T' would contain a curve $\mathfrak{c}'$ of minimal length β' freely homotopic to the boundary P' of T' and hence to P_j. Since T_j is contracting the greatest lower bound $\beta = \lim_{u \to \infty} \lambda_j(u)$ of the lengths of all curves on T_j freely homotopic to P_j would be less than β'. A sequence $\mathfrak{c}_\nu$ of such curves whose lengths β_ν tend to β would be unbounded (since it would otherwise contain a converging sub-sequence whose limit has minimal length β, see Section 5). On the other hand, $\mathfrak{c}_\nu$ must for $\beta_\nu < \beta'$ contain points of $T_j - T'$ which would imply $\beta_\nu \to \infty$.

A few *examples* will be welcome:

The curve $y = e^x$ rotated about the x-axis yields a surface R_1 which is, topologically, a cylinder. A simple closed geodesic polygon on R_1 which cannot be contracted to a point, bounds two tubes, of which one is contracting and the other expanding.

An ordinary half cylinder is an expanding tube.

The hyperboloid of one sheet has two expanding tubes. The curve $y = (1 + x^2)^{-1}$ rotated about the x-axis yields a surface R_2 which is, topologically, a cylinder. The curve $x = 0$ on R_2 is a closed geodesic and bounds two contracting tubes.

At the end of Section 30 we gave examples of a torus with constant negative curvature with an expanding tube, and also of one with a contracting tube.

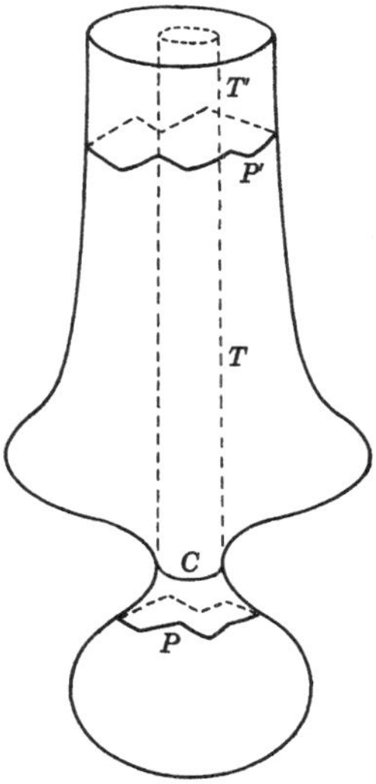

Fig. 47.

It is easier to describe an example for a bulging tube pictorially than by formulas. Figure 47 indicates a sphere with a bulging tube T bounded by P. Its sub-tube T' bounded by P' is contracting, but T is not, because it contains the curve C.

We return to the general theory. An expanding or bulging tube T_i contains a curve C_i freely homotopic to P_i of minimal length. Hence $R_i(u_i)$ may be chosen as C_i for any $u_i > C_i\, T_i$. It follows from Cohn-Vossen's Theorem that $R_i(u_i)$ is then either a closed geodesic or all its angles in $T_i(u_i)$ are concave. (43.3) leads to the following refinement of (43.10):

(43.13) ***Theorem.*** *If the tubes $T_1, \ldots, T_c$, $0 \leqslant c \leqslant k$, of R are contracting and the others are not, then for sufficiently large $u_{c+1}, \ldots, u_k$ and arbitrary positive $u_1, \ldots, u_c$*

$$\varepsilon(D_u) \leqslant 2\pi\, \chi(D_u) + \pi\, c, \tag{43.14}$$

with equality only if $c = 0$ and all $R_j(u_j)$ are closed geodesics.

If $c = k$, then for arbitrary $u_j > 0$

$$2\pi\, \chi(D_u) \leqslant \varepsilon(D_u) \leqslant 2\pi\, \chi(D_u) + \pi\, k.$$

We notice the corollary:

(43.15) *If R possesses no expanding tube, then $\varepsilon(D_u) \geqslant 2\pi\, \chi(D_u)$ for a suitable choice of tubes $T_1, \ldots, T_k$ and arbitrary positive u_j.*

For if the original tubes are not contracting, the bulging ones among them may be replaced by contracting sub-tubes.

We see from (43.14) that zero excess is compatible with the following cases: R is orientable, $\gamma = 0$ and $k = 1$, or $k = 2$, or $k = 3$ in which case $c \neq 3$. The combination $k = c = 4$ is impossible because $c > 0$ precludes equality; R is non-orientable, $\gamma = k = 1$. Hence,

(43.16) *A G-surface with finite connectivity and zero excess is topologically a torus, a one-sided torus, a plane, a cylinder, a Moebius strip, or a sphere with three holes, in which case all three tubes are contracting.*

44. Simple monogons, total excess. Surfaces with positive excess.

The last theorem deviates from the Riemannian case in that the sphere with three holes cannot carry a euclidean metric, see Section 30. In the Riemannian case the (single) angles of $R_j(u_j)$, $0 \leqslant j \leqslant c$, have for suitable positive u_j measures which are as close to π as desired (this is contained in (44.1)), so that the contribution of these $R_j(u_j)$ to $-\Sigma(\pi - \beta_i)$ in (43.3) applied to D_u is as small as desired. Since the other $R_j(u_j)$ contribute non-positive amounts, the term $\pi\, c$ in (43.14) may be replaced by an arbitrarily small positive number. This excludes the sphere with three holes from the surfaces with zero excess because $\chi(D_u) = -2\pi$.

The great arbitrariness of angular measure in the present case permits us to modify the Riemannian angular measure on a Riemannian surface so that it still satisfies our conditions α_1, α_2, but the contributions of the $R_j(u_j)$, $0 \leqslant j \leqslant c$, are as close to πc as desired for any choice of the u_j.

Angular measures uniform at π

We lack a property that insures that, in a uniform way, an angle cannot be nearly straight without having a measure close to π:

An angular measure is called "*uniform at* π" in the subset G of R if a non-decreasing positive function $\delta(\varepsilon) < 1$ and a positive function $\omega(p, \varepsilon) \leqslant \rho(p)/4$ defined for $0 < \varepsilon < \pi$ and $p \in G$ exist, such that the relations

$$0 < a_1 p = a_2 p < \omega(p, \varepsilon) \qquad \text{and} \qquad a_1 a_2/(a_1 p + p a_2) > 1 - \delta(\varepsilon)$$

imply $|a_1 p\, a_2| > \pi - \varepsilon$.

We may and will assume that $\delta(\varepsilon)$ increases and stays below $\frac{1}{2}$. For if the original $\delta(\varepsilon)$ does not satisfy these conditions, then $\delta(\varepsilon)(1 + \varepsilon)/2(2 + \varepsilon)$ will.

The uniformity is contained in the requirement that $\delta(\varepsilon)$ be independent of p. The definition is justified by:

(44.1) *The Riemannian angular measure on a Riemannian G-surface R is uniform at π on R.*

For obviously any $\delta(\varepsilon)$ less than 1-cos $(\varepsilon/2)$, for instance $\delta(\varepsilon) = 1 - \cos(\varepsilon/4)$, will satisfy our requirements, because

$$a_1 a_2/(a_1 p + p a_2) = \eta_\nu, \quad \eta_\nu \to \eta$$

and $a_1 p = p a_2 \to 0$ imply that $|a_1 p a_2| \to 2 \operatorname{Arc} \cos \eta$.

According to Cohn-Vossen, his theorem (43.9) may now be completed by:

(44.2) **Theorem.** *If the angular measure on the tube T_j is uniform at π, then for a suitable $v_j > 0$ the curve $R_j(v_j)$ is either a closed geodesic, or all angles of $R_j(v_j)$ in $T_j(v_j)$ are concave or $R_j(v_j)$ is a monogon, at whose vertex the angle in $T_j(v_j)$ is convex and has measure greater than $\pi - \varepsilon$.*

For a proof we put $\lambda_j(u) = \lambda(u)$ and consider the function

$$f(u) = \lambda(u) + \delta(\varepsilon)\, u, \qquad u \geqslant 1,$$

where $\delta(\varepsilon)$ is the function entering the definition of an angular measure uniform at π. According to (43.8) the function $\lambda(u)$ is continuous and non-increasing, hence $f(u)$ reaches a minimum for some $\nu_j \geqslant 1$. Then for $h > 0$

$$\lambda(\nu_j + h) + \delta(\varepsilon)(\nu_j + h) \geqslant \lambda(\nu_j) + \delta(\varepsilon)\,\nu_j,$$

hence

$$\delta(\varepsilon)\,h \geqslant \lambda(\nu_j) - \lambda(\nu_j + h) \qquad \text{for} \qquad h > 0. \tag{44.3}$$

If $R_j(\nu_j)$ is not a closed geodesic and its angles in $T_j(\nu_j)$ are not concave, then $R_j(\nu_j)$ has a single vertex p with a convex angle in $T_j(\nu_j)$. Choose points a_1, a_2 on the legs of this angle such that they lie on $R_j(\nu_j)$ and $a_1 p = a_2 p < \omega(p, \varepsilon)$. Consider the curve R' originating from $R_j(\nu_j)$ through replacing $T(a_1, p) \cup T(p, a_2)$ by $T(a_1, a_2)$. The distance of R' from P_j is at most $\nu_j + a_1 p$. Putting $h = a_1 p$ and denoting the length of R' by λ' we see that $\lambda(\nu_j + h) \leqslant \lambda'$, hence

$$\lambda(\nu_j) - (\nu_j + h) \geqslant \lambda(\nu_j) - \lambda' = 2\,h - a_1a_2 = 2\,h(1 - a_1a_2/2\,h)$$

and, by (44.3),

$$\delta(\varepsilon) > 1 - a_1a_2/(a_1 p + p\,a_2).$$

Because the angular measure is uniform at π we have $|a_1 p a_2| > \pi - \varepsilon$.

We saw in the beginning of this section that (44.2) implies the following addition to (43.16):

(44.4) *A sphere with three holes does not possess an angular measure which is uniform at π and has zero excess.*

Simple geodesic monogons

The metrizations of the plane which have interested us hitherto were principally those without conjugate points. If a G-plane is not straight, then questions like the existence of geodesics without multiple points, of simple monogons, etc. arise. It will be useful to have a simple example ready for the theorems proved below.

Let F_ε' be a cap of the unit-sphere in E^3 with area ε, $0 < \varepsilon \leqslant 2\pi$, bounded by a circle C, and denote by F_ε'' the part of the circular cone, enveloped by the tangent planes of the sphere along C, which is bounded by C and does not contain the apex. Then $F_\varepsilon = F_\varepsilon' \cup F_\varepsilon''$ is a surface whose integral curvature or total excess is ε, (44.13). For $\varepsilon \leqslant \pi$ all geodesics are simple (i. e., without multiple points) and none is closed, see (44.6). For $\pi < \varepsilon < 2\pi$ there are simple monogons, even very many, (44.9), but no simple closed geodesics,

(44.21 d). But $F'_{2\pi}$ is a hemisphere and $F''_{2\pi}$ a half cylinder, hence $F_{2\pi}$ possesses simple closed geodesics. For no ε does F_ε possess a straight line, (44.21 e) or closed geodesic with multiple points, (44.21 h).

A geodesic with multiple points contains at least one simple monogon. If D is a simply connected domain bounded by a simple monogon P, then $\chi(D) = 1$ and (43.3) yields

$$\varepsilon(D) = 2\pi - (\pi - \beta) = \pi + \beta, \tag{44.5}$$

where β is the angle of P in D at its vertex.

It is convenient to include simple closed geodesics among the simple monogons, and to denote an arbitrary point as vertex. (44.5) yields

(44.6) *The geodesics of a G-plane are all simple and not closed if there is no simple closed polygon bounding a compact domain of excess greater than* π.

This shows that the geodesics on F_ε are simple for $\varepsilon \leqslant \pi$. The problem arises to find conditions under which many geodesic monogons exist. The deepest theorem in this direction is (44.9) which is also due to Cohn-Vossen [2, p. 144]. We begin with a simple fact:

(44.7) *Let the tube T be bounded by the simple monogon P such that the angle at its vertex p (if any) in T is convex. Then every interior point q of T is vertex of a simple monogon which lies in the interior of T.*

Among all curves on T through q homotopic to P there is one, C, of minimal length. The argument for (43.9) shows that C is a geodesic polygon whose vertices, except for q, coincide with the vertices of P. But since p is the only vertex of P and the angle in T at p is convex, C cannot pass through p and have minimal length. As in the proof of (43.9), it is seen that C is simple.

The next statement is merely a lemma needed in the proof of (44.9).

(44.8) If the tube T is bounded by a simple closed monogon P with vertex p, then every point q in the interior of T which can be joined to p by a segment $T(p, q)$ which lies, except for p, in the interior of P and forms with P in T non-concave angles, is vertex of a simple monogon that contains no points of $P \cup T(p, q)$ except q.

As in the note following the proof (43.9) we consider T as cut along $T(p, q)$ obtaining a generalized tube bounded by a polygon P' consisting of P and $T(p, q)$ traversed twice. Among the curves in T' homotopic to P', or in the interior of T, homotopic to P and not crossing $T(p, q)$, there is a shortest arc, C. Clearly C is again a geodesic polygon, and its only conceivable vertex other than q is p. This is, however, impossible because neither of the angles which $T(p, q)$ forms with P in T is concave. C cannot contain any point of $T(p, q) \cup P - p - q$. The simplicity of C is easy to establish.

The result (44.6) might suggest that the existence of simply connected compact polygonal domains with excess greater than π suffices to insure the existence of monogons. It can be seen from examples that this is not correct. The following theorem of Cohn-Vossen shows in which direction the hypothesis must go, and then establishes the existence of surprisingly many monogons.

(44.9) **Theorem.** *In a G-plane R let a bounded set M and a positive η exist such that every simple closed polygon containing M in its interior bounds a domain $D \supset M$ with $\varepsilon(D) \geqslant \pi + 3\eta$. If the angular measure is uniform at π in R, then θ exists such that every point q with $qM \geqslant \theta$ is vertex of a simple geodesic monogon containing M in its interior.*

We first observe:

(44.10) *Any bounded set M of a G-plane R lies inside a simple closed geodesic polygon.*

For each point of $\overline{M}$ is interior point of a non-degenerate triangle. A finite number of these triangles covers M. Denote their union by U. Since U is compact it lies, with an auxiliary euclidean metric $e(x, y)$ in R, inside a circle $e(p, x) = N$. Therefore the boundary of the unbounded component of U is a simple closed geodesic polygon.

Applying (44.10) to the set M in the hypothesis of (44.9) yields a simple closed polygon P containing M in its interior. P bounds a tube T. Denote the length of P by ξ and choose θ' such that P and its interior lie in $S(M, \theta')$. We are going to prove (44.9) for $\theta = 3\,\xi\,\delta^{-1}(\eta) + \theta' + 1$.

Let $qM \geqslant \theta$. If p is a foot of q on P we know that a segment $T(p, q)$ lies except for q in the interior of T. Let $x(\tau)$, $0 \leqslant \tau \leqslant pq$, represent $T^+(p, q)$. For $0 \leqslant \tau \leqslant \xi\,\delta^{-1}(\eta) = \zeta$ we apply the note following the proof of (43.9) to $P_\tau = P \cup T(p, x(\tau))$ and $u = 1$, thus obtaining a simple closed polygon R_τ whose length $\varphi(\tau)$ is minimal among all curves homotopic to P and $T(p, x(\tau))$ traversed twice with distance at most 1 from P_τ. The length of the latter curve is $\xi + 2\tau$, hence

$$\varphi(\tau) \leqslant \xi + 2\tau \leqslant \xi + 2\zeta.$$

The distance of any point outside of $S(M, \theta)$ from P_τ is at least

$$\theta - \tau - \theta' \geqslant 2\zeta + 1 \geqslant \zeta + 2\xi + 1$$

because $\delta(\eta) \leqslant \frac{1}{2}$. Therefore any closed curve whose distance from P_τ is at most one and which contains points outside of $S(M, \theta)$ has length at least

$$2(\zeta + 2\,\xi + 1) - 2 > \varphi(\tau).$$

Therefore R_τ lies in $S(M, \theta)$.

Denote by E_τ the compact domain bounded by R_τ. If R_τ is for any τ a monogon for which the angle at its vertex in E_τ is concave or straight then it follows from (44.7) that every point of $R - E_\tau$, hence every point of $R - S(M, \theta)$, is vertex of a monogon containing E_τ and therefore M. We assume therefore that this happens for no τ.

Then all vertices are by Cohn-Vossen's Theorem vertices of P_τ and the angles at the proper vertices in E_τ are convex and coincide with vertices of P_τ. Moreover, there is at least one such common vertex. It suffices to show that for some τ_0 the point $x(\tau_0)$ is the only vertex, hence R_{τ_0} is a monogon. For since $T\big(x(\tau_0), p\big)$ lies inside or on R_{τ_0} the angles which $T\big(q, x(\tau_0)\big)$ forms with R_{τ_0} in the tube bounded by R_{τ_0} are non-concave. Hence (44.8) applied to q, R_{τ_0}, $x(\tau_0)$, $T\big(q, x(\tau_0)\big)$ instead of $q, P, p, T(p, q)$ yields then the existence of a simple monogon with vertex q containing E_{τ_0}, hence M.

If R_τ were for no τ a monogon with vertex $x(\tau)$ then R_τ would contain another vertex of P_τ which must be a point y of P.

Since R_τ lies in $S(M, \theta)$ and does not cover $T\big(x(\tau), p\big)$, it must intersect $T\big(q, x(\tau)\big)$ at a point z, hence

(44.11) $$\varphi(\tau) \geqslant 2\,zy \geqslant 2\,zp - 2\,p\,y \geqslant 2\,\tau - \xi.$$

If R_τ does not pass through $x(\tau)$, then

(44.12) $$\varphi(\tau + \sigma) \leqslant \varphi(\tau) \qquad \text{for} \qquad 0 < \sigma \leqslant x(\tau)\,R_\tau.$$

For R_τ has then distance at most 1 from $P_{\tau+\sigma}$. The assumption regarding the excess enters the discussion of the case where R_τ passes through $x(\tau)$. In that case denote by γ and γ', $\gamma \geqslant \gamma'$, the measures of the angles into which $T\big(x(\tau), p\big)$ decomposes the angle of R_τ in E_τ. One of the numbers may vanish but not both, since R_τ is simple. Since the angles of R_τ in E_τ are convex, we conclude from (43.2) and the hypothesis that

$$\pi + 3\,\eta \leqslant 2\pi - (\pi - \gamma - \gamma') \leqslant 2\,\gamma + \pi,$$

hence $\gamma \geqslant 3\,\eta/2$. Choose $a_1 = x(\tau + \sigma)$ and a_2 on the leg different from $T\big(x(\tau), p\big)$ of the angle with measure γ such that

$$a_1\,x(\tau) = \sigma = a_2\,x(\tau) \qquad \text{and} \qquad 0 < \sigma < \omega\big(x(\tau), \gamma\big),$$

where $\omega(x(\tau), \gamma)$ is the other function entering the definition of angular measure uniform at π.

Then by the definition of uniformity at π

$$\frac{a_1 a_2}{2\sigma} = \frac{a_1 a_2}{a_1 x(\tau) + x(\tau) a_2} \leqslant 1 - \delta(\gamma) \leqslant 1 - \delta(3\eta/2).$$

If the segment $T(x(\tau), a_2)$ on R_τ is replaced by $T(x(\tau), a_1) \cup T(a_1, a_2)$ then the resulting curve R' belongs to the class of curves whose length is minimized by $R_{\tau+\sigma}$. Therefore

$$\varphi(\tau + \sigma) \leqslant \text{length } R' = \varphi(\tau) + a_1 a_2 + a_1 x(\tau) - x(\tau) a_2$$
$$\leqslant \varphi(\tau) + 2\sigma(1 - \delta(3\eta/2)).$$

We see from (44.12) that the estimate

$$\varphi(\tau + \sigma) \leqslant \varphi(\tau) + 2\sigma(1 - \delta(3\eta/2)),$$

with sufficiently small positive σ holds for every τ.

The limit of a converging sequence of curves R_{τ_ν} with $\tau_\nu \to \tau_0$ has at most length $\liminf \varphi(\tau_\nu)$ [see (5.17)], hence

$$\varphi(\tau_0) \leqslant \liminf_{\tau \to \tau_0 -} \varphi(\tau).$$

The last two inequalities yield

$$\varphi(\tau) \leqslant \varphi(0) + 2\tau(1 - \delta(3\eta/2)) < \xi + 2\tau(1 - \delta(\eta)).$$

In particular, for $\tau = \zeta = \xi\,\delta^{-1}(\eta)$

$$\varphi(\zeta) < 2\zeta - \xi$$

which contradicts (44.11), and proves (44.9).

Total excess

The integral curvature of a set G on a Riemannian surface is the integral of the Gauss curvature over G. If G is not bounded, then the integral is improper and need not exist. In terms of excess we may define a "*total excess*" which equals integral curvature in the Riemannian case, as follows:

A *polygonal region* on a G-surface R is a compact domain whose boundary B is either empty or consists of a finite number of simple closed geodesic polygons.

Let G be a subset of R which is the union of an increasing sequence $\{D_\nu\}$ of polygonal regions (i.e., $G = \cup D_\nu$, $D_\nu \subset D_{\nu+1}$), such that $p_i \in D_{\nu_i+1} - D_{\nu_i}$ implies that $\{p_i\}$ has no accumulation point. If

$$\varepsilon(G) = \lim_{\nu \to \infty} \varepsilon(D_\nu)$$

exists for each such sequence $\{D_\nu\}$, $\pm\infty$ admitted, it is independent of the particular sequence $\{D_\nu\}$ and is called the *total excess* of G.

To see the independence it suffices to prove: If D is any polygonal region in G then $D \subset D_\nu$ for large ν. If this were not the case, then $p_i \in D - D_i$ would exist for every i. Since $p_i \in D \subset G$, there is a first subscript ν_i such that $p_i \in D_{\nu_i+1} - D_{\nu_i}$. But $\{p_i\}$ has an accumulation point because D is compact.

If G is bounded then the assumption that $\{p_i\}$ has no accumulation point implies $D_\nu = G$ from a certain ν on, so that the total excess of G coincides with the excess of G defined previously.

Except for the assumption of connectedness which could easily be avoided, the restriction to polygonal regions is necessary. For in Finsler spaces, even under strong differentiability hypotheses both on the metric and on the angular measure (or in Riemann spaces with other angular measures than the Riemannian), *the additive set function defined on simplicially decomposed sets by the excess can, in general, not be extended to a completely additive set function* on the Borel sets. Total excess can therefore not be regarded as an integral. The following trivial remark is useful:

(44.13) *If the excess of every triangle outside of a bounded subset of G is non-positive (non-negative) then $\varepsilon(G)$ exists.*

For then $\varepsilon(D_{\nu+1}) \leqslant \varepsilon(D_\nu)$ $\left(\varepsilon(D_{\nu+1}) \geqslant \varepsilon(D_\nu)\right)$ from a certain ν on.

Let T be a tube with boundary P. If $\varepsilon(T)$ exists it may be evaluated as follows: We determine on T a curve $R^1 = R(u^1)$, $u^1 > 0$, homotopic to P with distance from P not exceeding u^1 and of minimal length. We denote the subtube of T bounded by R^1 with T^1. Let T' be a subtube of T^1 whose boundary P' has distance greater than one from R^1. With T' we proceed exactly as with T obtaining a curve R^2 of minimal length homotopic to P' whose distance from P' is at most u^2, and denote the subtube of T bounded by R^2 with T^2.

If this construction is continued we obtain a sequence of tubes T^ν whose boundaries R^ν have the properties described in Cohn-Vossen's Theorem.

If D_ν denotes the domain bounded by T and T^ν, then D_ν has the properties required in the definition of total excess, hence

$$\lim \varepsilon(D_\nu) = \varepsilon(T).$$

The angles of R^ν in D_ν are either all convex, or R^ν is a closed geodesic, or R^ν has exactly one concave angle in D_ν. Since $\chi(D_\nu) = 0$ we find from (43.3)

$$\varepsilon(D_\nu) \leqslant -\sum(\pi - \beta_i) + \pi.$$

If T is expanding then for suitable u_ν the angles of R^ν in T^ν are concave, hence $\varepsilon(D_\nu) \leqslant -\Sigma(\pi - \beta_i)$. If T is contracting or bulging, then u_ν can be chosen such that the angle of R^ν in T^ν is convex, hence $\varepsilon(D_\nu) \geqslant -\Sigma(\pi - \beta_i)$. If, in addition, the angular measure is uniform at π in T, then u_ν can by (44.2) be determined such that the angle of R^ν in T^ν is greater than $\pi - 1/\nu$. Thus we obtain

(44.14) *If the tube T possesses a total excess and the angles in T of its boundary at the proper vertices are $\beta_1, \ldots, \beta_m$ then*

$$\varepsilon(T) \leqslant -\sum_{i=1}^{m}(\pi - \beta_i) \text{ if } T \text{ is expanding,}$$

$$-\sum_{i=1}^{m}(\pi - \beta_i) \leqslant \varepsilon(T) \leqslant -\sum_{i=1}^{m}(\pi - \beta_i) + \pi \text{ if } T \text{ is contracting or bulging.}$$

If the angular measure is uniform at π on T, then

$$\varepsilon(T) \leqslant -\sum_{i=1}^{m}(\pi - \beta_i)$$

with equality if T is not expanding.

If R is a G-surface which is topologically a compact manifold of genus γ, punctured at k points, we define the characteristic $\chi(R)$ by (43.2). Its total excess, if it exists, can be determined by treating each tube T_j as the tube T above. We then obtain from (43.14):

(44.15) **Theorem.** *If the G-surface R possesses a total excess and has genus γ and k holes, then*

$$\varepsilon(R) \leqslant 2\pi\,\chi(R) + k\pi.$$

$$\varepsilon(R) \geqslant 2\pi\,\chi(R)$$

if R has no expanding tubes. If the angular measure is uniform at π *on R then*

$$\varepsilon(R) \leqslant 2\pi\,\chi(R)$$

with the equality sign if R has no expanding tubes.

Another application of (44.14) is:

(44.16) *A tube T with non-negative excess is expanding if its angular measure is uniform at* π.

For otherwise T would contain a contracting subtube T' bounded by a monogon R' of length λ' whose angle in T' has measure $\beta' \leqslant \pi$. By (44.14)

$$0 \leqslant \varepsilon(T') \leqslant -(\pi - \beta').$$

This implies $\beta' = \pi$ and $\varepsilon(T') = 0$, so that R' is a closed geodesic and the excess on T' vanishes identically.

Consider a curve $R'(u)$ on T' of minimal length $\lambda'(u)$ homotopic to R' on T' and with distance at most u from R'. If $R'(u)$ does not coincide with R' it is disjoint from R' and applying (43.3) to the compact domain bounded by $R'(u)$ and R' we conclude from Cohn-Vossen's Theorem that $R'(u)$ is also a closed geodesic.

Consider the universal covering space $\overline{T}$ of T'. (This cannot be obtained from our general theory because T' is not a G-space, but there is no difficulty since T' is, topologically, a half cylinder.) The curves $\overline{R}$ over R' and $\overline{R}(u)$ over $R'(u)$ on $\overline{T}$ are geodesics without multiple points.

Any subarc of $\overline{R}$ is shortest connection of its endpoints in $\overline{T}$. Otherwise a geodesic arc $\overline{C}$ in $\overline{T}$ connecting two points $\overline{a}$, $\overline{b}$ on $\overline{R}$ would exist, which is shorter than the arc of $\overline{R}$ from $\overline{a}$ to $\overline{b}$. These two arcs bound together a simply connected domain $\overline{D}$ with two proper vertices, which contradicts $\varepsilon(\overline{D}) = 0$, see the proof of (43.6).

The argument used in the last part of the proof of (32.3) shows now that $\lambda'(u) \leqslant \lambda'$. Hence R' has minimal length among all curves homotopic to R' on T', and T' is, contrary to our assumption, not contracting.

We now prove another theorem of Cohn-Vossen which has an important consequence for planes with positive excess:

(44.17) *Let Q be an open Jordan curve on the G-surface R of the form* $\bigcup_{i=-\infty}^{\infty} T(a_i, a_{i+1})$, $a_i \neq a_{i+1}$, *where the* a_i *have no accumulation point and only a finite number of* a_i *are proper vertices. Assume moreover, that Q bounds on R* *a domain D homeomorphic to a half-plane and that each subarc of Q is a shortest connection of its endpoints in D. If the angular measure is uniform at π in D and $\varepsilon(D)$ exists then*

$$\varepsilon(D) \leqslant - \sum_{i=1}^{m} (\pi - \beta_i),$$

where β_i are the measures of the angles in D at the proper vertices of Q.

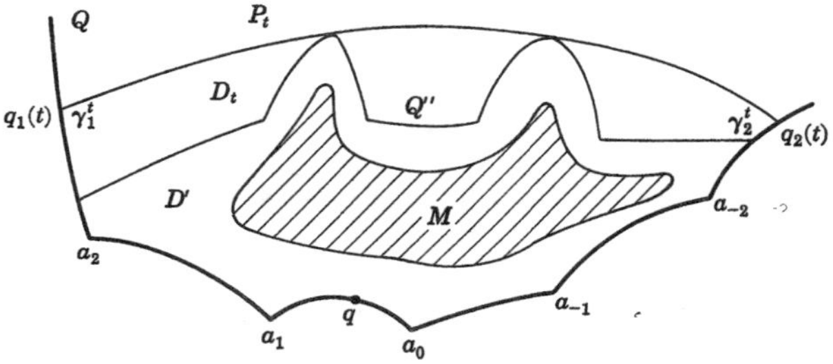

Fig. 48.

It follows from the definition of total excess that it suffices to prove: If an arbitrary bounded subset M of D and a positive ε are given, then a simple closed polygon D_s in D exists such that the domain bounded by D_s contains M and

$$\varepsilon(D_s) \leqslant - \sum_{i=1}^{m} (\pi - \beta_i) + 2\,\varepsilon. \tag{44.18}$$

We choose first an arc Q' of Q containing all proper vertices of Q and a simple geodesic polygon Q'' connecting the endpoints of Q' and otherwise in the interior of D, such that the domain D' bounded by $Q' \cup Q''$ contains M. For an arbitrary point $q \in Q'$ let $q_1(t)$, $q_2(t)$ be the two points on Q for which the subarcs of Q from q to $q_i(t)$ have length t, and let t' be so large that the union of the arcs contains Q'. Let P_t, $t \geqslant t'$, be the shortest connection of $q_1(t)$ and $q_2(t)$ in $(D - D') \cup Q''$. If $\lambda(t)$ denotes the length of P_t then the minimizing property of the subarcs of Q guarantees

$$\lambda(t) \geqslant 2\,t. \tag{44.19}$$

As in the proof of Cohn-Vossen's Theorem we see that P_t is — an obviously simple — geodesic polygon. Because Q' contains all vertices of Q, the points

$q_i(t)$ are the only common points of P_t and Q, and all proper vertices of P_t, if any, are such that the angle in the domain D_t bounded by P_t and the subarc of Q from $q_1(t)$ to $q_2(t)$ are convex. Denote the measure of the angle which P_t forms with Q at $q_i(t)$ in D_t by γ_i^t. Then by (43.3)

$$\varepsilon(D_t) \leqslant 2\pi - (\pi - \gamma_1^t) - (\pi - \gamma_2^t) - \sum (\pi - \beta_i) = \gamma_1^t + \gamma_2^t - \sum \pi - \beta_i.$$

We show that $s \geqslant t'$ exists such that $\gamma_i^s < \varepsilon$. Because of (44.18) the function $\lambda(t) - 2t + \delta(\varepsilon)t$ tends to infinity when $t \to \infty$. The triangle inequality implies the continuity of $\lambda(t)$, hence $\lambda(t)$ reaches a minimum for some value $s \geqslant t'$ of t. Then $\lambda(s+h) - 2(s+h) + \delta(\varepsilon)(s+h) \geqslant \lambda(s) - 2s + \delta(\varepsilon)s$ for $h > 0$, and

$$\lambda(s+h) - \lambda(s) \geqslant 2h - h\,\delta(\varepsilon) \qquad \text{for} \qquad h > 0. \tag{44.20}$$

Let $h > 0$ be less than $\min \omega(q_i(s), \varepsilon)$ and also such that the point b_i of P_s with $q_i(s)\,b_i = h$ lies on the side of P_s with endpoint $q_i(s)$.

If $a_i = q_i(s+h)$, then $|a_i\, q_i(s)\, b_i| = \pi - \gamma_i^s$ and $a_i\, q_i(s) = b_i\, q_i(s) = h$. The polygon P_{s+h} is at most as long as the polygon originating from P_s through replacing $T(b_i, q_i(s))$ by $T(b_i, a_i)$. Therefore

$$\lambda(s+h) \leqslant \lambda(s) + a_1 b_1 - h + a_2 b_2 - h.$$

This yields together with (44.20) that

$$2h - h\,\delta(\varepsilon) \leqslant a_1 b_1 - h + a_2 b_2 - h$$

or

$$\left(1 - \frac{a_1 b_1}{2h}\right) + \left(1 - \frac{a_2 b_2}{2h}\right) \leqslant \delta(\varepsilon)/2 < \delta(\varepsilon).$$

Since each of the terms on the left is non-negative, each is less than $\delta(\varepsilon)$, hence

$$|a_i\, q_i(s)\, b_i| = \pi - \gamma_i^s > \pi - \varepsilon,$$

which proves (44.18) and the theorem because $D_s \supset D' \supset M$.

SURFACES WITH POSITIVE EXCESS

We saw in (43.11) that a non-compact G-surface with positive excess is homeomorphic to the plane. Since $\varepsilon(R)$ exists, see (44.13), the preceding considerations contain many interesting results on these surfaces if their angular measure is, in addition, uniform at π. We formulate them in one theorem.

(44.21) ***Theorem.*** *Let R be a G-plane with non-negative, but not identically vanishing, excess and such that the angular measure is uniform at π. Then*

a) $\varepsilon(R) \leqslant 2\pi$.
b) *Two compact domains each bounded by a simple monogon have common interior points.*
c) *The exterior of a simple closed polygon is an expanding tube.*
d) *If the excess does not vanish identically outside any bounded domain, then R possesses no simple closed geodesic.*
e) *If $\varepsilon(R) \leqslant \pi$ then R possesses no simple monogons so that all geodesics are one-to-one continuous images of the real axis.*
f) *If $\varepsilon(R) > \pi$ then, given a bounded set N, a $\rho > 0$ exists such that every point q with $qN > \rho$ is vertex of a simple monogon containing N in its interior.*
g) *R possesses no straight line and no closed geodesic with multiple points.*
h) *If $x(\tau)$ represents a geodesic $\mathfrak{g}$ with multiple points, then $\mathfrak{g}$ contains exactly one simple monogon $\mathfrak{m}$. If $\mathfrak{m}$ is given by $x(\tau)$ for $\alpha \leqslant \tau \leqslant \beta$, $x(\alpha) = x(\beta)$, then each of the half geodesics represented by $x(\tau)$ for $\tau \leqslant \alpha$ and $\tau \geqslant \beta$ lies except for $\tau = \alpha$ or $\tau = \beta$ outside of the compact domain D bounded by $\mathfrak{m}$ and does not intersect itself.*

The last statement means $x(\tau') \neq x(\tau'')$ for $\tau' < \tau'' \leqslant \alpha$ and $\tau' > \tau'' \geqslant \beta$, but does not exclude $x(\tau') = x(\tau'')$, for $\tau' < \alpha < \beta < \tau''$. Clearly (44.15) implies a). The excess of a compact domain bounded by a simple monogon is by (44.5) greater than π, hence two such domains must have common interior points. c) follows from (44.16).

For the compact domain D bounded by a simple closed geodesic (44.5) yields $\varepsilon(D) = 2\pi$, hence $\varepsilon(R) = 2\pi$ and the excess vanishes outside of D. This proves d). The surface $F_{2\pi}$ constructed on page 294 shows that a simple closed geodesic may exist, if the excess vanishes outside of a bounded set.

To see f) we enclose N in a simple closed polygon which bounds a compact domain D' with $\varepsilon(D') = \pi + 3\eta$, $\eta > 0$. Such D' exist because there are by (44.10) domains D_ν bounded by a simple closed polygon and containing $S(N. \nu)$. By the definition of total excess $\lim \varepsilon(D_\nu) = \varepsilon(R) > \pi$. Hence D_ν may serve as D' for large ν. If D is any domain bounded by a simple closed polygon and containing D' then $\varepsilon(D) \geqslant \varepsilon(D')$ because excess is non-negative. Applying (44.9) to D as M yields a θ such that every point q with $qD \geqslant \theta$ is vertex of a simple monogon containing D in its interior. If $D \subset S(N, \theta_1)$ then $\rho = \theta + \theta_1$ will satisfy f).

A straight line Q on R and each of the half planes bounded by Q would satisfy the hypothesis of (44.17), but at least one of the half planes would have positive total excess, although Q has no proper vertices, which contradicts (44.17). The second part of g) is contained in h). h) is interesting because it furnishes far reaching information on the shape of the geodesics. Its proof is simple: $\mathfrak{g}$ contains at least one simple monogon $\mathfrak{m}$ which is not a closed geodesic. With the notation as under h) the angle of $\mathfrak{m}$ in D at $x(\alpha)$ is convex by (44.5) and b).

Therefore the point $x(\tau)$ lies outside of D for small $\tau - \beta > 0$. If there were a $\tau > \beta$ with $x(\tau) \in D$, then a first such τ, say τ_0 would exist.

It is impossible that $x(\tau') = x(\tau'')$ for $\beta \leqslant \tau' < \tau'' < \tau_0$ because if τ'' is the first τ following τ' with $x(\tau'') = x(\tau')$, then $x(\tau)$ represents for $\tau' \leqslant \tau \leqslant \tau''$ a monogon whose interior has no common point with D in contradiction to b). If τ_1 is determined by $\alpha < \tau_1 \leqslant \beta$ and $x(\tau_0) = x(\tau_1)$, then $x(\tau)$ represents for $\tau_1 \leqslant \tau \leqslant \tau_0$ a simple monogon $\mathfrak{m}_1$. The compact domain D_1 bounded by $\mathfrak{m}_1$ must by b) have a common point with D, and hence contain the points $x(\tau)$ for $\alpha \leqslant \tau \leqslant \tau_1$. But then the angle of $\mathfrak{m}_1$ at $x(\tau_1)$ in D_1 is concave which contradicts (44.5) and b).

The fact that $x(\tau)$ does not lie in D for $\tau > \beta$ precludes $x(\tau') = x(\tau'')$ for $\beta \leqslant \tau' < \tau''$, since this would again lead to a simple monogon whose interior has no common point with D.

Cohn-Vossen also proves that $x(0)x(\tau_\nu) \to \infty$ if $|\tau_\nu| \to \infty$. This is also true under the assumptions of (44.21). However, the proof which the author possesses rests on many case distinctions and is not publishable. Cohn-Vossen's proof uses properties of Gauss curvature and can therefore not be extended to the present case.

CHAPTER VI

HOMOGENEOUS SPACES

45. Introduction

The sixth and last chapter deals with spaces which are homogeneous in the sense that either the space itself, or at least its universal covering space, possesses a transitive group of motions. Thus the characterizations of the elliptic and Minkowskian geometries given in Chapter III could also have been placed here.

The elementary spaces, i. e., the euclidean, hyperbolic, and spherical spaces, have (with the exception of the circle) the property that the loci of the points equidistant from two distinct points, called *bisectors, are flat.* In Sections **46** and **47** *it is shown that this characterizes the elementary spaces* and that the spaces with constant curvature are the only spaces which have this property locally. Besides being of considerable historical interest these results form perhaps the most direct and conceptually simplest access to the elementary geometries, and they have many important applications.

Differentiability assumptions would reduce the major part of Section **47** to a few lines. Nothing shows the irrational attitude towards such assumptions more clearly than the fact that just this problem has always been treated without differentiability; the present version represents only a last step (due to H. G. Forder [1]) in a long series of efforts, beginning with Bolyai and Lobachevsky, to obtain explicit trigonometric formulae by purely synthetic arguments. We happen to deal with a field which developed as a continuation of Euclid's ideas into the foundations of geometry, where differentiability hypotheses are traditionally felt as alien and consequently rejected.

Among the many applications of the theorems on spaces with flat bisectors we mention in particular the solution, in Section **48**, of the *Helmholtz-Lie Problem* in several local and global forms, of which the simplest is the following: If for any two isometric point-triples of a G-space a motion exists which takes the first triple into the second, then the space is elementary. Another application characterizes the elementary surfaces as those in which an area for triangles exists that can be expressed in terms of the lengths of the sides.

The following Section **49** deals with *involutoric motions.* We discuss a few elementary properties, throw a passing glance at the — through E. Cartan's work — so interesting *symmetric spaces* and prove that a G-space which can be reflected in each lineal element is euclidean, hyperbolic, spherical, or elliptic.

The above formulation of the Helmholtz-Lie Problem suggests the problems of determining the G-spaces with *transitive and pairwise transitive groups of motions*. The latter means that for any two point pairs with equal distances a motion exists that takes the first pair into the second. Little is known concerning the first problem: If the space possesses a transitive abelian group of motions then a well-known result of Pontrjagin shows that it is, topologically, the product of a finite number of circles and straight lines and that its metric is Minkowskian. Here we reduce this result to the theory of spaces with non-positive curvature. We also characterize the two-dimensional Minkowski spaces through properties of triangle area, in particular those with symmetric perpendicularity by the property that triangle area is expressible in terms of base and altitude.

In order to determine all G-surfaces with transitive groups of motions we study in Section 51 the straight planes which possess all translations along two non-parallel lines. The case where these lines are asymptotes leads to a new geometry which we call *quasi-hyperbolic*. It has many interesting properties, some of which are listed under (46) in the Appendix.

The *G-surfaces with transitive groups of motions* are (Section 52) the Minkowski plane, the quasi-hyperbolic plane, the sphere, the elliptic plane, and the cylinder and torus with Minkowskian metrics. Our only general result on G-spaces with transitive groups of motions is that, when compact, they are topological manifolds and their groups of motions are Lie groups.

We then discuss examples of spaces with non-constant curvature and pairwise transitive groups of motions. The *hermitian elliptic and hyperbolic spaces* are treated in detail, other spaces more sketchily.

The spaces with pairwise transitive groups of motions have been determined by Wang [2, 3] and Tits [1]. They are the symmetric Riemann spaces of rank 1 with simple Lie groups as the identity components of their groups of motions.

This was proved in the compact case by Wang [3]. We derive his results in Section 54 by establishing directly the above mentioned properties. Tits [1] states the corresponding result for non-compact spaces, but does not give proofs; a complete exposition has not yet appeared. Therefore we restrict ourselvs here to the odd-dimensional case which had been settled previously by Wang [2]. We mention particularly the remarkable fact that in this case the euclidean, hyperbolic, spherical, and elliptic spaces are the only spaces with pairwise transitive groups of motions, whereas there are others for all even dimensions greater than two.

Also, for these results it suffices to assume that the space be finitely compact and M-convex, local prolongability and uniqueness of prolongation follow from pairwise transitivity. Even the convexity condition can be considerably relaxed, but we will not discuss this question here.

46. Spaces with flat bisectors I

Euclid describes in his Elements a curve as something which has length and no breadth (Γραμμὴ δὲ μῆκος ἀπλατές) and a surface as something that has only length and breadth (Ἐπιφάνεια δέ ἐστιν ὃ μῆκος και πλάτος μόνον ἔχει), and then proceeds to characterize the straight lines and planes among the curves and surfaces.

These definitions have been found objectionable already in antiquity. The concept of shortest connection leads readily to a satisfactory definition of lines, in fact, this is the underlying idea of the entire present book. The situation is not so simple for planes. Leibniz [1, p. 166] suggested defining a plane as the locus of points equidistant from two given points. This has many more implications than would appear at first sight.

We will denote the locus of the points x which have the same distance $ax = a'x$ from two given *distinct* points a, a' by $B(a, a')$ and call it briefly the "*bisector*" *of a and a'*. Leibniz' definition of plane can be considered as satisfactory only if the bisectors are flat, for only then do they have the characteristics of planes.

The bisectors are flat in the euclidean, hyperbolic, and spherical spaces, in the latter case only when the dimension exceeds one, which will be assumed. It is a fundamental fact that these elementary spaces are the only spaces with flat bisectors: *adding to the axioms of a G-space the single postulate that a bisector contains with any two points at least one segment connecting them characterizes the euclidean, hyperbolic, and spherical spaces,* see Theorem (47.4).

We are interested not only in these spaces but also in the spaces which are covered by them, that is, the spaces of constant curvature. We therefore prove a local theorem, which entails much manipulation with the triangle inequality, but this is amply justified by the applications of the theorem.

Local characterization of elementary geometry

Our principal result is:

(46.1) ***Bisector Theorem.*** *If for any five distinct points a, a', x, y, z in a sphere $S(p, \delta_p)$, $0 < \delta_p < \rho_1(p)$ of a G-space the relations $ax = a'x$; $az = a'z$ and $(x\, y\, z)$ imply $ay = a'y$ then $S(p, \delta_p)$ is isometric to a sphere in a finite dimensional euclidean, hyperbolic, or spherical space.*

The hypothesis means that $B(a, a')$ contains with any two points x, y every point of $T(x, y) \cap S(p, \delta_p)$. We precede the long proof by a trivial but useful remark on bisectors in general.

(46.2) *If a, a' are two distinct points in any G-space, then the midpoints of a and a' are the feet of a (or a') on $B(a, a')$.*

For if m is a midpoint of a and a', then $m \in B(a, a')$. If $x \in B(a, a')$ then

$$2\, a\, x = 2\, a'\, x = a\, x + x\, a' \geqslant a\, a' = 2\, a\, m = 2\, a'\, m,$$

so that m is a foot of a and of a' on $B(a, a')$. If x is also a foot, then it must be another midpoint of a and a'.

The proof of (46.1) consists of several steps of a different character. The first is a systematic use of the triangle inequality to produce smaller neighborhoods with additional properties.

Applications of the triangle inequality

Since $S(p, \delta_p) \supset S(q, \delta_p - p\, q)$ for $p\, q < \delta_p$ the sphere $S(q, \delta_p - p\, q)$ also satisfies the hypothesis of (46.1). Put

$$24\, \beta_q = \min\, (\delta_p - p\, q, \rho(q)), \quad q \in S(p, \delta_p).$$

It then follows readily from $|\rho(q) - \rho(r)| \leqslant qr$ that

(a) $$|\beta_q - \beta_r| \leqslant 24^{-1}\, q\, r.$$

We call the segment $T(x', y')$ a "*symmetric extension*" of the segment $T(x, y)$, $x \neq y$, if $T(x', y')$ contains $T(x, y)$ properly and has the same midpoint as $T(x, y)$. We now show — always under the hypothesis of (46.1) — that $B(a, a')$ not only contains points between two of its points, but also points on the prolongation:

(b) *If $x \neq y$ and $a \neq a'$ lie in $S(q, 3\,\beta_q)$ and $x, y \in B(a, a')$ then the symmetric extension $T(x', y')$ of $T(x, y)$ with $24\,\beta_q$ exists and lies on $B(a, a')$.*

Proof. Theorem (8.10) applied with $\nu = 6$ yields a symmetric extension of $T(x, y)$ with length $3\,\rho(q)/2 > 24\,\beta_q$; hence also a symmetric extension $T(x', y')$ of length $24\,\beta_q$. Let $(x'\,xy)$.

Assume (see Figure 49) some point z of $T(x', y')$ does not lie on $B(a, a')$. The hypotheses imply that z cannot belong to $T(x, y)$. Say $(x\,y\,z)$ and $a'\,z < az$. Since $T(a, y) \subset S(q, 6\,\beta_q)$ every point u of $T(a, y)$ satisfies the inequality

$$xu + yz\,\frac{u\,a}{y\,a} \leqslant xq + qu + yz < 3\,\beta_q + 6\,\beta_q + 12\,\beta_q = 21\,\beta_q.$$

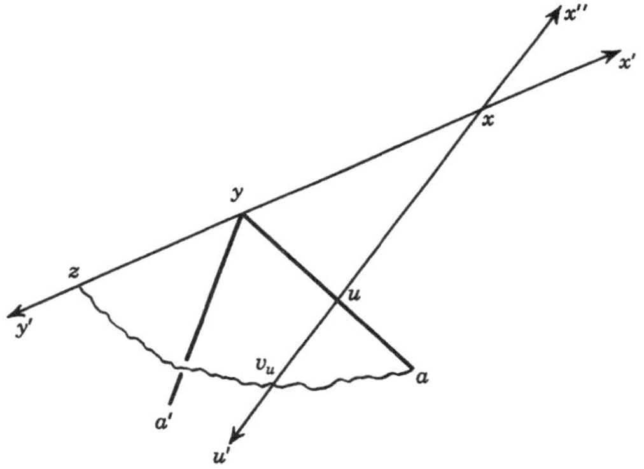

Fig. 49.

By (8.5) there is a symmetric extension $T(x'', u')$ of $T(x, u)$ with $(x\,u\,u')$ and length

$$2\,\rho(x) \geqslant 2\,\big(\rho(q) - qx\big) > 2\,\big(\rho(q) - \rho(q)/8\big) \geqslant 42\,\beta_q,$$

because

$$u\,x \leqslant u\,q + q\,x < 12\,\beta_q \leqslant \rho(q)/2 < 4\,\rho(x)/7.$$

Then $u'\,x > \rho(x) > 21\,\beta_q$. Hence $T(x, u')$ contains a point v_u with

$$x\,v_u = x\,u + y\,z\,\frac{u\,a}{y\,a}$$

and v_u depends continuously on u.

Now $v_a = a$ and $v_y = z$, hence $a'\,v_a > a\,v_a$ and $a'\,v_y < a\,v_y$ so that a point w on $T(a, y)$ exists for which $a'\,v_w = a\,v_w$. But $xq < 3\,\beta_q$, $w\,q < 6\,\beta_q$ and

$$v_w\,q \leqslant v_w x + x\,q < 21\,\beta_q + 3\,\beta_q = 24\,\beta_q.$$

Hence by hypothesis, the relations $x \in B(a, a')$ and $v_w \in B(a, a')$ imply $w \in B(a, a')$, but this is impossible because

$$a' w = a w = a y - y w = a' y - y w$$

and uniqueness of prolongation would imply $a = a'$.

Some of the precautions of this and the following proof will, a posteriori, prove unnecessary because it will be seen that the sphere $S(q, 3\beta_q)$ is convex. It is convenient to revive the notation m_{xy} for the midpoint of x and y introduced in Section 36. The next statement prepares for symmetry of perpendicularity:

(c) *Let a, a' be two distinct points in $S(q, 3\beta_q)$ and m their midpoint. If $x \in B(a, a')$, $m = m_{xx'}$, $0 < mx < 3\beta_q$ then $xa = x'a = xa' = x'a'$, hence $T(x, x') \subset B(a, a')$ and $T(a, a') \subset B(x, x')$.*

We choose c with $(x\, c\, m)$ so close to x that $c\, q < 3\beta_q$ and form the symmetric extension $T(x'', c')$ of $T(x, c)$ with (xcc') and length $24\beta_q$. Then by (46.2)

$$c' a \geqslant c' x - x a > 12\beta_q - 3\beta_q = 9\beta_q > x a > m a.$$

Because, for $y \in T(x, m)$,

$$y a \leqslant mx/2 + \max(ma, xa) \leqslant (ma + ax)/2 + xa < 2\, xa < 6\beta_q,$$

it follows that $(c'\, m\, x)$. Hence $T(c', m)$ contains a point x^* with $x^* a = x\, a$. By (b) the segment $T(x'', c')$ lies in $B(a, a')$ hence $x \in B(a, a')$ and

$$x a' = x^* a = x a \perp x a'.$$

Since $x q < 3\beta_q$, $T(a, a') \in S(q, 6\beta_q)$ and

$$x^* q < x^* a + aq = xa + aq \leqslant xq + 2\, aq < 9\beta_q,$$

we find $T(a, a') \in B(x, x^*)$; hence m must be the midpoint of x and x^*, so that $x^* = x'$ and $T(x, x') \subset B(a, a')$.

(d) *The spheres $K(q, \sigma)$ are strictly convex for $q \in S(p, \delta_p)$ and $0 < \sigma < 3\beta_q$. The set $S(q, 3\beta_q)$ is convex.*

It suffices by (20.8) to show that every point q of $S(p, \delta_p)$ has exactly one foot on any segment T that contains points of $S(q, 3\beta_q)$. If q had two different feet f, f' on T then $qf = qf' < 3\beta_q$ and $q \in B(f, f')$. If $m = m_{ff'}$ and $m = m_{qq'}$, it follows from (c) that $T(f, f') \subset B(q, q')$. Hence m would be the foot of q on $T(f, f')$.

Perpendicularity

We now select a definite sphere $U = S(z, \alpha)$ where $z \in S(p, \delta_p)$ and $0 < \alpha \leqslant 2\beta_z$. A segment with endpoints on $K(z, \alpha)$ lies by (d), except for its endpoints, in U. The segment without the endpoints will be called a line in U. Any two distinct points b, c in U lie on exactly one line $L(b, c)$. The line L in U is perpendicular to the set D (not necessarily entirely in U) at f if $f \in L \cap D$ and f is the foot on D of any point x on L. Because of (20.6) f is the only foot of x on D.

The symbol $B_u(a, a')$ will be used for the intersection $B(a, a') \cap U$, when it is not empty and at least one of the points a, a' lies in U. According to (b) the line $L(b, c)$, $b \neq c$, lies on $B_u(a, a')$ if b and c do.

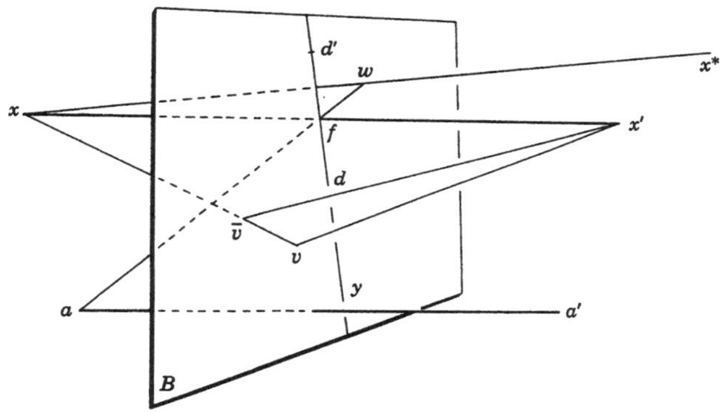

Fig. 50.

We can now establish the two facts on which the whole proof principally rests:

(e) *Let $f \in U$ be a foot of q on $B_u(a, a')$ where $q \neq f$ and $q \in U$. If x is any point of $L(q, f)$ different from f and $f = m_{xx'}$, then $B_u(a, a') = B_u(x, x')$.*

Proof. First consider the case where $(f\, x\, q)$. Then f is the unique foot of x on $B = B_u(a, a')$. Let y be any point of B different from f. Since f is also the unique foot of x on $L(y, f)$ this line contains two points d, d' with $(d\, f\, d')$ and $xd = xd'$. Let m be the midpoint of d, d' and $x''m = mx = x''x/2$. Then (b) and (c) yield $T(x, x'') \subset B(d, d')$ and $T(d, d') \subset L(f, y) \subset B_u(x, x'')$. Moreover, m is by (46.2) the foot of x on $B_u(x, x'')$, hence on $T(d, d')$, so that $m = f$ and $x' = x''$. Therefore $y \in B_u(x, x')$ and $B \subset B_u(x, x')$.

The same result follows now without the restriction $(q\,x\,f)$: If $\overline{x}$ is any point on $L(q, f)$ and $f = m[\overline{x}, \overline{x}']$, it follows from (b) that $T(\overline{x}, \overline{x}') \subset B(d, d')$, and from (c) that $T(d, d') \subset B(\overline{x}, \overline{x}')$, and from (e) that $y \in L(d, d') \subset B(\overline{x}, \overline{x}')$.

In order to show that $B = B_u(x, x')$ we first observe that x and x' lie on different sides of B, i. e., that $xa < xa'$ implies $x'a > x'a'$ and $xa < xa'$ implies $x'a < x'a'$. The relation $x'a = x'a'$ is impossible, because $x' \in B$, $f \in B$ and (b) would lead to $x \in B$. Assume for an indirect proof that $xa < xa'$ and $x'a < x'a'$. Let $(a\,f\,w)$ with sufficiently small fw. Then (see Figure 50)

$$a'w < a'f + fw = af + fw = aw.$$

If $x^* w = w x = xx^*/2$, then $x^* \to x'$ for $w \to f$, hence $x^*a < x^* a'$, for small fw. But then each of the segments $T(x, w)$ and $T(x^*, w)$ would intersect B, so that $T(x, x^*) \subset B$ and $x \in B$.

If now a point $v \in B_u(x, x') - B$ existed then $vx = vx'$ and $va > va'$. As we just saw, we may assume that $xa < xa'$. Hence $T(x, v)$ would contain a point $\overline{v}$ of $B \subset B_u(x, x')$, but

$$x\overline{v} = xv - v\overline{v} = x'v - v\overline{v} > x'\overline{v}.$$

Notice the following corollaries of (e), (46.2) and (c):

(f) If f is a foot of a point $x \in U - B_u(a, a')$ on $B_u(a, a')$ then $L(x, f)$ is a perpendicular to $B_u(a, a')$.

(g) If the line G in U is perpendicular to the line H in U, then H is perpendicular to G. If $G \cap H = f$, $y \in G - f$, and $f = m_{yy'}$, then $B_u(y, y')$ is the union of all lines perpendicular to G at f. It therefore depends only on G and f may be denoted by $B_u(G, f)$.

The second important fact referred to above is the following.

(h) *If $B = B_u(a, a')$ contains z then a foot f on $B(a, a')$ of a given point $x \in U - B$ lies in B and is unique. If $f = m_{xx'}$, then $x' \in U$. The involutoric mapping $\Omega(B)$ of U on itself which maps x on x' and every point of B on itself, maps any line in U that intersects B isometrically on a line intersecting B in the same point.*

Proof. A foot f of x on $B(a, a')$ lies in U because $xf \leqslant xz < \alpha$. It is the unique foot on $B(a, a')$ and hence on B, because by (a)

$$3\,\beta_x \geqslant 3\,(\beta_z - xz/24) \geqslant 3\,(\beta_z - \alpha/24) \geqslant 11\,\beta_z/4 \geqslant \alpha.$$

Hence the sphere $K(x, xf)$ is strictly convex, see (d). It follows from (e) that $B_u(x, x') = B_u(a, a')$ hence $z \in B_u(x, x')$ and $zx' = zx < \alpha$ or $x' \in U$.

Let G be a line in U not in B that intersects B in a point s and let $a, b \in G$ with (asb). As in the proof of (e) we see that a and b lie on different sides of B. Since a and a' lie on different sides and b and b' too, the points a', b' lie on different sides of B, so that the segment $T(a', b')$ intersects B in a point s'. The relation $B = B_u(a, a') = B_u(b, b')$ follows from (d) and yields

$$ab \leqslant as' + s'b = a's' + s'b' = a'b' \leqslant a's + sb' = as + sb = ab,$$

whence $ab = a'b'$ and $s = s'$.

If c is any point of G different from a, b, s then either (asc) or (csb). In the first case we just proved $(a's'c')$ and $ac = a'c'$, hence also $b\,c = b'c'$ which proves (h).

It is now clear that our aim will be to establish that $\Omega(B)$ not only preserves distances on lines intersecting B, but is an isometry of U on itself. This is done most easily by reducing the problem to two dimensions which will be our next step.

The existence of planes

We deduce from the last theorem:

(i) *U is the topological product of $B = B_u(a, a') \supset z$ and an open segment.*

For 1) each point of $U - B$ lies on exactly one perpendicular to B.

2) A point f of B lies on at most one perpendicular G to B. The existence of a second perpendicular H to B at f leads to a contradiction: If $q \in G - f$ and $x \in H - f$ lies sufficiently close to f, then $L(q, x)$ contains a point s of B. If $L = L(s, f)$ then $B(L, f) \supset H \cup G$, see (g), hence $B(L, f) \supset L(q, x) \supset y$ which is absurd.

3) There is a perpendicular to B at a given point f; it suffices to take a sequence $x_\nu \in U - B$ which tends to f. The perpendicular through x_ν to B will tend to the perpendicular to B at f. This also shows that the perpendicular to B at f depends continuously on f. The perpendiculars to B yield therefore a representation of U as the product of B and an open segment.

Now let z_1, z_2 be two points distinct from z and not on the same line with z. If $L(z_2, z)$ is not perpendicular to $L(z_1, x)$ then $L(z_1, x)$ with (z_2xz) contains for small xz a point u whose foot on $L(z, z_1)$ is z. If $L_1 = L(z_1, z)$, $L_2 = L(u, z)$ then L_1 and L_2 are perpendicular to each other.

Choose a maximal set of lines $L_1, L_2, \ldots$ which pass through z and are perpendicular to each other. If $q_i \in L_i$ and $q_iz = \alpha/2$, then $q_i\,q_k > \alpha/2$ because L_i is perpendicular to L_k. Hence the number of L_i is finite, say n ($\geqslant 2$). Then

$\bigcap_{i=1}^{n} B_u(L_i, z) = z$, because any line in the intersection would be perpendicular to all L_i. Also $\bigcap_{i=2}^{n} B_u(L_i, z) = L_1$. For if this intersection contained any point z_1' not on L_1, then as above, a point x on $T(z_1', z)$ and a line L' through a point u' on $L(z_1, x)$ would exist which is perpendicular to L_1. Since $u' \in B_u(L_i, z)$, $i \geqslant 2$, it follows that $L' \subset B_u(L_i, z)$, $i \geqslant 2$, so that L' would be perpendicular to all L_i.

We call a set Q in $S(z, \alpha)$ *plane* if it is homeomorphic to a plane and contains with any two points the whole line $L(x, y)$. If $x \in Q$ and $\eta \leqslant \alpha - zx$ then $Q \cap S(x, \eta)$ will be called a *disk* with center x and radius η.

Let q_2 be defined as above and $(q_2' z q_2,)$ $q_2' z = \alpha/2$. If $n = 2$ then $L_1 = B_u(q_2, q_2')$. $Q = U$ is by (i) the product of L_1 and a segment, hence a plane set and a disk.

If $n > 2$ put $Q = \cap B_u(L_i, z)$. Because $L_1 = B_u(q_2, q_2') \cap Q$, the line L_1 is the bisector of those points in Q which are equidistant from q_2 and q_2'. Since Q contains with any two points the line connecting them we can apply the general theory to Q and find from (i) that Q is the product of L_1 and an open segment, hence again, a plane set and a disk.

Now let $8\,\alpha_z = \beta_z$ so that by (a)

$$|\alpha_x - \alpha_y| \leqslant \varepsilon\, x\, y, \qquad \varepsilon = 1/192.$$

If $az < 6\,\alpha_z$ then

$$S(a, 2\,\beta_a) \supset S(z, 2\,\beta_a - a\,z) \supset S(z, 2\,\beta_z - az/12 - az) \supset S(z, 6\,\alpha_z).$$

Any three points a, b, c in $S(z, 6\,\alpha_z)$ lie in a plane set Q in $S(a, 2\,\beta_a)$ through a. The intersection $Q \cap S(z, 6\,\alpha_z)$ is a plane set in $S(z, 6\,\alpha_z)$ which contains a, b, c.

It is hardly necessary to mention that $S(z, 6\,\alpha_z)$ is — as a consequence of either (i) or Section 14 — homeomorphic to E^n.

Reflections

Let D be any disk with center x and radius $6\,\alpha_x$. If G and G' are perpendicular diameters of D, then the mapping $\Omega\ (B_u(G', x))$, where $U = S(x, 6\,\alpha_x)$, induces the mapping Ω_G of D on itself which leaves G pointwise fixed, maps each perpendicular to G in D on itself, any line in D intersecting G isometrically on a line in D through the same point of G. Hence Ω_G maps every disk in D with center x and radius $\eta < 6\,\alpha_x$ on itself. Our aim is to prove that *this mapping is isometric* for $\eta = \alpha_x$. The main point in the proof is the lemma:

(j) *If H and G are any diameters of the disk D and $H' = H\,\Omega_G$ then*

$$\Omega_H\,\Omega_G\,\Omega_{H'} = \Omega_G.$$

It is to be shown that the left side applied to an arbitrary point p of D yields the same point p' as the right side. Since this is obvious for $p \in H$ we assume $p \in D - H$ and put $q = p\,\Omega_H$. If $q' = q\,\Omega_G$ then clearly (see Figure 51)

$$q\,x = p\,x = p'x = q'x.$$

Choose t so close to x on H that the lines $L(p, t)$ and $L(q, t)$ intersect G in points a and b respectively. Ω_G maps $L(p, a)$ and $L(q, b)$ isometrically on the lines $L(p', a), L(q', b)$ which must pass through the image t' of t on H'. Therefore $q\,t = q't'$, $p\,t = p't'$ and, since also $pt = qt$ we conclude that $p't' = q't'$. Thus H' contains the two points t' and x of $B(p', q')$. It follows that

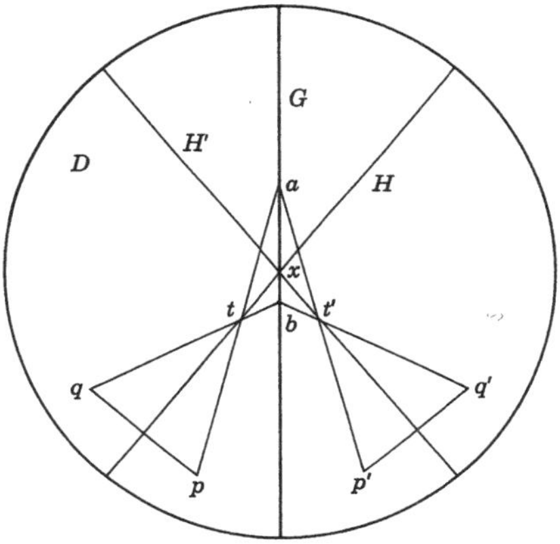

Fig. 51.

$$p\,\Omega_G = p' = q'\,\Omega_{H'} = q\,\Omega_G\,\Omega_{H'} = p\,\Omega_H\,\Omega_G\Omega_{H'}.$$

(k) *Let D^0 be a disk about z with radius $6\,\alpha_z$ and D the sub-disk of D^0 with center z and radius α_z. If G is a diameter of D^0 then the mapping Ω_G of D^0 on itself induces an isometry of D on itself.*

Since we already know that any line which intersects G is mapped isometrically on another line, it suffices to show that Ω_G sends any pair a, a_1 in D on a line H^0 in D^0 which does not intersect G into a pair $a' = a\,\Omega_G$ and $a_1\,\Omega_G$ with $aa_1 = a'a_1\,\Omega_G$.

It is more convenient to show that Ω_G maps the intersection $H = H^0 \cap D$ isometrically on another line in D.

We first use the triangle inequality to obtain some estimates needed later. Because $a'z = az$ and $|\alpha_z - \alpha_a| < \varepsilon\,\alpha_z$ and $|\alpha_z - \alpha_{a'}| < \varepsilon\,\alpha_z$ we have

$$(46.3) \qquad (1+\varepsilon)^{-1}\,\alpha_a < \alpha_z < (1-\varepsilon)^{-1}\,\alpha_a, \quad (1+\varepsilon)^{-1}\alpha_{a'} < \alpha_z < (1-\varepsilon)^{-1}\,\alpha_{a'}$$

hence

(1) $S(z, \alpha_z) \subset S(a, \alpha_z + a\,z) \subset S(a, 2\,\alpha_z) \subset S\left(a, 2\,\alpha_a(1-\varepsilon)^{-1}\right) \subset$

$S\left(z, 2\,(1-\varepsilon)^{-1}\,\alpha_a + \alpha_z\right) \subset S\left(z, (3+\varepsilon)\,(1-\varepsilon)^{-1}\,\alpha_z\right) \subset$

$S\left(a', (3+\varepsilon)\,(1-\varepsilon)^{-1}\alpha_z + a'\,z\right) \subset S\left(a', 4\,(1-\varepsilon)^{-1}\alpha_z\right) \subset S\left(a', 4\,(1-\varepsilon)^{-2}\alpha_{a'}\right) \subset S(a', 6\alpha_{a'})$.

We now choose any two distinct points u, v on G draw a circle $K = K(a, \eta)$ with $0 < \eta < \min\,(au, av, \alpha_z - az)$ and put $u' = T(a, u) \cap K$, $v' = T(a, v) \cap K$, compare Figure 52. We denote by A that arc of K which begins at u', ends at a point b of H and contains v', and by A' its subarc from u' to v'. Because K is convex these arcs have finite lengths. Choose b_1 so close to b on A that $0 < bb_1 < \lambda'/3$, where λ' is the length of A'.

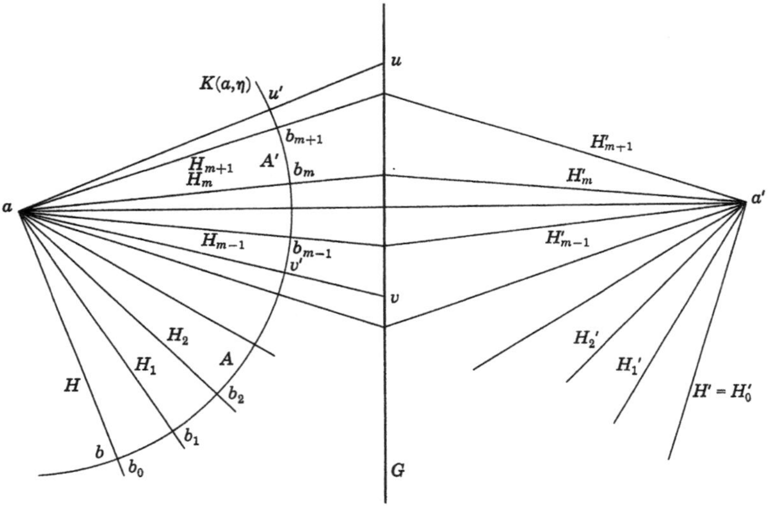

Fig. 52.

Because of (1) there is a disk D^a about a with radius $6\,\alpha_a$ which contains $S(z, \alpha_z)$. Denote by H^a the line in D^a containing H, by H_1^a the diameter of D^a through b_1, by Φ_1 the mapping $\Omega(H_1^a)$ of D^a on itself. Put $H_2^a = H^a\,\Phi_1$, $H_2^a \cap K = b_2$, $\Phi_2 = \Omega\,(H_2^a)$, $H_1^a\,\Phi_2 = H_3^a$, $H_3^a \cap K = b_3$ and generally

$$H_{i-1}^a\,\Phi_i = H_{i+1}^a, \quad b_i = H_i^a \cap K, \qquad \text{then} \qquad b_{i-1}\,b_i = b_i\,b_{i+1}.$$

Because of (j) we have

$$\Phi_{i-1}\Phi_i\Phi_{i+1} = \Phi_i \qquad \text{or} \qquad \Phi_{i-1} = \Phi_i\Phi_{i+1}\Phi_i \qquad (i = 2, \ldots, m).$$

We continue this process until we reach the first subscript ν for which $b_\nu \in A'$. Such a subscript exists because $ibb_1 = bb_1 + b_1b_2 + \ldots + b_{i-1}b_i$ is less than the length of the arc of A from b to b_i. The points $b_{\nu+1}$ and $b_{\nu+2}$ then also lie on A' because $b_ib_{i+1} < \lambda'/3$. Let m be the even one of the two numbers ν and $\nu + 1$.

The lines H^a_m and H^a_{m+1} intersect G because b_m and b_{m+1} lie on A'. Therefore Ω_G maps $H^a_m \cap D^0$ and $H^a_{m+1} \cap D^0$ isometrically on open segments H'_m and H'_{m+1} in D^0 which intersect at a'. Let $D^{a'}$ be the disk with center a' and radius $6\,\alpha_{a'}$, which contains D, and $H^{a'}_m$, $H^{a'}_{m+1}$ the diameters of $D^{a'}$ containing H'_m and H'_{m+1}. We define Φ'_m as the mapping $\Omega(H^{a'}_m)$ of $D^{a'}$ on itself, put $H^{a'}_{m-1} = H^{a'}_{m+1}\Phi'_m$, $\Phi'_{m-1} = \Omega(H^{a'}_{m-1})$, $H^{a'}_{m-2} = H^{a'}_m\,\Phi'_{m-1}$ and so on, until we reach Φ_1' and $H^{a'}_o = H^{a'}_2\,\Phi_1'$. We also put $H^{a'}_j \cap D = H'_j$. Because of (j) we have again

$$\Phi'_{i-1} = \Phi'_i\,\Phi'_{i+1}\Phi'_i, \qquad i = 2, \ldots, m.$$

Because of (l) the mapping Φ_i is defined in $S(z, \alpha_z)$ and maps this set on a set M in $S\left(a, 2\,(1-\varepsilon)^{-1}\alpha_a\right)$. Since this sphere lies in $S\left(z, (3+\varepsilon)\,(1-\varepsilon)^{-1}\alpha_z\right)$ the mapping Ω_G is defined in M and maps M on a set M' in $S\left(z, (3+\varepsilon)\,(1-\varepsilon)^{-1}\,\alpha_z\right)$, and the latter set lies in $S(a', 6\,\alpha_{a'})$ so that Φ_i' is defined in M'. Therefore $\Phi_i\,\Omega_G\,\Phi_i'$ is defined in $S(z, \alpha_z)$. We show that

$$\Phi_i\Omega_G\Phi'_i = \Omega_G \qquad \text{in} \qquad S(z, \alpha_z).$$

For $i = m$, $m + 1$ this is a consequence of (j). Assume the relation to be true for $i \geqslant k$, where $k \leqslant m$. Then

$$\Phi_{k-1}\,\Omega_G\Phi'_{k-1} = \Phi_k\,\Phi_{k+1}\,\Phi_k\,\Omega_G\Phi'_k\,\Phi'_{k+1}\,\Phi'_k = \Phi_k\,\Phi_{k+1}\,\Omega_G\Phi'_{k+1}\,\Phi'_k =$$
$$= \Phi_k\,\Omega_G\Phi'_k = \Omega_G.$$

Therefore

$$\Omega_G = \Phi_1\Phi_2\ldots\Phi_m\Omega_G\Phi'_m\Phi'_{m-1}\ldots\Phi'_1 \qquad \text{in} \qquad S(z, \alpha_z).$$

Now Φ_1 maps H^a isometrically on H^a_2 and Φ_2 leaves H^a_2 pointwise fixed, Φ_3 maps H^a_2 isometrically on H^a_4, and so forth. Altogether, $\Phi_1\ldots\Phi_m$ maps H^a isometrically on H^a_m, hence H isometrically on a subset H_m of H^a_m, and $H_m \subset H^a_m \subset D^0$ because $H_m \subset S(a, 2\,\alpha_z) \subset S(z, 6\,\alpha_z) \cap S(a, 6\,\alpha_a)$ by (l). Then Ω_G maps H_m isometrically on a subset H'_m of $H^{a'}_m$, and $\Phi'_m \ldots \Phi_1'$ maps H'_m isometrically on an open subsegment of $H^{a'}_0$. This proves (k).

We call Ω_G the reflection of $S(z, \alpha_z)$ in G.

47. Spaces with flat bisectors II

From the fact that Ω_G is an isometry we will now deduce the euclidean, hyperbolic, or spherical character of the metric in $S(z, \alpha_z/4)$ by establishing the trigonometric formulae for the right triangle. We first discuss congruence of triangles, then obtain the distinction between the three cases, and finally produce the formulae.

Angular measure, congruent triangles

With the previous notations let a, b be any two distinct points on the circle $K(z, \delta)$ where $0 < \delta < \alpha_z$, and A one of the two arcs of $K(z, \delta)$ from a to b. If c is an interior point of A, and d_1, d_2 on A are such that the arcs of A from a to d_1 and from d_1 to c as well as the arcs from c to d_2 and from d_2 to b have equal lengths, then the reflection of $S(z, \alpha_z)$ in $L(z, d_1)$ carries $K(z, \delta)$ into itself and a into c. The reflection in $L(z, d_2)$ carries c into b. The product of the two reflections is an orientation preserving isometry of $S(z, \alpha_z)$ on itself, hence a rotation of $S(z, \alpha_z)$ about z according to our definition in Section 42. We know from (42.7) that there is exactly one angular measure for the angles with vertex z, which is invariant under the rotations, and that this measure is proportional to the length of the arc on $K(z, \delta)$. (There is no restriction on the size of δ here since we know $K(z, \delta)$ to be convex.) We use *this angular measure* henceforth.

If A is a semicircle, i. e., (azb), and G is the perpendicular to $L(a, b)$ at z, then Ω_G interchanges the two arcs of A determined by its intersection with G. These therefore have equal lengths, and it follows that perpendicular lines form angles of measure $\pi/2$, which we call, of course, right angles.

A reflection of $S(z, \alpha_z)$ about a diameter which carries a into a' carries a circle $K(a, \varepsilon)$ with $0 < \varepsilon < \alpha_z - az$ into the circle $K(a', \varepsilon)$, and since the reflection preserves length of curves, an angle with vertex a will have the same measure as the angle with vertex a' corresponding to it under the reflection. Thus reflections and products of reflections, preserve angular measure.

Put $E = S(z, \alpha_z/4)$ and let abc, $a'b'c'$ be two triangles in E for which either $ab = a'b'$, $bc = b'c'$ and $ca = c'a'$ or $ab = a'b'$, $bc = b'c'$ and $|abc| = |a'b'c'|$. If $m = m_{aa'}$ then $mz < \alpha_z/4$ and from (46.3)

$$E \subset S(m, \alpha_z/4 + mz) \subset S(m, \alpha_z/2) \subset S\left(m, \alpha_m 2^{-1} (1 - \varepsilon)^{-1}\right).$$

Therefore the reflection Ψ of $S(m, \alpha_m)$ in the diameter perpendicular to the diameter carrying a and a' is defined in E and sends a into a'. Put $b'' = b\Psi$, $c'' = c\Psi$. Since $ab < \alpha_z/2$, $ac < \alpha_z/2$, it follows that the points b', c', b'', c''

lie in $S(a', \alpha_z/2) \subset S(a', \alpha_{a'} 2^{-1}(1-\varepsilon)^{-1}) \subset S(a', \alpha_{a'})$. Because of $a'b' = a'b''$ and $a'c' = a'c''$, both a reflection of $S(a', \alpha_{a'})$ in a diameter and a rotation of $S(a', \alpha_{a'})$ about a' exist which carry b'' into b', and one of these will carry c'' into c'.

The triangles abc and $a'b'c'$ are therefore *congruent* in the sense that all corresponding parts are equal. For instance, if (adc) and $(a'd'c')$ with $ac = a'c'$ then $bd = b'd'$ etc. It also follows exactly as in elementary geometry that the two triangles abc and $a'b'c'$ are congruent if $bc = b'c'$ and $|abc| = |a'b'c'|$, $|acb| = |a'c'b'|$.

Speaking of congruent triangles instead of motions has the advantage that we only use the triangles, hence avoid discussing the range of a motion with the concomitant, cumbersome use of the triangle inequality. We may continue using motions whenever convenient, without troubling about this question since we can always reformulate the procedure in terms of congruence.

Lambert Quadrilaterals

A convex quadrangle $abcd$ with three right angles is traditionally called a *Lambert quadrilateral*, because such quadrangles played an important role in Lambert's investigations on, what is now called, non-euclidean geometry. We imply in the notation that the angles at a, b, and c are right angles and that d is opposite to b. All our figures will lie in E. We show first

(47.1) *If abcd is a Lambert quadrilateral, then*

$$|adc| < \pi/2 \text{ if and only if } ad > bc \text{ (or } cd > ab)$$
$$|adc| = \pi/2 \text{ if and only if } ad = bc \text{ (or } cd = ab)$$
$$|adc| > \pi/2 \text{ if and only if } ad < bc \text{ (or } cd < ab)$$

Let the perpendicular to $L(a, b)$ at $m = m_{ab}$ meet $T(c, d)$ in n. The reflection in $L(m, n)$ maps c on the foot c' of n on $L(a, d)$. (The point c' might lie outside E, but this does not matter because it lies in $S(z, \alpha_z)$.) Obviously $c' = d$ or $ad = bc$ if and only if $|adc| = \pi/2$. If $(ac'd)$ then $bc < ad$ and $|adc| < \pi/2$, otherwise the perpendicular to $L(a, d)$ at d would intersect $T(c', n)$. If (adc') then $ad < bc$ and $|adc| > \pi/2$, otherwise the perpendicular to $L(a, d)$ at d would again intersect $T(c' n)$; see Figure 53.

We call a Lambert quadrilateral $abcd$ rectangular, acute or obtuse according to whether $|adc| = \pi/2$, $< \pi/2$, or $> \pi/2$. In order to distinguish the three elementary geometries we now prove:

(47.2) *Either all Lambert quadrilaterals are rectangular, or all are acute, or all are obtuse.*

The proof is rather long; the first and main step consists in showing: If one Lambert quadrilateral is rectangular then all are.

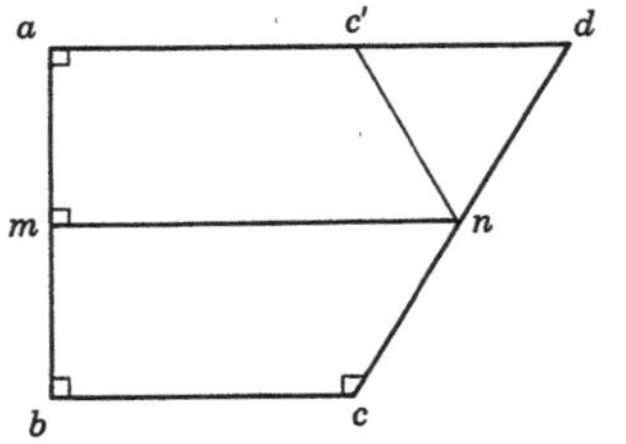

Fig. 53.

If $abcd$ is rectangular then — with the same notation as above — $c' = d$ implies that $L(m, n)$ is perpendicular to $L(c, d)$, hence $mbcn$ is rectangular and, by (47.1), $mn = ad = bc$. It is clear that repetition of this argument yields for any point x on $T(a, b)$ with $bx/ba = \mu\, 2^{-\nu}$ (μ, ν positive integers) that the perpendicular to $T(a, b)$ at x intersects $T(d, c)$ in a point x' for which $xbcx'$ is a rectangular Lambert quadrilateral and hence $xx' = bc$. Continuity shows the same for all $x \in T(a, b)$.

Applying this result to any $xbcx'$ shows that a perpendicular to $T(b, c)$ at a point y with (byc) intersects $T(x, x')$ at a point z for which $xbyz$ is a rectangular Lambert quadrilateral.

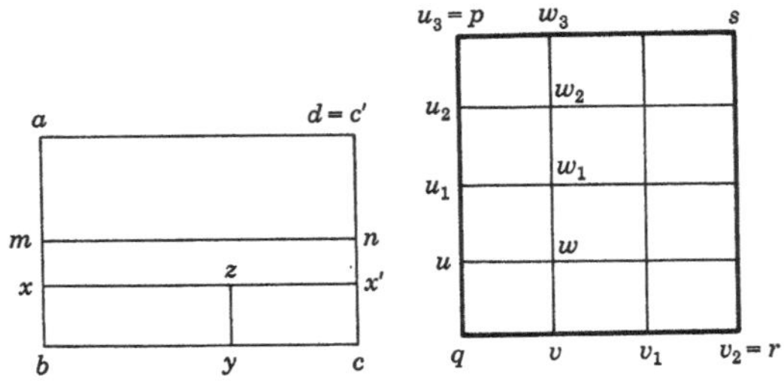

Fig. 54.

Now let $pqrs$ be any Lambert quadrilateral. Choose u with (puq) and v with (rvq) such that $qu < ab$, $qv < bc$ and qp/qu and qr/qv are positive integers. Let the perpendiculars to $T(p, q)$ at u and to $T(q, r)$ at v intersect at w. Choose $x \in T(a, b)$ with $bx = qu$ and $y \in T(b, c)$ with $by = qv$. If z is defined as above then the two Lambert quadrilaterals $uqvw$ and $xbyz$ are congruent, and since the latter is rectangular, $uqvw$ is too.

Reflection of $uqvw$ in $L(u, w)$ sends q, v into points u_1, w_1 with (quu_1), (vww_1) such that u_1uww_1 and u_1qvw_1 are rectangular Lambert quadrilaterals. Reflection in $L(u_1, w_1)$ carries u, w into points u_2, w_2 with (u_2u_1q), (w_2w_1v)

for which $u_2u_1w_1w_2$ *and* u_2qvw_2 are rectangular Lambert quadrilaterals. Since qp/qu is a positive integer, the point u_n will for some n coincide with p. Then (pw_ns) and $pqvw_n$ is rectangular. We now reflect $pqvw_n$ in $L(v, w_n)$ and see, proceeding as before, that $pqrs$ is rectangular.

It is now easy to finish the proof of (47.2). Let $abcd$ be an acute (obtuse) Lambert quadrilateral. If (axb) and (byc) denote by z the intersection of the perpendiculars to $L(a, b)$ at x and to $L(b, c)$ at y. If $xbyz$ were not acute (obtuse), then an obvious continuity argument would produce points $x' \in T(x, a)$ and $y' \in T(y, c)$ for which the corresponding Lambert quadrilateral $x'by'z'$ is rectangular; but then $abcd$ would be rectangular by the first part of this proof. Thus $xbyz$ is acute (obtuse).

If now $pqrs$ is any Lambert quadrilateral we choose u with (puq), and v with (qvr), x with (axb) and y with (byc) such that $qu = bx$, $qv = by$. Then the corresponding Lambert quadrilaterals $uqvw$ and $xbyz$ are congruent. Therefore $uqvw$ is acute (obtuse) and it follows that $pqrs$ is acute (obtuse).

THE FUNCTIONS $\varphi(\xi)$ AND $C(\theta)$

Rectangular, acute, and obtuse Lambert quadrilaterals correspond of course, to the euclidean, hyperbolic, and spherical geometries. Our next purpose is to actually derive these geometries from the mobility hypotheses, and we take the hyperbolic case as example, i. e., we assume that all Lambert quadrilaterals are acute.

The arguments in the spherical case are strictly analogous. In the euclidean case they are somewhat different but considerably simpler, we will briefly indicate its treatment later.

Let G and H be two lines perpendicular at w and Q one of the quadrants determined by G and H. If $p \in Q$ and x, y are the feet of p on G and H we put

$$px = \mu, \quad py = \lambda, \quad wx = \xi, \quad wy = \eta, \quad pw = \rho, \quad |pwx| = \theta.$$

The quantities μ and ξ for example, determine the others and if λ or ρ are considered as functions of μ and ξ we write $\lambda(\mu, \xi)$ or $\rho(\mu, \xi)$ to indicate this and use similar notations like $\theta(\xi, \rho)$ for other dependences.

Let $\xi_2 - \xi = \xi - \xi_1 > 0$, $\xi_1 \geqslant 0$, and put $x_i = (\xi_i, 0)$ (i. e., $\eta = 0$) $p_i = (\xi_i, \eta)$ (so that (p_1pp_2)), $\mu_i = p_i x_i$. Let the perpendicular to $L(p, x)$ intersect $L(p_i, x_i)$ in a_i. Finally let $p = m\,[p_1, p']$. Then the triangles a_1p_1p and $a_2p'p$ are congruent and since

$$|x_1a_1p| = |x_2a_2p| = |pa_2p'| < \pi/2 < |p_2a_2p|,$$

we have $(pp'p_2)$ and

(a) $$2\lambda = \lambda_1 + \lambda_2 - pp_2 + p_1p < \lambda_1 + \lambda_2.$$

Since $|x_1p_1y| = |a_1p_1p| = |a_2p'p| < \pi/2$ it follows that $p_2a_2 > p'a_2 = p_1a_1$ hence

$$2\mu < x_1a_1 + x_2a_2 = \mu_1 + \mu_2 + p_1a_1 - p_2a_2 < \mu_1 + \mu_2.$$

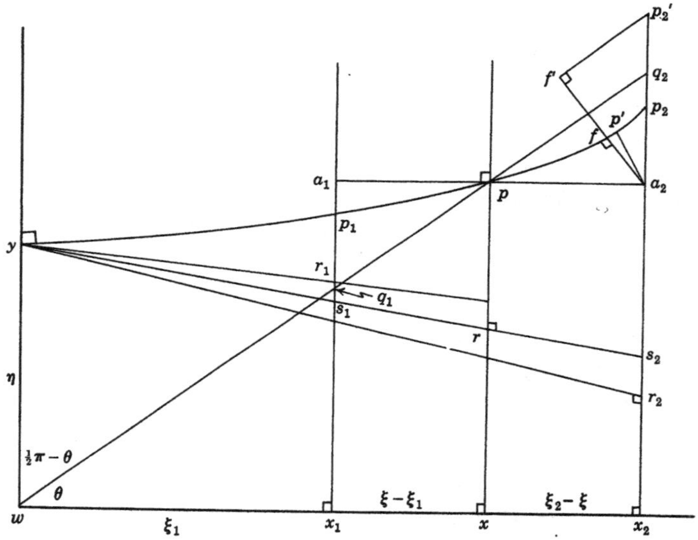

Fig. 55.

Thus we see: $\lambda(\xi, \eta)$ and $\mu(\xi, \eta)$ are, for fixed η (or ξ), *increasing convex functions of* ξ (or η). Hence $\lambda(\xi, \eta)$ possesses at $\xi = 0$ a right-hand partial derivative $\varphi(\eta)$ with respect to ξ:

$$\varphi(\eta) = \lim_{\xi \to 0+} \frac{\lambda(\xi, \eta)}{\xi} \geqslant 1$$

exists and, because of the inequality (a),

(b) $$\varphi(\eta) < \frac{\lambda(\xi, \eta)}{\xi} < \frac{\lambda(\xi', \eta)}{\xi'} \qquad \text{for} \qquad 0 < \xi < \xi'.$$

The function $\lambda(\xi, \eta)/\xi$ is for fixed $\xi > 0$ an increasing convex function of η which decreases monotonically to $\varphi(\eta)$ as ξ decreases. Therefore $\varphi(\eta)$ is a convex non-decreasing function of η, and as such continuous. By Dini's Theorem (see Courant [1, p. 106]) the convergence of $\lambda(\xi, \eta)/\xi$ to $\varphi(\eta)$ for $\xi \to 0+$ is uniform in η. (This also follows without using monotoneity from the convexity of $\lambda(\xi, \eta)/\xi$ as function of η.) Therefore

$$\text{(c)} \qquad \lim \frac{\lambda(\xi_\nu, \eta_\nu)}{\xi_\nu} = \varphi(\eta) \quad \textit{if} \quad \eta_\nu \geqslant 0,\ \eta_\nu \to \eta \quad \textit{and} \quad \xi_\nu \to 0+.$$

We will also need the information that $\varphi(\eta)$ is not constant. This may be seen as follows: let r, r_1, r_2 be the feet of y on $L(x, p)$, $L(x_1, p)$, $L(x_2, p)$ and let $L(y, r)$ intersect $L(x_i, p_i)$ at s_i. Then $rs_1 = rs_2$ hence

$$2\, yr = ys_1 + ys_2 > yr_1 + yr_2.$$

Therefore yr is a concave function of ξ whose right-hand derivative at 0 equals $\varphi(\eta)$ because of (c). Applying (b) we find

$$\varphi(\eta) \geqslant \frac{yr}{\xi} > \varphi(yr).$$

Now let, for fixed θ, the line $L(w, p)$ intersect $L(x_i, p_i)$ in q_i. Then $|wpx| < \pi/2$, and exactly the same arguments as above show that

$$pq_1 < pq_2 \qquad \text{and} \qquad a_1q_1 < a_2q_2;$$

thus we conclude that $\rho(\xi, \theta)$ and $\mu(\xi, \theta)$ are, for fixed θ, increasing convex functions of ξ. Hence the right-hand derivatives

$$C^{-1}(\theta) = \lim_{\xi \to 0+} \frac{\rho(\xi, \theta)}{\xi} \geqslant 1 \qquad \text{and} \qquad T(\theta) = \lim_{\xi \to 0+} \frac{\mu(\xi, \theta)}{\xi}$$

exist and are finite, moreover,

$$\text{(d)} \qquad C^{-1}(\theta) \leqslant \frac{\rho(\xi, \theta)}{\xi} < \frac{\rho(\xi', \theta)}{\xi'}, \qquad T(\theta) \leqslant \frac{\mu(\xi, \theta)}{\xi} < \frac{\mu(\xi', \theta)}{\xi'}$$
$$\text{if} \qquad 0 < \xi < \xi'.$$

The limit

$$S(\theta) = \lim_{\xi \to 0+} \frac{\mu(\xi, \theta)}{\rho(\xi, \theta)} = C(\theta)\, T(\theta)$$

also exists. Because $\varphi(0) = 1$ we conclude from (c) that $\lim \lambda(\xi, \theta)/\xi = 1$, hence

$$S(\pi/2 - \theta) = \lim_{\xi \to 0+} \frac{\lambda(\xi, \theta)}{\rho(\xi, \theta)} = \lim_{\xi \to 0+} \frac{\xi}{\rho(\xi, \theta)} = C(\theta) > 0.$$

We now keep x_1, x, x_2 fixed and let $\eta \to 0+$. Putting $\xi_2 - \xi = \xi - \xi_1 = h$ we have for $\eta \to 0+$

$$\frac{\mu(\xi_1, \eta)}{\eta} \to \varphi(\xi - h), \quad \frac{\mu(\xi, \eta)}{\eta} \to \varphi(\xi), \quad \frac{\mu(\xi_2, \eta)}{\eta} \to \varphi(\xi + h) \qquad \text{and}$$

$$\frac{x_1 a_1}{\mu} = \frac{x_2 a_2}{\mu} \to \varphi(h).$$

Moreover, with $\mu_i = \mu(\xi_i, \eta)$, $\mu = \mu(\xi, \eta)$,

$$\mu_1 + \mu_2 = x_1 a_1 - p_1 a_1 + x_2 a_2 + p_2 a_2$$

hence

$$\text{(e)} \qquad \frac{\mu_1 + \mu_2}{\eta} = \frac{2\, x_2\, a_2}{\mu} \frac{\mu}{\eta} + \frac{p_2 a_2 - p' a_2}{\eta}.$$

Because $|pp'a_2| < \pi/2$ the foot f of a_2 on $L(p, p_2)$ satisfies $(fp'p_2)$. Let $(a_2 p_2 p_2')$, $a_2 p'_2 = \delta$ where δ is any fixed positive number small enough for p_2' to lie in E, and let f' be the foot of p_2' on $L(a_2, f)$. Then by (d)

$$0 < \frac{p_2 a_2 - p' a_2}{\eta} < \frac{p_2 a_2 - a_2 f}{\eta} < \frac{p_2 f}{\mu_2} \frac{\mu_2}{\eta} < \frac{p_2 f}{p_2 a_2} \frac{\mu_2}{\eta} < \frac{p_2 f}{f a_2} \frac{\mu_2}{\eta} < \frac{p_2' f_2'}{a_2 f'} \frac{\mu_2}{\eta}$$

The line $L(p_1, p_2)$ tends for $\eta \to 0+$ to the line G, hence $|f'a_2 p_2'| \to 0$. Since δ is fixed $p_2 f'/a_2 f' \to 0$, $\mu_2/\eta \to \varphi(\xi_2)$. Therefore (e) yields for $\eta \to 0+$

$$\varphi(\xi + h) + \varphi(\xi - h) = 2\varphi(\xi)\,\varphi(h).$$

Since $\varphi(0) = 1$, $\varphi(\xi) \geqslant 1$, and $\varphi(\xi)$ is not constant and continuous, $\varphi(\xi)$ must have the form

$$\varphi(\xi) = \cosh k\,\xi \qquad \textit{with a suitable } k > 0,$$

see Picard [1, pp. 9—12].

The trigonometry of the right triangle

We are now ready for the principal step, the proof of the *Pythagorean Theorem*:

If abc is a triangle with a right angle at c, then $\cosh kab = \cosh kac \cosh kcb$, or $\varphi(ab) = \varphi(ac)\,\varphi(cb)$.

Let (aa_1c), (acc_1) and $a_1c_1 = ac$. Choose b_1 on the same side of $L(a, c)$ as b such that the triangle $a_1b_1c_1$ becomes congruent to abc. Then $T(a_1, b_1)$ and

$T(b, c)$ intersect at a point b_2. Let $(b_2a_1a_2)$ and $b_2a_2 = ab$. Choose c_2 on the same side of $L(a_2, b_2)$ as c (or c_1) such that $a_2b_2c_2$ is congruent to $a_1b_1c_1$ and hence to abc. Then

$$aa_1 = cc_1 \qquad \text{and} \qquad a_1a_2 = b_1b_2.$$

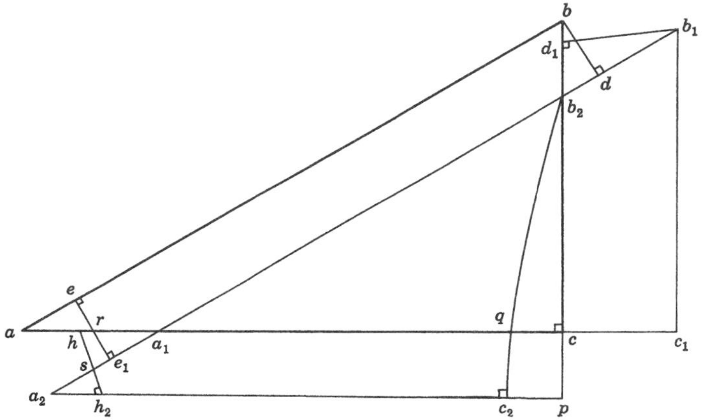

Fig. 56.

Put $T(b_2, c_2) \cap L(a, c) = q$ and $L(b, c) \cap L(a_2, c_2) = p$. The point p exists for small aa_1. In this order let d, d_1, e, e_1, h, h_2 be the feet of b on $L(a_1, b_1)$, b_1 on $L(b, c)$, $r = m_{aa_1}$ on $L(a, b)$, r on $L(a_1, b_1)$, $s = m_{a_1a_2}$ on $L(a, c)$, s on $L(a_2, c_2)$. Then as $a_1 \to a$

$$\lim \frac{b_1d_1}{a_1a_2} = \lim \frac{b_1d_1}{b_1b_2} = \lim \frac{bd}{bb_2} = S(|abc|) > 0.^1$$

Because of $|ear| = |e_1a_1r|$, $ar = a_1r$, $|aer| = |a_1e_1r| = \pi/2$ the triangles aer and a_1e_1r are congruent, hence (ere_1) and

$$\frac{ee_1}{aa_1} = \frac{er}{ar}, \quad \text{similarly} \quad \frac{hh_2}{a_1a_2} = \frac{h_2s}{a_2s},$$

so that

$$\lim \frac{ee_1}{aa_1} = \lim \frac{hh_2}{a_1a_2} = S(|bac|) > 0.$$

Then we obtain (using $aa_1 = cc_1$)

$$\text{(f)} \qquad \lim \frac{bd}{ee_1} = \lim \frac{bd}{bb_2}\,\frac{bb_2}{aa_1}\,\frac{aa_1}{ee_1} = \lim \frac{b_1d_1}{a_1a_2}\,\frac{bb_2}{aa_1}\,\frac{a_1a_2}{hh_2} = \lim \frac{b_1d_1}{cc_1}\,\frac{bb_2}{hh_2}.$$

But

$$cp = b_2p - b_2c > b_2c_2 - b_2c = bb_2 > b_2c_2 - b_2q = c_2q$$

and (c) yield

$$\varphi(ac) = \lim \frac{cp}{hh_2} = \lim \frac{c_2q}{hh_2} = \lim \frac{bb_2}{hh_2},$$

$$\varphi(ab) = \lim \frac{bd}{ee_1}, \quad \varphi(bc) = \lim \frac{b_1d_1}{cc_1},$$

so that the assertion follows from (f).

The rest is now fairly simple. In a triangle abc with a right angle at c let f be the foot on $L(a, b)$ of a point x with (cxa). Then (afb). Put

$$\alpha = kbc, \quad \beta = kca, \quad \gamma = kab, \quad \delta = kax,$$

$$\varepsilon = kaf, \quad \rho = kxf, \quad \theta = |bac|.$$

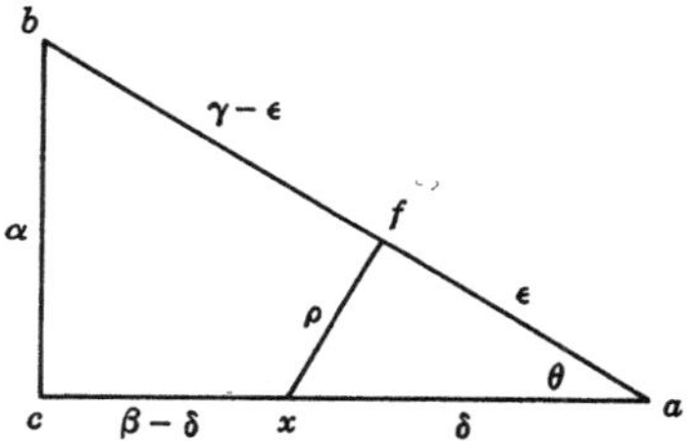

Fig. 57.

Then the Theorem of Pythagoras yields the two relations

$$\cosh \alpha \cosh (\beta - \delta) = \cosh kxb = \cosh \rho \cosh (\gamma - \varepsilon)$$
$$\cosh \alpha \cosh \beta \cosh \delta = \cosh \gamma \cosh \rho \cosh \varepsilon.$$

Subtracting the first equation from the second yields

$$\cosh \alpha \sinh \beta \sinh \delta = \cosh \rho \sinh \gamma \sinh \varepsilon$$

and division of this relation by the preceding one gives

$$\tanh \beta \tanh \beta = \tanh \gamma \tanh \varepsilon$$

or

$$\text{(g)} \qquad \frac{\tanh k\,ac}{\tanh k\,ab} = \frac{\tanh \varepsilon}{\tanh \delta} = \lim_{\varepsilon \to 0+} \frac{\tanh \varepsilon}{\tanh \delta} = \lim \frac{k\,af}{k\,ax} = C(\theta).$$

This implies that $C(\theta)$ is continuous for $0 \leqslant \theta \leqslant \pi/2$ if we put $C(0) = 1$, $C(\pi/2) = 0$.

It remains to identify $C(\theta)$ with $\cos \theta$. Let $0 < \delta < \theta$ and $\theta + \delta < \pi/2$. Consider an angle with vertex z, measure θ and legs R, R_1. Construct the two angles with measure δ, vertex z and R_1 as one leg. The perpendicular G to R_1 at a point y intersects R for small zy (because $\theta < \pi/2$) in a point p, and the other legs of the two angles with measure δ at points x, x' with $(px'y)$. Denote the feet of x, y, x' on R by u, v, u'. Then by (g):

$$C(|zpy|) = \frac{\tanh k\,pv}{\tanh k\,py} = \frac{\tanh k(pv + vu)}{\tanh k(py + xy)} = \frac{\tanh k(pv - vu')}{\tanh k(py - yx')}.$$

The line G converges for $y \to z$, hence $C(|zpy|)$ converges to a finite limit different from 0. Since $xy = yx'$ we conclude from the last relation that $uv/vu' \to 1$ for $y \to z$. Now

$$\frac{zv}{zx} = \frac{zv}{zy} \cdot \frac{zy}{zx} \to C(0)\,C(\delta)$$

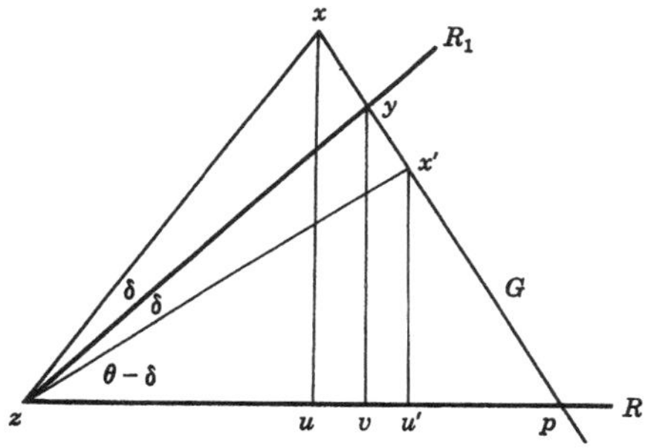

Fig. 58.

and

$$\frac{2\,zv}{zx} = \frac{zv}{zx} + \frac{zv}{zx'} = \frac{zu'}{zx'} + \frac{zu}{zx} + \frac{uv - u'v}{uv}\,\frac{uv}{zx'}.$$

Because $uv/vu' \to 1$ and $uv < zx'$, the last term tends to zero and we find

$$C(\theta + \delta) + C(\theta - \delta) = 2\,C(\theta)\,C(\delta).$$

Since $C(0) = 1$, $C(\pi/2) = 0$ and $C(\theta)$ is continuous it follows that $C(\theta) = \cos\theta$, see Picard [1, pp. 9—12]. The relations $S(\pi/2 - \theta) = C(\theta)$ and $T(\theta) = S(\theta)\,C^{-1}(\theta)$ prove $S(\theta) = \sin\theta$ and $T(\theta) = \tan\theta$.

The remaining relations between the parts of a triangle abc with a right angle at c follow immediately from the Theorem of Pythagoras, (g) and the trigonometric identities, and the oblique triangle reduces to the right triangle.

In the *case where all Lambert quadrilaterals are rectangular* we have, $\lambda(\xi, \eta) = \xi$, hence $\varphi(\eta) \equiv 1$. With the same notations as above the triangles $q_1 a_1 p$ and $q_2 a_2 p$ become congruent, hence

$$2\rho(\xi, \theta) = \rho(\xi_1, \theta) + \rho(\xi_2, \theta), \qquad 2\mu(\xi, \theta) = \mu(\xi_1, \theta) + \mu(\xi_2, \theta)$$

so that $\rho(\xi, \theta)$ and $\mu(\xi, \theta)$ are, for fixed θ, linear functions of ξ, and

$$C^{-1}(\theta) = \rho(\xi, \theta)/\xi, \qquad T(\theta) = \mu(\xi, \theta)/\xi, \qquad S(\theta) = \mu(\xi, \theta)/\rho(\xi, \theta).$$

The sum of the angles in a right triangle is π because a diagonal divides a Lambert quadrilateral into two congruent triangles.

If abc is a right triangle and f the foot of c on $L(a, b)$, then (afb) and $C(|abc|) = fb/bc = bc/ab$, hence

$$fb \cdot ab = bc^2, \text{ and similarly } \qquad af \cdot ab = ac^2$$

which yield $ab^2 = ac^2 + bc^2$.

The functional equation for $C(\theta)$ becomes quite trivial because $uv = vu'$ and no limit process is necessary.

The main theorem

It is now easy to complete the proof of Theorem (46.1). Consider a sphere $S(z, \alpha_z/4)$. Unless the space is two-dimensional there are three dimensional sets through z which contain with any two points the line connecting them. Let M be such a set. Any two planes P_1, P_2 in M through z intersect in a line L. The lines perpendicular to L at z in M form a plane set P. The intersections $P_i \cap P$ are two lines L_i, which determine in P two angular domains. Their angular bisectors G, H are perpendicular to each other and to L. Let a, a' be two points on H with midpoint z. The mapping $\Omega = \Omega(B(a, a'))$ of $S(z, \alpha_z/4)$ defined in (46.h) induces in M a mapping which leaves G and L fixed and carries L_1 into L_2. Since Ω maps every line which intersects L isometrically on a line intersecting L, it maps P_1 on P_2, and *the metrics in* P_1 *and* P_2, which we know to be euclidean, hyperbolic or spherical *must be the same.*

To see this it suffices to consider points $a \in L - z$, $b_i \in L_i - z$ with $b_1 z = b_2 z$. Then $b_2 = b_1 \Omega$, and if $ab_1^2 = az^2 + zb_1^2$, $\cosh k\, ab_1 = \cosh k\, az \cosh k\, zb_1$, or $\cos k\, ab_1 = \cos k\, az \cos k\, zb$, then the same relation holds for a, b_2, z. This implies that P_2 is euclidean if P_1 is, or is hyperbolic (spherical) with the same constant k if P_1 is. If the dimension of the space exceeds three and any two planes P_1, P_2 through z in $S(z, \alpha_z/4)$ are given which have only the point z in common, we take any line H_i in P_i through z. The three-dimensional flat subsets of $S(z, \alpha_z/4)$ through z containing $P_1 \cup H_2$ and $P_2 \cup H_1$ have the plane containing $H_1 \cup H_2$ in common. It follows from the preceding argument that the metrics in this plane are the same as in P_1 and P_2, so that they are the same in P_1 and P_2.

It follows that the same law of cosines applies to any triangle with one vertex at z, and this yields an isometric mapping of $S(z, \alpha_z/4)$ on a sphere in a euclidean, hyperbolic or spherical space.

This is, in particular, true for $S(p, \alpha_p/4)$ where p is the point occurring in (46.1). We map $S(p, \delta_p)$ on a sphere $S(p', \delta_p)$ in a euclidean, hyperbolic or spherical space by first mapping $S(p, \alpha_p/4)$ isometrically on $S(p', \alpha_p/4)$. Then for any radius $x(\tau)$ of $S(p, \delta_p)$, $0 \leqslant \tau < \delta_p$, (where $x(\tau)$ is part of a representation of a geodesic) the image $x'(\tau)$ is for $0 \leqslant \tau < \alpha_p/4$ part of a representation of a geodesic $x'(\tau)$. We map $x(\tau)$ for $0 \leqslant \tau < \delta_p$ on the point $x'(\tau)$. The mapping of $S(p, \delta_p)$ in the euclidean, in a hyperbolic or in a spherical space thus defined is isometric. The details for a proof of this obvious fact are found in the proof of (30.1).

A corollary of (46.1) and (30.1) is

(47.3) **Theorem.** *If every point p of a G-space R has a neighborhood $S(p, \delta_p)$, $\delta_p > 0$, in which for any given distinct points a, a', x, y, z the relations $ax = a'x$, $ay = a'y$ and $xz + zy = xy$ imply $az = a'z$, then the universal covering space of R is euclidean, hyperbolic, or spherical.*

Characterization of the Elementary Spaces

If the hypothesis of the last theorem is assumed in the large we obtain:

(47.4) **Theorem.** *If each bisector $B(a, a')$* (i. e., the locus $xa = xa'$) *of a G-space R contains with any two points x, y at least one segment $T(x, y)$, then the space is euclidean, hyperbolic, or spherical of dimension greater than* 1.

Notice the corollary

(47.5) *The euclidean, the hyperbolic, and the spherical spaces of dimension greater than one, are the only G-spaces in which the bisectors are flat.*

With the same method as for (46.b) we show first that under the hypotheses of (47.4) the bisectors are flat. Let $a \neq a'$. The set $B(a, a')$ is closed, hence finitely compact, and since it contains with any two points a segment connecting them, it is convex in Menger's sense. Uniqueness of prolongation is obvious because it holds in the whole space. To prove local prolongability it suffices to show that $B(a, a')$ contains with any segment $T(x, y)$, where $0 < xy < \rho(y)/2$ also a segment $T(x, y')$ with $xy = yy' = xy'/2$.

If a point z of $T(x, y')$ does not belong to $B(a, a')$ then $az < a'z$ say. Choose b with (yba') and $yb < \rho(y)/2$. Then $ba > ba'$ because $ba \geqslant ay - yb = a'y - yb = ba'$ and $ba = ba'$ would imply (yba) and hence $a = a'$.

For $u \in T(y, b)$ let v_u be defined by $v_b = b$ and, if $u \neq b$, by $(v_u ux)$ and $v_u u : zy = bu : by$. The point v_u exists because $v_u u \leqslant zy < \rho(y)/2$ and $xu < \rho(y)$. Also v_u depends continuously on u. Since $v_b a > v_b a'$ and $av_y = az < a'z = a'v_y$ there is a point w on $T(y, b)$ with $v_w \in B(a, a')$. The point w cannot lie on $B(a, a')$ by the same argument that proved $ba > ba'$. But this is impossible because $x, v_w \in B(a, a')$ and $T(x, v_w)$ is unique, and hence lies on $B(a, a')$.

Next we consider the case where the space is two-dimensional. Since $B(a, a')$, $a \neq a'$, is a G-space it is a straight line or a great circle, and since every geodesic is a bisector it follows from (31.3) that the space is homeomorphic to the plane, the sphere, or the elliptic plane, hence the metric is either euclidean, or hyperbolic, or spherical, or elliptic. The last case is impossible because a bisector in the elliptic plane consists of two great circles.

If the dimension of the space exceeds two then $S(p, \rho(p))$ is by (46.1) isometric to a sphere with radius $\rho(p)$ of an n-dimensional euclidean, hyperbolic, or spherical space, where $n \geqslant 3$ is finite. Take any plane P — in the same sense as previously — through p in $S(p, \rho(p))$ and let $L_1, \ldots, L_{n-2}$ be lines through p perpendicular to P and to each other. If $a_i \in L_i$ and $0 < a_i p = pa_i' = a_i a_i'/2 < \rho(p)$, then $B = \bigcap_{i=1}^{n-2} B(a_i, a_i')$ contains P and with any two points x, y the segment $T(x, y)$ whenever it is unique. By examining the set over B in the universal covering space, we find that B also contains at least one segment $T(x, y)$ when $T(x, y)$ is not unique. Since each of the $B(a_i, a_i')$ contains with two points x, y, $xy < \rho(y)/2$, the segment $T(x, y')$ with $y = m_{xy'}$, the same holds for B, which is therefore a two-dimensional G-space. If b, b' are two distinct points in B then $B \cap B(b, b')$ contains with any two points x, y the segment $T(x, y)$ when it is unique, and hence by continuity at least one segment $T(x, y)$ when it is not unique. We just proved that B is the euclidean plane, a hyperbolic plane or a sphere. This proves (47.4).

It should be observed that the results of Chapter IV need not be used to prove (47.4). The arguments of this and the preceding sections can be so arranged that they yield a complete proof of (47.4).

48. Applications of the Bisector Theorem. The Helmholtz-Lie Problem

The theorems of the last two sections have many important applications. A first type of application concerns questions of *mobility*. We begin by proving a few simple lemmas which establish the connection of mobility with the flatness of the bisectors.

Spheres permitting rotation

Concise formulation will be greatly facilitated by the definition:

The sphere $S(p, \delta)$ "*permits rotation*" if for any four points a, b, a', b' in $S(p, \delta)$ with $pa = pa'$, $pb = pb'$ and $ab = a'b'$ an isometry of $S(p, \delta)$ on itself exists which leaves p fixed and carries a into a' and b into b'. An isometry of $S(p, \delta)$ on itself is a motion of $S(p, \delta)$ considered as a metric space, and we will use this shorter term.

(48.1) *If* $S(p, \delta_p)$ *and* $S(q, \delta_q)$ *permit rotation and* $pq_1 = pq < \delta_p$ *then* $S\big(q_1, \min(\delta_q, \delta_p - pq)\big)$ *permits rotation.*

For let $q_1a_1 = q_1a_1' < \sigma = \min(\delta_q, \delta_p - pq)$, $q_1b_1 = q_1b_1' < \sigma$, and $a_1b_1 = a_1'b_1'$. Applying the hypothesis to the case where $a = b = q$ and $a' = b' = q_1$ we obtain a motion Φ of $S(p, \delta_p)$ which leaves p fixed and takes q_1 into q. The points $q_1, a_1, a_1', b_1, b_1'$ lie in $S(p, \delta_p)$. Therefore Φ carries them into points q, a, a', b, b' with $qa = qa' < \sigma$, $qb = qb' < \sigma$ and $ab = a'b'$. Since $S(q, \delta_q)$ permits rotation, $S(q, \sigma)$ does too. Therefore a motion Ψ of $S(q, \sigma)$ exists which sends q, a, b into q, a', b'. Then $\Phi\Psi\Phi^{-1}$ is an isometry of $S(q_1, \sigma)$ on itself which carries q_1, a_1, b_1 into q_1, a_1', b_1'.

(48.2) If $S(p, \delta_p)$ and $S(q, \delta_q)$ permit rotation and $pq < \min\big(\rho(p), \delta_p\big)$, $\beta = \min(\delta_q, \delta_p - pq, pq)$ and $x \in S(q, \beta)$ then $S(x, \beta - qx)$ permits rotation.

Proof. Since $pq < \rho(p)$ there is a point p' with $p'p = pq = p'q/2$. Let t_1 and t_2 be the midpoints of $T(p', p)$ and $T(p, q)$, respectively, and t any point distinct from t_1 and t_2 with distance $pt = pq/2$ from p. (t exists if the dimension exceeds 1.) Then $T(t_i, t) \subset S(p, pq)$. For any point $u \in T(t_1, t) \cup T(t, t_2)$ there is a point v with (puv) and $pv = pq$. As u traverses $T(t_1, t) \cup T(t, t_2)$ the point v traverses a Jordan arc A from p' to q which lies entirely on $K(p, pq)$.

The arc A intersects every sphere $K(q, \tau)$ with $\tau \leqslant \beta$ at some point y_τ because $p'q = 2\,pq > \beta$. By (48.1) the sphere $S\big(y_\tau, \min(\delta_q, \delta_p - pq)\big)$ and hence $S(y_\tau, \beta)$ permits rotation. If $x \in S(q, \beta)$ then $x \in K(q, \tau)$ for some $\tau < \beta$. Now (48.1) applied to q, y_τ, x instead of p, q, q_1 shows that $S(x, \beta - qx)$ permits rotation.

For brevity we call a sphere "*elementary*" if it is isometric to a sphere in a euclidean, hyperbolic or spherical space. The connection with (46.1) is now established by:

(48.3) *If* $S(p, \delta_p)$ *and* $S(q, \delta_q)$ *permit rotation,* $pq < \min(\rho(p)\ \delta_p)$ *and* $\beta = \min(\delta_q, \delta_p - pq, pq)$ *then* $S(q, \beta/3)$ *is elementary.*

Because of (46.1) it suffices to show: If a, a', x, y, z are five distinct points in $S(q, \beta/3)$ for which $ax = a'x$, $ay = a'y$ and (xzy) then $az = a'z$.

When $x \in S(q, \beta/3)$ then $2\beta/3 < \beta - qx$, hence $S(x, 2\beta/3)$ permits rotation by (48.2). Since $xa = xa' < 2\beta/3$, $xy = xy < \beta/3$ and $ay = a'y$ there is a motion Φ of $S(x, 2\beta/3)$ which carries x, a, y into x, a', y. Since also $xz < 2\beta/3$ and $yz < 2\beta/3$, the mapping Φ is defined at z and (xzy) implies $(x\Phi z\Phi y\Phi)$ or $(xz\Phi y)$ and $xz\Phi = xz$. Therefore $z\Phi = z$ and $az = a'z\Phi = a'z$.

A THEOREM OF F. SCHUR

Various well-known results of differential geometry can now be proved. A theorem of F. Schur [1], see also Cartan [2, p. 127], states that a Riemann space has constant curvature if two points p and q exist such that any two points a, b lie in a plane through p and in a plane through q. In this form the theorem cannot be extended to a non-Riemannian space, since every Desarguesian space satisfies the hypothesis. Here, the dimension of the space is, of course, assumed to exceed two.

An equivalent statement for Riemann spaces is this: if for any two plane elements dP_1 and dP_2 through p a motion exists which leaves p fixed and carries P_1 into P_2 and the same holds for q, then the space has constant curvature. The theorem is local as are almost all theorems in Riemannian geometry. To become applicable to two-dimensional and to Finsler spaces, the hypothesis must be modified to the effect that – in our terminology – there are spheres $S(p, \delta_p)$ and $S(q, \delta_q)$ which permit rotation. The assertion is that a certain neighborhood $S(p, v)$ of p say, is elementary. This cannot be true without restrictions on δ_p and δ_q. Clearly it is necessary that $q \in S(p, \delta_p)$ and $p \in S(q, \delta_q)$.

For let $f(r)$ be defined by

$$f(r) = \begin{cases} r^2 & \text{for} \quad 0 \leqslant r \leqslant 1 \\ 2r - 1 & \text{for} \quad r \geqslant 1, \end{cases}$$

then the surface $z = f(\sqrt{x^2 + y^2})$ in (x, y, z)-space has the property that for $p = (0, 0, 0)$ every sphere $S(p, \delta)$ permits rotation. Each point with $r > 1$ is interior point of a cone and has therefore a neighborhood isometric to a euclidean circular disk, but no neighborhood of p is elementary.

But even $p \in S(q, \delta_q)$ and $q \in S(p, \delta_p)$ are not sufficient. The ellipsoid $x^2 + y^2 + a^2z^2 = 1$, $a > 1$, provides an example: every sphere about $p = (0, 0, a^{-1})$ and $q = (0, 0, -a^{-1})$ permits rotation. This explains why the elegance of Schur's Theorem in its usual form must be sacrificed if the goal is a statement with complete hypotheses and a definite radius v for which $S(p, v)$ is elementary. Our version is:

(48.4) ***Theorem.*** *If $0 < pq < \min(\delta_p, \rho(p), \delta_q)$ and $S(p, \delta_p)$ and $S(q, \delta_q)$ permit rotation, then $S(p, v)$, $v = \min(\rho(p), \delta_p, 2\,pq)$ is isometric to a sphere in a euclidean, hyperbolic or spherical space.*

For a proof put $\beta = \min(\delta_q - pq, \delta_p - pq)/3$. By (48.3) the sphere $S(q, \beta)$ is elementary. Let $p = m_{qq'}$ and r not on the same segment with p and q so close to p that $V = T(q, r) \cup T(r, q') \subset S(p, \delta_p)$. For every point x on V there is (because $pq < \rho(p)$) a point y_x either with $(y_x x p)$ or on $T(p, x)$ such that $py_x = pq$.

Rotation about p carries $S(q, \beta)$ into $S(y_x, \beta)$. The number qy_x changes continuously with x from 0 to $2\,pq$ as x traverses V. For $qy_x < pq$ the sphere $S(y_x, \beta)$ lies in $S(q, \delta_q)$. If z is given with either $z \in T(p, q)$ or (pqz) and $qz < pq$, then $x \in V$ exists such that rotation about q carries $S(y_x, \beta)$ into $S(z, \beta)$.

Now a rotation about p will send a suitable one among these points z into a given point w of $S(p, v)$ because $v \leq \min(\delta_p, 2\,pq)$. Hence $S(w, \beta)$ is isometric to the elementary sphere $S(q, \beta)$. Since also $v < \rho(p)$ it follows as in previous proofs that $S(p, v)$ is elementary.

Notice the corollary:

(48.5) *If all spheres about two distinct points p, q of a straight space permit rotation, then the space is euclidean or hyperbolic.*

A slightly different application of (48.3) is

(48.6) *If $S(p, \delta)$, $0 < \delta < \rho(p)$ permits rotation and if for every τ, $0 < \tau < \delta$, a point q_τ with $pq_\tau = \tau$ and a $\delta_\tau > 0$ exist such that $S(q_\tau, \delta_\tau)$ permits rotation, then $S(p, \delta)$ is elementary.*

It follows from (48.3) that every point x in $S(p, \delta)$ except p has an elementary spherical neighborhood $S(x, \beta_x)$. If $\delta_\tau > pq$ for some τ then (48.6) follows immediately. If this is not the case let $0 < \tau < \rho(p)/3$, so that $\rho(q_\tau) > \tau$. We map $S(q_\tau, \beta_{q_\tau})$ isometrically on an elementary sphere and extend this mapping to a topological mapping of $S(q_\tau, \rho(q_\tau))$ on an elementary sphere U' such that each radius of $S(q_\tau, \rho(q_\tau))$ is mapped isometrically on a radius of U'.

By the argument used in the proof of (31.1) this mapping is locally isometric in the neighborhood of every point in $S(q_\tau, \rho(q_\tau)) \cap S(p, \delta)$ with the possible exception of p. Any segment $T(x, y)$ which does not pass through p is mapped on a curve of the same length, hence the distance of x and y equals the distance of their images. For reasons of continuity the mapping is isometric throughout.

The Helmholtz-Lie Problem

An immediate corollary of (48.6) is the important fact:

(48.7) **Theorem.** *If every point p of a G-space R has a neighborhood $S(p, \delta_p)$, $\delta_p > 0$, such that for any points a, a', b, b' in $S(p, \delta_p)$ with $ap = a'p$, $bp = b'p$ and $ab = a'b'$ an isometry of $S(p, \delta_p)$ on itself exists which leaves p fixed and takes a into a' and b into b' then the universal covering space of R is euclidean, hyperbolic or spherical.*

This theorem is the local version of the famous *Helmholtz-Lie Problem* whose history is briefly this: Riemann's lecture [1] of 1854 'Über die Hypothesen welche der Geometrie zu Grunde liegen" was answered in 1868 by Helmholtz [1]: "Über die Tatsachen die der Geometrie zu Grunde liegen". Helmholtz says that, while spaces of the general type discussed by Riemann are abstractly conceivable, physical space is distinguished by the *free mobility* of objects in it. A rigorous form of this property is this: If Φ is an isometric mapping of the set M on the set M' (in the same metric space), then Φ can be extended to a motion of the space.

Helmholtz stated correctly, and believed to have proved under certain differentiability assumptions, that the only spaces with this property are the euclidean, hyperbolic and spherical spaces. A serious gap in Helmholtz's proof was discovered by Lie [1, p. 437], who reformulated the problem and solved it by using his theory of transformation groups, which required strong analyticity assumptions. Weyl [1] treats the problem completely under reduced differentiability hypotheses. In all these investigations the number of points in the sets M to which the mobility property is applied depends on the dimension of the space, it is n or $n + 1$ for an n-dimensional space. Our last theorem shows that it suffices to consider point triples. This is particularly important in our case because we do not know whether an arbitrary G-space has finite dimension. On the other hand, G. Birkhoff [1, 2] showed, that *the mobility property applied to denumerable sets M can be used to deduce finite compactness from the much weaker assumption of completeness.*

The development of the foundations of geometry made differentiability assumptions appear altogether foreign to the Helmholtz-Lie Problem, and Hilbert [1, Appendix IV] eliminated them entirely in the two-dimensional case. Kolmogoroff [1] indicates without proof how Hilbert's idea can be generalized, but also works with $(n + 1)$-tuples instead of triples.

Recently Tits [1] indicated (with proofs forthcoming in the near future) how Kolmogoroff's approach can be completed, reducing at the same time the number of points from $n + 1$ to 3.

The work of Hilbert, Kolmogoroff and Tits uses much weaker hypotheses than ours (the modern results on Lie groups permit us to eliminate many of them). The space is not assumed to be a G-space, the groups of motions are shown to be isomorphic to those of the elementary spaces. It then follows that the elementary metrics provide the only M-convex metrizations. On the other hand, these methods definitely operate in the large and have hitherto not yielded theorems like our (48. 4, 6, 7). The latter results lead to the following solution of the Helmholtz-Lie Problem in the large within the framework of G-spaces:

(48.8) ***Theorem.*** *If a G-space possesses for any four points a, a', b, c with $ab = a'b$ and $ac = a'c$ a motion which leaves b and c fixed and carries a into a', then the space is euclidean, hyperbolic, or spherical.*

We first show: If p, a, b and p, a', b' are isometric triples then a motion of the given space R exists which leaves p fixed and carries a into a' and b into b'. For there is a motion Φ of R taking p, p, a into p, p, a'. If $b\Phi = b^*$ then $pb^* = pb = pb'$ and $a'b^* = ab = a'b'$. Therefore a motion Ψ exists which sends p, a', b^* into p, a', b' so that $\Phi\Psi$ carries p, a, b into p, a', b'. This means that every sphere permits rotation, hence the universal covering space of R is by (48.7) euclidean, hyperbolic or spherical. We next prove a lemma which will prove useful also on other occasions:

(48.9) *In a G-space R' let, for any two pairs a, b and a', b' with $ab = a' b' < \eta$, where η is a fixed positive number, a motion of R' exist which carries a into a' and b into b'. If $x(\tau)$ and $y(\tau)$ represent geodesics in R' then for any real α, τ_1, τ_2*

$$x(\tau_1)\, x(\tau_2) = y(\tau_1 + \alpha)\, y(\tau_2 + \alpha).$$

If R' is not compact then it is straight.

The number $\rho(p)$ is constant because there is a motion carrying p into an arbitrary other point. Let $0 < \beta < \min\,(\eta, \rho(p))$. There is a motion Φ, sending $x(0), x(\beta)$ into $y(\alpha), y(\alpha + \beta)$. Because $T\,(x(0), x(\beta))$ and $T\,(y(\alpha), y\,(\alpha + \beta))$

are unique, Φ takes $x(\tau)$ into $y(\tau+\alpha)$ for $0 \leqslant \tau \leqslant \beta$. But $x(\tau)\,\Phi$ represents a geodesic, hence $x(\tau)\,\Phi = y(\tau+\alpha)$ for $0 \leqslant \tau \leqslant \beta$ implies that this relation holds for all real τ. Since Φ is a motion, $x(\tau_1)\,x(\tau_2) = y(\tau_1+\alpha)\,y(\tau_2+\alpha)$ follows.

If R' is not compact then any point p is origin of a ray $x(\tau)$, $\tau \geqslant 0$ (see (22.1)), i. e., $x(\tau_1)\,x(\tau_2) = |\tau_1 - \tau_2|$ for $\tau_i \geqslant 0$. Let a, b be any two distinct points and let $y(\tau)$ be a geodesic which represents for $0 \leqslant \tau \leqslant ab$ a segment $T^+(a, b)$. By the first part of this proof $x(\tau_1)\,x(\tau_2) = y(\tau_1)\,y(\tau_2)$ for any τ_i in particular $(x(0)\,x(ab)\,x(2\,ab))$ implies $(aby(2\,ab))$. This shows $\rho(p) \equiv \infty$, so that the space is straight.

We now return to (48.8). Since the universal covering space of R is euclidean, hyperbolic or spherical it follows from (48.9) that R is euclidean or hyperbolic if it is not compact.

If R is compact, then it cannot be locally euclidean or hyperbolic. For closed geodesics exist and correspond in the hyperbolic case to axes of motions of the fundamental group and are countable in number. In the euclidean case the motions of the fundamental group correspond to sets of parallel lines. In either case there are both closed and non-closed geodesics, which contradicts (48.9).

This argument is not applicable in the spherical case because in an elliptic space all geodesics are isometric to each other. We take a plane P through q in a neighborhood $S(q, \beta)$ in the sense in which this word was used in Section 46. The geodesics containing the line in P through q form in R a set S with this property: If $\overline{P}$ lies over P in the universal covering space $\overline{R}$ of R then the great circles through the lines in $\overline{P}$ through the point $\overline{q}$ over q form a sphere $\overline{S}$ which lies over S. Hence S is, with its intrinsic metric, either a sphere or an elliptic plane. But from the way $\overline{S}$ is mapped on S it follows that the geodesics in S are great circles with the metric of R, hence S is also with the metric of R a sphere or an elliptic plane.

It is now easy to see that S cannot be an elliptic plane: If longitude θ, $0 \leqslant \theta \leqslant 2\pi$ and latitude φ, $-\pi/2 < \varphi < \pi/2$ are introduced as coordinates in $\overline{S}$, consider on $\overline{S}$ the points $a = (2\pi/3, 0)$, $b = (\pi/3, 0)$, $c = (0, 0)$, $a' = (\pi/6, \alpha)$, where $0 < \alpha < \pi/2$ and $\cos\alpha = 3^{-\frac{1}{2}}$. If 2β is the length of the great circle in R, and $d = (\pi/6, 0)$ then $bd = cd = \beta/6$ and $\cos(a'b'\pi/\beta) = \cos(a'c\,\pi/\beta) = \cos(bd\,\pi/\beta)\cos\alpha = 1/2$, hence $a'b = a'c = \beta/3$. If S were an elliptic plane and a, a', b, c lie over a_1, a_1', b_1, c_1 in S, then $a_1'b_1 = a_1'c_1 = \beta/3$ and $a_1b_1 = a_1c_1 = \beta/3$. Thus the triples a_1, b_1, c_1 and a_1', b_1, c_1 are isometric, but no motion can exist which carries the first triple into the second, because the first lies on a geodesic and the second does not. The last consideration shows how the assumptions of (48.8) must be modified in order to obtain a theorem which includes the elliptic spaces:

(48.10) *If for a suitable positive δ and for any four points a, a', b, c with $ab = a'b < \delta$, $ac = a'c < \delta$ a motion exists which leaves b and c fixed and carries a into a', then the space is euclidean, hyperbolic, spherical or elliptic.*

The proof runs along the lines of the last proof: first we see that, given p, a, b, a', b' with $pa = pa' < \delta/2$, $pb = pb' < \delta/2$ and $ab = a'b'$, a motion exists which takes p, a, b into p, a', b'. Therefore the space is by (48.7) locally euclidean, hyperbolic, or spherical. The lemma (48.9) shows exactly as above that in the first two cases the space is euclidean or hyperbolic. In the spherical case we see that, with the previous notations, S is a sphere or an elliptic plane. If S and S' are two such sets through q, i. e., through planes P and P' in $S(q, \beta)$ through q, then points a, b in P not collinear with q and a', b' in P' not collinear with q' exist such that $qa = qa' < \delta/2$, $qb = qb' < \delta/2$ and $ab = a'b'$. The motion which takes q, a, b into q, a', b' carries P into P' hence S into S'. Therefore S and S' are either both spheres or both elliptic planes which implies the assertion.

The last theorems suggest the question whether a G-space is euclidean, hyperbolic, spherical or elliptic if for any two pairs a,b and a',b' with $ab = a'b'$ a motion exists which carries a into a' and b into b'.

This is obviously so when the space is two-dimensional. On the other hand, the hermitian hyperbolic and elliptic spaces, see Section 53, are examples which show that *the answer is negative for all even dimensional spaces of dimension greater than two.*

Surprisingly enough the answer is affirmative for all odd-dimensional spaces. This is a beautiful result of Wang [2] which will be proved in Section 55.

Transversals of a triangle

The Bisector Theorem has various other unexpected consequences. In elementary geometry the length of a transversal of a triangle is determined by the lengths of the adjacent sides and the lengths of the segments into which it divides the third side. Our first aim is to show that this characterizes the elementary geometries.

In euclidean, hyperbolic, or spherical geometry there exists a function $\varphi(t_1, t_2, t_3, t_4)$ such that (qxr) (in the spherical case qr must be smaller than the diameter of the space) implies for any point p the relation

$$px = \varphi(pq, pr, xq, xr).$$

We take the *hyperbolic* case as an example. If f is the foot of p on the line through q and r and (qfx) say, (the case (fqx) is treated in an entirely analogous manner) we put

$$k \cdot pq = \alpha, \quad k \cdot pr = \beta, \quad k \cdot qx = \delta, \quad k \cdot rx = \varepsilon, \quad k \cdot qf = \nu, \quad k \cdot px = \xi$$

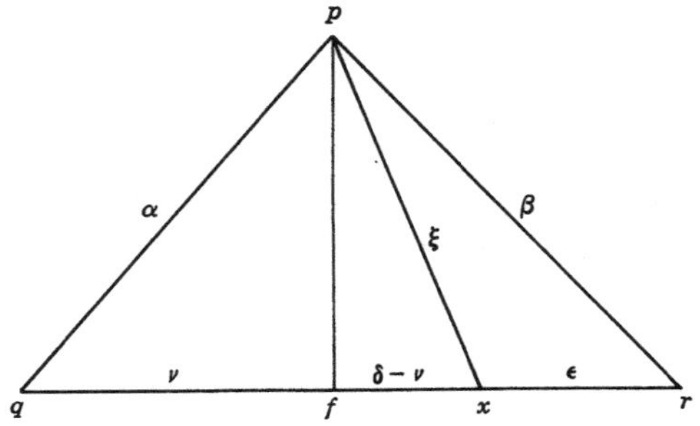

Fig. 59.

where k is the space constant. Then

$$\frac{\cosh \nu}{\cosh \alpha} = \frac{\cosh \delta \cosh \nu - \sinh \delta \sinh \nu}{\cosh \xi} = \frac{\cosh(\varepsilon + \delta)\cosh \nu - \sinh(\varepsilon + \delta) \sinh \nu}{\cosh \beta}$$

hence

$$\tanh \nu = \frac{\cosh \delta}{\sinh \delta} - \frac{\cosh \xi}{\cosh \alpha \sinh \delta} = \frac{\cosh(\varepsilon + \delta)}{\sinh(\varepsilon + \delta)} - \frac{\cosh \beta}{\cosh \alpha \sinh(\varepsilon + \delta)}$$

$$\cosh \xi \sinh (\varepsilon + \delta) = \cosh \alpha \cosh \delta \sinh (\varepsilon + \delta) - \cosh \alpha \cosh (\varepsilon + \delta) \sinh \delta$$
$$+ \cosh \beta \sinh \delta = \cosh \alpha \sinh \varepsilon + \cosh \beta \sinh \delta \qquad \text{or}$$

(48.11) $\cosh k\, px = (\cosh k\, pq \sinh k\, xr + \cosh k\, pr \sinh k\, xq)/\sinh k\, (qx + xr).$

The corresponding *spherical* formula is

(48.12) $\cos k\, px = (\cos k\, pq \sin k\, xr + \cos k\, pr \sin k\, xq)/\sin k(qx + xr)$ and holds only for $k \cdot qr < \pi$.

The *euclidean* relation is

(48.13) $px^2 = (pq^2 \cdot xr + pr^2 \cdot xq)\, (qx + xr)^{-1} - xq \cdot xr.$

We now show that the functions $\varphi(t_1, t_2, t_3, t_4)$ determined by (48.11, 12, 13) are the only possible general expressions for a transversal.

(48.14) ***Theorem.*** *In a G-space R let for every point z a neighborhood $S(z, \delta_z)$ and a function $\varphi_z(t_1, t_2, t_3, t_4)$ exist such that for any four points p, q, x, r with (qxr) in $S(z, \delta_z)$ the relation*

$$px = \varphi_z(pq, pr, xq, xr)$$

holds. Then R is locally euclidean, hyperbolic, or spherical.

For let p, q', q, r, x be five distinct points in $S(z, \delta_z)$ with (qxr) and $pq = p'q$, $pr = p'r$. Then

$$px = \varphi_z(pq, pr, xq, xr) = \varphi_z(p'q, p'r, xq, xr) = p'x.$$

This means that the hypotheses of (47.3) are satisfied and (48.14) is proved. The theorem in the large corresponding to (48.14) is

(48.15) *If in a G-space a function $\varphi(t_1, t_2, t_3, t_4)$ exists such that for any four points p, q, x, r with (qxr) the relation $px = \varphi(pq, pr, xq, xr)$ holds, then the space is euclidean or hyperbolic.*

For the same argument as above shows that $B(p, p')$ contains with any two points q, r every segment $T(q, r)$. We conclude from (47.4) that the space is euclidean, hyperbolic, or spherical. But in the spherical case a bisector does not contain every segment $T(q, r)$ when q and r are antipodes.

Area of a triangle in terms of its sides

A similar application of the Bisector Theorem shows that in two dimensions the elementary geometries are the only ones in which the area of a triangle is expressible in terms of its sides. If the triangle abc has sides $\alpha = bc$, $\beta = ca$, $\gamma = ab$ and we put as usual $s = (\alpha + \beta + \gamma)/2$, then in *euclidean* geometry Hero's formula states that the area $A(abc)$ of abc is given by

$$A^2(abc) = s(s-\alpha)(s-\beta)(s-\gamma).$$

In the hyperbolic and spherical geometry $A(abc)$ is proportional to the excess $\varepsilon(abc) = |abc| + |bca| + |cab| - \pi$. Therefore L'Huilier's formula (see Hessenberg [2, p. 134])

$$\tan^2 \frac{\varepsilon(abc)}{2} = \tan k \frac{s}{2} \tan k \frac{s-\alpha}{2} \tan k \frac{s-\beta}{2} \tan k \frac{s-\gamma}{2}$$

expresses the area of a *spherical* triangle in terms of its sides.

The corresponding formula for *hyperbolic* geometry is

$$\tan^2 \frac{\varepsilon(abc)}{2} = \tanh k \frac{s}{2} \tanh k \frac{s-\alpha}{2} \tanh k \frac{s-\beta}{2} \tanh k \frac{s-\gamma}{2},$$

but the following expression of Lobachevsky is more frequently found in the literature, see Liebmann [1, p. 129]

$$\sin^2 \frac{\varepsilon(abc)}{2} =$$

$$(e^{2s}-1)(e^{2s-2\alpha}-1)(e^{2s-2\beta}-1)(e^{2s-2\gamma}-1)(e^{\alpha}+1)^{-2}(e^{\beta}+1)^{-2}(e^{\gamma}+1)^{-2}.$$

Before we can show that these are the only possible general expressions for area in terms of the sides (except for a factor depending on the unit for area), we have to agree on the meaning of area. We need surprisingly little:

$A(abc)$ is called a "*triangle area*" in the subset M of a G-surface if

1) *$A(abc)$ is defined for all triples a, b, c in M, symmetric in a, b, c and non-negative* (real).

2) *$A(abc) = 0$, if and only if a, b, c lie on a segment.*

3) *If a, b, c, x lie in M and (bxc) then $A(abx) + A(axc) = A(abc)$.*

We now prove:

(48.16) **Theorem.** *If every point p of a G-surface R has a neighborhood $S(p, \delta_p)$, $\delta_p > 0$, in which a triangle area is defined and has the same value for isometric triples, then R is locally euclidean, hyperbolic or spherical.*

The formulation of this theorem does not exclude the possibility that $A(abc)$ has different values for a triangle abc in the overlap of two different neighborhoods $S(p, \delta_p)$, and the proof will show that it is not necessary to assume that the value be the same. On the other hand, it would be most unnatural to consider $A(abc)$ without this consistency assumption. Since the result implies moreover that the values of $A(abc)$ in the overlap of two triangles can differ only by a constant factor, we will agree to consider here, as well as in Theorem (50.6), only $A(abc)$ which are consistent, because then we obtain a reasonable area on the whole surface.

It suffices to show that $S(p, \varepsilon)$ with $\varepsilon = \min\left(\delta_p, \rho(p)/4\right)$ is elementary. Consider two distinct points a, a' in $S(p, \varepsilon)$ and two points c_1, c_2 $B(a, a')$ in $S(p, \varepsilon)$ which lie on different sides of the symmetric extension of $T(a, a')$ of length $\rho(p)$. The triangle inequality yields immediately that $T(c_1, c_2)$ intersects this extension in a point q with (aqa'). Let $m = m_{aa'}$. We are going to show that $q = m$. Assume $m \in T(q, a')$. Then the properties 1), 2), 3) of area and the hypothesis imply

$$A(ac_1q) + A(ac_2q) = A(ac_1c_2) = A(a'c_1c_2) = A(a'c_1q) + A(a'c_2q)$$

$$A(ac_1q) \leqslant A(ac_1q) + A(c_1qm) = A(ac_1m) = A(a'c_1m) \leqslant A(a'c_1q)$$

$$A(ac_2q) \leqslant A(ac_2q) + A(c_2qm) = A(ac_2m) = A(a'c_2m) \leqslant A(a'c_2q).$$

The first of these three relations implies equality in the last two, hence $A(c_1qm) = 0$ and $q = m$ by property 2).

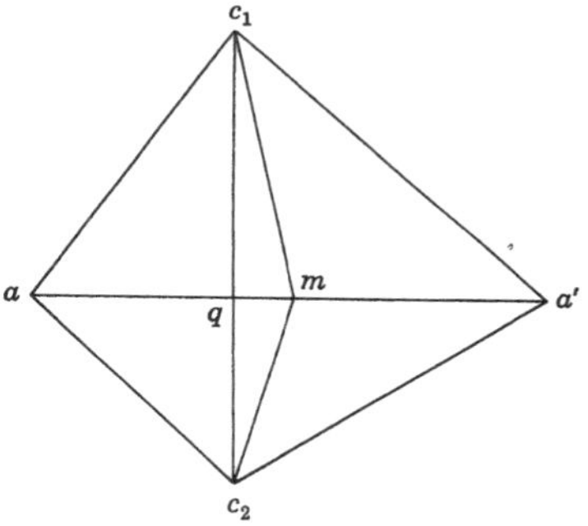

Fig. 60.

Let now c be any point in $S(p, \varepsilon)$ and on $B(a, a')$ different from q. Then either $T(c_1, c)$ or $T(c_2, c)$ intersects $T(a, a')$ and is therefore collinear with c_1 and q or with c_2 and q so that the points of $B(a, a')$ in $S(p, \varepsilon)$ lie on one segment. The assertion follows now fıom (46.1).

A triangle area in the large, i. e., defined for all point triples in a two-dimensional G-space is not a natural concept except when the space is straight or spherelike. A straight space satisfying the hypothesis of (48.16) is clearly euclidean or hyperbolic, and a spherelike space is, under the assumption of (48.16), spherical so that it is not necessary to require the existence of an area defined for all triples, although it can be verified in the euclidean and hyperbolic case, and with a little caution also in the spherical case, that such an area does exist.

In the elliptic case the lengths of the sides of a triangle do not necessarily determine the triangle uniquely, and no formula for the area $A(abc)$ in terms of the sides valid for all triangles exists.

49. Involutoric motions

Motions with period two play an important role in many parts of geometry. What is known about such motions in G-spaces forms the subject of this section. We begin with some observations on straight spaces and spaces of the elliptic type which are far from deep but amusing, because they show that phenomena with which we are well familiar in euclidean and non-euclidean geometry extend to much more general situations.

Reflections of Straight Spaces and Spaces of the Elliptic Type

We need some general facts on fixed points of motions.

(49.1) *The set F of the fixed points (if any) of a motion Φ of a straight space R is flat.*

For if Φ leaves the points a and b fixed, it leaves every point on the straight line through a and b fixed, which therefore lies in F. Since F is closed, see (4.2), it is with the metric of R itself a straight space, hence flat.

(49.2) *The set F of fixed points (if any) of a motion Φ of a space R of the elliptic type is the union of flat sets $F_1, F_2, \ldots, F_k$ such that a point in F_i is conjugate to every point of $F - F_i$. A point which is not fixed lies on at most one line containing two fixed points.*

If F contains two non-conjugate points a, b then it contains the whole great circle through a and b. This shows that F has the form $\cup F_i$, where each F_i contains with any two points the whole great circle, and any point of F_i is conjugate to any point of $F - F_i$. The number of F_i is finite because R is compact. Since F is closed each F_i is closed, so that each F_i is flat, and in fact again a space of the elliptic type.

A great circle which contains two fixed points goes under Φ into itself, a point lying on two great circles containing two fixed points goes, as unique intersection of the two great circles, into itself, hence is fixed.

A motion Φ is *involutoric* if $\Phi^2 = E$ but $\Phi \neq E$, where E is the identity.

(49.3) *An involutoric motion Φ of a straight space R is uniquely determined by the non-empty set F of its fixed points: If p is not in F, then the foot f of p on F is unique and $f = m[p, p\Phi]$.*

The motion Φ is called the "*reflection*" *of R in F.*

For because $\Phi \neq E$, there is a point $p \neq p\Phi$. The midpoint f of p and $p\Phi$ goes into itself because $p\Phi^2 = p$, so that F is not empty. Let g be any foot of p on F. Then the inequalities,

$$p\Phi f = pf \geqslant pg = p\Phi g, \qquad pf + fp\Phi = pp\Phi \leqslant pg + gp\Phi$$

show that $g = f$.

An involutoric motion of a space of the elliptic type need not have fixed points; the mapping which maps every point of a great circle on its antipode shows this already. Less trivially,

$$x'_1 = x_3, \quad x'_2 = x_4, \quad x'_3 = -x_1, \quad x'_4 = -x_2$$

defines a distance preserving mapping of the sphere $x_1^2 + x_2^2 + x_3^2 + x_4^2 = 1$ on itself and induces in the three-dimensional elliptic space which it covers a motion without fixed points.

(49.4) *If an involutoric motion Φ of a space R of the elliptic type has fixed points, then the set F of all fixed points is the union of two flat sets F_1 and F_2. Each point $p \in R - F$ lies on exactly one great circle $\mathfrak{g}_p$ intersecting both F_1 and F_2. The point $f_i = \mathfrak{g}_p \cap F_i$ is the unique foot of p on F_i, and f_1, f_2 are the midpoints of the two arcs of $\mathfrak{g}_p$ from p to $p\Phi$.*

Thus each of the sets F_1, F_2 determines Φ uniquely, and Φ is called the *"reflection" of R in F_1 or in F_2.*

By (49.2) the set F is the union of flat sets $F_1, F_2, \ldots$ such that every point of F_i is conjugate to every point in $R - F_i$. By hypothesis there is at least one set F_1. Let $p \subset R - F$, and let f be a foot of p on F_1. Then f is the unique foot of any point y with (pyf), and if $yf < \rho(f)$ then $(yfy\Phi)$ by the same argument as above. Hence Φ maps the great circle $\mathfrak{g}_p$ through p and f on itself (so that $p\Phi \in \mathfrak{g}_p$) and leaves therefore the point f_2 conjugate to f_1 on $\mathfrak{g}_p$ fixed. f_2 cannot lie in F_1 because this would imply $p \in F_1$, so that $R - F_1$ is not empty. It is also clear that f_1 and f_2 are the midpoints of the two arcs of $\mathfrak{g}_p$ from p to p.

A second foot of p on F_1 would yield two different lines through p which contain (each) two distinct fixed points in contradiction to (49.2). Hence F_1 determines Φ uniquely.

F cannot have more than two components because we proved that $p \in R - F$ lies on a great circle connecting a point of a given component F_i of F to a point of $R - F_i$, so that three or more components would lead to at least two distinct lines through p each containing two fixed points.

Symmetric spaces

We now turn to involutoric motions of general G-spaces. The motion Φ of the G-space R is called the *"symmetry" or "reflection" in the point p* if Φ is involutoric and p is its only fixed point in a certain sphere $S(p, \rho)$, $\rho > 0$.

An equivalent definition is: Φ is a motion and there exists a positive ρ such that $p = m[x, x\Phi]$ for $x \in S(p, \rho) - p$. This and (28.8) show that the symmetry in p is unique, which we anticipated by using the definite article.

If the symmetry in p exists for every point p of the G-space R then R is called *symmetric*. We notice

(49.5) *A symmetric space of the elliptic type is elliptic.*

For by (49.4) the component of the set of fixed points other than p consists of all conjugate points to p and is flat, so that (49.5) follows for spaces of dimension greater than two from Theorem (21.2). In the two-dimensional case we observe that the reflection in p is also the reflection in the line formed by the conjugate points to p. Then the reflection in every line exists and the assertion follows from Section 47.

There is no similar theorem for straight spaces. All Minkowski spaces are symmetric, the product $[E^{\nu} \times H^{\mu}]_2$ of a ν-dimensional euclidean space and a μ-dimensional hyperbolic space defined in Section 8 is an example of a symmetric Riemann space which does not have constant curvature, the hermitian hyperbolic spaces discussed in Section 53 provide other examples. The symmetric G-surfaces are determined in Theorem (52.8). Observe the following fact:

(49.6) *A symmetric G-space possesses a transitive group of motions.*

For if x, y are two arbitrary distinct points of the given space R and m is any midpoint of x and y, then the segments $T(x, m)$ and $T(y, m)$ are unique and form a segment $T(x, y)$, hence the symmetry in m carries x into y.

The symmetric Riemann spaces have been investigated by Cartan in several papers which are known for their depth, elegance, and importance for Lie groups; for a brief resumé and the literature see Cartan [3, Chapter IV]. It ought to be pointed out that Cartan uses the term symmetric space in two non-equivalent meanings. One coincides with ours, the other leads to spaces which might be called locally symmetric.

The G-space R is *locally symmetric* if every point p possesses a neighborhood $S(p, \rho)$, $0 < \rho < \rho(p)$ such that $p = m[x, x\Phi]$ and $p\Phi = p$ define an isometric mapping Φ of $S(p, \rho)$ on itself.

Obviously a symmetric space is locally symmetric, but not conversely. Every space of constant curvature is locally symmetric, but only very few of them possess transitive groups of motions: the Moebius strip with a euclidean metric does not possess a transitive group of motions because of the existence of a distinguished great circle, see Section 30. None of the surfaces with constant negative curvature except the hyperbolic plane is symmetric (see Theorem (52.8)). Possibly the universal covering space of a locally symmetric space is always symmetric[1]. Doubtless many of Cartan's results both on symmetric and on locally symmetric spaces can be carried over to G-spaces. This optimism is based on the following observations:

Assume that the G-space R is symmetric and compact. It follows from (49.6) and (52.4) that *the group Γ of all motions of R is a Lie group and that R is a topological manifold.* There is a Riemannian metrization R^* of R invariant under Γ, see Cartan [**3**, p. **43**]. Since the symmetry Φ of R in a point p is involutoric and has p as only fixed point in a certain sphere $S(p, \rho)$, the same holds for Φ as motion of R^*, hence *it is also the symmetry of R^* in p. Thus, in the compact case all results of Cartan on the structure of spaces which are symmetric (in the large) carry over to G-spaces.* We observe that, locally, the midpoints $m(x, y)$ of x and y for R and R^* coincide, hence the geodesics of R and R^* coincide and along each geodesic the distances in R and R^* differ only by a factor which in general depends on the geodesic. This will be so, for instance, on a torus with a Minkowskian metric. However, if the factors are known for the geodesics passing through one point, then they are known for all geodesics because every point can be moved into any other. The degree of freedom in the choice of the factor depends on the *structure of the socalled isotropic subgroup* of Γ, i. e., the subgroup of motions in Γ which leaves a given point p fixed.

Reflections in lineal elements

An involutoric motion Φ of a G-space R which leaves every point of a segment T with midpoint p and a positive length $2\,\rho < 2\,\rho(p)$ fixed, and has no other fixed points in $S(p, \rho)$ than those on T, is called the "*reflection*" *in* T. There cannot be more than one reflection in T. For if x is a point of $S(p, \rho/4)$ not on T then a foot f of p on T lies in $S(p, \rho/2)$ and we see as in the proof of (49.3) that f is unique and $f = m[x, x\Phi]$. This determines Φ in $S(p, \rho/4)$ hence in all of R.

The length of T is immaterial: A motion which leaves a non-degenerate segment pointwise fixed leaves the whole geodesic $\mathfrak{g}$ containing the segment pointwise fixed. Φ cannot have in $S\left(p, \rho_1(p)\right)$ other fixed points than those on $\mathfrak{g}$. For if q were a fixed point not on $\mathfrak{g}$ in $S\left(p, \rho_1(p)\right)$ then $T(p, q)$ would be pointwise fixed but contain points in $S(p, q)$ not on T.

To be definite we may use the lineal elements introduced in Section **9**. Whereas there are many types of symmetric spaces, there are only few which possess reflections in all lineal elements:

(49.7) **Theorem.** *A G-space R which can be reflected in each lineal element is euclidean, hyperbolic, elliptic or spherical.*

Proof. Let T be a segment with center p and length $2\,\rho(p)$ (a straight line if $\rho(p) = \infty$), q a point of $S\left(p, \rho(p)\right) - p$ and f its foot on T, moreover $f = m_{qq'}$.

If $r \in T$, $pr < \rho(f)/4$ then the reflection in T takes q into q' and leaves r fixed, so that $rq = rq'$. Also, if (qxf) then f is the unique foot of x on T and $f = m[x, x\Phi]$, hence $rx = rx\Phi$. This shows that f must be the foot of r on $T(q, q')$. Hence perpendicularity of segments is symmetric. If $f = m[r, r_0]$ and $y \in B(r, r_0) \cap S(f, \rho(f)/4) - f$ then the segment $T(x, f)$ will be perpendicular to T.

Let x, x' be two arbitrary distinct points and m a midpoint of x and x'. If (xym), $ym < \rho(m)/4$, and $m = m_{yy'}$ (so that $(my'x')$), choose a point $z \in B(y, y') - m$ with $z \in S(p, \rho(m)/4)$. Such a point z exists unless R has dimension 1. Then the reflection in $T(z, z')$, where $m = m_{zz'}$, takes y into y', hence x into x'. Thus the space possesses a transitive group of motions and $\rho(p)$ is constant. Put $\delta = \rho(p)/16$.

(a) If $0 < xy = x'y' < \delta$ then a motion exists that takes x into x' and y into y'.

We know there is a motion Φ' that takes x into x'. Let $y\Phi' = y'' \neq y'$. If $(y'x'y'')$ then the reflection Φ'' in a segment perpendicular to $T(y, y'')$ at x' will take y'' into y'. If the relation $(y'x'y'')$ does not hold then the reflection Φ'' in $T(x', m_{y'y''})$ will take y' into y'', in either case $\Phi'\Phi''$ satisfies (a). The main step is the lemma:

(b) *Let $q_1 \neq q_2$ and $\delta > q_1p = q_2p = \alpha > 0$, where p is the foot of q_1 and q_2 on the segment T with center p and length $8\,\delta$. If $m = m[q_1, q_2] \neq p$, then p is the foot of m on T.*

We know that $T(p, m)$ is perpendicular to $T_q = T(q_1, q_2)$. Let $x(\tau)$ represent T_q with $x(0) = m$ and $x(\beta) = q_1$, where $\beta = m\,q_1$, and let $y(\sigma)$ represent T with $y(0) = p$. Then

$$y(2\,\alpha)\,x(\tau) \geqslant 2\,\alpha - px(\tau) \geqslant \alpha > pm \qquad \text{for} \qquad |\tau| \leqslant \beta$$

$$y(\sigma)\,x(\pm\,\beta) \geqslant y(0)\,x(\pm\,\beta) > pm \qquad \text{for} \qquad |\sigma| \leqslant 2\,\alpha.$$

Therefore $y(\sigma)x(\tau)$ is as a function of σ and τ on the boundary of the rectangle $|\sigma| \leqslant 2\,\alpha$, $|\tau| \leqslant \beta$ everywhere greater than at $(0, 0)$ where it has the value pm. Consequently $y(\sigma)x(\tau)$ reaches a minimum at some interior point (σ', τ') of the rectangle. Therefore the segment $T' = T\big(y(\sigma'), x(\tau')\big)$ must be perpendicular to both T_q and T. We are using here and in the following argument obvious extensions of (20.9 f) to perpendicularity of segments.

The lemma will be proved by showing $\sigma' = \tau' = 0$. Let $\tau' \geqslant 0$. We extend $x(\tau)$ and $y(\sigma)$ to representations of geodesics. Then the reflection Φ

in T' takes q_2 into $q_2' = x(2\tau' + \beta)$, q_1 into $q_1' = x(2\tau' - \beta)$ and $p = y(0)$ into $p' = y(2\sigma')$. Moreover

$$pq_1' = pq_1 = pq_2 = p'q_2' = \alpha.$$

The relations

$$\alpha = p'x(2\tau' + \beta) = p'x(2\tau' - \beta) \qquad \text{and} \qquad 2\tau' - \beta \leqslant \beta \leqslant 2\tau' + \beta$$

imply $p'x(\beta) \leqslant \alpha$ with inequality unless $\tau' = 0$. Since p is the foot of $x(\beta) = q_1$ on T and $pq_1 = \alpha$ it follows that $p'x(\beta) \geqslant \alpha$, hence $p'x(\beta) = \alpha$ and $\tau' = 0$. Also $p = p'$ or $\sigma' = 0$.

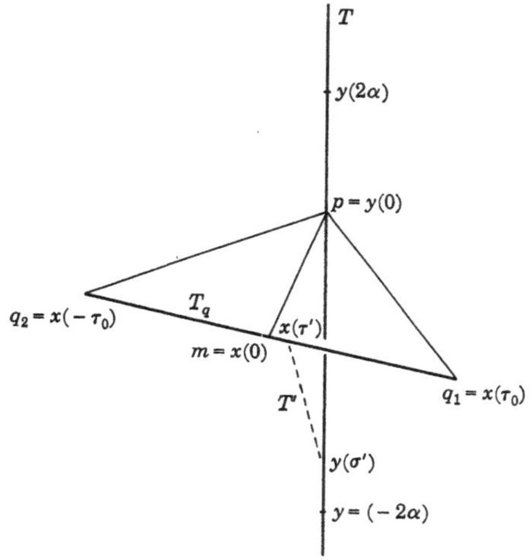

Fig. 61.

We use (b) to prove

(c) If ab, ac and $bc = bc'$ are positive and less than δ, and if $T(a, b)$ is perpendicular to $T(b, c)$ and to $T(b, c')$, then a motion exists that leaves a and b fixed and takes c into c'.

This is trivial for $c = c'$ and also for $b = m_{cc'}$, because in the latter case the reflection in $T(a, b)$ takes c into c'. If neither of these cases enters then $m = m_{cc'} \neq b$ and $T(b, m)$ is by (b) perpendicular to $T(a, a_0)$ where $b = m_{aa_0}$. The reflection Φ in $T(b, c)$ leaves b and c fixed and takes a into a_0, and the reflection Ψ in $T(b, m)$ takes a_0 into a, c into c' and leaves b fixed, hence $\Phi\Psi$ satisfies (c).

Theorem (49.7) will follow from (48.8) if we show

(d) *If* $ab = a'b' < \delta$, $bc = b'c' < \delta$, $ca = c'a' < \delta$ *then a motion* Ψ *exists that takes* a, b, c *into* a', b', c' *respectively.*

Let $bc \geqslant \max(ab, ac)$. The foot f of a on the segment T of length 8 δ with center b and containing $T(b, c)$ lies between b and c. For (bcf) or $c = f$ and symmetry of perpendicularity would imply $ba > bf \geqslant bc$. Since also $b'c' \geqslant \max(a'b', a'c')$ it follows that the foot f' of a' on T satisfies $(b'f'c')$.

Let $af \geqslant a'f'$. Since $bf + fc = b'f' + f'c'$ we have either $bf \geqslant b'f'$ or $cf \geqslant c'f'$. Assume $bf \geqslant b'f'$. Choose a_0 on $T(a, f)$ such that $a_0f = a'f'$ and b_0 on $T(f, b)$ such that $b_0f = b'f'$. By (a) and (c) there is a motion Ψ that takes a_0, f, b_0 into a', f', b' respectively. If $a\Psi = a''$, $b\Psi = b''$, $c\Psi = c''$, then $a''b'' \geqslant a'b'' \geqslant a'b'$ because $T(a', f')$ is perpendicular to $T(b'f')$. But $a''b'' = ab = a'b'$, hence $a'' = a'$, $b'' = b'$ and the relation $b''c'' = b'c'$ implies $c'' = c'$. Thus Ψ takes a, b, c into $a'b'c'$.

In analogy to locally symmetric spaces we may consider spaces for which reflections in segments exist locally. Without proof we state the following theorem which is quite easily established by using (48.3) and the idea of the preceding proof:

(49.8) *Let for every lineal element* T *with center* p *and length* $\sigma(p)$ *an involutoric isometry of* $S(p, \sigma(p)/2)$ *on itself exist which leaves the points on* T *and no others fixed. Then the space is locally euclidean, hyperbolic, or spherical.*

50. New characterizations of Minkowski spaces

The problem of determining all G-spaces with transitive groups of motions is, except for two dimensions, much too general to be solved with the present means of mathematics. Only the abelian case is simple.

Transitive abelian groups of motions

The translations of a Minkowskian Geometry form a transitive abelian group, and a result of Pontrjagin [1, p. 170] on more general spaces implies that every G-space with a transitive abelian group of motions must be locally Minkowskian. Since this result is of major importance for the theory of G-spaces it is desirable to establish it with the present methods. This can very easily be achieved with the help of our results on spaces with non-positive curvature.

(50.1) **Theorem.** *A G-space which possesses a transitive abelian group of motions is locally Minkowskian, and is, topologically, the product of a finite number of straight lines and circles.*

We first prove two trivial, but frequently useful, lemmas on motions in general metric spaces:

(50.2) *If Γ is a transitive abelian group of motions of the metric space R, and $\Phi \in \Gamma$, then $xx\Phi$ is independent of x. Hence no motion in Γ, except the identity, has fixed points.*

For if x and y are given, then a Ψ in Γ exists such that $y = x\Psi$ and

$$yy\Phi = x\Psi\, x\Psi\Phi = x\Psi\, x\Phi\Psi = xx\Phi.$$

(50.3) *A transitive group Γ of motions of a metric space R is simply transitive if and only if no motion in Γ, except the identity, has fixed points.*

For if a motion Φ in Γ different from the identity has a fixed point f and $p \neq f$, then a motion Ψ in Γ exists with $p\Psi = f$. Then $\Psi\Phi$ is different from Ψ and both motions take p into f so that Γ is not simply transitive.

Conversely let Γ be transitive, but not simply transitive. Then two different motions Φ, Ψ in Γ and two distinct points a, b exist such that $a\Phi = b$ and $a\Psi = b$. Then $\Phi\Psi^{-1}$ leaves a fixed and is not the identity.

Applying the lemmas to the hypothesis of (50.1) we see that the given transitive abelian group Γ is simply transitive, so that the motion in Γ which carries a into b may be denoted by $(a \to b)$. We observe first:

(50.4) If $a \neq c$ and the midpoint b of a and c is unique then $(a \to b)^2 = (a \to c)$. If $a, b, c \in S(p, \rho(p)/4)$ for a suitable p and $b = m_{ac}$, then $(a \to d)^2 = (a \to c)$ implies $d = b$.

For the proof of the first part put $\Phi = (a \to b)$ and $\Psi = (b \to c)$. Because of (50.2) we have

$$aa\Psi = bb\Psi = bc = ac/2,$$

$$a\Psi c = a\Psi a\Phi\Psi = aa\Phi = ab = ac/2$$

hence $a\Psi = m_{ac} = b$ or $\Psi = (a \to b)$ and $(a \to c) = \Phi\Psi = \Phi^2$.

Under the assumptions of the second part of (50.4) there is a point c' with $d = m_{ac'}$. Moreover d is the unique midpoint of ac'. The first part of (50.4) shows $(a \to c') = (a \to d)^2 = (a \to c)$, hence $c' = c$ and $d = b$.

If we can show that for any three points a, b, c in $S(p, \rho(p)/16)$ the relation

$$2\, m_{ab} m_{ac} = bc$$

holds, then it will follow from (39.12) that the space is locally Minkowskian. Put $b' = m_{ab}$, $c' = m_{ac}$, $(a \to b') = \Phi$, $(a \to c') = \Psi$, $(b' \to c') = \Omega$. Then $\Psi = \Phi\Omega$ and $\Phi\Omega\Phi = \Psi\Phi = \Psi\Omega^{-1}\Psi$ hence

$$d = a\Phi\Omega\Phi = a\Psi\Omega^{-1}\Psi.$$

Now by (50.4)

$$(b \to d) = (b \to a)\,(a \to d) = \Phi^{-2}\Phi\Omega\Phi = \Omega \quad \text{and}$$

$$(d \to c) = (d \to a)\,(a \to c) = \Psi^{-1}\Omega\Psi^{1}\Psi^{2} = \Omega.$$

Thus $(b \to c) = \Omega^2 = (b \to d)^2$ and (50.2) yields

$$bd = dc = b'c' \leqslant ab' + ac' \leqslant (ab + ac)/2 < \rho(p)/8, \qquad ad \leqslant ab + bd < \rho(p)/4$$

It follows from the second part of (50.4) that $d = m_{bc}$ hence $2\,b'c' = bc$.

The Moebius strip with a euclidean metric, compare Section 29, is an example which shows that not all locally Minkowskian spaces possess transitive abelian groups of motions. However, if $\bar{\Gamma}$ is the group of motions in the (Minkowskian) universal covering space $\bar{R}$ of R which lie over motions of Γ, then theorems (28.10) and (28.11) show that $\bar{\Gamma}$ is transitive and abelian. Hence $xx\bar{\Phi}$ is constant for any motion $\bar{\Phi}$ in $\bar{\Gamma}$, so that $\bar{\Phi}$ is a translation of $\bar{R}$. In particular, the motions of the fundamental group are translations, which implies that R is the product of a finite number of circles and straight lines.

This could also have been deduced from the fact that R is a group space (because Γ is simply transitive) and the general property of group spaces of having abelian fundamental groups.

Transitive groups of motions with maximal displacements

We saw in the last proof that $xx\Phi$ is constant for each Φ in Γ, and know from Theorem (32.5) that this condition is satisfied for a motion Φ of a straight space if there is a point z for which the displacement under Φ is maximal, i. e., $zz\Phi = \sup xx\Phi$. Thus we are led to conjecture:

(50.5) **Theorem.** *If a straight space R possesses a transitive group Γ of motions such that for each motion in Γ points with maximal displacement exist, then the space is Minkowskian.*

Since $xx\Phi$ is constant for each Φ in Γ no motion in Γ except the identity has fixed points. By (50.3) the group Γ is simply transitive on R, and the motion in Γ that carries a into b may again be denoted by $(a \to b)$.

If $a \neq b$ then $\Phi = (a \to b) \neq E$ and $xx\Phi = ab$ implies, trivially, $ab = \min_{x \in R} xx\Phi$. Hence, see (32.3), the line $\mathfrak{g}^+(a, b)$ is an oriented axis of $(a \to b)$ and for any point x, because of $xx\Phi = ab$, the line $\mathfrak{g}^+(x, x\Phi)$ is also an oriented axis of Φ. All these axes are by (32.4) parallel to each other. Since there is at most one parallel to a given line through a given point, each parallel to an axis of Φ is an axis of Φ, and the parallel axiom holds.

We show first that the space has the divergence property: If $y(\tau)$, $\tau \geqslant 0$, represents a ray $\mathfrak{r}$ and G is a line through $y(0)$ not containing the ray, then $y(\tau)\, G \to \infty$ for $\tau \to \infty$. For an indirect proof assume the existence of a sequence $\tau_\mu \to \infty$ with $y(\tau_\mu)\, G < M$. If f_μ is a foot of $y(\tau_\mu)$ on G, then $f_\mu \neq y(0)$ for large μ and $q_\mu = y(\tau_\mu)\,(f_\mu \to z)$ has z as foot on G. Choose a subsequence $\{\nu\}$ of $\{\mu\}$ for which $\{q_\nu\}$ converges to a point q [1]. The line $\mathfrak{g}\big(x(\tau_\nu), q_\nu\big)$ is an axis of $(f_\nu \to z)$ hence parallel to G, and tends therefore to the parallel G' to G through q. On the other hand, the definition of co-ray implies that $T\big(q_\nu, y(\tau_\nu)\big)$ tends to the co-ray from q to $\mathfrak{r}$, hence G' is also parallel to the line H carrying $\mathfrak{r}$. Since the concept of parallelism is symmetric G and H would be parallels to G' which is impossible.

The distance of two axes of the same motion, hence of parallels is bounded. Exactly as in Section 37 we deduce from the divergence property that two lines G_1 and G_2 are parallel if, and only if, x_1G_2 and x_2G_1 are bounded for $x_i \in G_i$.

We now show that Γ is abelian and thus reduce the present theorem to the preceding one. If $\Phi \neq E$ and $\Psi \neq E$ in Γ are given, we select an arbitrary point z and put $z\Phi = p$, $z\Psi = q$, so that $\Phi = (z \to p)$, $\Psi = (z \to q)$. The relation $\Phi\Psi = \Psi\Phi$ is obvious if z, p, q lie on one line. Assume therefore that $\mathfrak{g}(z, p) = G \neq H = \mathfrak{g}(z, q)$ and put $H' = H\Phi$. For $y \in H$ the relation $yy\Phi = zp$ shows that the distance of H and H' is bounded, hence H' is parallel to H and an axis of Ψ, so that $p\Psi$ is a point p' of H' with $zq = pp'$. On the other hand, $zq = z\Phi q\Phi$, hence $q' = q\Phi = p'$, so that $\Phi = (q \to q')$, $\Psi = (p \to q')$. It follows that

$$\Phi\Psi = (z \to p)\,(p \to q') = (z \to q') = (z \to q)\,(q \to q') = \Psi\Phi$$

which proves our theorem.

Characterization of Minkowskian Geometry by properties of area

When we discussed the geometries in which the area of a triangle can be expressed in terms of its sides, the reader probably expected to see the euclidean geometry characterized by the property that the area of a triangle is determined by base and altitude. It is a curious fact that this property does not single out euclidean geometry. We will show here that it characterizes the Minkowskian geometries with symmetric perpendicularity. There a triangle area $A(abc)$ exists which has the form $A(abc) = k \cdot bc \cdot a\mathfrak{g}(b, c)$ where k is a constant depending on the unit of area.

In Section 41 we introduced a volume in Finsler spaces, which yields for a Minkowski plane: The Minkowski area is proportional to the area in an arbitrary associated euclidean geometry. The factor of proportionality is so determined that the Minkowski area of the Minkowskian unit circle becomes π. Since the Minkowskian and euclidean distances are proportional on a fixed straight line, the Minkowskian area $A(abc)$ of a triangle shares with the euclidean area the property that $A(abc)$ depends only on a, bc and the line $\mathfrak{g}(b, c)$. We are going to prove that the existence of an area with this property characterizes Minkowskian geometry. It will then follow easily that perpendicularity is symmetric when $A(abc)$ depends only on $a\mathfrak{g}(b, c)$ and bc.

We remember that every point of a G-surface possesses arbitrarily small convex neighborhoods. A line $L(a, b)$ in such a neighborhood was the open segment containing a and b whose endpoints lie on the boundary of the neighborhood.

(50.6) ***Theorem.*** *If every point of a G-surface R possesses a convex neighborhood U_p in which a triangle area $A(abc)$ is defined and depends only on a, $L(b, c)$ and bc then R is locally Minkowskian.*

We assume that $A(abc)$ is defined consistently, see the remarks after Theorem (48.16), although we actually do not need this hypothesis.

If a, b, c are three non-collinear points of U_p, then the additivity property 3) of triangle area, and the fact that area is non-negative lead through standard arguments regarding the functional equation $f(u) + f(v) = f(u + v)$ (see Picard [1, pp. 1, 2]) to the relation

(a) $$A(axb) : A(axc) = xb : xc \qquad \text{for} \qquad x \in L(b, c).$$

Consider three non-collinear points a, b_1, b_2 in U_p. Let $m = m[b_1, b_2]$, (mqa) and put $c_1 = L(b_2, q) \cap T(b_1, a)$, $c_2 = L(b_1, q) \cap T(b_2, a)$, $n = T(a, m) \cap T(c_1, c_2)$. Then (anq). The relation (a) yields $A(ab_1m) = A(ab_2m)$ and $A(qb_1m) = A(qb_2m)$, hence

(b) $$A(qb_1a) = A(qb_2a)$$

so that by (a)

$$\frac{A(qb_1c_1)}{A(qc_1a)} = \frac{b_1c_1}{c_1a} = \frac{A(b_2b_1c_1)}{A(b_2c_1a)} = \frac{A(qb_1b_2) + A(qb_1c_1)}{A(qb_2a) + A(qc_1a)} = \frac{A(qb_1b_2)}{A(qb_2a)}$$

and similarly

$$\frac{A(qb_2c_2)}{A(qc_2a)} = \frac{b_2c_2}{c_2a} = \frac{A(qb_1b_2)}{A(qb_1a)}$$

Therefore $b_1c_1 : c_1a = b_2c_2 : c_2a$ and by (a) and (b)

(c) $$A(qb_1c_1) = A(qb_2c_2) \quad \text{and} \quad A(qc_1a) = A(qc_2a).$$

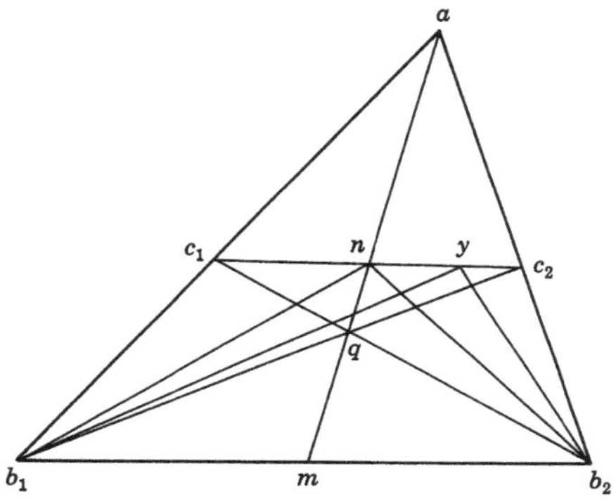

Fig. 62.

Now (a) yields

$$\frac{A(ac_1n)}{A(qc_1n)} = \frac{a\,n}{q\,n} = \frac{A(ac_2n)}{A(qc_2n)},$$

hence (c) implies $A(ac_1n) = A(ac_2n)$; therefore by (a)

(d) $$c_1n = c_2n.$$

We conclude from (a) and (d) that

$$\frac{A(ac_1b_2)}{A(an\,b_2)} = \frac{2\,A(a\,n\,c_2) + 2\,A(b_2n\,c_2)}{A(a\,n\,c_2) + A(b_2n\,c_2)} = 2$$

hence

(c) $$\frac{a\,n}{a\,m} = \frac{2\,A(a\,n\,b_2)}{2\,A(a\,m\,b_2)} = \frac{A(a\,c_1b_2)}{A(a\,b_1b_2)} = \frac{a\,c_1}{a\,b_1}.$$

Successive dyadic subdivisions of $T(b_1, b_2)$ and application of (d) and (e) yield

(f) *If* (b_1xb_2) *and* $T(a, x) \cap T(c_1, c_2) = y$ *then*

$$ay : ax = ac_1 : ab_1 \quad \text{and} \quad c_1y : c_2y = b_1x : b_2x.$$

This leads to

(g) If z_1 and z_2 are points of $L(b_1, b_2)$ then $A(z_1z_2y)$ is constant when y varies on $L(c_1, c_2)$.

For

$$\frac{A(ab_1b_2)}{A(c_1b_1b_2)} = \frac{ab_1}{c_1b_1} = \frac{ax}{xy} = \frac{A(ab_ix)}{A(yb_ix)} = \frac{A(ab_1b_2)}{A(yb_1b_2)}.$$

The general statement (g) follows from (a).

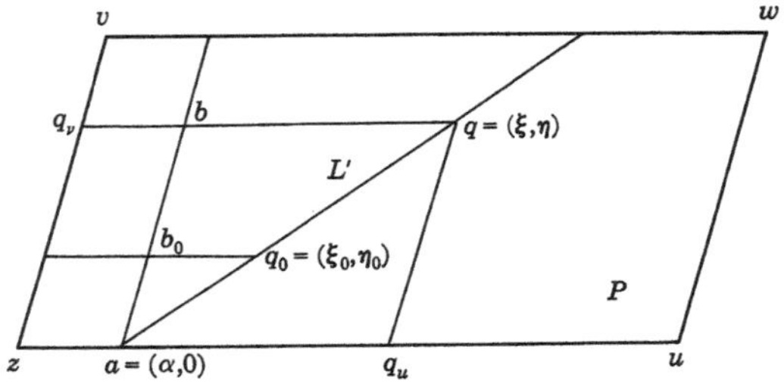

Fig. 63.

We call the line H parallel to the line L if for one pair of distinct points b_1, b_2 on L the area $A(yb_1b_2)$ is constant for $y \in L$. We just proved that $A(yz_1z_2)$ is constant for any pair of points z_1, z_2 of L. The relations (c) and (d) show that $A(c_1c_2b_1) = A(c_1c_2b_2)$ so that L is parallel to H if H is parallel to L. We also find

(h) *If H is parallel to L and K lies between H and L and is parallel to L, then it also is parallel to H. If for $i = 1, 2$, $a_i \in H$, $b_i \in K$, $c_i \in L$ and $(a_ib_ic_i)$ then $a_1c_1 : b_1c_1 = a_2c_2 : b_2c_2$.*

The proportion follows from (a) and this in turn implies that K is parallel to H.

If $T(a, c)$ and $T(b, d)$ are two non-collinear segments with the same center, then (h) shows that the opposite sides $L(a, b)$ and $L(c, d)$ as well as $L(a, d)$ and $L(b, c)$ are parallel. Thus every point in U_p, in particular p itself, is interior point of a parallelogram. In the interior P of any parallelogram $zuwv$, where w is opposite to z, coordinates ξ, η may be introduced as follows: Let the parallels to $L(z, v)$ and to $L(z, u)$ through a given point q in P intersect $T(z, u)$ at q_u and $T(z, v)$ at q_v. Then $\xi = zq_u$, $\eta = zq_v$ are the coordinates of q.

A line L which intersects P intersects the boundary of P. Let for instance $L \cap T(z, u) = a = (\alpha, 0)$. Choose any two distinct points $q = (\xi, \eta)$ and

$q_0 = (\xi_0, \eta_0)$ on $L' = L \cap P$ which are different from a. If $\xi_0 = \alpha$ then $\xi = \alpha$ and L' satisfies the equation $\xi = \alpha$. If $\xi_0 \neq \alpha$ let the parallel to $L(z, v)$ through a intersect the parallels to $L(z, u)$ through q and q_0 in b and b_0 respectively. Then (f) and (h) imply

$$\text{(i)} \qquad \frac{\xi - \alpha}{\xi_0 - \alpha} = \frac{a\,q}{a\,q_0} = \frac{a\,b}{a\,b_0} = \frac{\eta}{\eta_0}$$

so that *L' satisfies a linear equation.*

Besides the given distance $q_1 q_2$ we introduce in P the euclidean distance

$$e'(q_1, q_2) = [(\xi_1 - \xi_2)^2 + (\eta_1 - \eta_2)^2]^{\frac{1}{2}}.$$

Then (i) shows that $e'(q_1, q_2)$ and $q_1 q_2$ are proportional on a fixed line L' in P. The same holds then also for the more general euclidean distance

$$e(q_1, q_2) = [E(\xi_1 - \xi_2)^2 + 2F(\xi_1 - \xi_2)(\eta_1 - \eta_2) + G(\eta_1 - \eta_2)^2]^{\frac{1}{2}}.$$

We consider a system of neighborhoods in R which is formed by such parallelograms. In the overlap of two parallelograms we have two different euclidean metrics. If two parallel lines in one of the two parallelograms have common points with the second, then they lie also on parallel lines in the second parallelogram. To see this we do not need the consistency of the area $A(abc)$, we merely observe that, with the above notations, $L(b_1, b_2)$ is parallel to $L(c_1, c_2)$ if $(ac_i b_i)$ and $ac_1 : ab_1 = ac_2 : ab_2$.

Therefore the euclidean metric in the second parallelogram can be changed affinely, i. e., in the way in which $e'(q_1, q_2)$ was replaced by $e(q_1, q_2)$ such that the metrics coincide in the overlap. We thus remetrize R as a locally euclidean space with the same geodesics as R in such a way that the euclidean distance and the given distance differ on each geodesic locally only by a constant factor.

The universal covering space R carries then a euclidean metric in addition to the metric induced by R. The geodesics are in both metrics the euclidean straight lines, and the distances along the same straight lines are proportional. Therefore R is by (17.1) a Minkowski space which proves (50.6).

The assumption that $A(abc)$ be consistent induces a triangle area in $\overline{R}$ defined for all triples abc, which also depends only on a, $\mathfrak{g}(b, c)$ and bc. We omit the easy proof of this fact and rather turn to the uniqueness of area.

(50.7) *In a Minkowski plane there is, up to a constant factor, only one triangle area which depends only on a, $\mathfrak{g}(b, c)$ and bc.*

For let H, L be two straight lines intersecting at c. If $a \in H$ and $b \in L$, then (a) shows that $A(abc)$ is proportional to ac and bc so that

$$A(abc) = \varphi(H, L) \cdot ac \cdot bc, \tag{50.8}$$

where $\varphi(H, L) = \varphi(L, H) > 0$. We have to show that the area of one non-degenerate triangle abc determines the area of all other triangles. The value of $A(abc)$ determines with the above notation the value of $\varphi(H, L)$. If now K is any line not parallel to H or L, and not through c, then K intersects H and L in points x and y, and

$$A(cxy) = \varphi(H, L) \cdot xc \cdot yc = \varphi(K, L) \cdot xy \cdot cy;$$

hence $\varphi(H, L)$ determines $\varphi(K, L)$. We can now replace L by any line K' not through $K \cap L$ and not parallel to K or L and find in the same way that $\varphi(K, L)$ determines $\varphi(K, K')$. It is now clear that the value of $\varphi(H, L)$ determines $\varphi(G, G')$ for any pair G, G' of non-parallel lines.

Since we know that an area proportional to an associated euclidean metric satisfies the hypothesis of (50.7), we have proved that these are the only triangle areas $A(abc)$ depending only on a, $\mathfrak{g}(b, c)$, bc.

Area in terms of base and altitude

We now return to the question which led to the present discussion and show:

(50.9) *A straight plane possesses a triangle area $A(abc)$ which depends only on the distances bc and $a\mathfrak{g}(b, c)$, if and only if it is a Minkowski plane with symmetric perpendicularity.*

We know that the plane is Minkowskian and that

$$A(abc) = \varphi(H, L) \cdot ac \cdot bc \qquad \text{where} \qquad H = \mathfrak{g}(a, c), \qquad L = \mathfrak{g}(b, c).$$

Under the present assumption $\varphi(H, L)$ must have the same value $\alpha > 0$, whenever one of the lines H or L is perpendicular to the other.

If a pair of lines H, L existed such that H is perpendicular to L at c but L not to H, then a transversal to L at a point $b \neq c$ would intersect H in a point $a \neq c$ and

$$A(abc) = \alpha\, ac \cdot bc = \alpha\, ab \cdot bc$$

but $ab > bc$ because c is a foot of a on L.

It follows that the area of any triangle *abc* equals

$$(50.10) \qquad A(abc) = \alpha \cdot bc \cdot a\mathfrak{g}(b, c).$$

Conversely, it must be shown that in a Minkowski plane with symmetric perpendicularity such a formula always holds. Because of (50.8) it suffices to see this for "right" triangles where the legs of the right angle have length 1. This means, that in an associated euclidean geometry all these triangles have the same area, and follows directly from the construction of *all* Minkowskian geometries with symmetric perpendicularity given at the end of Section 17. A detailed discussion is found in the author's paper [6].

If anything in the relation (50.10) distinguishes the euclidean geometry from the Minkowskian geometries with symmetric perpendicularity, it can only be *the value of* α. Indeed, Petty [1] proves: *if area is normalized such that the unit circle has measure* π, *then* $\alpha = {}^1/_2$ *for the euclidean geometry and* $\alpha > {}^1/_2$ *for the other Minkowskian geometries with symmetric perpendicularity.*

51. Translations along two lines

A translation of a straight plane R along a straight line $\mathfrak{g}$, or with axis $\mathfrak{g}$, was defined in Section 32. Because of (10.11) there is for two points a, b of $\mathfrak{g}$ at most one translation along $\mathfrak{g}$ that takes a into b. If this translation exists for arbitrary a, b on $\mathfrak{g}$ we say that "*all translations along* $\mathfrak{g}$ *exist*". The results of Section 32 imply:

(51.1) *Let all translations* Φ *along the geodesic* $\mathfrak{g}$ *of the straight plane* R *exist. The parallels to* $\mathfrak{g}$ *coincide with the axes of* Φ *and are equidistant from each other. The other equidistant curves to* $\mathfrak{g}$ *(or its parallels) are convex curves which are concave toward* $\mathfrak{g}$.

Every point of R *has exactly one foot on* $\mathfrak{g}$ *(or any parallel to* $\mathfrak{g}$).

The last statement is not contained in Section 32 but is obvious: if p has two distinct feet f_1, f_2 on $\mathfrak{g}$ and $(f_1 f f_2)$ and the translation that takes f_1 into f takes p into p', then $T(p, f_2)$ would intersect $T(p', f)$ in a point q which has both f and f_2 as feet on $\mathfrak{g}$, which contradicts (20.6).

It may actually happen that some equidistant curves to $\mathfrak{g}$ are geodesics and others not, even in a Riemann space with non-positive curvature, see example following Theorem (37.12).

Also, the parallel axiom may hold and all translations along one line (and hence its parallels) may exist in a non-Desarguesian straight plane. This is shown by the example constructed at the end of Section 33.

Finally, all translations along $\mathfrak{g}$ may exist and all curves equidistant to $\mathfrak{g}$ may be straight lines in a Desarguesian plane which does not satisfy the Parallel Axiom. The strip $|y| < \pi/2$ of an (x, y)-plane metrized by

$$[(x_1 - x_2)^2 + (y_1 - y_2)^2]^{\frac{1}{2}} + |\arctan y_1 - \arctan y_2|$$

has these properties.

Thus the hypothesis that all translations along one line exist does not alone imply any deeper properties. This suggests investigating the implications of the requirement that all translations along two lines, which are not parallel, exist. It is to be expected that the case where the second line is an asymptote to the first is essentially different from the case where it is not.

Translations along a line and a non-parallel asymptote

We settle the first case by proving:

(51.2) ***Theorem.*** *In a straight plane R let $\mathfrak{a}^+$ be an oriented asymptote to $\mathfrak{g}^+$, but not $\mathfrak{a}^-$ to $\mathfrak{g}^-$. If all translations along $\mathfrak{g}$ and along $\mathfrak{a}$ exist, then the group $\mathfrak{G}$ generated by these translations is simply transitive on R.*

Let $x(\tau)$ represent $\mathfrak{g}^+$ and let $z(\sigma)$, $-\infty < \sigma < \infty$ represent the limit circle $K = K_\infty(\mathfrak{g}^+, x(0))$ in terms of the arclength with $z(0) = x(0)$. The asymptote to $\mathfrak{g}^+$ through a given point p of R intersect K in a point $z(\sigma)$ and p lies on one limit circle $K_\infty(\mathfrak{g}^+, x(\tau))$. In terms of these coordinates σ, τ the motions $(\sigma, \tau) \to (\sigma', \tau')$ in $\mathfrak{G}$ have the form

$$\sigma' = \sigma\,\delta^\beta + \alpha, \qquad \tau' = \tau + \beta$$

where δ is a fixed number between 0 and 1 and each pair of real numbers α, β determines a motion in $\mathfrak{G}$.

Proof. There can be no parallel $\mathfrak{g}_1$ to $\mathfrak{g}$ (different from $\mathfrak{g}$) on the same side Σ of $\mathfrak{g}$ as $\mathfrak{a}$. For $\mathfrak{g}^+$ and a suitable orientation $\mathfrak{g}_1^+$ of $\mathfrak{g}_1$ would both be asymptotes to $\mathfrak{a}^+$ which are not parallel to $\mathfrak{a}$ and would therefore have distance 0 from each other, see the proof of (32.14).

$\mathfrak{a}^+$ and any second asymptote $\mathfrak{b}^+$ to $\mathfrak{g}^+$ in Σ have distance 0 from $\mathfrak{g}^+$ Therefore there are points $b \in \mathfrak{b}^+$ and $a \in \mathfrak{a}^+$ which have the same distance from $\mathfrak{g}$. A suitable translation Ψ along $\mathfrak{g}$ carries a into b hence $\mathfrak{a}^+$ into $\mathfrak{b}^+$, so that all translations along $\mathfrak{b}$ exist and have the form $\Psi^{-1}\Phi\Psi$ where Φ is a translation along $\mathfrak{a}$, see (39.9). Applying the same argument to the side of $\mathfrak{a}$ which contains $\mathfrak{g}$ we find that all translations along all asymptotes to $\mathfrak{g}^+$ exist and are products of translations along $\mathfrak{a}$ and along $\mathfrak{g}$.

Since any asymptote can be carried into any other the group $\mathfrak{G}$ generated by the translations along $\mathfrak{a}$ and along $\mathfrak{g}$ is transitive in R. To see that $\mathfrak{G}$ is simply transitive it suffices, because of (50.3), to prove that a motion Φ in $\mathfrak{G}$ which has a fixed point p is the identity. Since every motion in $\mathfrak{G}$ maps the family F of oriented asymptotes to $\mathfrak{g}^+$ on itself, Φ leaves the line in F through p pointwise fixed. Because Φ preserves the orientation it is by (10.11) the identity.

General theorems on transformation groups (see Kerékjártó [2]) show now that $\mathfrak{G}$ must be representable as indicated under (51.2). But the general theorem does not contain the geometric interpretation of the coordinates σ, τ, and the proof of the general theorem is involved.

Since the lines in F have distance 0 from each other we conclude from (22.8, 9) that for any two asymptotes $\mathfrak{a}^+$, $\mathfrak{b}^+$ to $\mathfrak{g}^+$ and any point x

$$K_\infty(\mathfrak{a}^+, x) = K_\infty(\mathfrak{b}^+, x).$$

We therefore denote this limit sphere simply by $K_\infty(x)$. If the motion in $\mathfrak{G}$ which carries x into y is again denoted by $(x \to y)$ then $K_\infty(x)\,(x \to y) = K_\infty(y)$. If $y \in K_\infty(x)$ then $(x \to y)$ carries $K_\infty(x)$ into itself. Therefore $\mathfrak{G}$ contains as subgroup what may be called the translations along the $K_\infty(x)$. Each translation along one $K_\infty(x)$ is a translation along all the others. The proof of Theorem (32.12) shows that the $K_\infty(x)$ are convex curves, so that each subarc $K_\infty(u, v)$ between two points u, v of the same $K_\infty(x)$ has a finite length $\lambda(u, v)$, compare (25.3). This justifies the representation $z(\sigma)$ of $K_\infty(x(0))$ used in (51.2).

The translations along the $K_\infty(x)$ have the form

$$\sigma' = \sigma + \alpha, \qquad \tau' = \tau. \tag{51.3}$$

Every motion in $\mathfrak{G}$ is product of a translation along $\mathfrak{g}$ and a translation along K. It suffices therefore to find the expression for translations along $\mathfrak{g}$.

Let $\mathfrak{b}^+$ be any asymptote to $\mathfrak{g}^+$. For any $\tau_2 < \tau_1$ put $x(\tau_i) = x_i$ and put $K_\infty(x_i) \cap \mathfrak{b} = y_i$. The equidistant curves to $\mathfrak{g}$ through y_i and to $\mathfrak{b}$ through x_i intersect at a point r_i because the existence of perpendiculars to $\mathfrak{g}$ shows $x(\tau)\,\mathfrak{b} \to \infty$ for $\tau \to -\infty$. Then $(y_2 \to r_2)$ carries $\mathfrak{g}$ into itself and $\mathfrak{b}^+$ into the line $\mathfrak{c}^+$ in F through r_2. Hence $y_1(y_2 \to r_2)$ is a point r of $\mathfrak{c}$. Now $(y_2 \to r_2)\,(r_2 \to x_2)$ is a translation along $K_\infty(y_2)$, hence also along $K_\infty(y_1)$. Therefore $(r_2 \to x_2)$ must take r into the intersection of $K_\infty(y_1)$ with $\mathfrak{g}$, i.e., into x_1, and it follows that $r = r_1$.

Let $K_\infty(r_i)$ intersect $\mathfrak{g}$ in x_i' and $\mathfrak{b}$ in y_i'. Then

$$K_\infty(r_i, x_i')\,(r_i \to x_i) = K_\infty(y_i, x_i) = K_\infty(y_i', r_i)\,(r_i \to y_i)$$

hence

$$\lambda(y_i', r_i) = \lambda(r_i, x_i')$$

so that $\mathfrak{c}$ is the locus of the midpoints of the arcs $K_\infty(u, v)$ with $u \in \mathfrak{g}$ and $v \in \mathfrak{b}$. Repetition of this argument and continuity show:

(51.4) *The locus of the points w which divide $K_\infty(u, v)$, $u \in \mathfrak{g}$, $v \in \mathfrak{b}$ in a fixed ratio $\rho = \lambda(u, w) : \lambda(u, v)$ is an asymptote to $\mathfrak{g}^+$.*

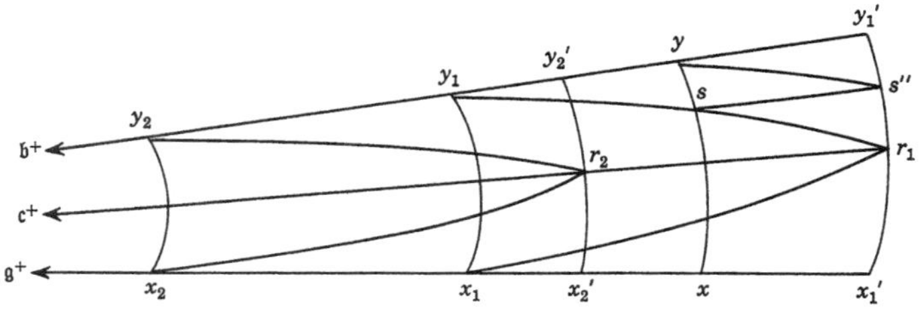

Fig. 64.

Moreover putting $\gamma = \tau_1 - \tau_1' = \tau_2 - \tau_2'$ we conclude from the arbitrariness of τ_1 and τ_2 that the arc of $K_\infty\bigl(x(\tau)\bigr)$ between $\mathfrak{b}$ and $\mathfrak{g}$ is for any τ twice as long as the arc intercepted by $\mathfrak{b}$ and $\mathfrak{g}$ on $K_\infty\bigl(x(\tau + \gamma)\bigr)$.

If $x = x(\tau_1 - \gamma/2)$ and $K_\infty(x) \cap \mathfrak{b} = y$ then $(x_1 \to x)$ sends y_1 into a point s of $K_\infty(x)$ and y into the intersection s'' of the line in F through s with $K_\infty(x_1')$. Then because of (51.4) and the last remark

$$\lambda(y_1, x_1) : \lambda(y, x) = \lambda(s, x) : \lambda(y, x) = \lambda(s'', x_1') : \lambda(y_1', x_1') = \lambda(y, x) : 2\,\lambda(y_1, x_1)$$

so that

$$\lambda\bigl(y, x(\tau_1 - \gamma/2)\bigr) = \sqrt{2}\,\lambda\bigl(y_1, x(\tau_1)\bigr).$$

Clearly the same method yields for any number ν of the form $\nu = k/2^l$, k, l integers,

$$\lambda\bigl(y, x(\tau_1 - \nu\gamma)\bigr) = 2^\nu \lambda\bigl(y_1, x(\tau_1)\bigr), \qquad \text{where} \qquad y = K_\infty\bigl(x(\tau_1 - \nu)\bigr) \cap \mathfrak{b}.$$

By continuity this extends to all real ν. If this relation is applied to $\tau_1 = 0$ and $\nu\gamma = -\beta$ it follows that the translation $\bigl(x(0) \to x(\tau)\bigr)$ along $\mathfrak{g}$ has the form

$$\sigma' = \sigma\, 2^{-\beta/\gamma}, \quad \tau' = \tau + \beta.$$

This proves together with (51.3) the second part of the Theorem (51.2).

A straight plane which satisfies the assumptions of (51.2) will be called "*quasi-hyperbolic*". The methods used at the beginning of Section **34** show readily that a quasi-hyperbolic plane satisfies the hyperbolic parallel axiom. A quasi-hyperbolic metric is not necessarily hyperbolic[1]. Some of its properties are listed in the Appendix. Here we only prove that it is, see Theorem (52.7), besides the Minkowskian geometry, the only metrization of the plane as a G-space with a transitive group of motions.

TRANSLATIONS ALONG TWO NON-ASYMPTOTIC LINES

The question which geometries possess all translations along two lines which are not asymptotes is answered as follows:

(51.5) ***Theorem.*** *If in the straight plane R all translations along two lines $\mathfrak{g}$ and $\mathfrak{h}$ exist, where $\mathfrak{h}$ is not an asymptote to either orientation of $\mathfrak{g}$, then R is Minkowskian or hyperbolic.*

Proof. It follows from (32.16) that $\mathfrak{g}$ cannot be an asymptote to either orientation of $\mathfrak{h}$. We may assume that $\mathfrak{g}$ and $\mathfrak{h}$ intersect. If they do not, and $x(\tau)$, $y(\sigma)$ represent $\mathfrak{g}$ and $\mathfrak{h}$ respectively, then $x(\tau)\mathfrak{h} \to \infty$ for $|\tau| \to \infty$ and $y(\sigma)\mathfrak{g} \to \infty$ for $|\sigma| \to \infty$. For if $y(\sigma)\mathfrak{g} < M$ for $\sigma > \sigma_0$ say (the convexity of the equidistant curves to $\mathfrak{g}$ implies that either $y(\sigma)\mathfrak{g} \to \infty$ for $\sigma \to \infty$ or $y(\sigma)\mathfrak{g} < M$ for $\sigma \geqslant 0$) and Φ_τ denotes the translation along $\mathfrak{g}$ which takes $x(0)$ into $x(\tau)$, then either $\lim_{\tau\to\infty} \mathfrak{h}\Phi_\tau$ or $\lim_{\tau\to-\infty} \mathfrak{h}\Phi_\tau$ would be a line $\mathfrak{h}'$ equidistant to $\mathfrak{g}$, and $\mathfrak{h}$ an asymptote to $\mathfrak{h}'$ hence to $\mathfrak{g}$.

Therefore there is a pair of values σ_0, τ_0 such that $x(\tau_0)y(\sigma_0) = \mathfrak{g}\,\mathfrak{h} > 0$. The equidistant curves C_g to $\mathfrak{g}$ through $y(\sigma_0)$ and C_h to $\mathfrak{h}$ through $x(\tau_0)$ have $\mathfrak{h}$ and $\mathfrak{g}$ as supporting lines. C_g and C_h cross each other twice. Let p be one of the intersections. The translation along $\mathfrak{g}$ which takes $y(\sigma_0)$ into p takes $\mathfrak{h}$ into a line $\mathfrak{h}^*$ along which all translations exist (compare the beginning of the preceding proof). Similarly the translations along $\mathfrak{h}$ which takes $x(\tau_0)$ into p takes $\mathfrak{g}$ into a line $\mathfrak{g}^*$ along which all translations exist. The lines $\mathfrak{g}^*$ and $\mathfrak{h}^*$ are not identical because C_g lies between $\mathfrak{h}^*$ and $\mathfrak{g}$ and C_h between $\mathfrak{g}^*$ and $\mathfrak{h}$.

We assume therefore that $\mathfrak{g}$ and $\mathfrak{h}$ intersect at a point z and that $z = x(0) = y(0)$. We take first the case where a parallel $\mathfrak{g}' \neq \mathfrak{g}$ to $\mathfrak{g}$ exists. Then $\mathfrak{g}'$ intersects $\mathfrak{h}$ in a point $y(\sigma')$, $\sigma' \neq 0$. If Ψ_σ denotes the translation along $\mathfrak{h}$ which takes $y(0)$ into $y(\sigma)$, then $\mathfrak{g}'\,\Psi_{\sigma'}$ is parallel to $\mathfrak{g}'$ and intersects $\mathfrak{h}$

in $y(2\sigma')$. It follows from (51.1) that every point of the plane lies on a parallel to $\mathfrak{g}$. We now observe:

(51.6) *A translation Ω of a straight plane along a line G takes a line H that intersects G into a line $H' = H\Omega$ which does not intersect H.*

For if $a = H \cap G$, then $b = H \cap H'$ would imply that $a\Omega$, b and $b\Omega$ lie on H', whereas the line through b and $b\Omega$ cannot intersect G because of (32.12).

This remark implies in the present case that $\mathfrak{g}_\sigma = \mathfrak{g}\Psi_\sigma$ is parallel to $\mathfrak{g}$: If $\mathfrak{g}_\sigma$ is the parallel to $\mathfrak{g}$ through a given point p and the translation Φ_τ takes $y(\sigma)$ into p then σ, τ are defined as the coordinates of p. In terms of these coordinates Φ_β has the form

$$\sigma' = \sigma, \qquad \tau' = \tau + \beta.$$

If $(\sigma, \tau)\,\Psi_\alpha = (\sigma'', \tau'')$, then obviously $\sigma'' = \sigma + \alpha$, moreover $\tau'' = \tau$ because of (51.1) and (32.4). Therefore the group of motions of R has $\sigma^* = \sigma + \alpha$, $\tau^* = \tau + \beta$ as subgroup and (50.1) shows that the geometry is Minkowskian.

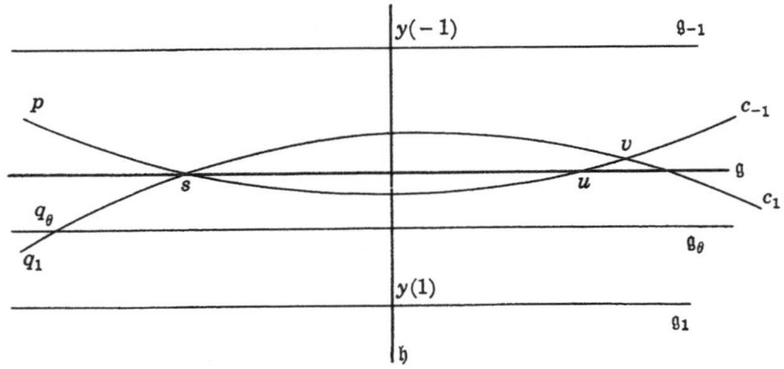

Fig. 65.

We now consider the case where no parallel $\mathfrak{g}' \neq \mathfrak{g}$ to $\mathfrak{g}$ exists and put again $\mathfrak{g}_\sigma = \mathfrak{g}\Psi_\sigma$. We know from (51.6) that $\mathfrak{g}_{\sigma'}$ and $\mathfrak{g}_\sigma$ do not intersect for $\sigma' \neq \sigma$. All translations along $\mathfrak{g}_\sigma$ exist. The $\mathfrak{g}_\sigma$ cover R because $x(\tau)\mathfrak{h} \to \infty$ for $|\tau| \to \infty$. Since Ψ_σ takes z into $y(\sigma)$ and a suitable translation along $\mathfrak{g}_\sigma$ takes $y(0)$ into an arbitrary point of $\mathfrak{g}_\sigma$ the plane R possesses a transitive group of motions.

No line $\mathfrak{g}_\sigma$ is asymptote to an other $\mathfrak{g}_{\sigma'}$. For then $\mathfrak{g}_\sigma$ and $\mathfrak{g}_{\sigma'}$ would satisfy the hypothesis of (51.2). ($\mathfrak{g}_\sigma$ and $\mathfrak{g}_{\sigma'}$ cannot be parallel because then $\mathfrak{g}$ and $\mathfrak{g}_{\sigma'-\sigma}$ would be parallel.) Hence $\mathfrak{g}_\sigma$ and $\mathfrak{g}_{\sigma'}$ have distance 0 from each other, and applying $\Psi_{\sigma'-\sigma}$ repeatedly would show that all $\mathfrak{g}_\sigma$ have distance 0 from each other and are therefore asymptotes to each other. Choose $\sigma_0 \neq 0$ such that

$y(\sigma_0)$ and z do not lie on the same limit circle with the $\mathfrak{g}_\sigma$ (properly oriented) as central rays. The group generated by the translations along $\mathfrak{g}_\sigma$ and $\mathfrak{g}_{\sigma'}$, contains then a translation Φ along a suitable $\mathfrak{g}_{\sigma''}$ which takes z into $y(\sigma)$. But then $\Phi\Psi_{-\sigma}$ would leave $\mathfrak{g}_{\sigma''}$ pointwise fixed. Since $\Phi\Psi_{-\sigma}$ preserves the orientation, it would be the identity, but Φ and Ψ_σ are translations along intersecting lines.

Choose $s \in \mathfrak{g}$ such that the equidistant curves $C_{\pm 1}$ to $\mathfrak{g}_{\pm 1}$ through s cross each other at s (this is so for all but at most one position of s on $\mathfrak{g}$). Let v be the second intersection of C_1 and C_{-1}. At least one of the curves C_1, C_{-1} intersects $\mathfrak{g}$ again, assume $u = C_{-1} \cap \mathfrak{g} - s$.

Let $\Omega_{\pm 1}$ be the translation along $\mathfrak{g}_{\pm 1}$ which carries s into v. Put $p = s\,\Omega_{-1}^{-1}$. Under $X_1 = \Omega_{-1}\Omega_1^{-1}$ the point p goes into a point $q_1 \neq p$ on C_1 and s goes into itself, so that X_1 is a rotation about s and not the identity.

If in this construction $\mathfrak{g}_1$ is replaced by $\mathfrak{g}_\theta$, $0 \leqslant \theta < 1$, and C_1 by the equidistant curve to $\mathfrak{g}_\theta$ through s (the other curves remaining as before), we obtain a rotation X_θ about s which takes p into a point q_θ that depends continuously on θ. Also X_θ is not always the same rotation, because for $\theta = 0$ the point q_θ lies on $\mathfrak{g}$.

Thus q_θ traverses an arc of the circle $K(s, p)$ and it follows that all rotations about s exist. Since R possesses a transitive group of motions, R can be freely rotated about each point and is therefore hyperbolic.

TRANSLATIONS ALONG TWO GREAT CIRCLES OF A SPHERELIKE SURFACE

A result similar to (51.5) which is however much easier to prove, holds for translations along two great circles of a spherelike surface S. A "*translation of S along a great circle*" $\mathfrak{g}$ is an orientation preserving motion of S which induces an orientation preserving mapping of $\mathfrak{g}$ on itself. It therefore maps each of the hemispheres S_1 and S_2 bounded by $\mathfrak{g}$ on itself.

Let Φ be a translation of S along $\mathfrak{g}$ which is not the identity. As a topological mapping of S_i on itself it has — by Brouwer's Fixed Point Theorem, see for instance Hurewicz-Wallman [1, p. 40][2] — at least one fixed point in the interior of S_i (because Φ has no fixed points on $\mathfrak{g}$). With any point the antipodal point stays also fixed. Hence Φ has exactly two fixed points. For if a second pair of antipodal points fixed under Φ existed, then the great circle through these four points would stay point-wise fixed, in particular its intersection with $\mathfrak{g}$.

If $p_i \in S_i$ are the fixed points of Φ and $\Phi^\nu \neq E$ or $\Psi^\mu = \Phi$ then they are also the fixed points of Φ^ν and Ψ. Therefore, if all translations along $\mathfrak{g}$ exist they all have the same fixed points p_1, p_2 and $p_1x = p_2x = p_1p_2/2$ for all $x \in \mathfrak{g}$. Also, all rotations about p_i exist and they coincide with the translations along $\mathfrak{g}$. It is now easy to prove:

(51.7) *If all translations along two different great circles of a spherelike surface S exist, then S is spherical.*

Proof. Choose the notations for the fixed points p_1, p_2 and q_1, q_2 of the translations along $\mathfrak{g}$ and $\mathfrak{h}$ such that p_1 and q_1 lie in the same hemisphere S_1 bounded by $\mathfrak{g}$. Then $\eta = p_1q_1 \leqslant \alpha = p_1p_2/2$.

The rotations about p_1 (or translations along $\mathfrak{g}$) take q_1 into all points of the circle $K(p_1, \eta)$, so that all rotations about each point of $K(p_1, \eta)$ exist. If $K(p_1, \eta)$ is revolved about q_1 it sweeps out the closed circular disk $\overline{S}(q_1, 2\eta) = S(q_1, 2\eta) \cup K(q_1, 2\eta)$. Hence all rotations about every point of $S(q_1, 2\eta)$ exist. The same holds for $\overline{S}(q_2, 2\eta)$. If $2\eta \geqslant \alpha$ then S can be freely rotated about each of its points and is spherical. If $2\eta < \alpha$ we continue: By revolving $\overline{S}(q_1, 2\eta)$ about p_1 we see that S can be freely rotated about every point of $\overline{S}(p_1, 3\eta)$. It is clear that a finite number of steps suffice to establish that all rotations about every point of S exist.

Since a surface of the elliptic type is homeomorphic to the projective plane, we cannot speak of orientation preserving motions. It is clear, however, that (51.7) contains the following fact proved by Gans [1]:

(51.8) *If for any pair p, q of points on either of the two great circles $\mathfrak{g}_1, \mathfrak{g}_2$ on a surface E of the elliptic type a motion of E exists which maps $\mathfrak{g}_i$ on itself and takes p into q, then E is elliptic.*

We did not require that the motion which takes $\mathfrak{g}_i$ into itself and p into q preserve the orientation of $\mathfrak{g}_i$, because the existence of such a motion follows from the arbitrariness of p and q. A similar remark applies to the preceding theorem.

52. Surfaces with transitive groups of motions

We observed already that Theorems (48. 8,10) lead naturally to the problems of determining the G-spaces with transitive groups of motions and those with *pairwise transitive* groups of motions, which means, by definition, that for any two pairs of points a, b and a', b' with $ab = a'b'$ a motion exists that takes a into a' and b into b'.

In this section we discuss the few facts known concerning the first problem beyond the abelian case treated in Theorem (50.1). From now on we will have to lean heavily on the theory of *topological and Lie groups*. However the knowledge required for the present section is quite moderate and the main theorem (52.9) does not use these theories at all.

MOTIONS OF COMPACT G-SPACES

We begin with the following corollary of (32.5):

(52.1) *For a compact G-space R and a motion $\Phi \neq E$ there exists a point z of maximal displacement, i. e., $\alpha = zz\Phi = \sup_{x \in R} xx\Phi$, and $zz\Phi^k = k\alpha$ if k is the smallest integer with $k\alpha \geqslant \rho(z)/2$.*

If $k > 1$ then a representation $x(\tau)$ of a geodesic exists such that $x(i\alpha) = z\Phi^i$, $i = 0, 1, 2, \ldots$, and $x(\tau)$ represents a segment for $i\alpha \leqslant \tau \leqslant (i+k)\alpha$.

The existence of z is obvious. Put $z_i = z\Phi^i$, so that $z_0 = z$. For $k = 1$ there is nothing to prove. If $k > 1$ then $k\alpha < \rho(z)$. It follows from (32.5) that (zz_1z_2), hence $(z_{i-1}z_iz_{i+1})$ for all integers i. Since $\rho(z_i) = \rho(z)$ the segment $T(z_{i-1}, z_{i+1})$ is unique and contains z_i. Now the existence of $x(\tau)$ is clear and $x(\tau)$ represents a segment for $i\alpha \leqslant \tau \leqslant (i+k)\alpha$ because $k\alpha < \rho(z_i)$. In particular $x(0)x(k\alpha) = zz\Phi^k = k\alpha$.

We agree to use as distance for two motions Φ, Ψ of a compact space R the number $\delta(\Phi, \Psi) = \sup_{x \in R} x\Phi x\Psi$. *Then $\delta(E, \Phi)$ equals the maximal displacement under Φ.* Before applying (52.1) to the proper subject of this section we notice the following interesting corollary:

(52.2) **Theorem.** *A one parameter group Γ of motions of a compact space possesses an orbit which is a geodesic.*

We remember from Section 29 that the orbit $0(p, \Gamma)$ of the point p under Γ consists of all points $p\Phi$ with $\Phi \in \Gamma$. We assume that Γ is represented in the standard form Φ_t with $\Phi_{t_1}\Phi_{t_2} = \Phi_{t_1+t_2}$ and choose $\varepsilon > 0$ such that $\delta(E, \Phi_t) < \inf_{x \in R} \rho(x)/2$ for $|t| < \varepsilon$.

For $0 < u < \varepsilon$ let z and z' be points of maximal displacement for Φ_u and $\Phi_{u/2}$ respectively. Then (32.5) and the choice of ε imply

$$z'z'\Phi_u = 2\,z'z'\Phi_{u/2} \geqslant 2\,zz\Phi_{u/2} = zz\Phi_{u/2} + z\Phi_{u/2}\,z\Phi_u \geqslant zz\Phi_u \geqslant z'z'\Phi_u$$

Thus z is also a point of maximal displacement for $\Phi_{u/2}$ and generally for $\Phi_{u2^{-n}}$. Moreover $(zz\Phi_{u/2}z\Phi_u)$ and generally $(zz\Phi_{u2^{-n-1}}\,z\Phi_{u2^{-n}})$. If $x(\tau)$ is the geodesic with $x(0) = z$ which represents the unique segment $T(z, z\Phi_u)$ for $0 \leqslant \tau \leqslant zz\Phi_u = \alpha'$ then (52.1) shows $x(i\alpha'\,2^{-n}) = z\Phi_{iu2^{-n}}$ for all i and non-negative n. Now a trivial continuity argument shows that $x(\alpha'\tau)x(0) = x(0)\Phi_{u\tau}$ for all τ, hence $x(\alpha\tau) = x(0)\,\Phi_\tau$ for $\alpha = \alpha'/u$, which expresses the assertion of (52.2) in a precise manner.

The geodesic $x(\tau)$ cannot have multiple points because then every point would be a multiple point. Considering the rotations of a two-dimensional sphere about a diameter we see that there need not be more than one orbit which is a geodesic. The rotations of $z = (x^2 + y^2)^{-1/2}$ about the z-axis show that the assumption of compactness is essential.

We now deduce from (52.1)

(52.3) ***Theorem.*** *A closed group of motions of a compact G-space is a Lie Group.*

Here we follow the usual agreement to consider finite groups as Lie groups. If Γ is the given group, then the spherical neighborhood of E with radius $\inf_{x \in R} \rho(x)/2$ contains, by Pontrjagin [1, Theorem 53], a subgroup Δ of Γ such that Γ/Δ is a Lie group. But Δ cannot contain any element other than the identity, because (52.1) shows that for any $\Phi \neq E$ a positive k exists such that $\delta(E, \Phi^k) \geqslant \inf \rho(x)/2$.

Before proceeding to other results we remember some standard definitions: With a point x and a group of motions Γ we associate (in any metric space) the *"isotropic group"* Γ_x of x which consists, by definition, of the motions in Γ which leave x fixed. If Γ contains a motion Ψ that takes x into y then $\Gamma_y = \Psi^{-1}\Gamma_x\Psi$ so that Γ_x and Γ_y are isomorphic. The group Γ is said to *"act effectively" on the orbit* $0(x, \Gamma)$ if the identity is the only element of Γ which leaves every point of $0(x, \Gamma)$ fixed. If Γ is transitive on the space R, then $0(x, \Gamma) = R$ and, as group of motions, Γ is automatically effective on R.

An example for non-transitive groups which act effectively on certain orbits and which will be used later is this: If in a G-space a group of motions has the property that $0(p, \Gamma_x) = K(x, \rho)$ for some $0 < \rho = px < \rho(x)$, then Γ_x acts effectively on $K(x, \rho)$. For if $\Phi \in \Gamma_x$ leaves every point p of $K(x, \rho)$ fixed, then it leaves every point of $T(x, p)$, hence every point of $S(x, \rho)$ fixed and must by (28.8) be the identity. But Γ_x is not effective on $0(x, \Gamma_x) = x$, or less trivially, if Γ is the group of all motions of an elliptic space, then Γ_x is transitive but not effective on the locus of conjugate points to x.

We now prove

(52.4) ***Theorem.*** *The group Γ of all motions of a compact G-space R is a Lie group. If Γ is transitive on R, then R is a topological manifold. If* $\dim R = n$ *then* $\dim \Gamma \leqslant n(n+1)/2$.

That Γ is a Lie group follows from (52.3). Hence Γ has finite dimension. Then for any point x

$$\dim \Gamma = \dim 0(x, \Gamma) + \dim \Gamma_x = \dim R + \dim \Gamma_x.$$

see Montgomery and Zippin [1, Corollary 3]. Hence $n = \dim R$ is finite. The relation $\dim \Gamma \leqslant n(n+1)/2$ is Theorem 9, l. c.. Since $R = 0(x, \Gamma)$ is locally connected it follows from Theorem 12, l. c., that R is a topological manifold.

G-SURFACES WITH TRANSITIVE GROUPS OF MOTIONS

The last theorem is all that is known in the general case about G-surfaces with transitive groups of motions, only in the two-dimensional case can all such spaces be determined.

Let R be a G-surface whose group Γ of motions is transitive. If the isotropic group Γ_x is infinite, it is trivial that it contains all rotations about x. In that case R is *euclidean, hyperbolic, spherical or elliptic.*

The universal covering space $\overline{R}$ of R is, topologically, either a sphere or a plane, and the group $\overline{\Gamma}$ of motions of $\overline{R}$ is transitive on $\overline{R}$, see (28.10). If $\overline{R}$ is a sphere then by (52.4) the group is a Lie group, and the only transitive Lie group acting on the sphere is the orthogonal group, see for instance the list in Section 10 of Mostow [1]. The metric is spherical because Γ_x is infinite. The only surface covered by a sphere is the elliptic plane.

If $\overline{R}$ is a plane and $\overline{\Gamma}_x$ is finite then $\overline{\Gamma}$ is a Lie group. This may be seen *either* by using the result of Gleason [1], that a finite dimensional locally compact topological group satisfying the second axiom of countability is a Lie group, when it has no subgroup of arbitrarily small diameter, and the older result of Montgomery and Zippin [2] that arbitrarily small groups cannot act on the plane.

We then conclude by inspecting the list of Mostow [1, pp. 626, 627] that $\overline{\Gamma}$ acting on a suitable (σ, τ)-plane *contains as subgroup one of the following two simply transitive subgroups*

(52.5) $$\sigma' = \sigma + \alpha, \qquad \tau' = \tau + \beta, \qquad -\infty < \alpha, \beta < \infty$$

or

(52.6) $$\sigma' = \sigma e^{-\beta} + \alpha, \qquad \tau' = \tau + \beta, \qquad -\infty < \alpha, \beta < \infty.$$

Or we may proceed in the following more elementary way for which the author is indebted to Prof. L. Zippin: The coset space $\overline{\Gamma}/\overline{\Gamma}_x$ is the plane.

Since $\bar{\Gamma}_x$ is finite, $\bar{\Gamma}$, as space, is a covering space of the plane. Since the plane is simply connected, $\bar{\Gamma}$ consists of a finite number of disconnected copies of the plane, in particular, the identity component of $\bar{\Gamma}$ is homeomorphic to the plane and simply transitive on the plane. v. Kerékjártó [2] proves with quite elementary means that then the identity component of $\bar{\Gamma}$ can be represented in one of the forms (52.5) or (52.6).

The first group is abelian and the metric of R is Minkowskian. The group of motions of R contains therefore at least the reflection

$$\sigma' = -\sigma + 2\alpha, \qquad \tau' = -\tau + 2\beta$$

in a given point (α, β), and it may contain other motions. For instance $\bar{\Gamma}_p$ may be any finite group of rotations about p of even order. It may also contain the reflection in a line through p. A complete discussion of all possibilities is easy but not worthwhile.

The group (52.5) can act transitively only on *the plane, the cylinder, and the torus* (see Mostow [1, p. 627]), and we know that each of these can be provided with an invariant Minkowskian metric. A metric argument, which shows that (52.5) cannot act, for instance, on a Moebius strip, although the latter can be provided with a euclidean metric, is the existence of a distinguished great circle, see Section 30.

There remains the discussion of the group (52.6), which we encountered in the last section in quasi-hyperbolic geometry. This geometry was defined as a straight plane which possesses all translations along a family of (or two) non-parallel asymptotes. We call these asymptotes and the corresponding limit circles "*distinguished*". Our main effort in this section will be the proof that any metrization of a (σ, τ)-plane as a G-space for which the transformations (52.6) are motions is quasi-hyperbolic, see Theorem (52.9). First we use this result to complete the list of all G-surfaces with transitive groups of motions.

The group (52.6) considered as transformation group, acts not only on the plane, but also on two types of cylinders which are obtained (see Mostow [1, p. 627]) by taking either a strip between two distinguished asymptotes and identifying points on them which lie on the same distinguished limit circle, or a strip between two distinguished limit circles and identifying points on them which lie on the same distinguished asymptotes. The geometric meaning of the parameters σ, τ in Theorem (51.2) shows at once that for these cylinders the transformations (52.6) are not all motions. Thus we are led to:

(52.7) **Theorem.** *The only G-surfaces with transitive groups of motions are: the plane with a Minkowskian or quasi-hyperbolic metric, the cylinder and torus with a Minkowskian metric, the sphere and the projective plane with a spherical metric.*

We notice that, with the exception of the quasi-hyperbolic plane, all surfaces in the list are symmetric spaces. For the cylinder and torus this follows from the relation $\Xi\Phi\Xi = \Phi$ which holds for any translation Ξ and any symmetry Φ of a Minkowski plane. If a quasi-hyperbolic plane is symmetric it is hyperbolic. For all translations along every line exist: if a, b are any two distinct points, then the reflection in $m[a, b]$ followed by the reflection in b is a translation along the line through a and b that takes a into b. By Theorem (51.5) the metric is hyperbolic. Since a symmetric space possesses a transitive group of motions, see (49.6), we deduce from (52.7):

(52.8) *The only symmetric G-surfaces are: the (metric) spheres, elliptic planes, hyperbolic planes, and the Minkowskian planes, cylinders, and tori.*

CHARACTERIZATION OF QUASI-HYPERBOLIC GEOMETRY

We now come to the theorem which we already used and which is the converse of the second part of Theorem (51.2)

(52.9) **Theorem.** *A (σ, τ)-plane which is metrized as a G-space P for which the transformations*

$$\Gamma : \sigma' = \sigma e^{-\beta} + \alpha, \qquad \tau' = \tau + \beta, \qquad -\infty < \alpha, \beta < \infty$$

are motions, is quasi-hyperbolic[1].

We introduce in the (σ, τ)-plane an auxiliary hyperbolic metric $h(a, b)$ for which $(0, \tau)$ represents a geodesic, $(\sigma, 0)$ a limit circle in terms of the arc length σ with $(0, \tau)$, $\tau \geqslant 0$ as central ray, and Γ is a group of motions. The orbits of the one-parameter subgroups of Γ will be called simply orbits. They are in the hyperbolic interpretation: the limit circles $\tau = \text{const.}$, the asymptotes orthogonal to these limit circles, and the equidistant curves to these asymptotes. There is exactly one orbit through two distinct points. (This follows from the simple transitivity of Γ or also because in a Poincaré model of hyperbolic geometry the orbits are the circles through one point on the absolute locus.)

We orient all orbits other than the limit circles, positively in the direction of increasing τ. Two orbits D_1, D_2 are equivalent if there is a motion in Γ taking D_1 into D_2. If $p_i \in D_i$ and $(p_1 \to p_2)$ is the motion in Γ taking D_1 into D_2 then $D_1(p_1 \to p_2) = D_2$. The families of equivalent orbits are:

1) The limit circles $\tau = \text{const.}$ 2) the asymptotes orthogonal to them 3) the families of curves equidistant to, on the same side of, and at the same hyperbolic distance from, the different asymptotes in 2). Each family of equivalent curves covers the plane simply.

We are now ready to prove our theorem:

(a) *Two geodesic arcs A_1, A_2 which have their endpoints on the same orbit D and have no other common point with D, lie on the same side of D.*

For let D_i be the curve equivalent to, but different from D, which is a supporting curve of A_i, i. e. contains at least one point p_i of A_i and A_i lies in the closed strip bounded by D and D_i. If A_1 and A_2 are on different sides of D then the two geodesic arcs A_2 and A_1 $(p_1 \to p_2)$ would both have p_2 as interior point without crossing each other at p_2 which contradicts (10.9).

(b) *If a proper subarc of an orbit D is a geodesic arc then D is a straight line and any geodesic arc that has two common points with D lies on D.*

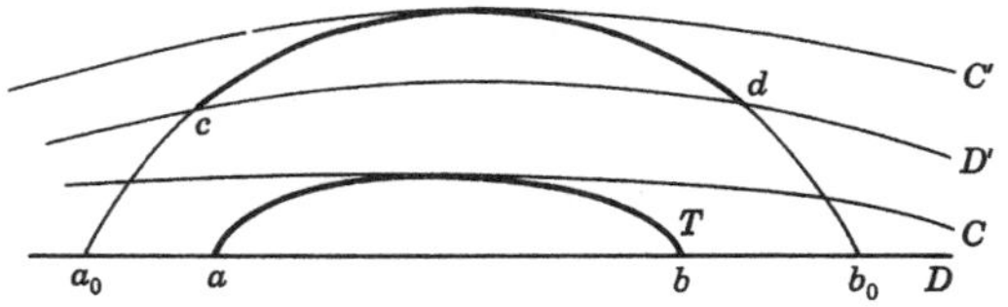

Fig. 66.

It is clear that D is a geodesic. If there were a geodesic arc A' that had two common points with D but did not lie on D, then A' would contain a subarc A with endpoints on D and completely on one side of D. The curve equivalent to but distinct from D, which is supporting line of A would then be a geodesic which does not cross A although it contains an interior point of A.

If D were not a straight line then two of its points could be connected by a segment A' not on D.

(c) *P is straight.*

It suffices to show that for a given segment $T = T(a, b)$ a segment $T(a', b')$ exists that contains T in its interior. Let D be the orbit through a and b. If $T \subset D$ then D is by (b) a straight line and the existence of $T(a', b')$ is obvious.

Otherwise a and b are by (b) the only common points of T and D. Choose two points a_0 and b_0 on D such that a and b are interior points of the arc D_0 of D with endpoints a_0 and b_0. By (a) a segment $T_0 = T(a_0, b_0)$ bounds with D_0 a domain which contains T. Let C be the hyperbolically equidistant curve to D and distinct from D which is supporting curve of T, and C' the curve equivalent to C which is supporting line of T_0 and lies on the other side of C from D. Finally denote by D' the image of D under a transformation Φ in Γ that takes C into C'. Then D' intersects T_0 in two points c and d, and $T\Phi$ is a segment with endpoints $a\Phi$, $b\Phi$ on D'. We can choose Φ such that the hyperbolic midpoints of the arcs of D from c to d and from $a\Phi$ to $b\Phi$ coincide.

The segment $T\Phi$ touches C'. If $a\Phi$, $b\Phi$ were interior points of the arc from c to d of D' then $T\Phi$ would either intersect the (unique) segment $T(c, d)$ twice or touch $T(c, d)$ at some point of C' without crossing it. For the same reason, c and d cannot lie between $a\Phi$ and $b\Phi$. Hence $a\Phi$ and $b\Phi$ must coincide with c and d, and since $T(c, d)$ is unique $T\Phi = T(c, d) \subset T_0$, hence $T_0\Phi^{-1}$ contains T in its interior.

We show next

(d) *At least one family of equivalent orbits consists of straight lines.*

It suffices to show that there is an orbit through a given point p which is a straight line, then all equivalent orbits are automatically straight lines.

Because a straight line coincides with an orbit or intersects it at most twice it suffices to show the existence of a $\Phi \neq E$ in Γ for which $(p\Phi^{-1}\, p\, p\Phi)$. The mapping

$$x' = x(x \to p)^2, \qquad x \in K = K(p, 1),$$

is a topological involutoric mapping of K on itself because

$$(x' \to x) = (x' \to p)\,(p \to x) = (p \to x')^{-1}\,(x \to p)^{-1} = (x \to p)^{-2}.$$

We show that for at least one point x of K the image x' coincides with the diametrically opposite point $\bar{x}$ to x.

Let K^+ be an orientation of K and u an arbitrary point of K. Unless $u' = \bar{u}$, traversing K^+ in the positive sense from u on, we either meet u' before $\bar{u}$ or $\bar{u}$ before u'. Take the first case. Then traversing K^+ in the positive sense from u' on, we meet $\bar{u}'$ before $(u')' = u$. For reasons of continuity there is an x for which $\bar{x} = x'$. This proves (d).

(e) *The family of limit circles among the orbits cannot consist of straight lines.*

For a translation along one of them is a translation along the others, so that they would form a family of parallel axes of these translations. A perpendicular H to one of these lines is perpendicular to all others. If it intersects the two limit circles L_1, L_2 of the family in p_1 and p_2, then $L_1(p_1 \to p_2) = L_2$ hence $H(p_1 \to p_2) = H$. Then all translations along H exist and it would follow from (51.5), that Γ is abelian (of the form $\sigma' = \sigma + \alpha$, $\tau' = \tau + \beta$).

Thus we see that at least one family F of equivalent hyperbolic equidistant curves consists of straight lines. Any two of these lines are, with their positive orientation, asymptotes to each other. For the hyperbolic distance $h(x, L')$ of a point x on any curve L in F from a second curve L' in F tends to 0 when x traverses L in the positive sense. Since P and the hyperbolic plane both have Γ as group of motions, the two metrics are uniformly equivalent, so that xL' also tends to 0. The lines L and L' cannot be parallel (in the sense of P), because L is not equidistant to L'. This proves the theorem.

It may be found disturbing that an arbitrary family F of hyperbolic equidistant curves functions as family of asymptotes instead of the family F^* formed by the orbits which are hyperbolic straight lines. But this cannot be helped, because the two cases are *equivalent: There is a topological mapping Ψ of the hyperbolic plane H (or P) on itself such that in the group $\Gamma^* : \{\Psi^{-1}\Phi\Psi\}$ with $\Phi \in \Gamma$ the translations along the curves in F become translations along the lines in F^**, and a translation along the limit circles remains a translation along the limit circles.

Ψ may be defined as follows: a given point p of H lies on exactly one curve L in F and L is equidistant to exactly one line L^* in F^*. We define $p^* = p\Psi$ as the hyperbolic foot of p on L^*. Clearly Ψ is a topological mapping and if K_∞ is a distinguished limit circle then $K_\infty\Psi$ is too, although a different one. If Φ takes L into itself then $\Psi^{-1}\Phi\Psi$ takes L^* into itself.

53. The hermitian elliptic and similar metrics

Before treating general spaces with pairwise transitive groups of motions, we discuss examples of such spaces which do not have constant curvature. The next section will show that they are more than just examples: In the compact case they represent together with the spherical and elliptic spaces all spaces with pairwise transitive groups of motions. For the sake of comparison we include the ordinary elliptic and hyperbolic spaces in our discussion. Let K denote the field of either *the real numbers, or the complex numbers or the (real) quaternions.*

The points of the *projective ν-space $P^\nu(K)$ over K* are the classes $[x] = [x_0, x_1, \ldots, x_\nu]$, other than $[0, 0, \ldots, 0]$, of $(\nu + 1)$-tuples $x = (x_0, x_1, \ldots, x_\nu)$

in K where two $(\nu+1)$-tuples x and y belong to the same class if and only if a $\lambda \neq 0$ in K exists such that

$$x_h = y_h \lambda, \qquad h = 0, 1, \ldots, \nu.$$

Since each element in a class determines the class uniquely we sometimes leave the brackets out.

If $|\xi|$ denotes the absolute value in the real and complex cases, and the norm of ξ if ξ is a quaternion, then $|\xi - \eta|$ is a distance in K which leads naturally to the following topology in $P^\nu(K)$:

$\lim_{m\to\infty} [x^m] = [x]$ means: for given $(\nu+1)$-tuples x^m in $[x^m]$ and x in $[x]$ there exist numbers λ_m in K such that

$$\lim_{m\to\infty} |x_h^m \lambda_m - x_h| = 0 \qquad \text{for} \qquad h = 0, \ldots, \nu.$$

If K stands for the real numbers then $P^\nu(K)$ is the ν-dimensional projective space P^ν. If K denotes the complex numbers, then $P^\nu(K)$ is a $2\,\nu$-dimensional space, the so-called *complex projective ν-space* $C^{2\nu}$. In case of the quaternions, $P^\nu(K)$ is a $4\,\nu$-dimensional space $Q^{4\nu}$, the socalled *right quaternion projective ν-space*. The term "right" refers to the fact that we defined the $(\nu+1)$-tuples of the form $x\lambda$ to constitute the class $[x]$. We could just as well have put the $(\nu+1)$-tuples λx into one class, which would have led to the left quaternion projective ν-space.

The projective, *complex*, *quaternion line* through two distinct points $[x]$, $[y]$ of the corresponding $P^\nu(K)$ is, by definition, the set of points of the form $[x\lambda + y\mu]$, $\lambda, \mu \in K$. The dimension of a line is 1, 2, 4 respectively. Because of the associative law of multiplication the point set forming a line is indeed uniquely determined by the points $[x]$ and $[y]$. Since Cayley numbers (see Section 55) do not satisfy the associative law, this definition of line cannot be applied to them, and there are various other reasons which preclude direct extension.

The ordinary elliptic and hyperbolic geometries are obtained as follows: for real x_i, y_i we put

$$(x, y)_\varepsilon = \varepsilon x_0 y_0 + x_1 y_1 + \ldots + x_\nu y_\nu, \qquad \varepsilon = \pm 1.$$

The *elliptic metric* is defined in all of P^ν by

$$\varepsilon(x, y) = \gamma \operatorname{Arc}\cos \frac{|(x, y)_1|}{[(x, x)_1 (y, y)_1]^{1/2}}, \tag{53.1}$$

where γ is a fixed positive number and Arc cos denotes, as before, the values of the inverse cosine in $[0, \pi]$.

The *hyperbolic metric* is defined for $(x, x)_{-1} < 0$, i. e., in the set $-x_0^2 + x_1^2 + \dots + x_\nu^2 < 0$ homeomorphic to the ν-dimensional euclidean space E^ν by

$$\delta(x, y) = \gamma \text{ Area cosh } \frac{|(x, y)_{-1}|}{[(x, x)_{-1} (y, y)_{-1}]^{1/2}} \tag{53.2}$$

where γ is again a fixed positive constant and Area cosh stands for non-negative values of the inverse hyperbolic cosine.

In both cases it is clear that the distance depends only on $[x]$ and $[y]$.

THE HERMITIAN ELLIPTIC AND HYPERBOLIC SPACES

In the case of $C^{2\nu}$ we use the hermitian froms

$$(x, y)_\varepsilon = \varepsilon\, \overline{x}_0 y_0 + \overline{x}_1 y_1 + \dots + \overline{x}_\nu y_\nu,$$

where $\overline{\alpha}$ is the conjugate complex number to α. Observe that

$$(y, x)_\varepsilon = \overline{(x, y)}_\varepsilon. \tag{53.3}$$

The hermitian elliptic metric is defined in all of $C^{2\nu}$ by (53.1). The hermitian hyperbolic metric is defined by (53.2) in the set $C_H^{2\nu}$: $(x, x)_{-1} < 0$, which is homeomorphic to $E^{2\nu}$. It is again obvious that $\varepsilon(x, y)$ and $\delta(x, y)$ depend only on $[x]$ and $[y]$ [1].

Of course we must prove that $\varepsilon(x, y)$ and $\delta(x, y)$ satisfy the axioms for a G-space. The only obvious property is the symmetry $\varepsilon(x, y) = \varepsilon(y, x)$ or $\delta(x, y) = \delta(y, x)$ which follows from (53.3).

The remaining properties are easily established by showing first that these spaces possess pairwise transitive groups of motions, where motion means for the present a transformation that preserves $\varepsilon(x, y)$ or $\delta(x, y)$. The proof rests on the following fact:

(53.4) *The transformation* $\Phi : x_i' = \sum_k a_{ik} x_k$, $i = 0, \dots, \nu$, *where the determinent* $D(a_{ik})$ *of the* a_{ik} *does not vanish, leaves* $(x, y)_\varepsilon$ *invariant* (i. e. $(x, y)_\varepsilon = (x', y')_\varepsilon$), *if and only if*

$$\varepsilon\, \overline{a}_{0i}\, a_{0k} + \sum_{j=1}^{\nu} \overline{a}_{ji}\, a_{jk} = 0 \quad \textit{for} \quad i \neq k,$$

(53.5)

$$\varepsilon\, \overline{a}_{00}\, a_{00} + \sum_{j=1}^{\nu} \overline{a}_{j0}\, a_{j0} = \varepsilon, \qquad \varepsilon\, \overline{a}_{0i}\, a_{0i} + \sum_{=1}^{\nu} \overline{a}_{ji}\, a_{ji} = 1,\ i > 0.$$

This follows by expanding $(x', y')_\varepsilon$ in terms of x_i and y_i and comparing coefficients.

If Φ leaves $(x, y)_\varepsilon$ invariant, then Φ^{-1} does too; this leads to the condition analogous to (53.5) with columns replaced by rows and shows that $|D(a_{ik})| = 1$.

It is clear that a Φ with (53.5) is a motion for $\varepsilon(x, y)$ or $\delta(x, y)$, but not every motion has this form. To prove pairwise transitivity we now show: *Given two pairs of points* $[x]$, $[y]$ *and* $[x']$, $[y']$ *(in* $C_H^{2\nu}$ *in the hyperbolic case) such that* $\varepsilon(x, y) = \varepsilon(x', y')$ *or* $\delta(x, y) = \delta(x', y')$ *then a* Φ *satisfying* (53.5) *exists that takes* $[x]$ *into* $[x']$ *and* $[y]$ *into* $[y']$.

Obviously it suffices to take the case where $x = (1, 0, \ldots, 0)$ and $y = (r_0, r_1, 0, \ldots, 0)$ where r_0, r_1 are real, $r_1 \neq 0$, and $(y, y)_\varepsilon = \varepsilon r_0^2 + r_1^2 = \varepsilon$. We normalize x' and y' such that $(x', x')_\varepsilon = \varepsilon$, $(y', y')_\varepsilon = \varepsilon$. Then $\varepsilon(x, y) = \varepsilon(x', y')$ or $\delta(x, y) = \delta(x', y')$ reduce to

$$|r_0| = |(x', y')_\varepsilon|$$

Since $(x' e^{i\theta}, y' e^{i\theta})_\varepsilon = (x', y')_\varepsilon$ we can normalize x' further such that $r_0 = \varepsilon (x', y')_\varepsilon$. It is, by the way, this very innocent looking second normalization which does not extend to triples. This explains why the hermitian spaces have only pairwise transitive groups of motions.

We determine successively the columns $(a_{0k}, a_{1k}, \ldots, a_{\nu k}) = a_k$ of Φ. We put

$$a_0 = x', \qquad a_1 = \frac{1}{r_1}(r_0 x' - y').$$

Then the second equation (53.5) is satisfied and also the first for $i = 0$, $k = 1$ because

$$(a_0, a_1)_\varepsilon = \Big(x', \frac{r_0}{r_1} x' - \frac{y'}{r_1}\Big) = \frac{r_0 \varepsilon}{r_1} - \frac{(x', y')_\varepsilon}{r_1} = 0.$$

Finally the third relation (53.5) holds for $i = 1$ because

$$(a_1, a_1)_\varepsilon = \frac{1}{r_1^2}(r_0 x' - y', r_0 x' - y')_\varepsilon =$$

$$= \frac{1}{r_1^2}(\varepsilon r_0^2 - r_0 (x', y')_\varepsilon - r_0 (y', x')_\varepsilon + \varepsilon) = \frac{\varepsilon - \varepsilon r_0^2}{r_1^2} = 1.$$

By the standard inductive method of forming orthonormal bases we can now find columns $a_2, a_3, \ldots, a_\nu$ such that the relations (53.5) hold.

As a corollary we observe

$$
(53.6)\qquad \begin{aligned} |(x, y)_1|^2 &\leqslant (x, x)_1 (y, y)_1, \\ |(x, y)_{-1}|^2 &\geqslant (x, x)_{-1} (y, y)_{-1} \quad \textit{if} \quad (x, x)_{-1} < 0, (y, y)_{-1} < 0 \end{aligned}
$$

with equality only when $[x] = [y]$ (we assume that neither is the zero class). For because of transitivity we may assume that $x = (1, 0, \ldots, 0)$. Then the inequalities reduce to the obvious

$$
\begin{aligned} |(x, y)_1|^2 &= |y_0|^2 \leqslant 1(|y_0|^2 + \ldots + |y_\nu|^2) \\ |(x, y)_{-1}|^2 &= |y_0|^2 \geqslant (-1)(-|y_0|^2 + |y_1|^2 + \ldots + |y_\nu|^2). \end{aligned}
$$

The inequalities (53.6) show that $\varepsilon(x, y)$ and $\delta(x, y)$ are defined for all pairs in $C^{2\nu}$ or $C_H^{2\nu}$ respectively and vanish only for $[x] = [y]$.

The triangle inequalities in whose proofs we may assume $\gamma = 1$, are equivalent to

$$
\cos \varepsilon(x, y) \cos \varepsilon(y, z) - \sin \varepsilon(x, y) \sin \varepsilon(y, z) \leqslant \cos \varepsilon(x, z)
$$

or

$$
\cosh \delta(x, y) \cosh \delta(y, z) - \sinh \delta(x, y) \sinh \delta(y, z) \geqslant \cosh \delta(x, z).
$$

It suffices to consider triples of the form

$$
\begin{aligned} &x = (\cos t, \sin t, 0, \ldots, 0), \quad 0 \leqslant t < \pi/2, y = (1, 0, \ldots, 0), \; z \text{ with } (z, z)_1 = 1 \\ &x = (\cosh t, \sinh t, 0, \ldots, 0), t > 0, y = (1, 0, \ldots, 0), \; z \text{ with } (z, z)_{-1} = -1. \end{aligned}
$$

Then the two triangle inequalities reduce to the obvious relations

$$
\begin{aligned} \cos t\, |z_0| - \sin t(|z_1|^2 + \ldots + |z_\nu|^2)^{\frac{1}{2}} &\leqslant |\cos t\, z_0 + \sin t\, z_1| \\ \cosh t\, |z_0| + \sinh t(|z_1|^2 + \ldots + |z_\nu|^2)^{\frac{1}{2}} &\geqslant |\cosh t\, z_0 + \sinh t\, z_1|. \end{aligned}
$$

It follows also that the equality sign can hold only when $z_2 = \ldots = z_\nu = 0$ and z_0/z_1 is real (if the points x, y, z are different). Then z lies on the complex line $Q_1 : x_2 = x_3 = \ldots = x_\nu = 0$ through x and y.

(53.7) *The equality sign holds in the triangle inequality only for points on a complex line.*

In the real case any point x^i on $x_2 = x_3 = \ldots = x_\nu = 0$ may be represented in the form $(\cos t_i, \sin t_i, 0, \ldots, 0)$ or $(\cosh t_i, \sinh t_i, 0, \ldots, 0)$. Hence

$$
\begin{aligned} \varepsilon(x^1, x^2) &= \gamma \operatorname{Arc\,cos} |\cos(t_1 - t_2)| = \gamma \min\{|t_1 - t_2|, \pi - |t_1 - t_2|\} \\ \delta(x^1, x^2) &= \gamma \operatorname{Area\,cosh} |\cosh(t_1 - t_2)| = \gamma\, |t_1 - t_2|. \end{aligned}
$$

In the first, that is the elliptic case, the line is therefore a great circle of length $\gamma\pi$, and in the hyperbolic case it is a straight line (γt is arc length).

THE METRICS IN C^2

Similarly, in order to prove that the hermitian spaces are G-spaces, we consider the geometries on a complex line in $C^{2\nu}$. The space C^2 consists of the points (x_0, x_1). All except $(0, 1)$ may be written as $(1, x_1)$. Thus C^2 becomes the complex plane, completed by one point, or the complex sphere in the terminology of the theory of complex variables. We are going to show that C^2 *is with the hermitian elliptic metric* (53.1) *isometric to a sphere in ordinary space of radius* $\gamma/2$. It suffices to take $\gamma = 1$.

Put $x_1 = \xi_1 + i\,\eta_1$ and consider the (ξ, η)-plane as belonging to a cartesian (ξ, η, ζ)-space. We map the sphere $S : \xi^2 + \eta^2 + \zeta^2 = 1$ stereographically on the (ξ, η)-plane by associating the point x_1 or $(\xi_1, \eta_1, 0)$ with the second intersection (ξ, η, ζ) of the line through $p = (0, 0, -1) \in S$ and $(\xi_1, \eta_1, 0)$ with S. A trivial calculation shows with $\xi_1^2 + \eta_1^2 = r^2$ that

$$\xi = \frac{2\,\xi_1}{1 + r^2} = \frac{x_1 + \bar{x}_1}{1 + x_1\bar{x}_1}, \qquad \eta = \frac{2\,\eta_1}{1 + r^2} = -\,i\,\frac{x_1 - \bar{x}_1}{1 + x_1\bar{x}_1},$$

$$\zeta = \frac{1 - r^2}{1 + r^2} = \frac{1 - x_1\bar{x}_1}{1 + x_1\bar{x}_1}.$$

The point p itself is mapped on the point $(0, 1)$ of C^2. If $y = (1, y_1)$ goes into (ξ', η', ζ') then

$$\cos \varepsilon(x, y) = \left[\frac{(1 + x_1\bar{y}_1)\,(1 + \bar{x}_1 y_1)}{(1 + x_1\bar{x}_1)\,(1 + y_1\bar{y}_1)}\right]^{1/2} = \left(\frac{1 + \xi\xi' + \eta\eta' + \zeta\zeta'}{2}\right)^2 = \cos\frac{\theta}{2}$$

if θ is the angle between the rays from the origin to (ξ, η, ζ) and (ξ', η', ζ'), hence $\varepsilon(x, y) = \gamma\,\theta/2$, which proves the assertion.

In order to reduce the hermitian hyperbolic metric in C_H^2 to a known form observe that in $-\bar{x}_0 x_0 + \bar{x}_1 x_1 < 0$ always $x_0 \neq 0$, hence every point may be written in the form $(1, x_1)$. The (transposed) conditions (53.5) become

$$-\bar{a}_{00}a_{10} + \bar{a}_{01}a_{11} = 0, \qquad -\bar{a}_{00}a_{00} + \bar{a}_{01}a_{01} = -1, \qquad -\bar{a}_{10}a_{10} + \bar{a}_{11}a_{11} = 1.$$

The last two imply $a_{00} \neq 0$, $a_{11} \neq 0$, then the first $a_{10}/a_{11} = \bar{a}_{01}/\bar{a}_{00}$. Putting this number equal to a, the last two relations give

$$|a_{00}|^2\,(1 - |a|^2) = |a_{11}|^2\,(1 - |a|^2) = 1$$

hence $|a| < 1$ and $a_{00}/a_{11} = e^{-i\varphi}$. The transformation then becomes

$$x_1' = \frac{a_{10} + a_{11}x_1}{a_{00} + a_{01}x_1} = e^{i\varphi}\,\frac{a + x_1}{1 + \bar{a}x_1}. \tag{53.8}$$

This is the general expression for a linear transformation of x_1, which carries the interior of the unit circle $|x_1| < 1$ into itself. Evaluating the hermitian hyperbolic distance δ for the points $(1, 0)$ and $(1, r)$, $0 \leqslant r \leqslant 1$, yields $\cosh^2 \delta/\gamma = (1 - r^2)^{-1}$, hence $\tanh \delta/\gamma = r$, or

$$\delta = \frac{\gamma}{2} \log \frac{1+r}{1-r} = \frac{\gamma}{2} \log R(0, r, 1, -1)$$

where R means cross ratio. Since cross ratio is invariant under linear transformations and (53.8) carries the diameter of $|x_1| < 1$ through 0 and r into the circle through the images x_1 and y_1 of 0 and r orthogonal to $|x_1| = 1$, we find as general formula for the hyperbolic hermitian distance in C_H^2

$$\delta[(1, x_1), (1, y_1)] = (\gamma/2)\,|\log R(x_1, y_1, x_1^*, y_1^*)|,$$

where x_1^* and y_1^* are the intersections with $|x_1| = 1$ of the circle orthogonal to $|x_1| = 1$ through x_1 and y_1. But this is the Poincaré model of the ordinary real hyperbolic plane.

Thus we have found:

(53.9) *C^2 is with the elliptic hermitian metric* (53.1) *isometric to a sphere of radius $\gamma/2$. The region $-\bar{x}_0 x_0 + \bar{x}_1 x_1 < 0$ is with the hyperbolic hermitian metric* (53.2) *isometric to a hyperbolic plane with the value $\gamma/2$ of the constant.*

THE METRICS IN $C^{2\nu}$

It is now easy to see the main properties of the higher dimensional hermitian spaces.

Because the two spaces possess pairwise transitive groups of motions, and two distinct points determine a complex line, any complex line can be moved into any other. Since the metric in $x_2 = x_3 = \ldots = x_\nu = 0$ is spherical with radius $\gamma/2$ or hyperbolic with constant $\gamma/2$ the same holds for every line.

It follows now from (53.7) that both hermitian spaces are G-spaces. The hyperbolic space is straight. The transformation $x_0' = x_0$, $x_i' = -x_i$ for $i > 0$ satisfies (53.5) and is therefore a motion. It is the symmetry in $(1, 0, \ldots, 0)$. Because the groups of motions are transitive, the symmetry in a given point exists. Thus we have found:

(53.10) *A $2\,\nu$-dimensional* ($\nu > 1$) *hermitian elliptic space is compact, symmetric, and possesses a pairwise transitive group of motions. Every complex line in the space is isometric to the sphere with radius $\gamma/2$. All geodesics are great circles of length $\gamma\pi$.*

This shows that Theorem (31.3) cannot be extended to even dimensional spaces of dimension greater than two.

(53.11) *A* 2ν*-dimensional* $(\nu > 1)$ *hermitian hyperbolic space is straight, symmetric, and possesses a pairwise transitive group of motions. The intersection of a complex line with the space is isometric to the hyperbolic plane with curvature* $-4\gamma^{-2}$.

Thus we see that for all even dimensions greater than two both compact and non-compact spaces exist which possess pairwise transitive groups of motions but do not have constant curvature.

THE MOTIONS OF HERMITIAN ELLIPTIC SPACES

In order to thoroughly understand at least one class of spaces with pairwise transitive groups of motions we discuss briefly the group of motions of an hermitian elliptic space R: the transformations $x_i' = \Sigma a_{ik} x_k$ which leave $(x, y)_1$ invariant, form by definition the unitary group. The determinant $D(a_{ik})$ has the form $e^{i\theta}$. If each a_{ik} is replaced by $a_{ik}\, e^{-i\theta/(\nu+1)}$ then the x_i' go into $x_i'\, e^{-i\theta/(\nu+1)}$ and the point $[x']$ does not change. Hence all the different projectivities among the unitary transformations are contained among those whose determinant equals 1. This subgroup is the so-called unitary unimodular group $\mathfrak{A}_\nu$.

But these are not the only motions of R: if $x' = \Sigma a_{ik} x_k$ leaves $(x, y)_1$ invariant, then $x_i' = \Sigma a_{ik} \bar{x}_k$ does too. (A transformation $x_i' = \Sigma a_{ik} \bar{x}_k$ with $D(a_{ik}) \neq 0$ is called an "*antiprojectivity*".) Now we have found all motions. For it follows from (53.10) that any motion of R carries a complex line into a complex line and is therefore either a projectivity or an antiprojectivity, see for instance Cartan [4]. The projectivities and antiprojectivities that leave $\varepsilon(x, y)$ invariant are readily seen to satisfy (53.5).

It is clear that no motion which is an antiprojectivity can lie in a continuous family of motions containing the identity: Therefore $\mathfrak{A}_\nu$, which is connected (and simply connected), see Weyl [1, pp. 194, 268, 269] represents the component Γ_0 of the identity of the group of motions of R. But $\mathfrak{A}_\nu$ does not represent Γ_0 in a one-to-one manner, because $x' = \lambda x$ with $\lambda = e^{-2i\pi/(\nu+1)}$, $i = 1, 2, \ldots, \nu+1$ is unimodular and corresponds to the identity in Γ_0. Clearly these are the only elements of $\mathfrak{A}_\nu$ which represent the identity of R, and they form the center $\mathfrak{Z}$ of $\mathfrak{A}_\nu$ (see Dieudonné, [1, p. 69]). The factor group $\mathfrak{A}_\nu/\mathfrak{Z}$ is isomorphic to Γ_0.

The matrix of a unitary transformation which leaves (0, . . ., 0, 1) fixed has the form

$$\begin{pmatrix} a_{00} & \cdot & \cdot & \cdot & a_{0,\nu-1} & 0 \\ a_{10} & \cdot & \cdot & \cdot & a_{1,\nu-1} & 0 \\ \cdot & \cdot & \cdot & \cdot & \cdot & \cdot \\ a_{\nu-1,0} & \cdot & \cdot & \cdot & a_{\nu-1,\nu-1} & 0 \\ 0 & \cdot & \cdot & \cdot & 0 & e^{-i\theta} \end{pmatrix}$$

where $M = (a_{ik})_{k=0,\ldots,\nu-1}^{i=0,\ldots,\nu-1}$ is a unitary matrix of order ν. If the transformation is unimodular then $D(M) = e^{i\theta}$, hence the subgroup of $\mathfrak{A}_\nu$ leaving (0, . . ., 0, 1) fixed is isomorphic to the direct product $\mathfrak{D}_1 \times \mathfrak{A}_{\nu-1}$ where $\mathfrak{D}_1$ is the group of rotations of the circle. The center $\mathfrak{Z}$ of $\mathfrak{A}_\nu$ lies in this subgroup. Therefore a subgroup of $\mathfrak{A}_\nu$ which leaves a given point of the space $C^{2\nu}$ fixed, i. e., an isotropic group, is isomorphic to $(\mathfrak{D}_1 \times \mathfrak{A}_{\nu-1})/\mathfrak{Z}$ and the space $C^{2\nu}$ is homeomorphic to the coset space $(\mathfrak{A}_\nu/\mathfrak{Z})/[(\mathfrak{D}_1 \times \mathfrak{A}_{\nu-1})/\mathfrak{Z}]$ or simply to $\mathfrak{A}_\nu/(\mathfrak{D}_1 \times \mathfrak{A}_{\nu-1})$.

The quaternion elliptic spaces

The elliptic and spherical spaces are, in the terminology of Cartan [1], the closed symmetric Riemann spaces whose group of motions (more precisely its identity component) is a simple Lie group and has type B D I and rank 1. The hermitian elliptic spaces are those of type A III, and rank 1, and the quaternion elliptic spaces are those of type C II and rank 1. Why one expects to find the compact spaces with pairwise transitive groups of motions among the *compact symmetric Riemann spaces with rank* 1 *and a simple Lie group as identity component of its group of motions* will be explained in the next section.

The spaces of type C II and rank 1 are defined as follows: The identity component of its group of motions is the symplectic unitary group $\mathfrak{C}_{\nu+1}$, which is defined as the group of homogeneous linear transformations of $2(\nu+1)$ complex variables $x = (x_0, \overset{*}{x}_0, x_1, \overset{*}{x}_1, \ldots, x_\nu, \overset{*}{x}_\nu)$ which leave invariant both the alternating form

$$(x_0\overset{*}{y}_0 - \overset{*}{x}_0 y_0) + (x_1\overset{*}{y}_1 - \overset{*}{x}_1 y_1) + \ldots + (x_\nu\overset{*}{y}_\nu - \overset{*}{x}_\nu y_\nu)$$

in two sets of variables x, y and the hermitian form

$$\bar{x}_0 x_0 + \bar{\overset{*}{x}}_0\overset{*}{x}_0 + \ldots + \bar{x}_\nu x_\nu + \bar{\overset{*}{x}}_\nu\overset{*}{x}_\nu. \tag{53.12}$$

The elements of the space are in Cartan's interpretation the transforms of the set

$$x_0 = \overset{*}{x}_0 = x_1 = \overset{*}{x}_1 = \ldots = \overset{*}{x}_{\nu-1} = 0$$

under $\mathfrak{C}_{\nu+1}$, that is, the elements are complex lines. The isotropic subgroup is then the subgroup of $\mathfrak{C}_{\nu+1}$ which leaves the above line fixed, it is isomorphic to $\mathfrak{C}_1 \times \mathfrak{C}_\nu$ (see l. c., p. 419). This is immediately verified by the same method that determined $\mathfrak{D}_1 \times \mathfrak{A}_{\nu-1}$ as isotropic subgroup in the hermitian elliptic spaces.

Instead of considering the elements of R as sets in a space with complex coordinates, we may interpret them as points in right quaternion projective space $Q^{4\nu}$. A quaternion may be written in the form

$$q = u + i\,u^* + j\,v + k\,v^* = (u + i\,u^*) + j(v - i\,v^*)$$

where u, u^*, v, v^* are real, $i^2 = j^2 = k^2 = -1$ and

$$i\,j = -j\,i = k, \qquad j\,k = -k\,j = i, \qquad k\,i = -i\,k = j.$$

The quaternion $\overline{q}$ conjugate to q is defined by

$$q = u - i\,u^* - j\,v - k\,v^* = (u - i\,u^*) - j(v - i\,v^*)$$

and $|q|^2 = q \cdot \overline{q} = u^2 + u^{*2} + v^2 + v^{*2}$.

If with the above notations we define

$$q_\mu = x_\mu + j\,\overline{x}^*_\mu, \qquad \mu = 0, \ldots, \nu,$$

then the hermitian form (53.12) becomes $\Sigma\overline{q}_\mu q_\mu$. The symplectic unitary transformations which leave $\Sigma\overline{q}_\mu q_\mu$ invariant may be written in the form of linear quaternion transformations

$$q_\mu{}' = \sum_i a_{\mu i}\, q_i.$$

We do not prove this here. The quaternion elliptic distance $\varepsilon(q, r)$ of two points $[q] = [q_0, \ldots, q_\nu]$ and $[r] = [r_0, \ldots, r_\nu]$ is defined in the whole right quaternion projective space $Q^{4\nu}$ by

$$\varepsilon(q, r) = \frac{|\Sigma\overline{q}_\mu r_\mu|}{(\Sigma\overline{q}_\mu q_\mu)^{1/2}\,(\Sigma\overline{r}_\mu r_\mu)^{1/2}}$$

If we translate this back into complex numbers it becomes obvious that this distance remains invariant under $\mathfrak{C}_{\nu+1}$.

We saw above that the elliptic hermitian metric in two dimensions is isometric to the two-dimensional spherical geometry. Algebraically, this comes from the fact that $\mathfrak{A}_1$ and $\mathfrak{B}_1$ are isomorphic, where $\mathfrak{B}_\nu$ denotes generally the orthogonal group in $2\nu + 1$ real variables. Similarly $\mathfrak{C}_2$ and $\mathfrak{B}_2$ are isomorphic, hence it is not surprising that the metric $\varepsilon(q, r)$ in Q^4 is isometric to a 4-dimensional spherical space. This will also follow from a general theorem proved in the next section, see (54.2). Thus the geometry on each quaternion line is spherical, and all geodesics are great circles of the same length.

That the space possesses a pairwise transitive group of motions follows from the fact that it has rank 1: each geodesic is orbit of a one parameter group of motions and each geodesic can be transformed into any other by a motion of the space, see Cartan l. c.

The Cayley elliptic plane

In addition to the classes of compact symmetric Riemann spaces of rank 1 which exist for all n (spherical and elliptic), for all even n (hermitian elliptic), and for all n divisible by 4 (quaternion elliptic) there is a *single 16-dimensional symmetric closed Riemann space of rank* 1 (see Cartan [1, pp. 466]), which has the exceptional simple group $\mathfrak{F}_4$ as identity component of its group of motions. The isotropic subgroup is the group $\mathfrak{B}_4$. The geodesics are all great circles of the same length, and through any point p and a point which has maximal distance from p there are ∞^7 geodesics which form a set isometric to S^8.

This geometry is closely related to the Cayley numbers (see Section 55). It corresponds to an elliptic metrization of a projective "plane" formed with these numbers and may therefore be called the Cayley elliptic plane. As mentioned above, there is no direct analogue to the definitions in hermitian or quaternion projective spaces, because the Cayley numbers do not obey the associative law. We may think of one point p as determined by its isotropic subgroup Γ_p (which is isomorphic to $\mathfrak{B}_4$). If q is a second point and Φ a motion of the space which takes p into q, then $\Gamma_p\Phi$ is the totality of all motions (in $\mathfrak{F}_4$) which take p into q. Thus the points may be considered as the cosets of $\mathfrak{F}_4$ modulo $\mathfrak{B}_4$. Another group Γ_l isomorphic to $\mathfrak{B}_4$ may be considered as the group which carries a certain line into itself (for all this compare Freudenthal [1]), the cosets $\Gamma_l\Psi$ are the other lines. That the point $\Gamma_p\Phi$ and the line $\Gamma_l\Psi$ are incident is equivalent to $\Gamma_p\Phi \cap \Gamma_l\Psi \neq 0$.

The "plane" thus obtained shares with the ordinary projective plane the properties that two points lie on exactly one line and that two lines intersect in exactly one point. Moreover, the standard construction (compare Section 13) determines the *fourth harmonic point uniquely. However, Desargues' Theorem does not hold,* as one might have expected from the non-associativity of the Cayley numbers. This does not contradict the implications listed at the beginning of Section 13, because a "line" is 8-dimensional and does not satisfy the order relations.

The non-validity of Desargues' Theorem explains why we cannot construct Cayley "spaces" and thus obtain similar geometries for all dimensions n divisible by 8. Of course, the exceptional character of the group $\mathfrak{F}_4$ explains this already.

This exhausts the list of *all closed symmetric Riemann spaces of rank* 1 *with a simple Lie group as group of motions,* see Cartan [1, pp. 466, 467]. We mention that Cartan obtains these spaces not under the assumption of symmetry in the large, but under the weaker hypothesis of local symmetry, see Section 49. In Riemann spaces local symmetry is equivalent to the requirement that the curvature belonging to a two-dimensional element be invariant under parallel displacement of this element.

54. Compact spaces with pairwise transitive groups of motions

In contrast to the few results on spaces with transitive groups of motions the G-spaces with pairwise transitive groups of motions have all been determined, all compact and the odd-dimensional non-compact spaces by Wang [2, 3], the even-dimensional non-compact spaces by Tits [1]. Since proofs for the results of Tits are not yet available we restrict ourselves to Wang's papers.

General remarks

We notice first that all results on these spaces hold under an apparently much weaker condition:

(54.1) *If a group of motions Γ of a G-space R has the property that for some positive $\delta(p, q) < \rho(p)$ and any two pairs a, b and a', b' with $ab = a'b' = \delta(a, a')$ a motion exists which takes a into a' and b into b' then Γ is pairwise transitive on R.*

Observe first that the assumption implies transitivity of Γ on R, hence $\rho(p)$ is constant. If now $ab_1 = a'b_1' > \delta(a, a')$ then segments $T(a, b_1)$ and $T(a', b_1')$ contain points b, b' respectively for which $ab = a'b' = \delta(a, a')$. There is a motion Φ in Γ that takes a into a' and b into b'. Since b_1 and b_1'

are the only points x, x' which satisfy the relations $bx = b'x' = ab_1 - \delta(a, a')$, (abx), $(a'b'x')$ the motion Φ takes b_1 into b_1'.

If $ab_1 = a'b_1' < \delta(a, a')$ then we determine b and b' by (ab_1b), $(a'b_1'b')$ and $ab = a'b' = \delta(a, a')$ which is possible because $\delta(a, a') < \rho(a) = \rho(a')$. There is a motion Φ in Γ that takes a into a' and b into b'. Because $ab = a'b' < \rho(a) = = \rho(a')$ the points b_1 and b_1' are the only solutions x, x' of

$$ax = a'x' = ab_1, \quad (axb), \text{ and } (a'x'b')$$

so that $b_1\Phi = b_1'$.

It will turn out that in the compact case not all the assumptions for a G-space are needed, it suffices to know that the space is M-convex, i. e., the last two axioms for a G-space, see Section 8 may be omitted. This rests on two facts, the first of which is the following:

(54.2) *Let R and R^* be two equivalent metrizations of the same set S as finitely compact, M-convex space. Let R possess a pairwise transitive group Γ of motions, which is also a group of motions for R^*. Then the metrics of R and R^* differ only by a constant factor* (so that Γ is also pairwise transitive on R^*).

For a proof we denote the distance of two points a, b in S as points of R by ab and as points of R^* by $\delta(a, b)$. If $ab = a'b'$ then Γ contains a motion Φ of R which takes a into a' and b into b'. Since Φ is also a motion of R^* it follows that $\delta(a, b) = \delta(a', b')$. Thus there is a function $f(\tau)$ such that

$$\delta(a, b) = f(ab), \qquad \text{with} \qquad f(0) = 0, \quad f(\tau) > 0 \qquad \text{for} \qquad \tau > 0.$$

If R is compact then there is a pair c, d whose distance equals the diameter β of R, and since a segment $T(c, d)$ exists (see Section 6) the range of $f(\tau)$ is the interval $0 \leqslant \tau \leqslant \beta$. If R is not compact then it is, as a finitely compact space, unbounded and segments of arbitrary length exist, so that $0 \leqslant \tau < \infty$ is the range for $f(\tau)$.

Let $x(\tau)$ represent a segment in R for $0 \leqslant \tau \leqslant \beta$, where β is the diameter of R in the compact case, and an arbitrary positive number in the non-compact case. Then

$$f(\tau) = \delta\big(x(0), x(\tau)\big)$$

and $f(\tau)$ is continuous because $\tau_\nu \to \tau_0$ implies $x(\tau_\nu) \to x(\tau_0)$, and the equivalence of the two metrics yields

$$f(\tau_\nu) = \delta\big(x(0), x(\tau_\nu)\big) \to \delta\big(x(0), x(\tau_0)\big) = f(\tau_0).$$

We also have

$$(54.3)\qquad f(\tau_2)-f(\tau_1)\leqslant f(\tau_2-\tau_1)\quad \text{if}\quad \tau_2>\tau_1$$

because

$$\delta\big(x(0),x(\tau_2)\big)-\delta\big(x(0),x(\tau_1)\big)\leqslant\delta\big(x(\tau_1),x(\tau_2)\big)=f(\tau_2-\tau_1).$$

The function $f(\tau)$ increases. For if $\tau_2>\tau_1$ then on a segment T^* in R^* that connects $x(0)$ to $x(\tau_2)$ the function $x(0)y$, $y\in T^*$, must attain the value τ_1 at some interior point y^* of T^*. Hence

$$f(\tau_2)=\delta\big(x(0),x(\tau_2)\big)>\delta\big(x(0),y^*\big)=f(\tau_1).$$

It follows that $\delta(a,b)=\delta(a',b')$ only if $ab=a'b'$, so that Γ is also pairwise transitive as group of motions of R^*. The inverse function $g(t)$ of $f(\tau)$ is defined, and since now the relation between R and R^* is entirely symmetric, the relation

$$g(t_2)-g(t_1)\leqslant g(t_2-t_1)\quad \text{for}\quad t_2>t_1$$

analogous to (54.3) must hold. On the other hand, if $g(t_i)=\tau_i$ or $t_i=f(\tau_i)$, then the last inequality yields, because $f(\tau)$ increases,

$$f(\tau_2-\tau_1)\leqslant t_2-t_1=f(\tau_2)-f(\tau_1),$$

which proves with $f(0)=0$ and $f(\tau)>0$ for $\tau>0$ that $f(\tau)$ has for $0\leqslant\tau\leqslant\beta$ the form $k\tau$, $k>0$. Since, in the non-compact case, β was arbitrary the assertion follows.

The Riemannian character of the metric in the compact case

We now prove:

(54.4) *A compact, M-convex metric space R which possesses a pairwise transitive group of motions is a Riemann space.*

If R is a G-space the assertion follows at once: By (52.4) the group Γ of all motions of R is a Lie group, and a well known result of Cartan [3, p. 43] states that there is a Riemannian metrization of R invariant under Γ. By the preceding result the metric of R differs from this Riemannian metric only by a constant factor and is therefore itself Riemannian.

The extension of this result to M-convex spaces rests on the following result of Wang [3]. It may be omitted by those interested only in G-spaces.

(54.5) **Theorem.** *If the compact M-convex metric space R possesses a pairwise transitive group Γ of motions, then every non-discrete, closed, normal subgroup Δ of Γ is transitive on R.*

We show first: given a point p and a positive ε' there is a positive $\varepsilon < \varepsilon'$ such that $K(p, \varepsilon) \subset 0(p, \Delta)$.

The orbit $0(p, \Delta)$ is either finite or dense in itself. Because Δ is normal $0(p\Psi, \Delta) = 0(p, \Delta)\,\Psi$ for any Ψ in Γ. Since Γ is transitive we conclude first that $0(p, \Delta)$ cannot consist of one point, since then $0(p\Psi, \Delta)$ would equal $p\Psi$ and Δ would consist of the identity only. Also $0(p, \Delta)$ cannot consist of a finite number of distinct points $p = p_1, p_2, \ldots, p_k$, $k > 1$. For since Δ is non-discrete, there is an element Φ in Δ such that $0 < \delta(E, \Phi) < m = \min_{i \neq j} p_i p_j$, where $\delta(E, \Phi) = \sup_{x \in R} xx\Phi$. Since Φ transforms $0(p, \Delta)$ into itself it must leave every p_i fixed. Because $0(p\Psi, \Delta) = 0(p, \Delta)\,\Psi$ for $\Psi \in \Gamma$ and $p_i\Psi p_j\Psi = p_i p_j \geqslant m$ the motion Φ also leaves every point of $0(p\Psi, \Delta)$ fixed, and Γ being transitive, Φ would be the identity, which contradicts $\delta(E,\Phi) > 0$.

Since $0(p, \Delta)$ is dense in itself it contains a point x with $0 < px < \varepsilon'$. Put $px = \varepsilon$ and let $y \in K(p, \varepsilon)$. Because Γ is pairwise transitive it contains a motion Ψ with $p\Psi = p$ and $x\Psi = y$. The relation $x \in 0(p, \Delta)$ means the existence of a Φ in Δ such that $p\Phi = x$. Then

$$y = p\Psi^{-1}\Phi\Psi \in 0(p, \Delta)$$

because Δ is normal, hence $K(p, \varepsilon) \subset 0(p, \Delta)$.

For the proof of (54.5) we define for any set M in R

$$0(M, \Delta) = \bigcup_{\Phi \in \Delta} M\Phi = \bigcup_{x \in M} 0(x, \Delta).$$

Then

$$0(M \cup M', \Delta) = 0(M, \Delta) \cup 0(M', \Delta).$$

By the first part of this proof there is for an arbitrary point p a decreasing sequence of positive ε_i with $\varepsilon_i \to 0$ and $K(p, \varepsilon_i) \subset 0(p, \Delta)$.

If

$$S_i = S(p, \varepsilon_i), \qquad \overline{S}_i = S_i \cup K(p, \varepsilon_i)$$

then $0(\overline{S}_i, \Delta)$ is closed since Δ is closed and $0(S_i, \Delta)$ is open because S_i is open. But

$$0(\overline{S}_i, \Delta) = 0(S_i, \Delta) \cup 0\big(K(p, \varepsilon_i), \Delta\big) \subset (S_i, \Delta) \cup 0(p, \Delta) = 0(S_i, \Delta)$$

so that $0(\overline{S}_i, \Delta) = R$ because R is connected.

Let x be any point of R. Since $x \in 0(S_i, \Delta)$, the sets $0(x, \Delta)$ and $\overline{S}_i$ intersect for each i, so that $0(x, \Delta) \cap \overline{S}_i$ is a decreasing sequence of non-empty compact sets. The intersection is not empty and is therefore the point p. Thus $p \in 0(x, \Delta)$ or $x \in 0(p, \Delta)$ which shows that Δ is transitive on R.

We now turn to the proof of (54.4): there is a decreasing sequence of closed normal subgroups $\Delta_1, \Delta_2, \ldots$ of Γ whose intersection is the identity and such that Γ/Δ_i is a Lie group (see Pontrjagin [1, Theorem 53], where subgroups of topological groups are by definition closed). The diameter of Δ_i is therefore from a certain subscript j on, smaller than the diameter β of R. It follows from (54.5) that Δ_i is discrete for $i \geqslant j$, hence finite. Put $\eta = \min_{\Phi \in \Delta_j - E} \delta(E, \Phi)$. The group Δ_{i_0} lies for a sufficiently large $i_0 > j$ in the neighborhood $\delta(E, \Psi) < \eta$ of E, and thus consists of E alone. Therefore $\Gamma/\Delta_{i_0} = \Gamma$ is a Lie group. The remainder of the proof runs in the same way as for G-spaces.

Under the assumptions of (52.4) spheres $K(p, \rho)$ with a positive radius $\rho < \rho(p)$ (notice that $\rho(p)$ is constant) are homeomorphic to the $(n-1)$-dimensional sphere if the space has dimension n.

Determination of all compact spaces with pairwise transitive groups of motions

The problem of finding all G-spaces with pairwise transitive groups of motions was completely solved in the compact case by Wang [3] who showed that they are the symmetric Riemann spaces of rank 1 whose groups of motions have simple Lie groups as identity components. However, Wang does not establish these properties directly, he obtains his result, so to say, by eliminating all other possibilities. Here we will aim directly at these properties.

Let R be a compact Riemann space of dimension $n > 2$ with a pairwise transitive group of motions. Denote by Γ' the group of all motions of R. We decompose the proof into several steps

(a) *If Γ_1 is a non-discrete normal subgroup of Γ' then every geodesic is orbit of a suitable one-parameter subgroup of Γ_1.*

All geodesics of R are closed and without multiple points.

Let Φ_τ be a one-parameter subgroup of Γ_1 (which may coincide with Γ') which is isomorphic to $\mathfrak{D}_1{}^1$. By (52.2) there is at least one geodesic $x(\tau)$ which is an orbit of Φ_τ: for a suitable $\alpha > 0$ we have $x(0)\,\Phi_\tau = x(\alpha\tau)$. Since Γ' is pairwise transitive it contains for a given geodesic $y(\tau)$ a motion Ψ such that

$x(\tau)\,\Psi = y(\tau)$. Because Γ_1 is a normal subgroup of Γ' the motion $\Psi^{-1}\Phi_\tau\Psi$ lies in Γ_1 and $y(0)\Psi^{-1}\Phi_\tau\Psi = x(0)\Phi_\tau\Psi = x(\alpha\tau)\ \Psi = y(\alpha\tau)$, hence $y(\tau)$ is also an orbit of a one-parameter subgroup of Γ_1.

Because Φ_τ is isomorphic to $\mathfrak{D}_1$ the geodesic $x(\tau)$ is closed and as orbit it cannot have multiple points, see (9.4). The same holds then for $y(\tau)$.

We notice in passing that this implies the transitivity of Γ_1 on R, a result proved above for M-convex spaces.

(b) *The identity component Γ of Γ' is pairwise transitive on R.*

Since Γ is a normal subgroup of Γ' it is, as we just saw, transitive on R. Let Δ' be the isotropic subgroup of Γ' belonging to an arbitrary but fixed point p. Because Δ' acts transitively on $K = K(p, \rho)$, $0 < \rho < \rho(p)$, which is homeomorphic to S^{n-1} its identity component Δ acts transitively on K and lies in Γ.

If now $ab = a'b' > 0$, let $x(\tau)$, $x'(\tau)$ represent segments of length β, where β is the diameter of R (compare (48.9), p. 337) with $x(0) = a$, $x(ab) = b$, $x'(0) = a'$, $x'(ab) = b'$. There are motions Φ, Ψ in Γ with $a\Phi = p$, $p\Psi = a'$. If $x(\rho)\,\Phi = a''$, $x'(\rho)\,\Psi^{-1} = b''$ then a'', b'' lie on K, hence a suitable Ω in Δ takes a'' into b''. Then $\Phi\Omega\Psi$ lies in Γ and takes a into a' and b into b'.

(c) *Γ is simple.*

If Γ were not simple it would contain a proper normal non-discrete subgroup Γ_1 of lower dimension. The group Γ has a covering group $\bar{\Gamma}$ which is the direct product of compact semi-simple groups and of a compact commutative group (see Pontrjagin [1, p. 285]). The totality of elements in $\bar{\Gamma}$ which lie over elements of Γ_1 is a normal subgroup $\bar{\Gamma}_1$ of $\bar{\Gamma}$ and $\bar{\Gamma}_1$ is a direct factor of $\bar{\Gamma}$.

Any element of a one-parameter subgroup $\bar{\Phi}_\tau$ of $\bar{\Gamma}$ isomorphic to $\mathfrak{D}_1$ which does not lie in $\bar{\Gamma}_1$ commutes with every element in $\bar{\Gamma}_1$. Let Φ_τ be the image of $\bar{\Phi}_\tau$ in Γ. Then Φ_τ has an orbit $x(\alpha\tau) = x(0)\,\Phi_\tau$ which is a closed geodesic. Put $z = x(0)$. Let $y(\tau)$ represent any geodesic in R different from $x(\tau)$ with $y(0) = z$. By (a) the geodesic $y(\tau)$ is orbit of a one parameter subgroup Ψ_τ of Γ_1: $y(\beta\tau) = z\Psi_\tau$. The motions Φ_τ and Ψ_σ commute for any τ, σ. The subset W of R consisting of the points $z\Phi_\tau\Psi_\sigma$ is then intrinsically isometric to a torus with a euclidean metric. The proof is the same as that of Theorem (50.1). "*Intrinsically isometric*" means, that, while the distances in W are only locally the same as in R, the set U will in general not be flat, but the geodesics on W are geodesics in R. Clearly W contains open geodesics which contradicts (a).

(d) *The rank of R equals* 1.

This is contained in the proof just given: The rank of the space (to be clearly distinguished from the rank of Γ) is the maximal dimension of a torus which can be generated in the way W was generated above. We need not go into the exact definition: the rank is at least 1, and as soon as it is greater than 1, there are manifolds in the space which have dimension greater than 1 and which are intrinsically tori with a euclidean metric, see Cartan [1].

(e) *The local spheres.*

A Riemann space is locally euclidean and a Finsler space is locally Minkowskian. In many cases it is useful to think of the local euclidean or Minkowskian geometry as imbedded in the given space E (we did this essentially in the proof of (41.12)):

Let p be any point of E. In $U = S(p, \rho(p))$ we define a new distance as follows: if a, b are two distinct points in U, consider the segments $x(\tau)$, $y(\tau)$, $0 \leqslant \tau \leqslant \rho(p)$, with $x(0) = y(0) = p$, which contain a and b respectively. We put

$$\mu(a, b) = \lim_{\tau \to 0+} \frac{x(\tau/pa)\, y(\tau/pb)}{\tau}.$$

In case $a = p$ we put $\mu(p, b) = pb$. Then $\mu(a, b)$ is under the usual (actually under very weak) differentiability hypotheses a Minkowskian metric; it is closely related to the introduction of normal coordinates and may be called the "*normal Minkowski metric belonging to* p".

It is clear that any motion of E which leaves p fixed induces a motion of $S(p, \rho(p))$ with the metric $\mu(a, b)$ and is an affinity of the affine space (obtained by extension of $S(p, \rho(p))$ for which the Minkowskian lines are the affine lines.

Although not needed for the present proof we notice the following consequence for later use:

(54.6) *If in a Finsler space E the motions which leave p fixed are transitive on $K = K(p, \rho)$, $0 < \rho < \rho(p)$, then the normal Minkowskian metric at p is euclidean.*

For then the motions of E which leave p fixed induce a group of affinities which leave p fixed and are transitive on K; hence K is by (16.11) an ellipsoid.

(f) *The space is symmetric.*

We return to the present case. Since Γ is pairwise transitive its isotropic subgroup Δ belonging to p is transitive on K. Therefore Δ induces a transitive group Δ_E of motions of the euclidean sphere K of dimension $n-1$. But all transitive groups of motions of spheres are known from the two papers Montgomery and Samelson [1] and Borel [1]. They are the following:

If $n-1$ is even Δ_E must be the group $\mathfrak{B}_{(n-1)/2}$ of all proper rotations of K, with the exception of $n-1=6$ where in addition to $\mathfrak{B}_3$, the exceptional group $\mathfrak{G}_2$ is a transitive group of motions. This group will be discussed in detail in the next section, where we will show that it is pairwise transitive on S^6. Since $\mathfrak{B}_{(n-1)/2}$ is also pairwise transitive, we see that for all odd n the group Δ_E contains for any two two-dimensional planes through p in the normal euclidean space at p a motion that takes the first plane into the second. The space R has therefore constant curvature and is locally symmetric, hence it is *spherical or elliptic*. This is, of course, also obvious without using symmetric spaces.

If $n-1$ is odd then there is a much greater variety of groups Δ_E that act transitively on K. They have the form $\mathfrak{H}$, $\mathfrak{D}_1 \times \mathfrak{H}$, $\mathfrak{C}_1 \times \mathfrak{H}$, where $\mathfrak{H}$ may be of the following types:

(a) If $n \equiv 2 \bmod 4$ then $\mathfrak{H} = \mathfrak{D}_{n/2}$ or $\mathfrak{H} = \mathfrak{A}_{n/2-1}$.

(b) If $n \equiv 0 \bmod 4$ then we have $\mathfrak{H} = \mathfrak{C}_{n/4}$ besides (a).

(c) If $n = 8$ then $\mathfrak{H} = \mathfrak{B}_3$ in addition to (a) and (b).

(d) If $n = 16$ then $\mathfrak{H} = \mathfrak{B}_4$ in addition to (a) and (b).

We met all these groups before except $\mathfrak{D}_\nu$ which is the (proper) orthogonal group in 2ν variables. Now $\mathfrak{D}_{n/2}$ and $\mathfrak{C}_{n/4}$ contain the reflection of K in p, or the antipodal mapping of K on itself. The same is true for $\mathfrak{A}_{n/2-1}$ if $n/2$ is even or $n \equiv 0 \bmod 4$, and also for $\mathfrak{B}_4$ because it occurs as isotropic group in the Cayley plane which is symmetric. (This argument is far fetched but it saves us the trouble of studying the action of $\mathfrak{B}_4$ on S^{15}.) In all these cases the space R is therefore symmetric.

There remain only the cases of $\mathfrak{B}_3$, $\mathfrak{B}_3 \times \mathfrak{D}_1$, $\mathfrak{B}_3 \times \mathfrak{C}_1$ acting on S_7 and $\mathfrak{A}_{n/2-1}$ acting on S^{n-1} for $n \equiv 2 \bmod 4$, for we know from our discussion of the hermitian elliptic spaces that $\mathfrak{A}_{n/2-1} \times \mathfrak{D}_1$, hence also $\mathfrak{A}_{n/2-1} \times \mathfrak{C}_1$ (which does not really occur) contains the antipodal mapping. To show that, apart from one exception, these cases cannot occur we use the fact that the group Γ of motions of R is simple and hence is one of the groups $\mathfrak{A}_\nu$ of dimension $\nu(\nu+2)$, $\mathfrak{B}_\nu$ of dimension $\nu(2\nu+1)$, $\mathfrak{C}_\nu$ of dimension $\nu(2\nu+1)$, $\mathfrak{D}_\nu$ of dimension $\nu(2\nu-1)$, or of the exceptional groups $\mathfrak{G}_2$, $\mathfrak{F}_4$, $\mathfrak{E}_6$, $\mathfrak{E}_7$, $\mathfrak{E}_8$ with $\dim \mathfrak{G}_2 = 14$, $\dim \mathfrak{F}_4 = 54$, $\dim \mathfrak{E}_6 = 78$, $\dim \mathfrak{E}_7 = 133$, $\dim \mathfrak{E}_8 = 248$. The subscripts denote everywhere the rank of the corresponding group.

Although many cases can be eliminated a priori on the basis of our knowledge concerning these groups, comparison of the dimensions proves so simple that we use exclusively this method. We have

$$\dim \Gamma = \dim R + \dim \Delta = n + \dim \Delta_E.$$

If $\Delta_E = \mathfrak{A}_\nu$ then $2\nu + 1 = n - 1$, because $\mathfrak{A}_\nu$ acts on $S^{2\nu+1}$ and $2\nu = n - 2 \equiv 0$ mod 4 so that ν is even.

$$\dim \Gamma = 2\nu + 2 + \nu(\nu + 2) = \nu^2 + 4\nu + 2 = K_\nu.$$

We notice that $K_\nu + 2$ is a square. Since $\dim \mathfrak{A}_\mu + 2 = \mu(\mu + 2) + 2$ is never a square, Γ cannot be an $\mathfrak{A}_\mu$.

The same remark eliminates for Γ all the exceptional groups except $\mathfrak{G}_2$. In the case of $\mathfrak{G}_2$ we find from $\dim \mathfrak{G}_2 = 14$ that $\nu = 2$, $n = 6$. This is the exceptional case mentioned above, but we will see in the next section that the geometry is spherical or elliptic.

In the remaining cases we use that the rank μ of Γ must at least equal the rank ν of $\mathfrak{A}_\nu$. This leads for $\mathfrak{B}_\mu$ and $\mathfrak{C}_\mu$ to

$$\mu(2\mu + 1) = \nu^2 + 4\nu + 2 \qquad \text{with} \qquad \mu \geqslant \nu.$$

But for $\mu = 2, 3$ the number $\mu(2\mu + 1) + 2$ is no square and

$$2\mu^2 + \mu > \nu^2 + 4\nu + 2, \qquad \text{for} \qquad \mu \geqslant 4.$$

The argument is similar for $\mathfrak{D}_\mu$. There we have to satisfy

$$\mu(2\mu - 1) = \nu^2 + 4\nu + 2, \qquad \text{with} \qquad \mu \geqslant \nu$$

and observe that $\mu(2\mu - 1) > \nu^2 + 4\nu + 2$ if $\mu \geqslant 6$ and that $\mu(2\mu - 1) + 2$ is no square for $\mu = 2, 3, 4, 5$.

Finally we turn to $\mathfrak{B}_3$ acting on S^7. There $\dim \Gamma = n + \dim \mathfrak{B}_3 = 29$. The dimensions of $\mathfrak{A}_\nu$, $\mathfrak{B}_\nu$, $\mathfrak{C}_\nu$, $\mathfrak{D}_\nu$ are not prime numbers which excludes $\mathfrak{B}_3$ alone. $\mathfrak{B}_3 \times \mathfrak{D}_1$ is also impossible because then $\dim \Gamma = 30$ and 30 does not have the form $\nu(2\nu + 1)$, $\nu(2\nu - 1)$ or $\nu(\nu + 2)$. Similarly $\mathfrak{B}_3 \times \mathfrak{C}_1$ leads to $\dim \Gamma = 32$ which does not have any of these three forms. In the case of the exceptional groups we notice that none has the dimension 29, 30, or 32.

Thus, anticipating the fact regarding $\mathfrak{G}_2$, and using the list of Cartan [1] quoted in the last section we have proved Wang's result:

(54.7) ***Theorem.*** *A compact M-convex space with a pairwise transitive group of motions is spherical, elliptic, hermitian elliptic, quaternion elliptic or the Cayley elliptic plane.*

In none of these spaces is the geodesic through two points unique except in the elliptic spaces. Therefore we have the corollary:

(54.8) *The elliptic spaces are the only compact G-spaces with pairwise transitive groups of motions in which the geodesic through two points is unique.*

55. Odd-dimensional spaces with pairwise transitive groups of motions

There are various reasons why the non-compact spaces are much more difficult to treat; one is that the groups of motions are not compact, and we know much less about non-compact Lie groups. Another reason, which follows from this, is that we cannot establish the Riemannian character of the metric a priori, for this conclusion was essentially based on the result of Cartan [3, p. 43] which holds only for compact spaces. But even if we know that the metric is Riemannian there is no simple analogue to the conclusion that the rank of the space must equal 1, because the tori become now euclidean planes, and all geodesics are congruent.

If a Finsler space has a pairwise transitive group of motions, then we conclude from (54.6) that it is Riemannian and from Myers-Steenrod [1] that its group of motions is a Lie group. If the space has an odd dimension then the local euclidean spheres are even dimensional and we conclude exactly as in the compact case that the space has constant curvature (using the result on $\mathfrak{G}_2$ which is derived at the end of this section). Therefore we have here *a case where differentiability hypotheses mean a very considerable simplification.* The reader who is satisfied with this case may omit everything preceding the discussion of Cayley numbers.

The space is a manifold

We know that a non-compact G-space with a pairwise transitive group of motions is straight. We are going to show that any such space is a manifold, provided its dimension is finite[1]. We observe first

(55.1) *A sphere $K(p, \rho)$ with $0 < \rho < \rho(p)$ of a G-space of dimension greater than* 1 *is arcwise connected and locally arcwise connected.*

If x, y are any two distinct not diametrically opposite points of $K(p, \rho)$ define x', y' by $(px'x)$ and $(py'y)$ and $px' = py' = \rho/3$. Then $T(x', y') \subset S(p, \rho)$. Because x and y are not diametrically opposite no point $u' \in T(x', y')$ coincides with p, so that there is a point u with $(pu'u)$ and $pu = \rho$. The points u form obviously an arc from x to y on $K(p, \rho)$. If x and y are diametrically opposite we merely have to choose any third point z of $K(p, \rho)$, which exists because the dimension is greater than 1, and connect x with z and z with y.

The continuity properties of geodesics imply that for a given $\varepsilon > 0$ a number $\delta > 0$ exists such that for any two points x, y with $xy < \delta$ the above constructed arc lies in $S(x, \varepsilon)$.

Keeping (52.4) in mind we now prove:

(55.2) *A finite dimensional*[1] *non-compact G-space with a pairwise transitive group of motions is a topological manifold.*

Consider a sphere $K = K(p, \rho)$ (here $\rho > 0$ is arbitrary because the space is straight). The isotropic subgroup Γ_p of the group of all motions of the space acts transitively and, as we saw in the last section, effectively on K which is finite dimensional, compact, connected and locally connected. It follows from Montgomery-Zippin [1, Theorem 11 and Corollary 6′] that Γ_p is a Lie group, hence K is a manifold.

The set $U = S(p, \rho) - p$ is the topological product of K and a straight line, hence a manifold. It follows that R is a manifold, and if $\dim R = n$, then $\dim K = n - 1$.

ODD-DIMENSIONAL SPACES

For odd dimensions we now establish the following analogue to Theorem (54.7) which is found in Wang [2]:

(55.3) **Theorem.** *A G-space whose dimension is finite*[1] *and odd (or two) and which possesses a pairwise transitive group of motions is euclidean, hyperbolic, spherical or elliptic.*

Because of (48.9) and Theorem (54.7) we may assume that the space is straight. In addition we assume, of course, $\dim R = n > 2$. We begin by showing that the spheres $K = K(p, \rho)$ are homeomorphic to S^{n-1}. We know already from the preceding proof that K and $U = S(p, \rho) - p$ are manifolds. Since $S(p, \rho)$ can be contracted to p along its radii, U has for $i < n - 1$ the same homotopy groups π_i as $S(p, \rho)$ so that $\pi_i(U) = 0$ for $i < n - 1$. Because U is the topological product of K and a straight line we conclude, that

$$\pi_i(K) = 0, \qquad i = 1, 2, \ldots, n-2.$$

Thus K is a simply connected homology sphere of the even dimension $n - 1$. It follows now from another result of Wang [1] that K is homeomorphic to S^{n-1}.

With the notation of the proof of (55.2) the identity component Γ_p^0 of Γ_p is a connected compact Lie group acting effectively and transitively on K. Since $\dim K = n-1$ is even, Γ_p^0 is by Wang [1] isomorphic either to the group $\mathfrak{D}_{(n-1)/2}$ of the rotations of the $(n-1)$-sphere or to the exceptional group $\mathfrak{G}_2$ which acts on S^6.

But Γ_p^0 is not only isomorphic, but equivalent to $\mathfrak{D}_{(n-1)/2}$ or $\mathfrak{G}_2$, for $\mathfrak{D}_{(n-1)/2}$ see Montgomery-Samelson [1, Lemma 1], for $\mathfrak{G}_2$ see Wang [1, Lemma 6]. Equivalence is defined as follows:

The topological transformation groups Γ and Δ acting on the same space M are *equivalent* if a topological mapping Ψ of M on itself exists such that $\Psi^{-1}\Gamma\Psi = \Delta$, i. e., Δ is identical with the set of mappings $\Psi^{-1}\Phi\Psi$ with $\Phi \in \Gamma$. Isomorphic groups acting on the same space are not always equivalent.

We know that $\mathfrak{D}_{(n-1)/2}$ is with the ordinary spherical distance on S^{n-1} pairwise transitive on S^{n-1}, and we are going to prove the same for $\mathfrak{G}_2$ and S^6. Taking the latter fact at the moment for granted and denoting $\mathfrak{D}_{(n-1)/2}$ or $\mathfrak{G}_2$ by Γ we now finish the proof of (54.7) as follows:

Since Γ_p^0 is equivalent to Γ there is a topological mapping Ψ of K on itself such that $\Psi^{-1}\Gamma_p^0\Psi = \Gamma$. If $\sigma(x, y)$ is a spherical distance on K invariant under Γ and we define $\varepsilon(x, y) = \sigma(x\Psi, y\Psi)$ then $\varepsilon(x, y)$ is invariant under Γ_p^0 because for $\Phi \in \Gamma_p^0$

$$\varepsilon(x\Phi, y\Phi) = \sigma(x\Psi\Psi^{-1}\Phi\Psi, y\Psi\Psi^{-1}\Phi\Psi) = \sigma(x\Psi, y\Psi) = \varepsilon(x, y).$$

The metric $\varepsilon(x, y)$ is, of course, also spherical and has Γ_p^0 as pairwise transitive group of motions. Now Γ_p^0 is also a group of motions with respect to the metric xy on K given by the original space R. We cannot apply (54.2) because xy is as metric on K not convex. But we conclude again that a function $f(\tau)$ exists such that

$$xy = f(\varepsilon(x, y)), \qquad f(\tau) = 0, \qquad f(\tau) > 0 \quad \text{for} \quad \tau > 0.$$

If $xy = x'y'$ only for $\varepsilon(x, y) = \varepsilon(x', y')$ then Γ_p^0 will also be pairwise transitive on K for the metric xy.

We have to show that $f(\tau)$ takes every value only once. If $\varphi = f(\tau) = f(\sigma)$ with $\tau > \sigma > 0$, then for a given point $q \in K = K(p, \rho)$ the spheres $H_\tau : \varepsilon(q, x) = \tau$ and $H_\sigma : \varepsilon(q, x) = \sigma$ on K would both lie on $qx = \varphi$. The sphere H_σ separates p from H_τ on K. The cone V formed by the rays with origin p and containing a point x of H_σ separates H_τ from q. Therefore a segment $T(q, y)$ with $y \in H_\tau$ must cross V at some point u. Let x be the point of H_σ on the generator of V through u. Then the two triangles pqx and pqy would satisfy the relations

$px = py = \rho$, $qx = qy = \varphi$ and $T(q, y)$ would intersect $T(p, x)$ or its prolongation beyond x at u, which is impossible according to the proof of (10.11).

Thus we see that Γ_p^0 is pairwise transitive on K. It follows readily that for any two pairs a, b and a', b' with $ab = a'b'$ a motion in Γ_p^0 (now a motion of R) exists that takes a into a' and b into b'. For the segments from p to a, b, a', b' or their prolongations intersect K in points $\bar{a}, \bar{b}, \bar{a}', \bar{b}'$ respectively. If $\bar{a}\bar{b} > \bar{a}'\bar{b}'$ then the segments $T(p, x)$ with $x \in T(a, b)$ or their prolongations beyond x would intersect K in an arc from a to b which contains a point a^* with $a^* \bar{b} = \bar{a}'\bar{b}'$. There is a motion in Γ_p^0 that takes $\bar{a}'$ into a^* and $\bar{b}'$ into $\bar{b}$. It takes b' into b and a' into a point a'', for which $T(a, b)$ intersect $T(p, a'')$ or its prolongation beyond a''. The two triangles pba and pba'' are impossible for the same reason as before. Thus $\bar{a}\bar{b} = \bar{a}'\bar{b}'$ and the motion in Γ_p^0 that takes $\bar{a}$ into $\bar{a}'$ and $\bar{b}$ into $\bar{b}'$ also takes a into a' and b into b'.

Thus R satisfies the hypothesis of Theorem (48.10) and Theorem (55.3) is proved except for the assertion regarding the pairwise transitivity of $\mathfrak{G}_2$ on S^6.

The Cayley numbers

The group $\mathfrak{G}_2$ is closely connected to the Cayley numbers. We list their properties, as far as they are needed here and refer for proofs to Freudenthal [1], quoted as F, which is not very accessible but concentrates just on the questions which interest us here.

While Cayley arrived at his numbers more or less by experimenting, we can now obtain them axiomatically together with the real numbers, complex numbers and quaternions, which elucidates the algebraic side of the results obtained in the last section.

We consider a *linear algebra* $\boldsymbol{H}$ over the real numbers, i. e., a finite dimensional linear vector space over the real numbers, in which *a product* xy is defined for any two elements in $\boldsymbol{H}$, lies in $\boldsymbol{H}$, and satisfies the following *distributive law*

$$(rx + sy)(r'x' + s'y') = rr' \cdot xx' + rs' \cdot xy' + sr' \cdot yx' + ss' \cdot yy',$$

where r, s, r', s' are real numbers and x, y, x', y' elements of $\boldsymbol{H}$.

We also assume that $\boldsymbol{H}$ has a *unit element* e_0 such that $xe_0 = e_0 x = x$ for any x in $\boldsymbol{H}$, but we assume neither the commutative nor the associative law for multiplication.

Finally we postulate the existence of a *real, symmetric inner product* (x, y) which is defined for any two elements of $\boldsymbol{H}$, is linear: $(rx + sy, z) = r(x, z) +$

$+ s(y, z)$ for real r, s and any x, y, z in $\boldsymbol{H}$, is positive definite: $(x, x) > 0$ for $x \neq 0$, and satisfies $|xy| = |x| \cdot |y|$, where $|x| = +(x, x)^{1/2}$.

These postulates imply, see F, p. 5.,

(a) $(ax, ay) = (xa, ya) = (a, a)(x, y)$ *for any* x, y, a *in* $\boldsymbol{H}$.

If the dimension of $\boldsymbol{H}$ is n, and we form an orthonormal base $(e_0, e_1, \ldots, e_{n-1})$ (notice that the first element is the unit) which means

(b) $(e_i, e_j) = 0$ *if* $i \neq j$, $(e_i, e_i) = 1$,

then the e_i satisfy the relations

(c) $e_i^2 = -e_0$ *if* $i > 0$, $e_ie_j = -e_je_i$ *if* $i \neq j$, $i, j > 0$

(d) $e_i(e_ix) = -x$ *if* $i > 0$

(e) $e_i(e_je_k) = -e_j(e_ie_k) = e_k(e_ie_j)$ *if* $i \neq j$, $i \neq k$ $i, j, k > 0$.

The base may in addition be chosen such that each product e_ie_j is, except possibly for sign, again an e_k. The advantage is a multiplication table which allows us to read off the product of two arbitrary numbers

$$x = \sum_{i=0}^{n-1} r_ie_i \quad \text{and} \quad y = \sum_{i=0}^{n-1} s_ie_i.$$

For any orthonormal base

$$(x, y) = \sum_{i=0}^{n-1} r_is_i$$

is an inner product (but there may be others). All properties are obvious, except perhaps $|xy| = |x| \cdot |y|$ which follows from (a) and (b):

$$|xy|^2 = \left(\sum (r_ie_i)\, y, \sum (r_je_j)\, y\right) = \sum r_ir_j(e_iy, e_jy)$$
$$= \sum r_ir_j(e_i, e_j)\, |y|^2 = \sum r_i^2\, |y|^2 = |x|^2\, |y|^2.$$

This relation is an identity of the form

$$\sum_{i=0}^{n-1} x_i^2 \sum_{i=0}^{n-1} y_i^2 = \sum_{i=0}^{n-1} z_i^2,$$

where the z_i are bilinear forms in the x_i and y_i. It was shown by Hurwitz [1] with matrix theory that such an identity can exist only for $n = 1, 2, 4$, or 8.

Thus the only systems $\boldsymbol{H}$ satisfying all our conditions are the real numbers, the complex numbers, the quaternions, and the Cayley numbers. A very elegant proof of Hurwitz' Theorem is found in F, pp. 7, 8.

The quaternions may be written in the standard form

$$x = r_0 + r_1 i + r_2 j + r_3 k$$

with multiplication table

$$i^2 = j^2 = k^2 = -1$$

$$ij = k, \quad ji = -k, \quad jk = i, \quad kj = -i, \quad ki = j, \quad ik = -j$$

and this is, except for trivial variations, the only multiplication table.

The *Cayley numbers* may be written in the form

$$x = r_0 + \sum_{i=1}^{7} r_i e_i,$$

but there are non-trivial variations between the different multiplication tables. We describe one *process of forming a basis* which will prove useful later:

We choose e_1 and e_2 arbitrarily such that $(e_0, e_1) = (e_0, e_2) = (e_1, e_2) = 0$ and $|e_1| = |e_2| = 1$. We put

(f) $$e_3 = e_1 e_2$$

then $|e_3| = |e_1| \cdot |e_2| = 1$ and by (a)

(g) $$(e_3, e_1) = (e_1 e_2, e_1 e_0) = (e_2, e_0) = 0.$$

Similarly $(e_3, e_2) = 0$. Moreover $(e_3, e_0) = (e_1 e_2, e_0) = \big(e_1(e_1 e_2), e_1 e_0\big) = -(e_2, e_1) = 0$. Here we use (d) for e_1, although we have no complete base, which is permissible because e_0, e_1 can always be completed to some orthonormal base. We choose e_4 arbitrarily with $(e_4, e_i) = 0$ for $i < 4$ and $|e_4| = 1$. The remaining elements of the base are now defined by

(h) $$e_5 = e_1 e_4, \qquad e_6 = -e_2 e_4, \qquad e_7 = e_3 e_4.$$

Clearly $|e_5| = |e_6| = |e_7| = 1$ and, as in (g), e_5 for example is orthogonal to any e_i that contains e_1 or e_4 as factor. The remaining orthogonality relations are proved as above:

$$(e_5, e_0) = (e_1 e_4, e_0) = \big(e_1(e_1 e_4), e_1 e_0\big) = -(e_4, e_1) = 0$$

and exactly the same for (e_5, e_2), (e_5, e_3). e_6 and e_7 are treated in the same way or as

$$(e_6, e_3) = (e_4e_2, e_1e_2) = (e_4, e_1) = 0.$$

The other products e_ie_j can now be calculated, for instance using (e):

$$e_1e_6 = -e_1(e_2e_4) = -e_4(e_1e_2) = -e_4e_3 = e_7$$

$$e_2e_5 = e_2(e_1e_4) = -e_4(e_1e_2) = e_7$$

$$e_3e_5 = e_3(e_1e_4) = -e_1(e_3e_4) = -e_1e_7 = -e_1(e_1e_6) = e_6.$$

We chose these three products because we now have the complete list of basic relations used by Wang [2] and Dickson [1, p. 14], (the list in F is different) although the last three relations are really superfluous because they follow from (f) and (h).

THE GROUP $\mathfrak{G}_2$

An automorphism Φ of the Cayley algebra leaves each real number fixed, preserves the value of the norm $|x|$ (see F, p. 17) and therefore induces a topological mapping of the set:

(55.4) $$u = \sum_{i=1}^{7} r_ie_i \quad \text{with} \quad \sum_{i=1}^{7} r_i^2 = 1$$

on itself which in turn determines Φ uniquely. Thus the group of automorphisms of the Cayley algebra may be regarded as a topological transformation group of the six-dimensional sphere $\sum_{i=1}^{7} r_i^2 = 1$ on itself. The identity component of this group is $\mathfrak{G}_2$, see F, pp. 17, 18.

An automorphism Φ preserves the euclidean distance $|x-y|$ since $(x-y)\Phi = x\Phi - y\Phi$ and Φ preserves the norm, hence Φ also preserves (x, y) and the spherical distance Arccos (x, y) on S^6; this follows from $(x-y, x-y) = 2(1-(x,y))$. To complete the proof of (54.7) we must show that $\mathfrak{G}_2$ is pairwise transitive on S^6. Let x, y and x', y' with $(x, y) = (x', y')$ be given. The fact that the points lie on S^6 implies

$$(x, 1) = (x', 1) = (y, 1) = (y', 1) = 0.$$

Put $x = u_1$, $x' = u_1'$. If $y = \pm x$, then $y' = \pm x'$ and we complete $1, u_1$ and $1, u_1'$ to bases $(1, u_1, \ldots, u_7)$, $(1, u_1', \ldots, u_7')$ by the definite process described above. Then the multiplication table for the two bases are identical, hence

$$\Phi : r_0 + \sum_{i=1}^{7} r_i u_i \to r_0 + \sum_{i=1}^{7} r_i u_i'$$

defines automorphism of the Cayley algebra that sends x into x' and y into y'.

If y is not dependent on x we put

$$u_2 = \frac{y - (x, y)\, x}{|y - (x, y)\, x|}, \qquad u_2' = \frac{y' - (x', y')\, x'}{|y' - (x', y')\, x'|}$$

Then $(u_2, u_1) = (u_2, 1) = (u_2', u_1') = (u_2', 1) = 0$, $|u_2| = |u_2'| = 1$ and we can again extend $1, u_1, u_2$ and $1, u_1', u_2'$ to complete bases by the above process because the first two elements e_1, e_2 were arbitrary (except for orthonormality).

If Φ is defined as before it will again be an automorphism. To make sure that Φ is in $\mathfrak{G}_2$ we ought to ascertain that the vectors $u_1, \ldots, u_7$ and $u_1', \ldots, u_7'$ determine the same orientation of S^6, but we can avoid this problem by observing that $-u_1', \ldots, -u_7'$ determines the opposite orientation from $u_1', \ldots, u_7'$ and has exactly the same multiplication table.

Because $(x, y) = (x', y')$ and

$$\lambda^2 = \big(y - (x, y)\, z,\, y - (x, y)\, x\big) = 1 - (x, y)^2$$

it follows that

$$y = (x, y)\, u_1 + \lambda u_2, \qquad y' = (x, y)\, u_1' + \lambda u_2'$$

hence $x\Phi = x'$ and $y\Phi = y'$.

This finishes the proof of (55.3) and also the proof of (54.7) in the case where $\mathfrak{G}_2$ acts on S^6 in the local euclidean geometry of a seven-dimensional Riemannian space.

To complete the proof of (54.7) we have to discuss the exceptional case where $\mathfrak{G}_2$ is the group of motions of a 6-dimensional Riemann space and $\mathfrak{A}_2$ is its isotropic subgroup. We obtained the combination $\mathfrak{G}_2$, $\mathfrak{A}_2$ simply by counting parameters. That $\mathfrak{A}_2$ is really the isotropic subgroup of $\mathfrak{G}_2$ as transformation group of S^6, may be seen as follows: We consider the automorphisms of the Cayley algebra $\boldsymbol{H}$ which leave e_7 fixed. Let $\boldsymbol{H}'$ be the subspace of $\boldsymbol{H}$ spanned by $e_1, \ldots, e_6$. For any element $x = \sum_{i=1}^{6} r_i e_i$ of $\boldsymbol{H}$ we introduce complex coordinates z_1, z_2, z_3 by

$$z_1 = -r_1 + ir_6, \qquad z_2 = -r_2 + ir_5, \qquad z_3 = -r_3 + ir_4,$$

where we use the same multiplication table as above. If $x = (z_1, z_2, z_3)$ and $y = (w_1, w_2, w_3)$ are elements in $\boldsymbol{H}'$ then

$$xy = -\alpha \cdot 1 - \beta e_7 + (-(\bar{z}_2\bar{w}_3 - \bar{z}_3\bar{w}_2), -(\bar{z}_3\bar{w}_1 - \bar{z}_1\bar{w}_3), -(\bar{z}_1\bar{w}_2 - \bar{z}_2\bar{w}_1))$$

where α and $i\beta$ are, respectively, the real and imaginary parts of

$$z_1\bar{w}_1 + z_2\bar{w}_2 + z_3\bar{w}_3.$$

Thus, a unitary unimodular transformation of $\boldsymbol{H}'$ with the coordinates z_1, z_2, z_3 on itself induces a transformation of $\boldsymbol{H}$ which preserves the product hence an automorphism of $\boldsymbol{H}$, and we see that the isotropic subgroup of $\mathfrak{G}_2$ as transformation group of S^6 contains $\mathfrak{A}_2$ as subgroup.

Conversely, an automorphism Φ of $\boldsymbol{H}$ which leaves e_7 fixed induces a real linear transformation of $\boldsymbol{H}'$ on itself. This follows from the relations $(e_7x)\,\Phi = e_7\Phi x\Phi = e_7(x\Phi)$ for $x \in \boldsymbol{H}'$ and

$$e_7e_1 = e_6, \quad e_7e_2 = e_5, \quad e_7e_3 = e_4, \quad e_7e_4 = -e_3, \quad e_7e_5 = -e_2, \quad e_7e_6 = -e_1$$

which are contained in the above multiplication table. Also, Φ preserves the norm $z_1\bar{z}_1 + z_2\bar{z}_2 + z_3\bar{z}_3$, therefore the transformation induced by Φ lies in $\mathfrak{A}_2$, so that $\mathfrak{A}_2$ is the isotropic subgroup of $\mathfrak{G}_2$.

There is no group different from $\mathfrak{G}_2$ and locally isomorphic to $\mathfrak{G}_2$, whereas there is a group locally isomorphic to, and different from, $\mathfrak{A}_2$. Corresponding to these two cases the coset space $\mathfrak{G}_2/\mathfrak{A}_2$ is, topologically, the six-dimensional sphere or projective space. To see that the metric of the space is spherical it suffices to prove:

If the sphere S^6 is metrized as a G-space R such that a group Γ isomorphic to $\mathfrak{G}_2$ is a group of motions of R, then R is spherical.

But this is clear; for we just saw that $\mathfrak{G}_2$ may be interpreted as a subgroup of the rotations of S^6. Since $\mathfrak{G}_2$ and Γ are equivalent we produce, by the process described above, a spherical metric invariant under Γ, and Γ is pairwise transitive for this spherical metric. Now (54.2) shows that the metric of R is also spherical.

According to Tits [1], in even-dimensions the hermitian and quaternian hyperbolic spaces and a hyperbolic analogue to the Cayley elliptic plane are, besides the euclidean and hyperbolic spaces, the only non-compact spaces with pair-wise transitive groups of motions.

Finite dimensionality is no longer assumed, but is deduced from the recent results of Yamabe on Lie groups. Moreover, Tits also omits the last two axioms for a G-space and indicates that M-convexity may be replaced by a weaker hypothesis. As mentioned before, proofs for these statements have not yet appeared.

Appendix: Problems and Theorems

We conclude the book by listing some problems, many of which were already mentioned in the text, and a few theorems which either indicate possible further developments of the theory of Finsler spaces or elucidate points not fully discussed in the text. P, 15 means a problem and T, 15 a theorem connected with material treated in Section 15.

Finite compactness is not a topological concept, but it is easy to characterize the spaces which possess finitely compact metrizations:

(1) T, 1 A Hausdorff space can be metrized as a finitely compact space if and only if it is locally compact and has a countable base. See Vaughan [1], also Busemann [2, p. 205].

(2) P, 7 Are all geodesics in a finitely compact, M-convex metric space R straight lines, if R satisfies the Axioms of local prolongability with $\rho(p) \equiv \infty$?

(3) P, 8 Does $\rho_1(p) \equiv \infty$ imply $\rho(p) \equiv \infty$? Or in words: In a simply connected space does uniqueness of the shortest connection imply the absence of conjugate points?

(4) P, 10 Is a circle $K(p, \delta)$, $0 < \delta < \rho(p)$, in a two-dimensional G-space rectifiable?

The problem whether every G-space is a topological manifold is at present, inaccessible. The following problems which go in this direction may be accessible.

(5) P, 10 Is a three-dimensional G-space a topological manifold?

(6) P, 10 Is a G-space locally homogeneous in the sense of Montgomery [1]? This paper shows that local homogeneity has strong topological implications for finite dimensional spaces.

(7) P, 10 Has every G-space a finite dimension?

(8) P, 10 Does the property of domain invariance (see Section 58) hold in every G-space? The answer is affirmative if the dimension is finite (communication from Prof. D. Montgomery).

(9) P, 11 To determine the systems of curves which occur in metrizations of the projective plane as a space of the elliptic type.

(10) P, 14 Can a two-dimensional Desarguesian space of the elliptic type be imbedded in a three-dimensional Desarguesian space?

The discrepancy between the sufficient conditions for a G-space given in Section 15 and necessary conditions is very great. It seems doubtful whether a considerable reduction can be obtained by the classical methods. The methods of Busemann and Mayer [1] may prove helpful, although this paper does not aim at uniqueness, but at differentiability properties.

(11) P, 15 Find weaker conditions on the integrand $F(x, \xi)$ than in Section 15 which imply that the space is a G-space. To find necessary and sufficient conditions is probably not a reasonable problem.

The same problem may be approached from a different angle, namely, by imposing conditions on a G-space which make it a Finsler space. Two results are known in this direction, see Busemann [1, Chapter II p. 63 and p. 70] of which the first is this:

(12) T, 15 For $a, b < \sigma(a)$ (Section 9) denote by $\mathfrak{x}(a, b)$ the segment with origin a through b of length $\sigma(a)$. Let the following hold: if a_ν, b_ν, c_ν tend to p such that $\mathfrak{x}(a_\nu, b_\nu)$ and $\mathfrak{x}(a_\nu, c_\nu)$ converge, but to different segments, and if $\lim a_\nu b_\nu / a_\nu c_\nu$ exists (∞ admitted), then $\mathfrak{x}(b_\nu, c_\nu)$ converges, and $\lim (b_\nu a_\nu + a_\nu c_\nu)/b_\nu c_\nu = 1$ if and only if $\lim \mathfrak{x}(b_\nu, a_\nu) = \lim \mathfrak{x}(b_\nu, c_\nu)$ or $\lim \mathfrak{x}(a_\nu, c_\nu) = \lim \mathfrak{x}(b_\nu, c_\nu)$.

Then a Minkowskian metric $m(x, y)$ (equivalent to xy) can be introduced in $S(p, \rho(p))$ such that $m(p, x) = px$ and $x_\nu y_\nu / m(x_\nu, y_\nu) \to 1$ if $x_\nu \to p$, $y_\nu \to p$ and $(x_\nu p + p y_\nu)/x_\nu y_\nu$ is bounded.

(13) T, 18 A Hilbert geometry with symmetric perpendicularity is hyperbolic, Kelly and Paige [1].

(14) T, 18 In a Hilbert geometry let for any two opposite rays $\mathfrak{r}, \mathfrak{s}$ (i. e., $\mathfrak{r}$ and $\mathfrak{s}$ have only their origin in common and lie on the same line) the locus $\alpha(\mathfrak{r}, x) = \alpha(\mathfrak{s}, x)$ (see Section 22) be flat. Then the metric is hyperbolic.

In all spaces considered in this book the distance is symmetric ($xy = yx$). Many problems remain significant, or can be reworded so as to become significant, for spaces with non-symmetric distances. Only the simple results corresponding to our Sections 1 to 8, 22, 27, 28 have been extended to non-symmetric spaces, called E-spaces, in Busemann [3].

(15) P, 20–55 A systematic investigation of non-symmetric spaces by the present methods.

Since (6.9) is a best estimate even in straight spaces there is in general for two distinct points a, b no line through b whose distance from a exceeds $ab/2$.

(16) T, 20 Given two points a, b in a straight plane, there is a line through b with distance at least $ab/2$ from a.

(17) P, 20 Given a line in a straight space of dimension greater than 1, are there always points with arbitrarily large distance from the line?

(18) T, 21 In a two-dimensional space of the elliptic type there is for every point p exactly one geodesic $\mathfrak{g}_p$ with maximal distance from p. The mapping $p \to \mathfrak{g}_p$ is topological, Busemann [2].

(19) P, 22 Do two straight lines in a straight space coincide, if they have representations $x(\tau)$, $y(\tau)$ which satisfy

$$\lim_{\tau \to \infty, \sigma \to \infty} x(\tau)\, y(\sigma) = \lim_{\tau \to -\infty, \sigma \to -\infty} x(\tau)\, y(\sigma) = 0 \ ?$$

For general G-spaces the answer would be negative, e. g., for the surface

$$z = [1 + x^2 + y^2]^{-1}.$$

(20) P, 23 Does symmetry of the asymptote relation imply transitivity in straight spaces of dimension greater than 2?

(21) P, 24 Are Theorems (24.6) and (24.10) true if the symmetry condition is omitted from the parallel axiom? (It is used only in (24.4).) More generally, does the second part of the parallel axiom imply the first?

(22) T, 24 and 39 Let R_λ denote the G-space originating from the G-space R by multiplying all the distances by $\lambda > 0$. If R_λ is isometric to R for all $\lambda > 0$ or only for all λ in an interval of positive length, then R is Minkowskian.

One proof is obtained by using Theorem (39.12).

(23) P, 24 To find a simple characterization of some Desarguesian space other than the Minkowskian, hyperbolic and elliptic spaces (for instance through a metric property of their planes as in (24.7)).

The following problem goes in the same direction (compare (14)):

(24) P, 24 To determine the straight spaces in which the loci $\alpha(\mathfrak{r}, x) = \alpha(\mathfrak{s}, x)$ are flat for any two opposite rays $\mathfrak{r}, \mathfrak{s}$.

If the dimension exceeds 2 and the spheres are differentiable the Desarguesian character follows as in the proof of (24.6).

(25) P, 25 Is a straight plane Minkowskian when its limit circles are straight lines?

(26) T, 26 The parallel axiom holds in a straight plane if the loci of the points equidistant to a given line consist of two straight lines.

There is a remarkable example of Funk [2] which shows that this is not necessarily the case when the distance is not symmetric, even if the plane is Desarguesian.

(27) P, 27 To find conditions (other than (27.16)), in particular conditions applying to an ordinary cylinder, under which a non-compact space has the property that every locally isometric mapping of the space on itself is a motion.

(28) P, 31 Is an odd-dimensional G-space in which all geodesics are straight lines or great circles, necessarily straight, sphere-like or of the elliptic type?

The hermitian elliptic spaces discussed in Section 53 show that the answer would be negative for even dimensions greater than 2. If it should be positive in odd dimensions, a proof is at present hopelessly difficult.

Owing to the topological character of M. Morse's methods a great number of his results on the calculus of variations in the large, see Morse [2], extend to G-spaces. Similarly the methods of Lustermik and Schuirelmann should prove applicable to G-spaces.

(29) P, 32 To extend the methods of the calculus of variations in the large to G-spaces.

(30) P, 33 On a (two-dimensional) torus without conjugate points, is each non-closed geodesic everywhere dense?

(31) P, 33 To find necessary and sufficient conditions for curve systems on a cylinder or compact surfaces of higher genus to serve as geodesics in metrizations of the surfaces without conjugate points.

It was mentioned in the text that the results of Nielsen (Section **34**) have been extended in many directions to more general Riemannian surfaces, principally by Morse and Hedlund. Much of this theory holds for G-surfaces.

(32) P, 34 To extend to G-surfaces the work of Morse and Hedlund concerning geodesics on Riemannian surfaces.

The methods of Morse [1] apply without change to G-surfaces. In fact. some proofs can be considerably simplified by extending the concept of axial motion to general G-spaces and applying the present methods. The central result in Morse [1] is the following:

Let a Poincaré model of a compact orientable G-surface (not necessarily without conjugate points) be defined as in Section **34**. For any two distinct points of the unit circle there is at least one straight line in P with these points as endpoints, and a constant β exists such that any straight line $\mathfrak{g}$ in P and the hyperbolic straight line $\mathfrak{g}_h$ with the same endpoints satisfy $\mathfrak{g}_h \subset S(\mathfrak{g}, \beta)$ and $\mathfrak{g} \subset S(\mathfrak{g}_h, \beta)$.

The corresponding result for the torus is found in Hedlund [1] and can — with the same simplifications as above — be extended to G-surfaces.

(33) P, 34 Does the Divergence Property hold in a straight space which is covering space of a compact space?

(34) P, 36 Is a Hilbert geometry with negative curvature hyperbolic?

(35) P, 37 Is the asymptote relation transitive or symmetric in a straight space with convex capsules?

(36) P, 40 In a G-space with non-positive curvature let $x(\tau)$, $y(\tau)$ represent geodesics which are asymptotic of some type, and are rays for $\tau \geqslant 0$. Are these rays co-rays to each other? Does the converse hold in spaces of finite connectivity?

(37) T, 40 A cylinder with non-positive curvature is simply covered by a suitable family of straight lines.

A cylinder with negative curvature either carries exactly one great circle $\mathfrak{g}$ and is simply covered by the straight lines perpendicular to $\mathfrak{g}$, or it contains no closed geodesic and its straight lines are all asymptotes to each other, Busemann [5, p. 300].

(38) P, 36–41 To investigate G-spaces with non-negative or positive curvature, where these concepts are defined analogously to (36.1, 3). In particular, some of the results of Preissmann [1] on Riemann spaces with positive curvature should prove extendable.

(39) P, 41 In an n-dimensional space with non-positive curvature and suitable differentiability properties, are $|S(p, \sigma)|_n$ (or even $|S(p, \sigma)|_n - \varkappa^{(n)} \sigma^n$) and $|K(p, \sigma)|_{n-1}$ convex functions of σ for $0 \leqslant \sigma < \rho(p)$?

The answer is affirmative for Riemann spaces, see Beckenbach [1]. This paper deals only with surfaces, but its methods generalize.

(40) T, 44 Every point of a G-plane with positive excess and angular measure uniform at π lies on a geodesic without multiple points.

A proof for the Riemannian case, not extendable to G-surfaces, is found in Cohn-Vossen [2].

(41) T, 48 If all spheres about two distinct points of a space of the elliptic type permit rotation, then the space is elliptic.

(42) T, 48 Let every point of a G-space have neighborhood $S(z, \delta_z)$ such that any quadruple of points in $S(z, \delta_z)$ is isometric to a quadruple of points in a euclidean, hyperbolic or spherical space (which, theoretically, may depend on z). Then the space is locally euclidean, hyperbolic, or spherical.

This is an immediate consequence of (48.14). Uniqueness of prolongation follows obviously from the hypothesis on $S(z, \delta_z)$ and may be omitted from the axioms of a G-space. For further considerable reductions concerning the problem in the large, see Blumenthal [1].

(43) P, 49 Is the universal covering space of a locally symmetric space symmetric?

(44) P, 49 To extend Cartan's theory of symmetric spaces to G-spaces.

(45) P, 50 If a G-space possesses a transitive group of motions such that for each motion Φ in the group the distance $xx\Phi$ is independent of x, then the space is locally Minkowskian.

(46) T, 51 The following properties of quasihyperbolic geometry are found in Busemann [7]:

The line element of any quasihyperbolic geometry can be written in the form:

$$F(dx, dy)\, y^{-1}, \quad -\infty < x < \infty, \quad y > 0.$$

Here $F(x, y)$ satisfies the conditions 1) 2) 3) in (17.17), is differentiable except at (0,0), and the tangents of the curve $F(x, y) = 1$ at its intersections with $x = 0$ touch this curve at one point only. Conversely, every line element of this form yields a quasihyperbolic geometry.

The hyperbolic parallel axiom holds. A Desarguesian geometry is hyperbolic. The circles are not necessarily convex, but each circle contains an arc greater than a semi circle which is convex. Therefore the space does not have convex capsules, still less negative curvature.

Any angular measure which is invariant under motions and continuous is non-extendable. There is (under a mild additional hypothesis) always exactly one angular measure invariant under motions for which the total excess exists. For this measure the excess of a triangle is negative and proportional to the area.

These facts show clearly, how the different aspects of curvature in Riemann spaces may become entirely dissociated in Finsler spaces. They contain, moreover, the following important negative result, whose precise formulation is found in Busemann [7]:

(47) T, 51 and 44 There is no universal angular measure for two-dimensional Finsler spaces which satisfies the Gauss-Bonnet Theorem.

Literature

Alexandroff, P. and H. Hopf [1] Topologie, Berlin 1935.

Beckenbach, E. F. [1] Some convexity properties of surfaces of negative curvature, Am. Math. Monthly 55 (1948), 285—301.

Behrend, Felix [1] Über die kleinste umschriebene und die größte einbeschriebene Ellipse eines konvexen Bereichs, Math. Ann. 115 (1938), 379—411.

Bertini, Eugenio [1] Einführung in die projektive Geometrie mehrdimensionaler Räume, Wien 1924.

Birkhoff, Garret [1] Metric foundations of geometry, Proc. Nat. Acad. Sci. U.S.A. 27 (1941), 402—406.

[2] Metric foundations of geometry I, Trans. Am. Math. Soc. 55 (1944), 465–492.

Blaschke, Wilhelm [1] Räumliche Variationsprobleme mit symmetrischer Transversalitätsbedingung, Ber. Sächs. Akad. Wiss. Leipzig 68 (1916), 50—55.

[2] Vorlesungen über Differentialgeometrie I, 3rd edition, Berlin 1929.

Bliss, G. A. [1] Lectures on the calculus of variations, Chicago 1946.

Blumenthal, L. M. [1] Theory and applications of distance geometry, Oxford 1953.

Bolza, Oscar [1] Lectures on the calculus of variations, Chicago 1904.

Bonnesen, T. and W. Fenchel [1] Theorie der konvexen Körper, Erg. der Math. III 1, Berlin 1934 and New York 1948.

Borel, Armand [1] Le plan projectif des octaves et les sphères comme espaces homogènes, C.R. Acad. Sci. Paris 230 (1950), 1378—1380.

Busemann, Herbert [1] Metric methods in Finsler spaces and in the foundations of geometry, Ann. Math. Study No. 8, Princeton 1942.

[2] On spaces in which two points determine a geodesic, Trans. Am. Math. Soc. 54 (1943), 171—184.

[3] Local metric geometry, Trans. Am. Math. Soc. 56 (1944), 200—276.

[4] Intrinsic area, Ann. of Math. 48 (1947), 234—267.

[5] Spaces with non-positive curvature, Acta Math. 48 (1947), 234—267.

[6] The foundations of Minkowskian geometry, Comm. Math. Helv. 24 (1950), 156—186.

[7] Quasihyperbolic Geometry, to appear in Rend. Conti Cire. Mat. Palermo

Papers which have been integrally taken over into [1] or the present book are not listed.

Busemann, H. and P. Kelly [1] Projective geometry and projective metrics, New York 1953.

Busemann, H. and W. Mayer [1] On the foundations of calculus of variations, Trans. Am. Math. Soc. 49 (1941), 173—198.

Cartan, Élie [1] Sur certaines formes remarquables des géometries à groupe fondamental simple, Ann. Normale Sup 3rd series 44 (1927), 345—467.

[2] Leçons sur la géométrie des espaces de Riemann, Paris 1928.

[3] La théorie des groupes finis et continus et l'analysis situs, Mem. Sci. Math. No 42, Paris 1930.

[4] Leçons sur la géométrie projective complèxe, Paris 1931.
Cohn-Vossen, Stefan [1] Kürzeste Wege und Totalkrümmung auf Flächen, Compositio Math. 2 (1935), 63—133.
[2] Totalkrümmung und geodätische Linien auf einfach zusammenhängenden, offenen, vollständigen Flächenstücken, Mat. Sbornik N.S. 1 (1936), 139—164.
Courant, Richard [1] Differential and integral calculus II, New York 1937.

Dickson, L. E. [1] Linear Algebras, Cambridge Tract No 16, London 1930.
Dieudonné, Jean [1] Sur les groupes classiques, Actual. Sci. No 1040, Paris 1948.
Douglas, Jesse [1] Solution of the inverse problem of the calculus of variations, Trans. Am. Math. Soc. 50 (1941), 71—128.
Duschek, A. and W. Mayer [1] Lehrbuch der Differentialgeometrie II, Leipzig 1930.

Finsler, Paul [1] Über Kurven und Flächen in allgemeinen Räumen, Dissertation Göttingen 1918 and Basel 1951.
Forder, G. H. [1] Coordinates in geometry, Auckland University College, Bulletin No. 41, 1953.
Freudenthal, Hans [1] Oktaven, Ausnahmegruppen und Oktavengeometrie, Math. Inst. Rijksuniv. Utrecht 1951.
Funk, Paul [1] Über die Geometrien, in denen die Geraden die kürzesten sind, und die Äquidistanten zu einer Geraden wieder Geraden sind, Monatshefte Math. Phys. 37 (1930), 153—158.

Gans, David [1] Axioms for elliptic geometry, Can. Jour. Math. 4 (1952), 81—92.
Gleason, A. M. [1] Groups without small subgroups, Ann. of Math. 56 (1952), 193—212.
Green, L. W. [1] Manifolds without conjugate points, Trans. Am. Math. Soc. 76 (1954), 529—546

Hadamard, Jacques [1] Les surfaces à courbures opposées et leur lignes géodésiques, Jour. Math. Pur. Appl. 5th series 4 (1898), 27—73.
Hamel, Georg [1] Über die Geometrien, in denen die Geraden die Kürzesten sind, Math. Ann. 57 (1903), 221—264.
Hardy, G. H., J. E. Littlewood and G. Polya [1] Inequalities, 2nd ed., Cambridge 1952.
Hausdorff, Felix [1] Grundzüge der Mengenlehre, Leipzig 1914.
[2] Mengenlehre, 3rd ed., Berlin and Leipzig 1935.
Hedlund, G. A. [1] Geodesics on a two-dimensional Riemannian manifold with periodic coefficients, Ann. of Math. 33 (1932), pp. 719—739.
Helmholtz, H. von [1] Über die Tatsachen, die der Geometrie zugrunde liegen, Wissensch. Abhandl. vol II, 618—637.
Hessenberg, Gerhard [1] Grundlagen der Geometrie, Berlin and Leipzig 1930.
[2] Ebene und sphärische Trigonometrie, Berlin und Leipzig 1934.
Hilbert, David [1] Grundlagen der Geometrie, 7th ed., Leipzig 1930.
[2] Mathematische Probleme, Nachr. Akad. Wiss. Göttingen, Math. Phys. 1900, 253—297.
Hopf, Eberhard [1] Closed surfaces without conjugate points, Proc. Nat. Acad. Sci. U.S.A. 34 (1948), 47—51.
Hurewicz, W. [1] Sur la dimension des produits Cartésiens, Ann. of Math. 36 (1935), 194–197
Hurewicz, W. and H. Wallman [1] Dimension theory, Princeton 1941.

Hurwitz, Adolf [1] Über die Komposition quadratischer Formen von beliebig vielen Variabeln, Nachr. Akad. Wiss. Göttingen Math. Phys. 1898, 309—316.

Jessen, Börge [1] To Sätninger om konvekse Punktmängder, Mat. Tidsskrift B 1940, 66—70.

Kelly, P. J. and L. J. Paige [1] Symmetric perpendicularity in Hilbert geometries, Pacific Jour. Math 2 (1952), 319—322.

Kerékjártó, B. von [1] Vorlesungen über Topologie I, Berlin 1923.

[2] Geometrische Theorie der zweigliedrigen kontinuierlichen Gruppen, Abh. Math. Sem. Hamburg 8 (1930), 107—114.

Koebe, Paul [1] Riemannsche Mannigfaltigkeiten und nichteuklidische Raumformen, Sitz.Ber. Preuss. Akad. Wiss. 1927, 164—196; 1928, 345—442; 1929, 414—457; 1930, 304—364, 505—541; 1931, 506—534.

Kolmogoroff, Alexander [1] Zur topologisch-gruppentheoretischen Begründung der Geometrie, Nachr. Akad. Wiss. Göttingen Math. Phys. 1930, 208—210.

Leibniz, G. W. [1] Mathematische Schriften, zweite Abteilung I, Berlin 1849.

Lefschetz, Solomon [1] Algebraic Topology, Am. Math. Soc. Pub. No. 27, New York 1942.

Lie, Sophus [1] Theorie der Transformationsgruppen, 3. Abschnitt, Leipzig 1893.

Liebmann, Heinrich [1] Nichteuklidische Geometrie, Leipzig 1905.

Menger, Karl [1] Untersuchungen über allgemeine Metrik I, II, III, Math. Ann. 100 (1928), 75—163.

Montgomery, Deane [1] Locally homogeneous spaces, Ann. of Math. 52 (1950), 261—271.

Montgomery, D. and H. Samelson [1] Transformation groups of spheres, Ann. of Math. 44 (1943), 454—470.

Montgomery, D. and L. Zippin [1] Topological transformation groups, Ann. of Math. 41 (1940), 778—791.

[2] Compact abelian transformation groups, Duke Math. Jour. 4 (1938), 363—373.

Morse, Marston [1] A fundamental class of geodesics on any closed surface of genus greater than one, Trans. Am. Math. Soc. 26 (1924), 25—60.

[2] The calculus of variations in the large, Am. Math. Soc. Pub. No. 18, New York 1934.

Morse, M. and G. Hedlund [1] Symbolic dynamics, Am. Jour. Math. 60 (1938), 815—866.

[2] Manifolds without conjugate points, Trans. Am. Math. Soc. 51 (1942), 362—386.

Mostow, G. D. [1] The extensibility of local Lie groups of transformations and groups on surfaces, Ann. of Math. 52 (1950), 606—636.

Myers, S. B. and N. E. Steenrod [1] The group of isometries of a Riemann manifold, Ann. of Math. (1939), 400—416.

Nielsen, Jakob [1] Om geodätiske Linier i lukkede Manigfoldigheder med konstant negativ Krumning, Mat. Tidsskrift B 1925, 37—44.

[2] Untersuchungen zur Topologie der geschlossenen zweiseitigen Flächen, Acta Math. 50 (1927), 189—358.

Pedersen, F. P. [1] On spaces with negative curvature, Mat. Tidsskrift B 1952, 66—89.

Petty, C. M. [1] On the geometry of the Minkowski plane, to appear in Rivista di Matematica della Universitá di Parma.

Picard, Emile [1] Leçons sur quelques équations fonctionnelles, Paris 1928.

Pontrjagin, L. [1] Topological groups, Princeton 1939.

Preissmann, Alexandre [1] Quelques propriétés globales des espaces de Riemann, Comm. Math. Helv. 15 (1943), 175—216.

Radon, J. [1] Über eine besondere Art ebener konvexer Kurven, Ber. Sächs. Akad. Wiss. Leipzig 68 (1916), 131—134.

Riemann, Bernhard [1] Über die Hypothesen, welche der Geometrie zugrunde liegen, Mathematische Werke, Leipzig 1892, 272—287.

Schur, Friedrich [1] Über den Zusammenhang der Räume konstanten Krümmungsmaßes mit den projektiven Räumen, Math. Ann. 27 (1886), 537—567.

Seifert, H. and W. Threlfall [1] Lehrbuch der Topologie, Leipzig 1934.

Tits, Jacques [1] Etude de certains espaces métriques, Bull. Soc. Math. Belgique (1952), pp. 44—52.

Vaughan, H. E. [1] On locally compact metrizable spaces, Bull. Am. Math. Soc. 43 (1937), 532—535.

Veblen, O. and J. W. Young [1] Projective geometry vol II, Boston 1918.

Wang, Hsien-Chung [1] A new characterization of spheres of even dimension, Nederl. Akad. Wet. Proc. 52 (1949), 838—845.

[2] Two theorems on metric spaces, Pacific Jour. Math. 1 (1951), 473—480.

[3] Two-point homogeneous spaces, Ann. of Math. 55 (1952), 177—191.

Weyl, Hermann [1] Mathematische Analyse des Raumproblems, Berlin 1923.

[2] The classical groups, Princeton 1939.

Whitehead, J. H. [1] The Weierstrass E-function in differential metric geometry, Quarterly Jour. Oxford Ser. 4 (1953), 291—296.

Notes to the Sections

Preface

1) The introductions to the individual chapters provide a good survey of the results.
2) More precisely: the author knows of no Riemannian theorem which is definitely not a special case of a theorem on Finsler spaces and has been proved without differentiability. However, most probably some of A. D. Alexandrov's results are of this type, which emphasizes their unique position in differential geometry.
3) At the moment the question of analytical versus geometric methods is largely vacuous, because for many of the present results on Finsler spaces no analytical tools are available.

Section 2

1) See Busemann [3].
2) See Blumenthal [1].
3) We use "denumerable" for the cardinal number of the integers and "countable" for denumerable or finite, but not empty.

Section 3

1) See Busemann [3, p. 206].

Section 5

1) Theorem (5.14) will not be used.

Section 8

1) See Hardy, Littlewood and Pólya [1, p. 30].

Section 11

1) See for instance Bolza [1].

Section 12

1) That the surface is a straight G-space follows immediately from the Gauss-Bonnet Theorem. That the theorem of Desargues cannot hold follows from (13.1) and Beltrami's Theorem (15.1).

Section 13

1) Possibly (2) is a consequence of (1). A complete proof would entail many case distinctions and seems hardly worthwhile.
2) This theorem and (14.1) were stated without proof in a slightly less complete form in Busemann [1]. Complete proofs of the theorems as stated there, and different from the present proofs, were given by D. B. Dekker in his (unpublished) Master's Thesis 1943.
3) For details of this construction compare the general theory of covering spaces developed in Chapter IV, in particular Section 31.
4) See Section 16.

Section 14

1) See Section 16.
2) See Hessenberg [1, pp. 103—106].
3) See Bertini [1, p. 55].

Section 15

1) See for instance Duschek and Mayer [1].
2) See Blaschke [2, p. 127] where references to the literature are found, and for higher dimensions Cartan [2], where the significance of the "axiome du plan" is thoroughly discussed.
3) Blaschke [2, p. 101].
4) l. c. 3) p. 119.
5) l. c. 3) § 60.
6) Cartan [2, p. 120].

Section 16 1) For the properties of convex sets quoted here and in the following pages see Bonnesen and Fenchel [1].

2) This fact will not be used and is mentioned only for comparison with (16.4) and a better understanding of (17.21) and (24.6).

3) If C is differentiable then $f'(\alpha)$ exists and is continuous. In that case the assertion is trivial because the hypothesis implies $f'(\alpha)/f(\alpha) = f'(\alpha + \pi)/f(\alpha + \pi)$. This method can be applied to the general case by using a fact which the present proof circumvents. The convexity of C implies that the difference quotients $(f(\alpha + h) - f(\alpha))\, h^{-1}$ are bounded for sufficiently small $|h|$. The hypothesis and (16.5) show that $f'(\alpha)/f(\alpha) = f'(\alpha + \pi)/f(\alpha + \pi)$ except for a countable set of α's. A standard argument from real variables shows then that $f(\alpha) \equiv c\, f(\alpha + \pi)$, and $c = 1$ because this holds for at least one value of α.

4) See Behrend [1]; Loewner communicated the result orally.

Section 17 1) This fact was discovered by Radon [1].

Section 18 1) See Hilbert [1, Anhang I].

2) See Bonnesen and Fenchel [1, Section 27].

3) This is automatically the case when the space is straight.

Section 19 1) Levi-Civita's parallelism is, of course, a generalization of parallelism in euclidean spaces. However, we are here interested in the analogue to "points at infinity".

2) See last section.

Section 20 1) For the definition of peakless and strictly peakless functions see Section 18.

Section 21 1) Compare Busemann and Kelly [1, p. 46 and p. 248].

Section 23 1) It turns out that transitivity is nowhere used. The question, whether transitivity follows from symmetry in higher dimensions, is open.

2) See, for instance, Hilbert [1, Anhang III].

3) See Busemann and Kelly [1, p. 162].

Section 24 1) For the literature on this theorem see Jessen [1].

Section 25 1) There is apparently no other word which is generally understood and describes these properties. The danger that the reader might be confused by the various uses of the term "convex" seems remote.

Section 27 1) See Busemann and Kelly [1, p. 304].

2) See l. c. 1) p. 193.

Section 30 1) For the definitions of $K_\infty(\mathfrak{r}, p)$ and $\alpha(\mathfrak{r}, x)$ see Section 22.

2) See, for instance, Busemann and Kelly [1, p. 149].

3) R' is an n-sheeted covering space if exactly n distinct points of R' lie over the same point of R.

Section 31 1) For an exact formulation see Alexandroff and Hopf [1, p. 64].

Section 33 1) The lines $x =$ const. are considered as rational lines with slope ∞.

2) Compare the proof of (11.2).

3) It is, however, not known, whether I, II, III, and IV imply V.

Section 34 1) A straight Riemann space whose curvature is bounded away from $-\infty$ has the divergence property, see Green [1]. It follows that a straight Riemannian plane which is covering space of a compact surface always has the divergence property.

Section 35 1) The results on G-spaces with non-positive curvature are found in Busemann [5]. That non-positive curvature is a stronger condition than convex capsules, and the extension of many theorems to spaces with convex capsules are found in Pedersen [1].

Section 43 1) Cohn-Vossen [1] calls a contracting tube a Schaft and uses Kelch for both bulging and expanding tubes. The latter are distinguished as eigentliche Kelche.

Section 47 1) $|cpb_2| < \pi/2$ because $|pb_2b_1| + |b_2b_1c_1| + |b_1c_1p| = 3\pi/2$, hence (apc) and $|a_1b_2c| > |a_1b_2p| = |abc|$. Thus $|a_1b_2c|$ only approaches $|abc|$ from above and $\lim b_1d_1/b_1b_2 = \lim bd/bb_2$ requires justification. The argument leading to (d) yields without essential changes that

$$[\rho(\delta, \xi') - \rho(\theta, \xi')]\, \xi'^{-1} \geqslant [\rho(\delta, \xi) - \rho(\theta, \xi)]\, \xi^{-1}$$

for $\pi/2 > \delta > \theta$ and $0 < \xi < \xi'$. This leads readily to $\rho(\delta, \xi)/\xi \to C^{-1}(\theta)$ and $\mu(\delta, \xi)/\xi \to T(\theta)$ for $\delta \to \theta +$ and $\xi \to 0 +$.

Section 49 1) It is easily seen that this is so, when the universal covering space is straight, in particular therefore, if the curvature is negative. — While the present book was in print, Kostant proved this conjecture for all Riemann spaces.

Section 50 1) q is different from z because of (22.23).

Section 51 1) This follows from Theorem (52.9).

2) This may also be proved without using Brouwer's theorem by referring to an elementary theorem on surfaces of the elliptic type, see (18) in the Appendix.

Section 52 1) Replacing the δ in Theorem (51.2) by e^{-1} amounts to replacing the arclength on $\mathfrak{g}^+$ by a parameter proportional to the arclength.

Section 53 1) A detailed discussion of the hermitian metrics in C^6 is found in Cartan [4].

Section 54 1) The existence of such subgroups is wellknown (in standard terminology the rank of Γ_1 is at least 1) and may be verified by forming the closure of any one-parameter subgroup. This group is abelian and hence the direct product of groups isomorphic to $\mathfrak{D}_1$.

Section 55 1) According to Tits [1] the assumption of finite dimensionality is superfluous.

Index

A CATALOG OF SELECTED

DOVER BOOKS

IN SCIENCE AND MATHEMATICS

Astronomy

BURNHAM'S CELESTIAL HANDBOOK, Robert Burnham, Jr. Thorough guide to the stars beyond our solar system. Exhaustive treatment. Alphabetical by constellation: Andromeda to Cetus in Vol. 1; Chamaeleon to Orion in Vol. 2; and Pavo to Vulpecula in Vol. 3. Hundreds of illustrations. Index in Vol. 3. 2,000pp. 6⅛ x 9¼.
Vol. I: 23567-X
Vol. II: 23568-8
Vol. III: 23673-0

EXPLORING THE MOON THROUGH BINOCULARS AND SMALL TELESCOPES, Ernest H. Cherrington, Jr. Informative, profusely illustrated guide to locating and identifying craters, rills, seas, mountains, other lunar features. Newly revised and updated with special section of new photos. Over 100 photos and diagrams. 240pp. 8¼ x 11. 24491-1

THE EXTRATERRESTRIAL LIFE DEBATE, 1750–1900, Michael J. Crowe. First detailed, scholarly study in English of the many ideas that developed from 1750 to 1900 regarding the existence of intelligent extraterrestrial life. Examines ideas of Kant, Herschel, Voltaire, Percival Lowell, many other scientists and thinkers. 16 illustrations. 704pp. 5⅜ x 8½. 40675-X

THEORIES OF THE WORLD FROM ANTIQUITY TO THE COPERNICAN REVOLUTION, Michael J. Crowe. Newly revised edition of an accessible, enlightening book recreates the change from an earth-centered to a sun-centered conception of the solar system. 242pp. 5⅜ x 8½. 41444-2

A HISTORY OF ASTRONOMY, A. Pannekoek. Well-balanced, carefully reasoned study covers such topics as Ptolemaic theory, work of Copernicus, Kepler, Newton, Eddington's work on stars, much more. Illustrated. References. 521pp. 5⅜ x 8½.
65994-1

A COMPLETE MANUAL OF AMATEUR ASTRONOMY: Tools and Techniques for Astronomical Observations, P. Clay Sherrod with Thomas L. Koed. Concise, highly readable book discusses: selecting, setting up and maintaining a telescope; amateur studies of the sun; lunar topography and occultations; observations of Mars, Jupiter, Saturn, the minor planets and the stars; an introduction to photoelectric photometry; more. 1981 ed. 124 figures. 26 halftones. 37 tables. 335pp. 6½ x 9¼.
42820-6

AMATEUR ASTRONOMER'S HANDBOOK, J. B. Sidgwick. Timeless, comprehensive coverage of telescopes, mirrors, lenses, mountings, telescope drives, micrometers, spectroscopes, more. 189 illustrations. 576pp. 5⅝ x 8¼. (Available in U.S. only.)
24034-7

STARS AND RELATIVITY, Ya. B. Zel'dovich and I. D. Novikov. Vol. 1 of *Relativistic Astrophysics* by famed Russian scientists. General relativity, properties of matter under astrophysical conditions, stars, and stellar systems. Deep physical insights, clear presentation. 1971 edition. References. 544pp. 5⅝ x 8¼. 69424-0

Chemistry

THE SCEPTICAL CHYMIST: The Classic 1661 Text, Robert Boyle. Boyle defines the term "element," asserting that all natural phenomena can be explained by the motion and organization of primary particles. 1911 ed. viii+232pp. 5⅜ x 8½. 42825-7

RADIOACTIVE SUBSTANCES, Marie Curie. Here is the celebrated scientist's doctoral thesis, the prelude to her receipt of the 1903 Nobel Prize. Curie discusses establishing atomic character of radioactivity found in compounds of uranium and thorium; extraction from pitchblende of polonium and radium; isolation of pure radium chloride; determination of atomic weight of radium; plus electric, photographic, luminous, heat, color effects of radioactivity. ii+94pp. 5⅜ x 8½. 42550-9

CHEMICAL MAGIC, Leonard A. Ford. Second Edition, Revised by E. Winston Grundmeier. Over 100 unusual stunts demonstrating cold fire, dust explosions, much more. Text explains scientific principles and stresses safety precautions. 128pp. 5⅜ x 8½. 67628-5

THE DEVELOPMENT OF MODERN CHEMISTRY, Aaron J. Ihde. Authoritative history of chemistry from ancient Greek theory to 20th-century innovation. Covers major chemists and their discoveries. 209 illustrations. 14 tables. Bibliographies. Indices. Appendices. 851pp. 5⅜ x 8½. 64235-6

CATALYSIS IN CHEMISTRY AND ENZYMOLOGY, William P. Jencks. Exceptionally clear coverage of mechanisms for catalysis, forces in aqueous solution, carbonyl- and acyl-group reactions, practical kinetics, more. 864pp. 5⅜ x 8½. 65460-5

ELEMENTS OF CHEMISTRY, Antoine Lavoisier. Monumental classic by founder of modern chemistry in remarkable reprint of rare 1790 Kerr translation. A must for every student of chemistry or the history of science. 539pp. 5⅜ x 8½. 64624-6

THE HISTORICAL BACKGROUND OF CHEMISTRY, Henry M. Leicester. Evolution of ideas, not individual biography. Concentrates on formulation of a coherent set of chemical laws. 260pp. 5⅜ x 8½. 61053-5

A SHORT HISTORY OF CHEMISTRY, J. R. Partington. Classic exposition explores origins of chemistry, alchemy, early medical chemistry, nature of atmosphere, theory of valency, laws and structure of atomic theory, much more. 428pp. 5⅜ x 8½. (Available in U.S. only.) 65977-1

GENERAL CHEMISTRY, Linus Pauling. Revised 3rd edition of classic first-year text by Nobel laureate. Atomic and molecular structure, quantum mechanics, statistical mechanics, thermodynamics correlated with descriptive chemistry. Problems. 992pp. 5⅜ x 8½. 65622-5

FROM ALCHEMY TO CHEMISTRY, John Read. Broad, humanistic treatment focuses on great figures of chemistry and ideas that revolutionized the science. 50 illustrations. 240pp. 5⅜ x 8½. 28690-8

Engineering

DE RE METALLICA, Georgius Agricola. The famous Hoover translation of greatest treatise on technological chemistry, engineering, geology, mining of early modern times (1556). All 289 original woodcuts. 638pp. 6¾ x 11. 60006-8

FUNDAMENTALS OF ASTRODYNAMICS, Roger Bate et al. Modern approach developed by U.S. Air Force Academy. Designed as a first course. Problems, exercises. Numerous illustrations. 455pp. 5⅜ x 8½. 60061-0

DYNAMICS OF FLUIDS IN POROUS MEDIA, Jacob Bear. For advanced students of ground water hydrology, soil mechanics and physics, drainage and irrigation engineering, and more. 335 illustrations. Exercises, with answers. 784pp. 6⅛ x 9¼. 65675-6

THEORY OF VISCOELASTICITY (Second Edition), Richard M. Christensen. Complete, consistent description of the linear theory of the viscoelastic behavior of materials. Problem-solving techniques discussed. 1982 edition. 29 figures. xiv+364pp. 6⅛ x 9¼. 42880-X

MECHANICS, J. P. Den Hartog. A classic introductory text or refresher. Hundreds of applications and design problems illuminate fundamentals of trusses, loaded beams and cables, etc. 334 answered problems. 462pp. 5⅜ x 8½. 60754-2

MECHANICAL VIBRATIONS, J. P. Den Hartog. Classic textbook offers lucid explanations and illustrative models, applying theories of vibrations to a variety of practical industrial engineering problems. Numerous figures. 233 problems, solutions. Appendix. Index. Preface. 436pp. 5⅜ x 8½. 64785-4

STRENGTH OF MATERIALS, J. P. Den Hartog. Full, clear treatment of basic material (tension, torsion, bending, etc.) plus advanced material on engineering methods, applications. 350 answered problems. 323pp. 5⅜ x 8½. 60755-0

A HISTORY OF MECHANICS, René Dugas. Monumental study of mechanical principles from antiquity to quantum mechanics. Contributions of ancient Greeks, Galileo, Leonardo, Kepler, Lagrange, many others. 671pp. 5⅜ x 8½. 65632-2

STABILITY THEORY AND ITS APPLICATIONS TO STRUCTURAL MECHANICS, Clive L. Dym. Self-contained text focuses on Koiter postbuckling analyses, with mathematical notions of stability of motion. Basing minimum energy principles for static stability upon dynamic concepts of stability of motion, it develops asymptotic buckling and postbuckling analyses from potential energy considerations, with applications to columns, plates, and arches. 1974 ed. 208pp. 5⅜ x 8½. 42541-X

METAL FATIGUE, N. E. Frost, K. J. Marsh, and L. P. Pook. Definitive, clearly written, and well-illustrated volume addresses all aspects of the subject, from the historical development of understanding metal fatigue to vital concepts of the cyclic stress that causes a crack to grow. Includes 7 appendixes. 544pp. 5⅜ x 8½. 40927-9

Mathematics

FUNCTIONAL ANALYSIS (Second Corrected Edition), George Bachman and Lawrence Narici. Excellent treatment of subject geared toward students with background in linear algebra, advanced calculus, physics, and engineering. Text covers introduction to inner-product spaces, normed, metric spaces, and topological spaces; complete orthonormal sets, the Hahn-Banach Theorem and its consequences, and many other related subjects. 1966 ed. 544pp. 6⅛ x 9¼. 40251-7

ASYMPTOTIC EXPANSIONS OF INTEGRALS, Norman Bleistein & Richard A. Handelsman. Best introduction to important field with applications in a variety of scientific disciplines. New preface. Problems. Diagrams. Tables. Bibliography. Index. 448pp. 5⅜ x 8½. 65082-0

VECTOR AND TENSOR ANALYSIS WITH APPLICATIONS, A. I. Borisenko and I. E. Tarapov. Concise introduction. Worked-out problems, solutions, exercises. 257pp. 5⅝ x 8¼. 63833-2

THE ABSOLUTE DIFFERENTIAL CALCULUS (CALCULUS OF TENSORS), Tullio Levi-Civita. Great 20th-century mathematician's classic work on material necessary for mathematical grasp of theory of relativity. 452pp. 5⅝ x 8¼. 63401-9

AN INTRODUCTION TO ORDINARY DIFFERENTIAL EQUATIONS, Earl A. Coddington. A thorough and systematic first course in elementary differential equations for undergraduates in mathematics and science, with many exercises and problems (with answers). Index. 304pp. 5⅜ x 8½. 65942-9

FOURIER SERIES AND ORTHOGONAL FUNCTIONS, Harry F. Davis. An incisive text combining theory and practical example to introduce Fourier series, orthogonal functions and applications of the Fourier method to boundary-value problems. 570 exercises. Answers and notes. 416pp. 5⅜ x 8½. 65973-9

COMPUTABILITY AND UNSOLVABILITY, Martin Davis. Classic graduate-level introduction to theory of computability, usually referred to as theory of recurrent functions. New preface and appendix. 288pp. 5⅜ x 8½. 61471-9

ASYMPTOTIC METHODS IN ANALYSIS, N. G. de Bruijn. An inexpensive, comprehensive guide to asymptotic methods–the pioneering work that teaches by explaining worked examples in detail. Index. 224pp. 5⅜ x 8½ 64221-6

APPLIED COMPLEX VARIABLES, John W. Dettman. Step-by-step coverage of fundamentals of analytic function theory–plus lucid exposition of five important applications: Potential Theory; Ordinary Differential Equations; Fourier Transforms; Laplace Transforms; Asymptotic Expansions. 66 figures. Exercises at chapter ends. 512pp. 5⅜ x 8½. 64670-X

INTRODUCTION TO LINEAR ALGEBRA AND DIFFERENTIAL EQUATIONS, John W. Dettman. Excellent text covers complex numbers, determinants, orthonormal bases, Laplace transforms, much more. Exercises with solutions. Undergraduate level. 416pp. 5⅜ x 8½. 65191-6

TENSOR CALCULUS, J.L. Synge and A. Schild. Widely used introductory text covers spaces and tensors, basic operations in Riemannian space, non-Riemannian spaces, etc. 324pp. 5⅜ x 8¼. 63612-7

ORDINARY DIFFERENTIAL EQUATIONS, Morris Tenenbaum and Harry Pollard. Exhaustive survey of ordinary differential equations for undergraduates in mathematics, engineering, science. Thorough analysis of theorems. Diagrams. Bibliography. Index. 818pp. 5⅜ x 8½. 64940-7

INTEGRAL EQUATIONS, F. G. Tricomi. Authoritative, well-written treatment of extremely useful mathematical tool with wide applications. Volterra Equations, Fredholm Equations, much more. Advanced undergraduate to graduate level. Exercises. Bibliography. 238pp. 5⅜ x 8½. 64828-1

FOURIER SERIES, Georgi P. Tolstov. Translated by Richard A. Silverman. A valuable addition to the literature on the subject, moving clearly from subject to subject and theorem to theorem. 107 problems, answers. 336pp. 5⅜ x 8½. 63317-9

INTRODUCTION TO MATHEMATICAL THINKING, Friedrich Waismann. Examinations of arithmetic, geometry, and theory of integers; rational and natural numbers; complete induction; limit and point of accumulation; remarkable curves; complex and hypercomplex numbers, more. 1959 ed. 27 figures. xii+260pp. 5⅜ x 8½. 42804-4

POPULAR LECTURES ON MATHEMATICAL LOGIC, Hao Wang. Noted logician's lucid treatment of historical developments, set theory, model theory, recursion theory and constructivism, proof theory, more. 3 appendixes. Bibliography. 1981 ed. ix+283pp. 5⅜ x 8½. 67632-3

CALCULUS OF VARIATIONS, Robert Weinstock. Basic introduction covering isoperimetric problems, theory of elasticity, quantum mechanics, electrostatics, etc. Exercises throughout. 326pp. 5⅜ x 8½. 63069-2

THE CONTINUUM: A Critical Examination of the Foundation of Analysis, Hermann Weyl. Classic of 20th-century foundational research deals with the conceptual problem posed by the continuum. 156pp. 5⅜ x 8½. 67982-9

CHALLENGING MATHEMATICAL PROBLEMS WITH ELEMENTARY SOLUTIONS, A. M. Yaglom and I. M. Yaglom. Over 170 challenging problems on probability theory, combinatorial analysis, points and lines, topology, convex polygons, many other topics. Solutions. Total of 445pp. 5⅜ x 8½. Two-vol. set.
Vol. I: 65536-9 Vol. II: 65537-7

INTRODUCTION TO PARTIAL DIFFERENTIAL EQUATIONS WITH APPLICATIONS, E. C. Zachmanoglou and Dale W. Thoe. Essentials of partial differential equations applied to common problems in engineering and the physical sciences. Problems and answers. 416pp. 5⅜ x 8½. 65251-3

THE THEORY OF GROUPS, Hans J. Zassenhaus. Well-written graduate-level text acquaints reader with group-theoretic methods and demonstrates their usefulness in mathematics. Axioms, the calculus of complexes, homomorphic mapping, *p*-group theory, more. 276pp. 5⅜ x 8½. 40922-8

Math–Decision Theory, Statistics, Probability

ELEMENTARY DECISION THEORY, Herman Chernoff and Lincoln E. Moses. Clear introduction to statistics and statistical theory covers data processing, probability and random variables, testing hypotheses, much more. Exercises. 364pp. 5⅜ x 8½. 65218-1

STATISTICS MANUAL, Edwin L. Crow et al. Comprehensive, practical collection of classical and modern methods prepared by U.S. Naval Ordnance Test Station. Stress on use. Basics of statistics assumed. 288pp. 5⅜ x 8½. 60599-X

SOME THEORY OF SAMPLING, William Edwards Deming. Analysis of the problems, theory, and design of sampling techniques for social scientists, industrial managers, and others who find statistics important at work. 61 tables. 90 figures. xvii +602pp. 5⅜ x 8½. 64684-X

LINEAR PROGRAMMING AND ECONOMIC ANALYSIS, Robert Dorfman, Paul A. Samuelson and Robert M. Solow. First comprehensive treatment of linear programming in standard economic analysis. Game theory, modern welfare economics, Leontief input-output, more. 525pp. 5⅜ x 8½. 65491-5

PROBABILITY: An Introduction, Samuel Goldberg. Excellent basic text covers set theory, probability theory for finite sample spaces, binomial theorem, much more. 360 problems. Bibliographies. 322pp. 5⅜ x 8½. 65252-1

GAMES AND DECISIONS: Introduction and Critical Survey, R. Duncan Luce and Howard Raiffa. Superb nontechnical introduction to game theory, primarily applied to social sciences. Utility theory, zero-sum games, n-person games, decision-making, much more. Bibliography. 509pp. 5⅜ x 8½. 65943-7

INTRODUCTION TO THE THEORY OF GAMES, J. C. C. McKinsey. This comprehensive overview of the mathematical theory of games illustrates applications to situations involving conflicts of interest, including economic, social, political, and military contexts. Appropriate for advanced undergraduate and graduate courses; advanced calculus a prerequisite. 1952 ed. x+372pp. 5⅜ x 8½. 42811-7

FIFTY CHALLENGING PROBLEMS IN PROBABILITY WITH SOLUTIONS, Frederick Mosteller. Remarkable puzzlers, graded in difficulty, illustrate elementary and advanced aspects of probability. Detailed solutions. 88pp. 5⅜ x 8½. 65355-2

PROBABILITY THEORY: A Concise Course, Y. A. Rozanov. Highly readable, self-contained introduction covers combination of events, dependent events, Bernoulli trials, etc. 148pp. 5⅜ x 8¼. 63544-9

STATISTICAL METHOD FROM THE VIEWPOINT OF QUALITY CONTROL, Walter A. Shewhart. Important text explains regulation of variables, uses of statistical control to achieve quality control in industry, agriculture, other areas. 192pp. 5⅜ x 8½. 65232-7

History of Math

THE WORKS OF ARCHIMEDES, Archimedes (T. L. Heath, ed.). Topics include the famous problems of the ratio of the areas of a cylinder and an inscribed sphere; the measurement of a circle; the properties of conoids, spheroids, and spirals; and the quadrature of the parabola. Informative introduction. clxxxvi+326pp; supplement, 52pp. 5⅜ x 8½. 42084-1

A SHORT ACCOUNT OF THE HISTORY OF MATHEMATICS, W. W. Rouse Ball. One of clearest, most authoritative surveys from the Egyptians and Phoenicians through 19th-century figures such as Grassman, Galois, Riemann. Fourth edition. 522pp. 5⅜ x 8½. 20630-0

THE HISTORY OF THE CALCULUS AND ITS CONCEPTUAL DEVELOPMENT, Carl B. Boyer. Origins in antiquity, medieval contributions, work of Newton, Leibniz, rigorous formulation. Treatment is verbal. 346pp. 5⅜ x 8½. 60509-4

THE HISTORICAL ROOTS OF ELEMENTARY MATHEMATICS, Lucas N. H. Bunt, Phillip S. Jones, and Jack D. Bedient. Fundamental underpinnings of modern arithmetic, algebra, geometry, and number systems derived from ancient civilizations. 320pp. 5⅜ x 8½. 25563-8

A HISTORY OF MATHEMATICAL NOTATIONS, Florian Cajori. This classic study notes the first appearance of a mathematical symbol and its origin, the competition it encountered, its spread among writers in different countries, its rise to popularity, its eventual decline or ultimate survival. Original 1929 two-volume edition presented here in one volume. xxviii+820pp. 5⅜ x 8½. 67766-4

GAMES, GODS & GAMBLING: A History of Probability and Statistical Ideas, F. N. David. Episodes from the lives of Galileo, Fermat, Pascal, and others illustrate this fascinating account of the roots of mathematics. Features thought-provoking references to classics, archaeology, biography, poetry. 1962 edition. 304pp. 5⅜ x 8½. (Available in U.S. only.) 40023-9

OF MEN AND NUMBERS: The Story of the Great Mathematicians, Jane Muir. Fascinating accounts of the lives and accomplishments of history's greatest mathematical minds–Pythagoras, Descartes, Euler, Pascal, Cantor, many more. Anecdotal, illuminating. 30 diagrams. Bibliography. 256pp. 5⅜ x 8½. 28973-7

HISTORY OF MATHEMATICS, David E. Smith. Nontechnical survey from ancient Greece and Orient to late 19th century; evolution of arithmetic, geometry, trigonometry, calculating devices, algebra, the calculus. 362 illustrations. 1,355pp. 5⅜ x 8½. Two-vol. set. Vol. I: 20429-4 Vol. II: 20430-8

A CONCISE HISTORY OF MATHEMATICS, Dirk J. Struik. The best brief history of mathematics. Stresses origins and covers every major figure from ancient Near East to 19th century. 41 illustrations. 195pp. 5⅜ x 8½. 60255-9

Physics

OPTICAL RESONANCE AND TWO-LEVEL ATOMS, L. Allen and J. H. Eberly. Clear, comprehensive introduction to basic principles behind all quantum optical resonance phenomena. 53 illustrations. Preface. Index. 256pp. 5⅜ x 8½. 65533-4

QUANTUM THEORY, David Bohm. This advanced undergraduate-level text presents the quantum theory in terms of qualitative and imaginative concepts, followed by specific applications worked out in mathematical detail. Preface. Index. 655pp. 5⅜ x 8½. 65969-0

ATOMIC PHYSICS: 8th edition, Max Born. Nobel laureate's lucid treatment of kinetic theory of gases, elementary particles, nuclear atom, wave-corpuscles, atomic structure and spectral lines, much more. Over 40 appendices, bibliography. 495pp. 5⅜ x 8½. 65984-4

A SOPHISTICATE'S PRIMER OF RELATIVITY, P. W. Bridgman. Geared toward readers already acquainted with special relativity, this book transcends the view of theory as a working tool to answer natural questions: What is a frame of reference? What is a "law of nature"? What is the role of the "observer"? Extensive treatment, written in terms accessible to those without a scientific background. 1983 ed. xlviii+172pp. 5⅜ x 8½. 42549-5

AN INTRODUCTION TO HAMILTONIAN OPTICS, H. A. Buchdahl. Detailed account of the Hamiltonian treatment of aberration theory in geometrical optics. Many classes of optical systems defined in terms of the symmetries they possess. Problems with detailed solutions. 1970 edition. xv+360pp. 5⅜ x 8½. 67597-1

PRIMER OF QUANTUM MECHANICS, Marvin Chester. Introductory text examines the classical quantum bead on a track: its state and representations; operator eigenvalues; harmonic oscillator and bound bead in a symmetric force field; and bead in a spherical shell. Other topics include spin, matrices, and the structure of quantum mechanics; the simplest atom; indistinguishable particles; and stationary-state perturbation theory. 1992 ed. xiv+314pp. 6⅛ x 9¼. 42878-8

LECTURES ON QUANTUM MECHANICS, Paul A. M. Dirac. Four concise, brilliant lectures on mathematical methods in quantum mechanics from Nobel Prize–winning quantum pioneer build on idea of visualizing quantum theory through the use of classical mechanics. 96pp. 5⅜ x 8½. 41713-1

THIRTY YEARS THAT SHOOK PHYSICS: The Story of Quantum Theory, George Gamow. Lucid, accessible introduction to influential theory of energy and matter. Careful explanations of Dirac's anti-particles, Bohr's model of the atom, much more. 12 plates. Numerous drawings. 240pp. 5⅜ x 8½. 24895-X

ELECTRONIC STRUCTURE AND THE PROPERTIES OF SOLIDS: The Physics of the Chemical Bond, Walter A. Harrison. Innovative text offers basic understanding of the electronic structure of covalent and ionic solids, simple metals, transition metals and their compounds. Problems. 1980 edition. 582pp. 6⅛ x 9¼. 66021-4

HYDRODYNAMIC AND HYDROMAGNETIC STABILITY, S. Chandrasekhar. Lucid examination of the Rayleigh-Benard problem; clear coverage of the theory of instabilities causing convection. 704pp. 5⅜ x 8¼. 64071-X

INVESTIGATIONS ON THE THEORY OF THE BROWNIAN MOVEMENT, Albert Einstein. Five papers (1905–8) investigating dynamics of Brownian motion and evolving elementary theory. Notes by R. Fürth. 122pp. 5⅜ x 8½. 60304-0

THE PHYSICS OF WAVES, William C. Elmore and Mark A. Heald. Unique overview of classical wave theory. Acoustics, optics, electromagnetic radiation, more. Ideal as classroom text or for self-study. Problems. 477pp. 5⅜ x 8½. 64926-1

PHYSICAL PRINCIPLES OF THE QUANTUM THEORY, Werner Heisenberg. Nobel Laureate discusses quantum theory, uncertainty, wave mechanics, work of Dirac, Schroedinger, Compton, Wilson, Einstein, etc. 184pp. 5⅜ x 8½. 60113-7

ATOMIC SPECTRA AND ATOMIC STRUCTURE, Gerhard Herzberg. One of best introductions; especially for specialist in other fields. Treatment is physical rather than mathematical. 80 illustrations. 257pp. 5⅜ x 8½. 60115-3

AN INTRODUCTION TO STATISTICAL THERMODYNAMICS, Terrell L. Hill. Excellent basic text offers wide-ranging coverage of quantum statistical mechanics, systems of interacting molecules, quantum statistics, more. 523pp. 5⅜ x 8½. 65242-4

THEORETICAL PHYSICS, Georg Joos, with Ira M. Freeman. Classic overview covers essential math, mechanics, electromagnetic theory, thermodynamics, quantum mechanics, nuclear physics, other topics. xxiii+885pp. 5⅜ x 8½. 65227-0

PROBLEMS AND SOLUTIONS IN QUANTUM CHEMISTRY AND PHYSICS, Charles S. Johnson, Jr. and Lee G. Pedersen. Unusually varied problems, detailed solutions in coverage of quantum mechanics, wave mechanics, angular momentum, molecular spectroscopy, more. 280 problems, 139 supplementary exercises. 430pp. 6½ x 9¼. 65236-X

THEORETICAL SOLID STATE PHYSICS, Vol. I: Perfect Lattices in Equilibrium; Vol. II: Non-Equilibrium and Disorder, William Jones and Norman H. March. Monumental reference work covers fundamental theory of equilibrium properties of perfect crystalline solids, non-equilibrium properties, defects and disordered systems. Total of 1,301pp. 5⅜ x 8½. Vol. I: 65015-4 Vol. II: 65016-2

WHAT IS RELATIVITY? L. D. Landau and G. B. Rumer. Written by a Nobel Prize physicist and his distinguished colleague, this compelling book explains the special theory of relativity to readers with no scientific background, using such familiar objects as trains, rulers, and clocks. 1960 ed. vi+72pp. 23 b/w illustrations. 5⅜ x 8½. 42806-0 $6.95

A TREATISE ON ELECTRICITY AND MAGNETISM, James Clerk Maxwell. Important foundation work of modern physics. Brings to final form Maxwell's theory of electromagnetism and rigorously derives his general equations of field theory. 1,084pp. 5⅜ x 8½. Two-vol. set. Vol. I: 60636-8 Vol. II: 60637-6